SOILS —OF THE— PAST

TITLES OF RELATED INTEREST

Aeolian geomorphology
W. G. Nickling (ed.)

Cathodoluminescence of geological materials
D. J. Marshall

The changing climate
M. J. Ford

Chemical fundamentals of geology
R. Gill

Deep marine environments
K. Pickering *et al.*

A dynamic stratigraphy of the British Isles
R. Anderton *et al.*

Environmental change and tropical geomorphology
I. Douglas and T. Spencer (eds)

Environmental chemistry
P. O'Neill

Environmental magnetism
F. Oldfield and R. Thompson

Experiments in physical sedimentology
J. R. L. Allen

Geomorphology and soils
K. Richards *et al.* (eds)

Hillslope processes
A. D. Abrahams (ed.)

The history of geomorphology
K. J. Tinkler (ed.)

Image interpretation in geology
S. Drury

Introduction to theoretical geomorphology
C. Thorn

Karst geomorphology and hydrology
D. C. Ford and P. W. Williams

Mathematics in geology
J. Ferguson

Paleopalynology
A. Traverse

Pedology
P. Duchaufour (translated by T. R. Paton)

Petrology of the sedimentary rocks
J. T. Greensmith

A practical approach to sedimentology
R. C. Lindholm (ed.)

Principles of physical sedimentology
J. R. L. Allen

Quaternary environments
J. T. Andrews (ed.)

Quaternary paleoclimatology
R. S. Bradley

Rocks and landforms
J. Gerrard

Sedimentary structures
J. Collinson and D. Thompson

Sedimentology: process and product
M. R. Leeder

Trace fossils
R. Bromley

Volcanic successions
R. A. F. Cas and J. V. Wright

SOILS —OF THE— PAST

An introduction to paleopedology

G. J. RETALLACK
University of Oregon, Eugene

Boston
UNWIN HYMAN
London Sydney Wellington

Unwin Hyman, Inc.,
8 Winchester Place, Winchester, Mass. 01890, USA

Published by the Academic Division of
Unwin Hyman Ltd
15/17 Broadwick Street, London W1V 1FP, UK

Allen & Unwin (Australia) Ltd,
8 Napier Street, North Sydney, NSW 2060, Australia

Allen & Unwin (New Zealand) Ltd,
in association with the Port Nicholson Press Ltd,
Compusales Building, 75 Ghuznee Street, Wellington 1, New Zealand

First published in 1990

Library of Congress Cataloging-in-Publication Data

Retallack, Gregory J. (Gregory John), 1951–
Soils of the past: an introduction to paleopedology/by Gregory J. Retallack.
p. cm.
Includes bibliographical references.
ISBN 0-04-551128-4 (alk. paper)
ISBN 0-04-445757-X (pbk.)
1. Paleopedology. I. Title.
QE473.R473 1990
552′.5—dc20 89-37143
CIP

British Library Cataloguing in Publication Data

Retallack Gregory J.
Soils of the past.
1. Palaeosoils
I. Title
551.7
ISBN 0-04-551128-4
ISBN 0-04-445757-X

Typeset in 10 on 12 point Palatino by Fotographics (Bedford) Ltd
and printed in Great Britain by Cambridge University Press

To
Ken and Wendy Retallack,
for letting me be

Preface

Landscapes viewed from afar have a timeless quality that is soothing to the human spirit. Yet a tranquil wilderness scene is but a snapshot in the steady stream of surficial change. Wind, water and human activities reshape the landscape by means of gradual to catastrophic and usually irreversible events. Much of this change destroys past landscapes, but at some times and places, landscapes are buried in the rock record. This work is dedicated to the discovery of past landscapes and their life through the fossil record of soils. A long history of surficial changes extending back almost to the origin of our planet can be deciphered from the study of these buried soils, or paleosols. Some rudiments of this history, and our place in it, are outlined in a final section of this book. But first it is necessary to learn something of the language of soils, of what happens to them when buried in the rock record and which of the forces of nature can be confidently reconstructed from their remains. Much of this preliminary material is borrowed from soil science, but throughout emphasis is laid on features that provide most reliable evidence of landscapes during the distant geological past.

This book has evolved primarily as a text for senior level university courses in paleopedology: the study of fossil soils. It is not the usual view of this subject from the perspectives of soil science, Quaternary research or land use planning. It is rather the view of an Earth historian and paleontologist. Compared to the elegant outlines of a fossil skull or the intricate venation on a fossil leaf, fossil soils may at first appear unprepossessing subjects for scientific investigation. These massive, clayey and weathered zones are fossils in their own way. Their identification within a classification of modern soils presupposes particular past conditions, in the same way as the lifestyle that can be inferred from modern relatives of a fossil species of skull or leaf. Particular features of paleosols also may reflect factors in their formation in the same way as ancient diet can be inferred from the shape of fossil teeth or former climate from the marginal outline of a fossil leaf. This book is an exploration of the idea that paleosols are trace fossils of ecosystems.

Examples in this book are drawn largely from my own work on fossil soils, some of it not yet published elsewhere. Theoretical concepts have been borrowed more widely from allied areas of science including geomorphology, coal petrography, plant ecology, astronomy and soil science, to name a few. The fossil record of soils is a new focus for integrating existing knowledge about land surfaces and their biota. Paleopedology remains an infant discipline, hungry for theory and data of the most elementary kinds. This book is one attempt to partially quell the growing pains.

Gregory J. Retallack,
Eugene, Oregon, 1989

Acknowledgments

This book on paleosols would be slender indeed without extensive borrowing of facts, experiments, ideas and inspiration from allied areas of science. I have been fortunate to be able to draw upon the wise counsel of prominent sedimentologists (J. R. L. Allen, A. Basu, D. R. Lowe, R. M. H. Smith and L. J. Suttner), paleontologists (R. Beerbower, A. K. Knoll, J. W. Schopf and P. Shipman), geochemists (G. G. Goles, J. M. Hayes, H. D. Holland and W. Holser) and soil scientists (P. W. Birkeland, S. W. Buol, L. D. McFadden and P. F. McDowell). Among the emerging cadre of paleopedologists concerned with rocks older than Quaternary, it is a pleasure to acknowledge stimulating discussions with D. E. Fastovsky, M. J. Kraus, W. R. Sigleo and V. P. Wright. Last, and certainly not least, many of my ideas have been reshaped by students at the University of Oregon (E. A. Bestland, D. P. Dugas, C. R. Feakes, P. R. Miller, J. A. Pratt, S. C. Radosevich, G. S. Smith and G. D. Thackray). They gave real meaning to the Socratic dictum that the unexamined life is not worth living.

Photographs, specimens and other illustrative materials were generously and promptly provided by H. J. Anderson, J. B. Adams, H. P. Banks, T. M. Bown, J. Gray, R. Greeley, M. J. Kraus, S. C. Morris, NASA Space Science Center, C. J. Percival and V. P. Wright. Others graciously acquiesced in my adaptation of their published work. For several fine photographs (Figs. 3.5, 3.6, 21.1, and 21.3) I thank Sean Poston. For reams of accurate word processing, I am indebted to C. D. Bonham and K. Fletcher. The writing would not have proceeded nearly so much to my satisfaction without the happy home life created by Diane, Nicholas and Jeremy.

We are grateful to the following individuals and organizations who have kindly given permission for the reproduction of copyright material (figure numbers in parentheses):

© 1952 University of Chicago Press (3.13); Table 4.1 part reproduced from Bodman & Mahmud, *Soil Science* **33**, 363–74, © 1932 Williams and Wilkins; Figure 4.9 reproduced from Stephenson, *Soil Science* **107**, 470–79, © 1969 Williams and Wilkins; Figure 9.1 reproduced with permission from L. R. Holdridge *et al.*, *Forest Environments in Tropical Life Zones*, © 1971 Pergamon Press PLC; Figure 9.2 reproduced with permission from *Annual Review of Earth and Planetary Sciences* 7, © 1979 Annual Reviews Inc.; Figure 10.2 reproduced by permission from Perry & Adams, *Nature* **276**, 489–91, © Macmillan Magazines Ltd; Figure 10.3 reproduced from

Wright, *Geological Journal* **19**, by permission of John Wiley & Sons © 1984; Figure 10.4 reproduced from J. M. Anderson *et al.*, *Palaeoflora of Southern Africa: Molteno Formation (Triassic)* Vol. 1, 1983 by permission of A. A. Balkema; © 1969 University of Chicago Press (11.2, 11.3); Blackwell Scientific Publications (11.7, 16.2D); Figure 13.3 reproduced from Gile *et al.*, *Soil Science* **101**, 347–60, © 1966 Williams & Wilkins; Figure 15.5 reproduced with permission from Heiken *et al.*, *Proc. 4th Lunar Conference 1* and *Proc. 7th Lunar Conference 1,* © 1973 and 1976 Pergamon Press PLC; Figure 15.6 reproduced with permission from McKay *et al.*, *Proc. 5th Lunar Science Conference 1,* © 1974 Pergamon Press PLC; Figures 16.1C, D, F, H and 16.2C reproduced by permission from J. W. Schopf (ed.), *Earth's Earliest Biosphere: its Origin and Evolution,* © 1983 Princeton University Press; © American Association for Advancement of Science (16.1G); Figure 16.2A reproduced from Ronov, *Geochemistry International* **4**, 713–37, by permission of John Wiley & Sons © 1964; © 1985 American Association for the Advancement of Science (16.2E).

Table of Contents

List of tables

PART ONE

SOILS AND PALEOSOLS

1 Paleopedology

Paleopedology is the study of ancient soils, and is derived from an ancient Greek word (πεδον, πεδου) for ground. Soils of the past, either buried within sedimentary sequences or persisting under changed surface conditions, are the main subject matter of paleopedology. In this book, it is seen as an historical perspective on soil genesis and as a way of reconstructing the geological history of land surfaces on Earth. Soils, like organisms, sediments, and surface environments, have changed over the past 4500 million years of recorded Earth history.

The concept of fossil soils can be traced back to Scottish physician James Hutton (1795). His insistence on arguing past causes from those which can be observed today was a prerequisite to making a connection between soils of today and those of the distant geological past. Thus red rocks along angular unconformities that he had discovered along the River Jed and at Siccar Point, southeast of Edinburgh, were regarded as comparable to surface soils and sediments on the modern landscape (Fig. 1.1). "From this it will appear, that the schistus mountains or vertical strata of indurated bodies had been formed, and had been wasted and worn in natural operations of the globe, before horizontal strata were begun to be deposited in these places . . ." (Hutton 1795, Vol. 1, p. 438). In considering French accounts of similar phenomena, Hutton refers to the "*sol mort rouge*," or "dead red soil," along unconformities in the mining districts of Germany (Hutton 1795, Vol. 1, p. 445). Such phrases cannot be taken literally. The red, brecciated, and conglomeratic material was labeled "soil" in a very broad sense, including both sediment and soil. German miners termed it "dead" because it lacked the ore-bearing veins of the underlying deformed rocks rather than because it was viewed as a former living soil. These ideas were reiterated in John Playfair's (1802) *Illustrations of the Huttonian theory of the Earth* which, because of its conciseness and clarity of expression, was more influential than Hutton's original two volumes. Playfair also cited a 1799 record of a fossil forest in Lincolnshire, now covered by tidal flat sediments. This is the oldest record of a Quaternary fossil soil.

The oldest record of buried soils within consolidated sedimentary rocks

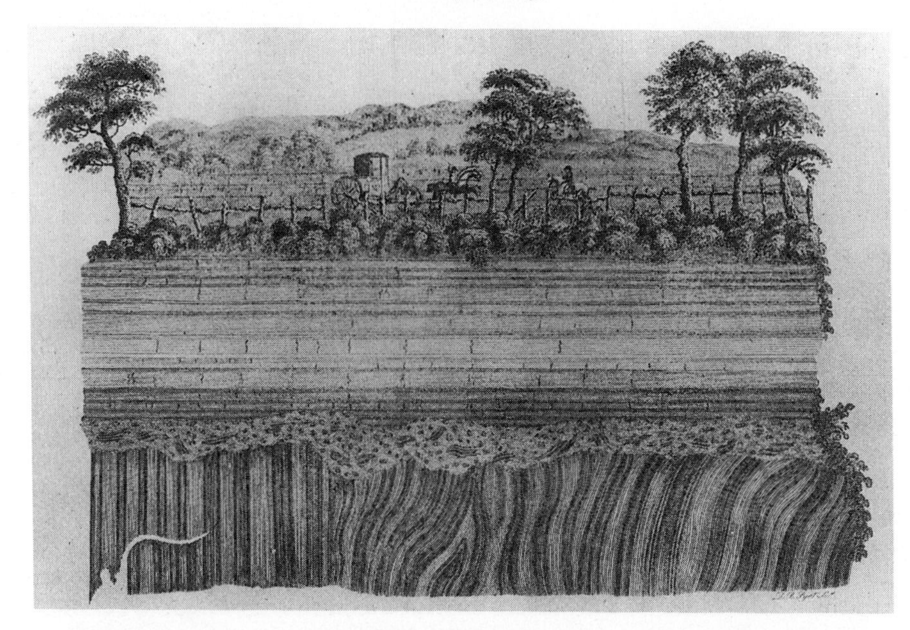

Figure 1.1 Angular unconformity between Early Silurian, Hawick Rocks, and Late Devonian, Upper Old Red Sandstone along the River Jed in southeastern Scotland (from Hutton 1795).

was the "dirt beds" (Fig. 1.2) and fossil stumps reported in latest Jurassic limestones of the Dorset Coast by Webster (1826) and popularized in William Buckland's (1837) "Bridgewater Treatise." A variety of other fossil forests were subsequently discovered, and their stumps and associated fossil plants described, but little was made of their substrates as fossil soils. Well publicized examples discovered in the late 19th century include the Eocene "fossil forests" of Yellowstone National Park in the United States and the Carboniferous stumps of tree-lycopods excavated at Clayton (Yorkshire) and in Victoria Park (Glasgow) in the British Isles. A comprehensive summary of these early discoveries of pre-Quaternary fossil soils can be found in the introductory chapters of Albert C. Seward's (1898) monumental work *Fossil plants*. He appreciated the significance of fossil soils as evidence for the immensity of geological time and as indicators of past worlds. Study of the paleosols themselves had to await the development of soil science.

During the late 19th century, buried soils also were recognized within surficial deposits of loess and till. These "weathered zones," "forest zones," and "soils," as they were variously termed, were found in the Russian Plain by Feofilatkov (in the 1870s, as recounted by Polynov 1927), in the midcontinental United States by McGee (1878), and in New Zealand by Hardcastle (1889). By the turn of the century such observations had been

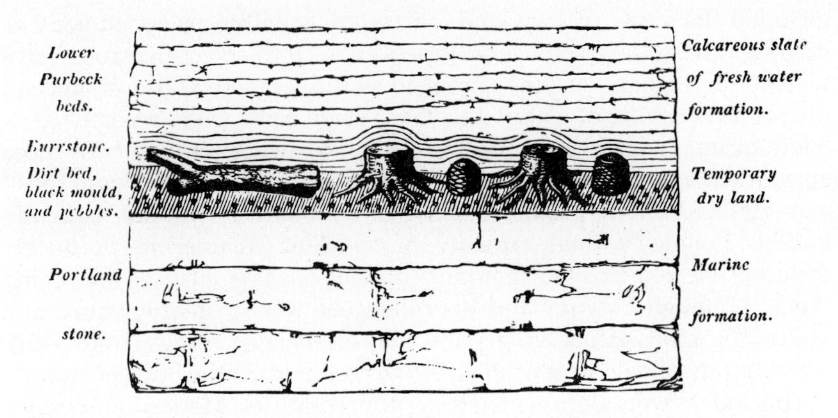

Section of the Dirt-bed in the Isle of Portland, shewing the subterranean remains of an ancient Forest. De la Beche.

Figure 1.2 "Dirt beds" (paleosols) in a stratigraphic section through latest Jurassic (Tithonian), Purbeck Formation on the Isle of Portland, Dorset, England (from Buckland 1837).

used for stratigraphic subdivision of glacial deposits (Chamberlain 1895).

The origin of paleopedology as a discrete field of inquiry can be traced back not so much to these early observations of buried soils, but to the late 19th century development of soil science. Since classical times, soils have been studied from the point of view of plant nutrition. It was not until 1862 that the Saxon scientist Friedrich A. Fallou first published the term "pedologie" for the study of soil science, as opposed to what he termed "agrologie," or practical agricultural science. The foundations of modern soil science were laid by Vasily V. Dokuchaev (1883) with a detailed account of the dark, grassland soils of the Russian Plain. This monograph demonstrated that soils could be described, mapped and classified in a scientific fashion. Furthermore, their various features could be related to environmental constraints, of which climate and vegetation were considered especially important. By the early part of the 20th century there was an established scientific tradition of research on soil geography, classification, and genesis in Russia, as summarized in the influential general works of K. D. Glinka (1927, 1931).

In the course of these early Russian investigations, certain soils were found to be anomalous in that their various features did not fit the general relationship between soil type and their climate and vegetation. It had long been suspected that these were very old soils, perhaps products of past environments. In 1927, Boris B. Polynov summarized Soviet observations of this kind. His short paper, which introduced the term paleopedology, can be considered to be the foundation of this branch of science. Polynov included the study of four kinds of materials within paleopedology as follows. "Secondary soils" encompass those formed by two successive weathering regimes such as grassland soils degraded by the advance of forests after retreat of glacial ice. "Two-stage soils" were recognized to have an upper horizon of recent origin, but deeper horizons of more ancient vintage, such as the lower horizons of an older soil. "Fossil soils" were defined as soil profiles developed on a surface and subsequently buried. Polynov's final category of "ancient weathering products" included the redeposited remnants of soils such as laterites and china clays. Although Polynov and his colleagues were primarily concerned with surface soils of Quaternary age, they established a logical framework for the study of paleosols of all ages that continues in the Soviet Union.

These ideas were slow to penetrate other countries. Modern soil science in North America can be traced back to Eugene W. Hilgard's (1892) monograph on the relationship of soil and climate. This, and the first nationwide mapping and classification of North American soils by Milton Whitney (1909), were largely independent of comparable research done by Russian soil scientists. The work of Curtis F. Marbut, which culminated in a monumental soil survey of the United States (1935), was strongly influenced by Soviet soil science, especially the general works of K. D. Glinka. Paleopedology also was introduced into North America through a Soviet connection. Constantin C. Nikiforoff completed doctoral studies at the University of St. Petersburg in pre-revolutionary Russia, but by 1943 was a scientist with the US Soil Conservation Service when he published a short essay outlining the role and scope of paleopedology. A supporting study of actual paleosols by Kirk Bryan and Claude Albritton (1943) made clear the practical application of such studies.

Ideas on the classification and origin of soils have been especially useful for studies of Quaternary stratigraphy and geomorphology (Valentine & Dalrymple 1976). Such studies are now conducted in most parts of the world, coordinated to some extent by a Commission on Paleopedology established in 1965 at the 7th Congress of the International Association for Quaternary Research in Denver, USA. An early result of the commission's activities was the publication of recommendations for recognizing and classifying paleosols in a volume of research papers edited by Dan H. Yaalon (1971). Mapping units for Quaternary paleosols have been incorporated into official stratigraphic codes (e.g., American Commission on Stratigraphic Nomenclature 1961, North American Commission on Stratigraphic Nomenclature 1982). Modern research on Quaternary paleosols can be found in books and journals on soil science, geography, archaeology, and Quaternary research (Birkeland 1984).

In contrast to a steady level of interest in Quaternary paleosols, studies of older paleosols have been slow to gain momentum. In many sequences now known to contain them in abundance, paleosols were not recognized or their features were explained as diagenetic phenomena. As with 19th century accounts, little was made of those few cases explicitly recognized as paleosols, e.g., by Barrell (1913), Collins (1925), Allen (1947), and Thorp & Reed (1949). Beginning in the 1960s interest in pre-Quaternary paleosols has been increasing on several fronts. The study of paleosols is especially compatible with the overall aims of sedimentology to reconstruct ancient environments and geological processes (Andreis 1981). Studies on Devonian fluvial rocks by John R. L. Allen (1973, 1974) have done much to popularize the use of paleosols for reconstructing paleoclimate and rates of sediment accumulation. Another influential early work by R. J. Dunham (1969) revealed that paleosols can be evidence of subaerial exposure in ancient marine reef limestones such as the Permian examples of the Guadalupe Mountains in Texas and New Mexico. These are regarded as pioneering papers by a new generation of professional sedimentologists spawned by the petroleum industry. Many of these scientists are finding and reporting pre-Quaternary paleosols in sedimentological and geological journals (Wright 1986a).

Paleontological research has always been concerned with reconstructing past biotas, their ecology, and the ways in which they are preserved in the rock record. Paleosols can be regarded as both trace fossils of past ecosystems and as preservational environments for many kinds of fossils (Retallack 1976, 1977). Paleontologically oriented reports on paleosols are appearing now in a variety of journals concerned with paleoecology and other paleontological subjects (Retallack 1988b).

A final area of geological sciences now contributing to paleopedology is geochemistry. The chemical study of weathering by Samuel S. Goldich (1938), from which was derived his well known mineral stability series, dealt with Cretaceous paleosols in Minnesota. The use of paleosols as indicators of atmospheric conditions and the nature of weathering processes in the very distant geological past has been explored in pioneering studies by R. P. Sharp (1940), A. V. Sidorenko (1963), G. E. Williams (1968) and S. M. Roscoe (1968). Such studies are now more common in journals and books concerned with geochemistry and Precambrian geology (Holland 1984, Retallack 1986a).

Although paleopedology has been a recognized scientific field for about 60 years, its growth has been modest. Few scientists yet regard themselves primarily as paleopedologists. Few conferences and no journals or societies are entirely devoted to paleopedology. For these reasons the subject has special charm and promise.

Fossil soils may be used as stratigraphic markers in the ongoing inventory and mapping of the geological resources of this and other planets. Particular features of paleosols may aid in locating especially valued resources. For example, changes in the degree of development and mineral content of paleosols reflect former time for formation and degree of waterlogging, and can be used to guide exploration for petroleum, coal and uranium ores. Coal and uranium accumulated in parts of the landscape where groundwater was poorly oxygenated. Uranium is more likely to be found in paleosols that are variegated and that formed on the margins of uplands of uraniferous granites. Coal is associated with paleosols that are gray and formed in swamps, rather than paleosols that are red and were formerly well drained. Paleochannel sandstones, located by following lateral trends in paleosol development and waterlogging, can be local petroleum reservoirs.

Fossil soils also provide historical validation for theories about how soils form. The geological history of soils can be viewed as a long-term natural experiment in which many fundamental conditions of soil formation, such as vegetation and atmospheric composition, have changed. Information from fossil soils will strengthen ideas about how soils form and how they should be classified. Because such ideas form the basis for much agricultural and engineering activity, it is all the more important that they have a firm scientific basis.

Finally, fossil soils are evidence for reconstructing past terrestrial ecosystems and environments. They can be used to bring particular times and places into sharper focus as evidence independent of associated fossils and sedimentary structures. They are a record of the evolution of ecosystems and of their interaction with environments on land, and provide a perspective on our place on Earth.

2 Soils on and under the landscape

Difficulties in defining "fossil soil" arise not so much from its fossil nature as from confusion over what is meant by soil. A fossil soil or buried paleosol, like other kinds of fossils, is the remains of an ancient soil buried by later deposits. Paleosols also may be at the surface but no longer actively forming in exactly the same way. The word "soil" on the other hand, like other commonly used words such as love and home, means different things to different people. For farming purposes, soil is fertile, loose, tillable ground. In engineering specifications, soil is any material that can be excavated without recourse to quarrying or blasting. By both of these practical definitions, unaltered sediments, such as dune sands or flood silt, are regarded as soil, yet some extremely altered soil materials, such as hard laterites, are not.

To some soil scientists, soil is the medium in which vascular plants take root. This narrow definition includes rock crevices supporting large plants, yet it excludes rocks colonized by plant-like microbes and rootless plants such as mosses and liverworts. It also excludes hard-setting parts of soils and parts of soils below the level of roots.

Taking an even wider view, what are we to call Lunar, Venusian, and Martian land surfaces that have been altered in place by surficial processes? The Moon and Venus are lifeless and Mars probably is also. Should calling their altered land surfaces soils necessarily imply that life is present? Or should an alternative label such as regolith imply that they are barren of life? Similar problems beset the naming of fossil soils older than Ordovician, before the advent of large land plants. If one takes the view that soils are a medium of plant growth, then both the terms Precambrian soil and Precambrian regolith beg the question of the antiquity and nature of life on land. Like the question of life on Mars, the origin of life on land is an important scientific issue in its own right and should not be confused by semantic considerations.

As paleopedologists, a wider definition of soil is needed. In this book I take soil to be material forming the surface of a planet or similar body and altered in place from its parent material by physical, chemical, or biological processes. This is close to what Nikiforoff (1959) had in mind in proposing that soil be regarded as the "excited skin of the subaerial part of the Earth's crust." Soil can be envisaged as a zone of interaction between the atmosphere and crust of a planetary body. A flux of energy from the sun, water, snow or living creatures continually alters rocks or sediments into what we call soil.

Some technical terms for soils

Soils are complex zones of interaction between sediment or solid rock and the ecosystem or atmosphere. Because of the varying levels of interaction down from the land surface, it is usual to study soils in profiles that show the layering of alteration, or soil horizons. Soil profiles naturally exposed in cliffs and river banks may be altered by weathering at the cliff face. They are best observed in freshly exposed material such as trenches and cores. Similarly, fossil soils are best studied by excavation.

Solum is a technical word for that part of the soil profile most altered by soil processes (Fig. 2.1). In many cases it is riddled with roots of plants.

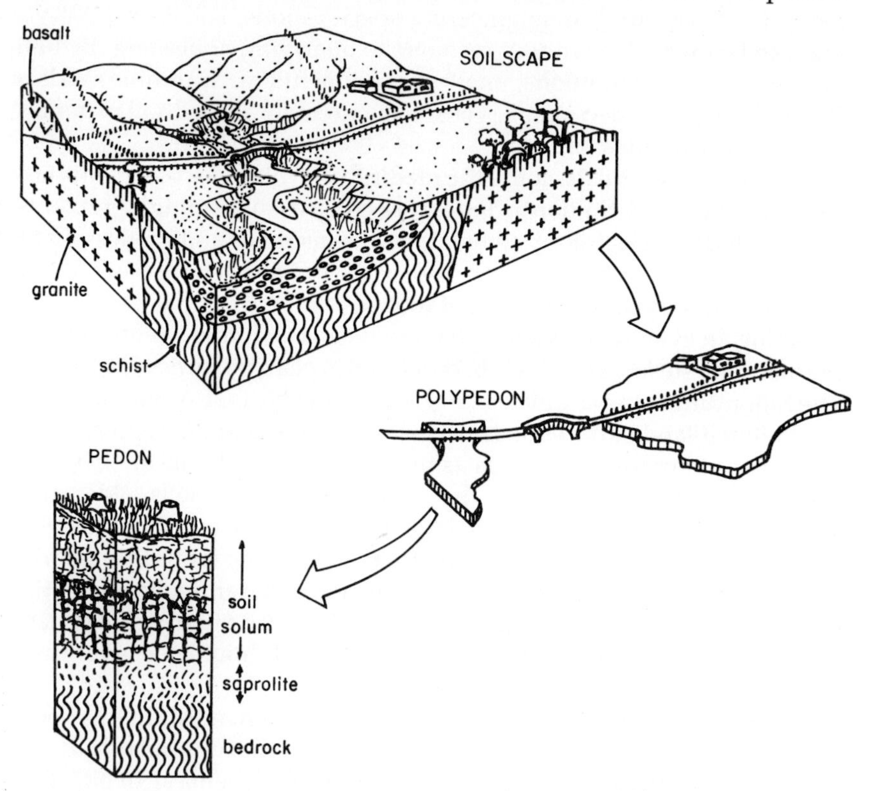

Figure 2.1 Some technical terms for soils and their relationship to landscapes.

The soil solum also may be dark, red, clayey, or massive, and so is very different from its underlying parent material. Weathered material between the solum and underlying sediment or bedrock has a mix of soil and inherited features, and is called saprolite. This material may be soft, oxidized, clayey or otherwise altered like a soil, but not to the same extent as the soil solum. Some saprolites show clearly the bedding, schistosity, and deformation of parent metamorphic rock. In the Piedmont region of the southeastern United States, there is a standing joke about this difference. Newcomers to the area have been surprised when told to sample a saprolite that, rather than ringing to the crack of the hammer as would be expected from what appears to be a strongly foliated metamorphic rock, the soft clay swallows the hammer head. The distinction between saprolite and solum is a matter of degree of alteration.

Soils on the landscape

Soils blanket most of the landscape, except for areas covered by rivers or lakes or areas freshly uncovered by erosion and human excavation. Because conditions of sunshine, moisture, and other soil-forming factors vary in different parts of the landscape, so do the soils that form there. The fundamental unit of soil is a column of soil material, or pedon, of the kind that could be dug out of the wall of a trench. Soils vary in such complex ways that few pedons are exactly alike. Some pedons are sufficiently similar that they are recognised as a discrete kind of soil, different from others in the area. One or more of these similar pedons covering an area of ground is called a polypedon. These are like tiles in a mosaic of soils over the landscape. The assemblage of polypedons mantling the landscape is sometimes called a soilscape.

Soils take time to form until interrupted by a cover of sediment or by removal through erosion. In river valleys, for example, deposits of sand and silt left behind by an especially powerful flood, or the rubble of a landslide, may cover soils and drive off or destroy plants and animals (Fig. 2.2). Deep burial is a common way in which paleosols are formed. The covering deposits provide a surface for recolonization by plants and animals on which soil begins to form anew. Soon the bare surface is dry and cracked and small plants are taking root. With the advent of plants, come worms and other burrowing invertebrates,and also herbivorous mammals. In humid regions, these early successional plants and animals are followed by shrubs and ultimately the kind of woodland and soil that existed before is reconstituted.

As a soil is burrowed, penetrated by roots, and otherwise altered, the ripple marks and bedding planes of the original alluvium are progressively destroyed. Such sedimentary relicts persist in many weakly developed

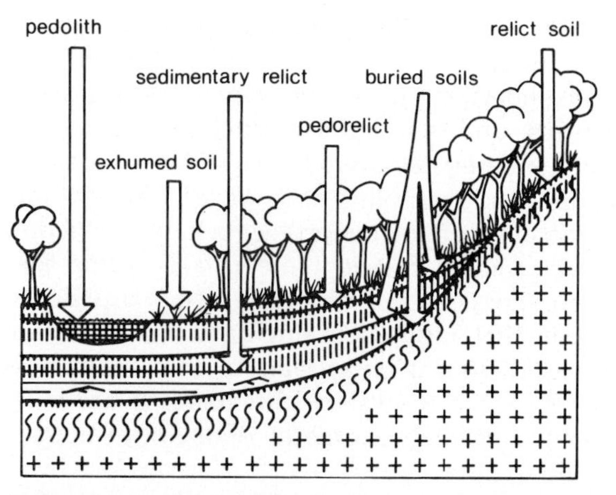

Figure 2.2 Technical terms for fossil soils (paleosols) and their relation to sediments (from Retallack 1983a, reprinted with permission from the Geological Society of America).

bottomland soils and paleosols. Their persistence is a reflection of the degree to which the soil has been able to form and does not negate the identification of these materials as soils or paleosols. It is not difficult to recognize paleosols developed during times of peace and quiet between deposition of thick sedimentary layers.

Sequences of paleosols can become very difficult to decipher, however, if the intervening sedimentary layers are too thin to separate them effectively (Schaetzl & Sorenson 1987). A common situation that may cause confusion is the overprinting of the upper horizon of an older soil by the development of a lower horizon of a younger soil at a slightly higher level. The older near-surface horizon, then, is not genetically related to the younger subsurface horizon. From the point of view of the younger soil, it can be considered to be a pedorelict of the older soil. Another common kind of pedorelict are nodules or clasts of pre-existing soils incorporated in sediments on which later soils have formed. Distinguishing between these older nodules or clasts and nodules or clods of the younger soil may be difficult unless they have sharp, ferruginized, or truncated boundaries.

Some sedimentary layers are very distinctive because they are composed entirely of a particular kind of soil material. This kind of bed, with clasts of soil mineralogy and appearance but sedimentary organization, may be called a pedolith in the extended sense suggested by Gerasimov (1971). This term was originally coined (by Erhart 1965) for redeposited laterites of Tertiary age. These remain a good example because such locally derived soil material forms brightly colored red beds very distinct from enclosing alluvium. The term is only useful for such clear cases, where soil-derived sediments are distinct, because most sediment ultimately is derived from soils and so is pedolithic in some sense.

Quaternary paleosols

By current estimates (Palmer 1983), the Quaternary geological period is the past 1.6 million years. The time spans required to form soils are thousands to millions of years, so those of Quaternary age are important evidence for the way in which soils form. The study of factors in soil formation usually involves a carefully constrained analysis of soils and paleosols of varying age or situation. In the earthquake-prone area of the Transverse Ranges of California, around Santa Paula (Rockwell et al. 1985a,b), river terraces are uplifted and tilted by folding that is still continuing (Fig. 2.3A). The youngest surfaces are those nearest the streams where they are still disturbed by annual floods. Slightly older terraces, dated by radiocarbon and other means, are at higher levels. This stepped landscape includes successively older surfaces at higher levels and these in turn bear progressively better developed soils (Fig. 2.3B). Many such sequences of alluvial terraces and their soils have been studied in order to document changes in soil formation with time (Harden 1982a). Another favorite landscape for such studies is areas around the terminus of retreating glaciers (Burke & Birkeland 1979). This kind of research provides basic information about the way in which soils form, which is vital to the interpretation of older paleosols.

Studies of Quaternary soils and paleosols reveal clearly what a complex thing a soil is and how many factors enter into their formation (Johnson &

Figure 2.3 Tectonically tilted Pleistocene (about 160 000–200 000 years old) terraces disrupted at Culbertson Fault (Rockwell *et al.* 1985a, b) on left skyline and the Tertiary rocks of Santa Paula Ridge in central Timber Canyon north of Santa Paula, California (A), and a thick, well developed soil (a Typic Palexeralf in the US taxonomy) on the central tilted terrace (B): soil horizons are indicated by standard shorthand and tape is 6 ft (2 m) long.

Watson-Stegner 1987). Not all these complications are relevant to the interpretation of older paleosols. Among these are human impacts such as increased incidence of fires, forest clearance, and paving. Some of the difficulty in understanding Quaternary paleosols also has been exacerbated by focusing study on those found in outcrops or shallow trenches rather than in deep boreholes. Paleosols of uplifted terraces or stable continental regions are likely to have been influenced by a greater variety of weathering regimes than those subsiding below the zone of weathering shortly after formation. Both problems are apparent from desert soils of Africa and Australia, which are red, thick and deeply weathered. These soils are in part relicts of Miocene and earlier forests which grew under a more humid climate (McFarlane 1976, Senior & Mabbutt 1979). In areas of granitic basement rocks of Precambrian age, these deeply weathered materials have persisted at the surface through many changes of environment. The true soils of deserts are not red, clayey and deeply weathered, but thin, sandy and little altered from the color of their parent materials, as can be seen from soils of eastern Oregon, southern California, Nevada, the Middle East and Central Asia.

Paleosols at major unconformities

Many unconformities show evidence of paleosols. Not all do, because some have been scoured clean by fluvial or marine erosion before being covered by later sediment. Examination of geological maps for unconformities remains a productive method for locating paleosols, especially in Precambrian rocks, where they are difficult to recognize otherwise. Paleosols at major geological unconformities include thick, well differentiated layers of rock enriched in ferric oxide (laterite), alumina (bauxite), silica (silcrete), or calcium carbonate (calcrete). The origin of these distinctive materials is a complex issue for at least two reasons (Goudie 1973). First, they take so long to form that conditions originally encouraging their formation are almost certain to have changed in some way before their burial and preservation. Second, these are all indurated and weather resistant materials that can withstand subsequent erosional events. A brief consideration of some of the leading theories of the formation of one kind of duricrust serves to illustrate some of these complexities.

Laterites are thought not to form within the soil solum, but within deeper and thicker zones of saprolite below (McFarlane 1976). Especially appropriate sites for the accumulation of ferric oxides to such a concentration are places on the side of plateaus where groundwaters enriched in iron dissolved in swamps within depressions on the plateau are oxidized within a zone of seepage around the plateau margin. Lateritic profiles may reach considerable thickness in such geomorphic positions (Figs. 2.4 &

2.5). Such zones of concentrated ferric oxide are soft and easily excavated within the ground but, once exposed to air, they become indurated like a brick. Their excavation and drying for construction stone on the Indian subcontinent are the source of their name from the Latin *later* (*lateritis* in the genitive) for brick. Indurated laterites may armor hillsides against further erosion or may persist as pebbles of conglomeratic material similar to pisolitic or nodular original laterite. Once formed, laterites are very persistent.

From this brief account of laterites, they can be seen to involve more than just soil formation. Erosional landscape lowering, reorganization of pre-existing soil horizons, changing flow of groundwater, and progressive modification of the landscape also occur. The overlapping effects of so many processes make the interpretation of such old surfaces of weathering difficult and controversial. Interpretation of buried examples is confounded by additional difficulties. Along major erosional unconformities it is difficult to be sure that the entire pre-existing profile has been preserved. Modern duricrusts form the surface of many landscapes because the soil under which they formed has been eroded. Many paleosols at major unconformities are likely to represent saprolite or other deep layers rather than the surface solum (Schau & Henderson 1983). A second difficulty with unconformities is the way in which they juxtapose

Figure 2.4 Laterite (dark lower zone) and mottled zone (variegated middle zone) of a Miocene paleosol (Plinthic Paleudult of US taxonomy) overlain unconformably by sand capped with a modern soil (Hapludoll of US taxonomy) at north end of Long Reef beach near Sydney, Australia (laterites discussed by Faniran 1971).

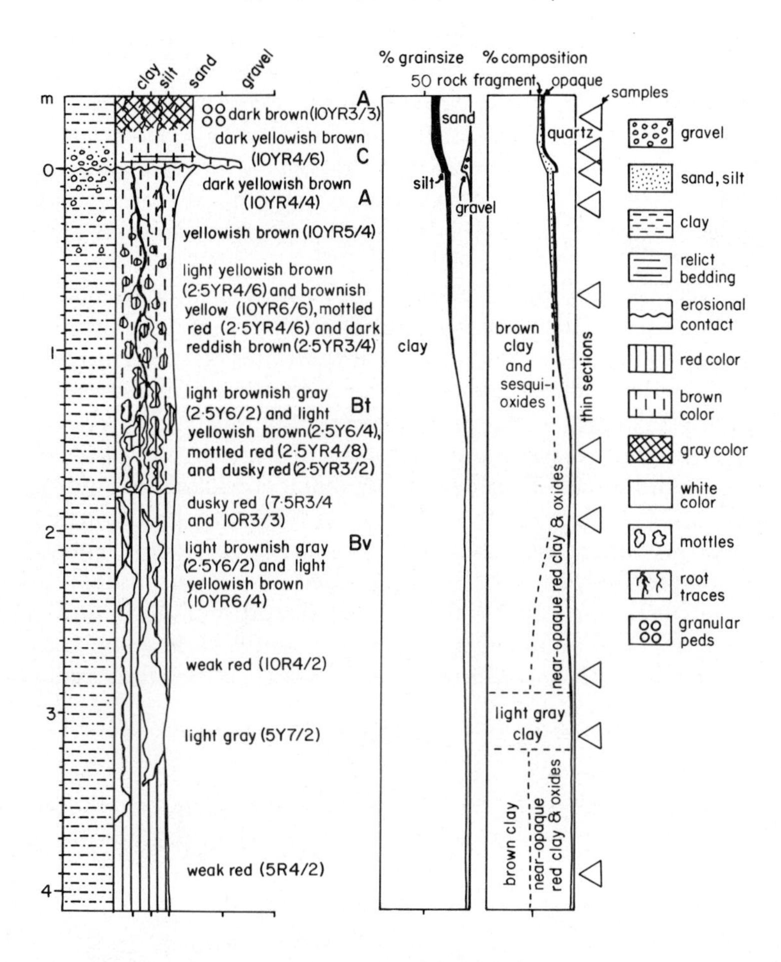

Figure 2.5 Interpretative stratigraphic section, grain size distribution, and mineral composition of Miocene paleosol and modern soil of Figure 2.4.

materials of very different chemical and physical characteristics. Commonly clayey impermeable paleosols are overlain by gravelly or sandy, permeable fluvial deposits. Passage of groundwater through overlying sediments could substantially alter the underlying paleosol with effects becoming less marked downward from the unconformity in a manner difficult to distinguish from former soil formation (Pavich & Obermeier 1985). Examples of formerly well drained paleosols which are mineralized with sulfide or uranium minerals characteristic of reducing environments (Retallack 1983b) are indications that such modifications may occur.

Despite these problems, paleosols at major unconformities often present soil formation so extreme as to be unmistakable. The accumulated

Figure 2.6 Alluvial sediments consisting mostly of superimposed paleosols of Oligocene age in the Pinnacles area of Badlands National Park, South Dakota, USA (from Retallack 1983a, reprinted with permission of the Geological Society of America).

alteration of ages is not easily erased by later events of lesser duration. As evidence of the antiquity and geological history of deep weathering and of duricrusts, they are of interest in themselves.

Paleosols in sedimentary sequences

Paleosols are abundant in some sedimentary successions (Fig. 2.6). In many cases paleosols have masqueraded under non-genetic terms such as red beds, variegated beds, tonstein, ganister, and cornstone. Unlike conglomerates and sandstones, which are readily identifiable as hardened gravel and sand, these other rock types do not resemble modern kinds of sediment. Their similarity to modern soils was unrecognized as long as they were compared to sediments and rocks rather than soils.

Ganisters, for example, are rocks found in Euramerican Carboniferous coal measures. The word was coined by British miners for hard, silicified quartz sandstone which is so chemically and physically inert that it is used for lining furnaces. Ganisters commonly are penetrated by carbonaceous root traces and underlie coal seams. Only recently has it been recognized that these were upper horizons of moderately well drained soils. Their quartz-rich composition was produced in part by destruction of easily weathered associated minerals and their silicification by diagenetic mobilization of accumulated plant opal (Retallack 1977).

Another kind of rock that seemed puzzling from a sedimentary perspective is cornstone. This rock is riddled with yellowish nodules of calcium or magnesium carbonate, irregular to rounded in shape and several centimeters in diameter. Marine rocks commonly contain calcareous, sideritic, and other kinds of nodules presumed to have formed after burial of the sediment (Coleman 1985), so cornstone also was thought to be produced by burial diagenesis. Some of the classical cornstone sequences of the Old Red Sandstone in Britain contain fossil freshwater fish but no marine fossils. From this it could be argued that they formed after burial, but in lake sediments. Careful evaluation of the associated sandstones revealed that cornstones were more commonly associated with rocks thought to have formed in ancient rivers and that they resemble calcareous nodules of modern soils of dry climates (Steel 1974, Allen 1986a). Carbonate nodules formed by burial diagenesis are less complexly cracked, less micritic and do not show displacive fabrics. The differences between calcareous nodules of marine rocks and of paleosols are now sufficiently well established that nodular paleosols can be recognized as evidence of low sea level within sequences of shallow marine limestones (Wright 1982).

Several general problems with the study of paleosols in sedimentary successions are apparent from these two examples. Foremost among these is their alteration upon deep burial. Cornstone was most easily understood because it is indurated and similar in appearance to modern soil nodules. Not all modern soils are colored as brightly as red beds, nor are modern soils hard and flinty like ganisters. The likelihood of some changes after burial should not be taken to mean that all features of these rocks are late diagenetic. Careful attention to relationships with unquestionably original features, such as fossil root traces and burrows, may allow discrimination between original and burial diagenetic portions of a paleosol. A second problem is caused by the confused boundaries of some paleosols in sedimentary successions due to overlapping of successive paleosols. Erosion and redistribution of soil material also can be confusing. These complications can obscure the expression of soil horizons or other features that would be more obvious indications of paleosols. A third problem has been the application of inappropriate conceptual models to the interpretation of these rocks. Many non-marine sedimentary rocks include evidence of soil formation in addition to sedimentation. Each needs to be considered in interpreting geological history.

Despite these problems, paleosols in sedimentary rocks are promising because of evidence for paleoenvironments that they contain. Evidence from fossil soils can not only be used to validate interpretations based on other lines of enquiry, but can also be used to frame new kinds of interpretations. For example, fossil soils can be used as evidence of former vegetation against which the degree of adaptation of limbs and teeth of associated fossil vertebrates can be assessed. Fossil soils also can provide records of rainfall, vegetation, and other factors controlling sedimentation against which the style of paleochannels and past fluvial processes can be assessed. Because of their unique problems and potential, the study of paleosols in ancient sedimentary successions is developing a research tradition of its own, distinct from that of paleosols at major unconformities and from that of Quaternary paleosols and soils.

3 Features of fossil soils

Compared with cross-bedded sandstone or coarsely crystalline granite, paleosols at first sight may seem massive and featureless. Despite these nondescript first impressions, paleosols do have distinctive features. For the most part these are characteristics also found in modern soils, yet many paleosols are no longer loose, cracked, and at the land surface. Important differences result from compaction and alteration upon burial which change many of the diagnostic chemical properties of modern soils such as pH, Eh and base saturation. Thus, identification of paleosols can be a problem for both geologists and soil scientists alike.

There are three main kinds of features by which paleosols may be recognized in the field and from laboratory studies: root traces, soil horizons, and soil structure. Using these and other observations, it is generally possible to distinguish paleosols from unaltered sedimentary deposits, volcanic flows, or zones altered by faulting. Paleosols can be altered by groundwater, hydrothermal activity, or metamorphism, and in these cases often have a mix of features that are difficult to disentangle. Alteration of paleosols after burial is the subject of a later chapter. This one is concerned with criteria to determine whether a rock is part of a paleosol or something else entirely.

Root traces

Fossil roots or root traces are one of the best criteria for recognition of paleosols in sequences of sedimentary rocks. They are evidence that plants once lived in it and that, regardless of its other features, it was once a soil. For example, a gray shale with clear bedding planes may look like an ordinary sedimentary deposit, but a few fossil root traces penetrating it mean that it was once a soil. The fossil record of roots is extensively documented and will be reviewed after considering some general aspects of fossil roots useful for identifying paleosols.

The top of a paleosol can be recognized as the surface from which root traces emanate (Fig. 3.1). Concentrations of other trace fossils such as

Figure 3.1 Calcareous rhizoconcretions emanating from the top of a mangal paleosol (Aquept) in the Late Eocene, Birket Qarun formation near Madwar el Bighal, Fayum depression, Egypt (photograph courtesy of Thomas L. Bown).

burrows can also be used because they record periods of reduced or no deposition during which sediment was extensively modified at the surface. There are situations when sedimentation keeps pace with burrowing and vegetative growth, but irregularities in depositional processes are such that perfect balance between depositional disturbance and vegetative colonization is seldom attained. Usually there are zones with more than the usual density of root traces and burrows that can be interpreted as horizons close to the top of a paleosol.

Under favorable circumstances, the original organic matter of a fossil root may be well preserved. Even if only a trace of roots is preserved by an infill of clay or calcite, there are several distinctive features by which they can be recognized. Unlike other trace fossils such as burrows, most root traces taper and branch downward. They also are very irregular in width. Large, near-vertical root traces characteristically have a concertina-like outline because of compaction of surrounding sediments. Outward flexures of the concertina are located at large lateral roots extending out into the matrix. Despite these characteristics the distinction between root traces and burrows is not always easy. Root mats may spread laterally over hardpans or around nodules. Some kinds of roots, such as the pneumatophores of mangroves, branch upwards and out of the soil (Jenik 1978).

Furthermore, a range of soil invertebrates, especially ants, termites, and worms, form complex branching burrow systems (Ratcliffe & Fagerstrom 1980) that may be irregular and partly collapsed in places. The distinction between root traces and burrows is further blurred by soil invertebrates such as cicadas, which burrow around and into roots to feed on them. This practice may have obscured the distinction between root traces and burrows in paleosols as old as Triassic (Retallack 1976). Other structures which could be confused for root traces include gas escape structures (Neumann-Mahlkau 1976) and tubular masses of soil fused by lightning strikes (Essene & Fisher 1986). The latter, called fulgurites, are lumpy masses of glass with exotic high-temperature minerals completely different from ordinary soil matrix. Gas escape structures, such as those forming the conduit to sand volcanoes in alluvium covering methane-generating organic matter, are not so copiously branched or pervasive as root traces. In most cases, wispy tubular structures forming an irregular, dense network within non-marine rocks are root traces.

One limitation on the use of root traces for recognizing paleosols is that they have not definitely been found in rocks much older than Silurian when the first vascular land plants appeared (Retallack 1985). There are burrows of invertebrates in paleosols as old as Ordovician (Retallack & Feakes 1987). For paleosols older than mid-Ordovician, root traces and burrows are of no use for identifying ancient soils.

Kinds of roots

Fossil root traces are most easily recognized when their original organic matter is preserved. Paleobotanical research has now unearthed fossil examples of most of the major kinds of root now found (Stewart 1983). Roots are downward-growing plant axes, with numerous fine branches or rootlets (Fig. 3.2). Both roots and rootlets are anatomically simple and similar in a wide variety of plants, unlike the anatomical diversity seen in aerial parts of plants. Usually a central cylinder (stele) of elongate woody cells (tracheids) is separated by a zone (cortex) of equidimensional fleshy cells (parenchyma) from a tough outer rim of thick-walled cells (epidermis). The central woody cylinder and tough outer rim withstand decay longer than the intervening zone of soft cells. Some partly decayed root traces may show a central dark woody streak and a carbonaceous epidermis separated by a zone filled with mud, calcite, or other minerals where the cortex has decayed (Retallack 1976). With further decay even the epidermal and stelar organic matter is replaced with other materials, but a concentric pattern of replacement may remain.

Root hairs are individual elongate epidermal cells found in zones near the tips of fine rootlets. Because of their increased surface area compared with that of older parts of root systems, they are especially significant in

Figure 3.2 Fossil root and rootlets from the Miocene, Molalla Formation on High Hill near Scotts Mills, Oregon, USA. Their original organic matter has been weakly ferruginized (Retallack specimen R261).

gathering water and nutrients from the soil. They are so small and delicate that they are only preserved under the exceptional circumstances of cellular permineralization (Retallack 1988a). Some Silurian and Devonian land plants, like living mosses and liverworts, lacked true roots. Their unicellular rhizoids performed a similar function. These also are preserved only under exceptional circumstances (Kidston & Lang 1917).

Various kinds of roots are distinguished by their pattern of branching and anatomical structure (Raven *et al.* 1981). Many plants have a single, thick vertical root or tap root, as in Devonian *Eddya* (Beck 1967). Carrots and parsnips are familiar modern plants that have tap roots modified into large underground structures for the storage of carbohydrates. Another kind of root system is seen in living grasses (Gramineae) and quillworts (*Isoetes*). These have fibrous roots radiating from a thickened stem base known as a corm or rhizophore, as in Triassic *Pleuromeia* (Retallack 1975). If the roots arise from the stem of a plant rather than its base, they are called adventitious roots. These may arise from rhizomes, which are stems lying in or along the ground. They also may anchor stems scrambling above ground (runners or stolons) as in modern strawberries and Carboniferous *Callistophyton* (Rothwell 1975). Adventitious roots also form prop or stilt roots connecting erect stems and branches to the ground, as in Cretaceous *Weichselia* (Alvin 1971). In tree ferns, such as Carboniferous *Psaronius*

(Morgan 1959), a very weak stem and leaf bases are completely enclosed by a mass of fine adventitious roots. They may look like tree trunks, but these masses of roots and leaf bases are best called false stems.

A variety of specialized structures of roots also are known from the fossil record. For example, tubers are underground storage organs branching from roots and rhizomes such as the common potato and Cretaceous *Equisetites* (Rushforth 1971). Some plants of waterlogged soils have rootlets which extend vertically into the air and these can be fossilized by continued deposition in the swamp and intertidal habitats where such roots are common (Whybrow & McClure 1981). These peg roots (pneumatophores) may play a role in allowing access to air for root respiration. For similar reasons, plants of waterlogged habitats may have thin-walled openings to the inside of the root (aerophores) or spongy parenchymatous tissue (aerenchyma). The most obvious of these aerating adaptations are large hollow cavities (lumina) found in roots of some swampland plants, such as Permian *Vertebraria* (Schopf 1982). Rooting structures such as these not only indicate the existence of paleosols, but are evidence of particular soil conditions.

From the known fossil record of roots, most kinds of rooting structures have been in existence since Carboniferous time. Extinct woody plants whose aerial parts were very different from those alive today, showed surprisingly modern kinds of roots. Presumably this is because functional constraints on root evolution have been more important than phylogenetic

Figure 3.3 Tabular root system of a large, extinct aborescent lycopod (*Stigmaria ficoides*) of Early Carboniferous age (Namurian) in the Lower Limestone Coal Group, Victoria Park, Glasgow, Scotland. The scale bar is 1 m for foreground only.

constraints. Useful paleoecological interpretations can be made from fossil roots by comparison with modern studies of root ecology and arrangement.

Patterns of root traces

When digging for fossil root traces it is useful to consider their arrangement because this may provide evidence for former drainage, vegetation types, and originally indurated parts of a paleosol (Sarjeant 1983). Because roots need oxygen in order to respire, they seldom penetrate permanently waterlogged parts of soils. Laterally spreading (or tabular) root systems are characteristic of plants growing in swampy ground (Jenik 1978) and are common among fossil stumps in sedimentary rocks of lowland environments (Fig. 3.3). On the other hand, well drained paleosols may be deeply penetrated by root traces. Under wooded grassland, deeply penetrating and stout roots of trees and shrubs are scattered among a diffuse network of fine (less than 2 mm diameter) grass roots (Van Donselaar-ten Bokkel Huinink 1966). In vegetation of drier climates, the pattern of roots becomes shallower and more irregular as the vegetation becomes more sparse and clumped. Under tall grass prairie, a network of fine roots may extend up

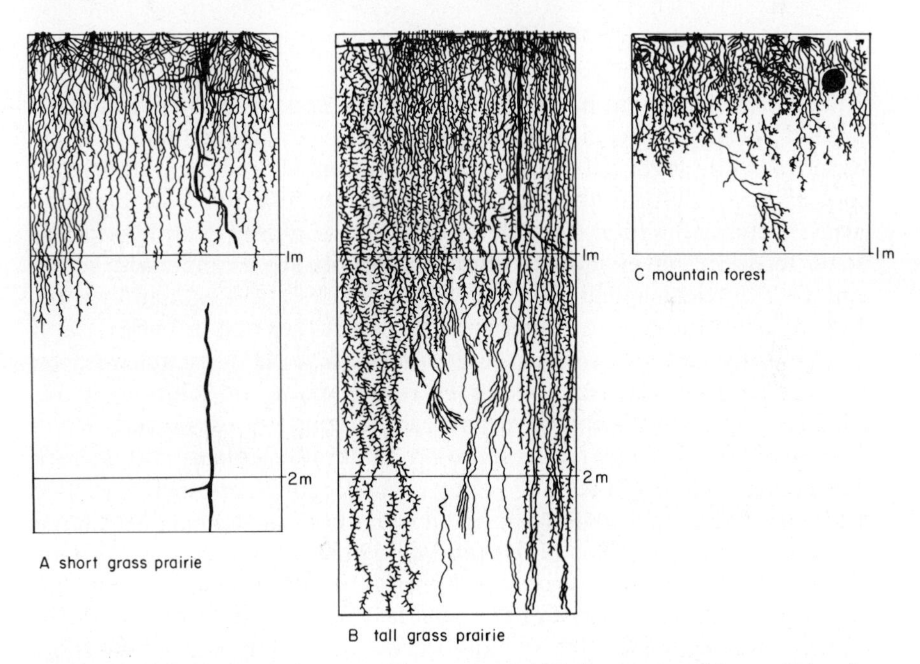

Figure 3.4 Scale drawings of excavated root systems of (A) short grass prairie near Colorado Springs, Colorado, USA, (B) lowland tall grass prairie near Lincoln, Nebraska, USA, and (C) mountain forest near Pikes Peak, Colorado (after Weaver 1919, 1920, with permisson of the Carnegie Institute of Washington).

to 2 m below the surface (Fig. 3.4). Under short grass prairie, grass roots are clumped under individual tussocks and interspersed with tubers and other rooting structures of desert perennials. Documentation of rooting patterns in modern soils is a laborious process involving digging, erecting a supportive net for the excavated roots, and then carefully washing them out (Weaver 1919, 1920). Such research provides a useful data base against which fossil root systems can be compared.

Patterns of root traces also are clues for distinguishing original features of paleosols from those formed after burial. Root traces tend to run around hard parts of soils such as pebbles, nodules, and cemented horizons (Haasis 1921). Abundant subhorizontal root traces deep within a profile are the best line of evidence for recognizing originally lithified horizons (fragipans and duripans) in paleosols now entirely lithified. Similarly, the avoidance of nodules by roots and burrows closely approaching them is evidence that the nodules were an original part of the soil. Not all root traces avoid nodules because in their early stages of formation nodules may be unindurated chemical segregations (Gile *et al.* 1966). With time they become indurated and better defined and may preserve root traces and other soil features within them. As one of the clearest primary features of paleosols, root traces are a guide to other original features of former soils.

Rhizoconcretions

Rhizoconcretions form in soils because of the special local environment created by roots. Water is taken in by roots because of the tendency for soil water to dilute their cell sap across semipermeable cell membranes (by osmosis) and because the whole plant maintains a negative pressure (water potential) within thin, water-conducting tubes (xylem) by loss of water from the leaves (transpiration). Nutrients are taken in with water and their uptake is enhanced by a variety of substances exuded by roots and their surrounding mucigel zone (rhizosphere) rich in bacteria and fungi. Many nutrient cations (Ca^{2+}, Mg^{2+}, K^{+}, Na^{+}) are released by displacement with hydrogen ions (H^{+}) in mildly acidic solutions maintained by organic acids and by carbonic acid arising from dissolved carbon dioxide of microbial and root respiration. Other nutrients such as iron are dissolved in a reduced state (Fe^{2+}) by organic reductants such as caffeic acid or are fixed in particularly favorable molecular sites (chelated) by large organic molecules such as EDTA (ethylene diamine tetracetic acid). This is not to say that the rhizosphere is always or uniformly acidic or reducing as was once thought. Most of the time it is near neutral in pH and Eh, allowing for normal aerobic respiration and nutrient uptake of both roots and associated microbes (Richards 1987). Conditions can change over short periods of time. Heavy rainfall may cause temporary waterlogged, reducing and acidic conditions. Long periods of nutrient starvation may induce dramatic increases in the production of exuded reductants or chelates over periods of a few hours (Olsen *et al.* 1981). The net effect of root action is thus to deplete the adjacent soil of nutrients. Zones of depletion are not especially prominent in soils because of the continued elongation of roots. The most actively adsorbing part is the zone of root hairs just behind the growing point of the root. This root apex may elongate at rates of more than 6 cm/day. About 2 cm/day is considered typical (Russell 1977). Because they are so transitory, detection of the effects of fossil rhizospheres is problematic.

Nevertheless, the effects of roots can be striking. Soils rich in calcium carbonate are widespread in desert regions where rainfall is insufficient to leach it from the soil, and also in coastal dunes and beaches in wetter regions where the sand includes numerous grains made of broken seashells. With repeated cycles of wetting and drying in such friable, sandy soils, the root margins become alternately wet and acidic (thus dissolving carbonate) and then dry and alkaline (thus precipitating carbonate). Under these conditions, roots may become heavily encrusted with concentric

Figure 3.5 Calcareous rhizoconcretions exhumed from Holocene coastal sand dunes north of Wanda Beach, New South Wales, Australia. Scale in millimeters (Retallack specimens P2606A, C, D).

layers of very fine-grained low-magnesian calcite (Fig. 3.5). These calcareous rhizoconcretions may become so thick and unyielding that the root dies and the hole remaining is filled with other materials (Esteban & Klappa 1983). When plugged and lacking organic matter in the center, they may superficially resemble the fused siliceous tubes created in soils by lightning strike, and for this reason calcareous rhizoconcretions have been called "pseudofulgurites." Similarly, iron mobilized in the drab ferrous state from minerals within the wet rhizosphere may be oxidized to yellow or red ferric oxides near the roots to form ferruginous rhizoconcretions. Calcium carbonate and iron oxyhydrates are the most common materials encrusting roots in soils, but a variety of other substances also form soil rhizoconcretions and nodules (Brewer 1976).

Drab-haloed root traces

A common feature of root traces in paleosols is a bluish or greenish gray halo extending out into the paleosol matrix (Fig. 3.6). Such drab haloed root traces can be formed in several different ways. Two of these are uncommon in paleosols and can readily be recognized using textural

Figure 3.6 Fossil root traces (2 mm diameter central streak of pale yellow, 5Y7/4), drab halo (light gray, 5Y7/2), and ferruginized matrix (dark red, 2.5YR3/6) from the subsurface (Bt) horizon of a paleosol of early Oligocene age in the basal John Day Formation, near Clarno, Oregon, USA. Scale in millimeters (Retallack specimen R417; paleosol undescribed).

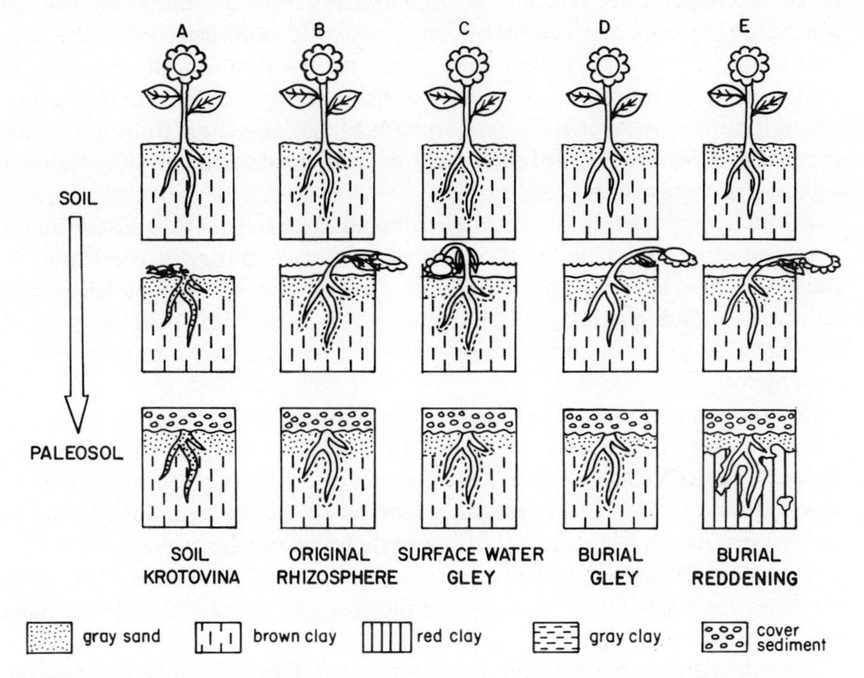

Figure 3.7 Processes and products in the formation of drab-haloed root traces and superficially similar features: (A) krotovinas, (B) remnants of the rhizosphere, (C) surface water gley, (D) reduction of soil by anaerobic decay of buried roots, and (E) dehydration of ferric oxyhydrate mottles.

relationships. First, drab root traces and burrows can form in red beds or horizons when drab material from a higher horizon is washed down into a lower horizon (Fig. 3.7). Such structures are called krotovinas, tonguing, or glossic features. The boundary between drab and red material of a krotovina is a sharp discontinuity between materials of different texture, unlike the diffuse contact in most drab haloes.

A second kind of drab halo forms around root traces in clayey, periodically waterlogged soils. Anaerobic bacterial activity in stagnant water around roots, burrows, and cracks in the soil can cause chemical reduction of those surfaces, leaving the interiors of the soil clods oxidized as before. Such surficial reduction may also result in mineralization with birnessite, pyrite or sphaerosiderite. These features of surface water gley are found in heavy-textured lowland soils with impermeable subsurface horizons. Such variegated soils have drab-haloed root traces accompanied by drab burrows and soil cracks. Surface water gleying does not explain soils in which only the root traces have drab haloes.

A third possible origin for drab-haloed root traces is as a result of reduction and mobilization of iron within the rhizosphere. However, drab-haloed root traces are seldom found in modern, well drained soils, but are abundant in paleosols. The highly reducing conditions implied by the halo would be too poor in oxygen for normal respiration of the deeply penetrating roots commonly haloed in this way. Furthermore, water and nutrient uptake is most active near the tips of rootlets where there are zones of root hairs, whereas drab haloes around fossil roots commonly are widest where the roots are largest.

A fourth explanation for drab-haloed root traces is that they are reduced areas of anaerobic bacterial decay of organic matter buried within paleosols. This burial gley origin of drab-haloed mottles is likely for paleosols in which the whole surface horizon is drab, from dispersed organic matter there, and in which there are comparable amounts of total iron in both drab and red areas.

The contrast between drab-haloed root mottles and surface horizons and the reddish remainder of many paleosols may have been enhanced by dehydration of yellow and brown ferric oxyhydrates to brick-red hematite during deep burial. However, it is unlikely that the original oxidation occurred at depth, especially within nearly impermeable clayey paleosols. This well established diagenetic change (Walker 1967) may have enhanced the color contrast of drab-haloed root mottles, but is unlikely to have been their cause.

If drab-haloed root mottles can be regarded as reflecting the former rhizosphere or anaerobic decay of buried organic matter, then the drab-haloed root traces represent the last crop of plants in the paleosol. Former roots and their rhizospheres decay rapidly once they die in well drained soils. Drab-haloed root systems are especially useful in distinguishing between forest, woodland, and wooded grassland of the past from paleosols (Retallack 1983b).

Soil horizons

The exact nature of soil horizons depends on conditions under which the soil formed and can be modified substantially upon burial in paleosols. There are, however, some general features of horizons useful for recognizing them even in highly altered paleosols. Horizons of paleosols are distinct from many kinds of geological layering in that the top of the uppermost horizon of a paleosol is usually truncated sharply, whereas boundaries between lower horizons and underlying parent material commonly are gradational (Fig. 3.8).

Exceptions to the general rule of diffuse contacts below the sharp top of paleosols are common enough to deserve special consideration. Some lowland soils receive thin increments of sediment through which vegetation continues to grow. Their tops are overthickened by sedimentation

Figure 3.8 Sharp upper contact and gradational lower contact of soil horizons in three paleosols: (A) Long Reef clay paleosol (probably a Hapludult) of Triassic age (Scytho-Anisian) in the Bald Hill Claystone at Long Reef, New South Wales, Australia [scale bar graduated in inches and paleosols discussed by Retallack (1977)] and (B) modern grassland soil (upper left only) and two comparable Chogo clay paleosols (probably Haplustolls) of Miocene age (about 14 million years) in the Fort Ternan Beds in the main excavation of Fort Ternan National Monument, Kenya [hammer for scale and paleosols discussed by Retallack (1986c)].

rather than sharply defined by erosional truncation. In these cases there may be several zones of denser rooting and burrowing that can be taken as surfaces formed during breaks in sedimentation. The one closest to the top of a paleosol can be taken as its surface and others above that as the tops of additional very weakly developed, younger soils.

Other exceptions to the generally diffuse boundaries of paleosol horizons are sharp contacts occasionally found within profiles. For the most part these are relict beds from sedimentary parent materials not yet obliterated by soil formation. Associated sedimentary features such as ripple marks or load casts allow confident identification of relict bedding. There also may be erosional surfaces within profiles where a pre-existing paleosol has been substantially eroded and soil development proceeded on an additional layer of sediment. Such cases may be difficult to detect in the field if the erosional contact has been obscured by subsequent soil formation. Thin lag deposits of weather-resistant pebbles such as quartz may be clues to complexities of this type of soil and paleosol. Stone lines are common in very ancient soils of stable geological settings (Johnson *et al.* 1987). Without such field indications, the true complexity of a paleosol may not become apparent until petrographic or chemical studies reveal discontinuities.

Kinds of horizons

Although paleosol horizons are varied, a mental image of common kinds of soil horizons can be useful in recognizing them in the field. Some kinds of paleosol horizons are so striking that they have attracted specific geological names. Cornstones (Fig. 3.8B), for example, are nodular calcareous horizons (or Bk in the shorthand of soil science) and ganisters are silicified, near-surface, sandy horizons (or E horizons of soil science). Successions of paleosols with gray–green, organic surface (A) horizons and red to purple, clayey subsurface (Bt) horizons (Fig. 3.8A) may form strikingly scenic sequences as gaudy as a barber pole or candy cane. Some Precambrian paleosols have surface (A) horizons of a distinctive lime-green color. Some kinds of paleosol horizons may prove to be extinct, but for most the horizon nomenclature of soil science is appropriate.

The variety of horizons found in modern soils are labeled with a shorthand system of letters and numbers (such as A and Bt). Laboratory studies may force one to change the designation of a horizon, but it is best to attempt to work out the nature of paleosol horizons in the field. Field assessment of horizons may determine how a paleosol is sampled and also its interpretation and identification in a modern classification of soils. The field nomenclature for horizons is descriptive rather than genetic in orientation (Guthrie & Witty 1982). Horizon nomenclature has remained relatively stable over the years, although some changes have recently been proposed (Table 3.1).

Horizons are defined on the basis of the materials that comprise them. For buried soils, it also is useful to have an idea of typical thicknesses of different kinds of horizons in modern soils. Most surface horizons of soils (A and E) are less than 50 cm thick and subsurface (Bt and Bk) horizons are less than 2 m thick (Fig. 3.9). Unaltered sediments, duricrusts, and saprolites, on the other hand, may be much thicker and reflect the operation of processes other than soil horizon differentiation.

The different kinds of soil horizons reflect varied conditions of soil formation, but the degree of expression of horizons can be related to the intensity and duration of soil formation. For example, clayey subsurface (Bt) horizons form at the expense of bedding of alluvial parent materials most rapidly under forest vegetation in warm humid climates. Calcareous subsurface (Bk) horizons develop in less deeply weathered soils in dry climates (Birkeland 1984). The nature of paleosol horizons is therefore a guide to the former conditions under which they formed. The genetic significance of horizons has also been recognized in soil classifications,

Table 3.1 Descriptive shorthand for labelling paleosol horizons.

Category	New term	Description	Old term
Master horizons	O	Surface accumulation of organic materials (peat, lignite, coal) overlying clayey or sandy part of soil	O
	A	Usually has roots and a mixture of organic and mineral matter; forms the surface of those paleosols lacking an O horizon	A
	E	Underlies an O or A horizon and appears bleached because lighter colored, less organic, less sesquioxidic or less clayey than underlying material	A2
	B	Underlies an A or E horizon and appears enriched in some material compared to both underlying and overlying horizons (because darker colored, more organic, more sesquioxidic or more clayey) or more weathered than other horizons	B
	K	Subsurface horizon so impregnated with carbonate that it forms a massive layer	K
	C	Subsurface horizon, slightly more weathered than fresh bedrock: lacks properties of other horizons, but shows mild mineral oxidation, limited accumulation of silica, carbonates, soluble salts or moderate gleying	C
	R	Consolidated and unweathered bedrock	R
Gradations between master horizons	AB	Horizon with some characteristics of A and of B, but with A characteristics dominant	A3
	BA	As above, but with B characteristics dominant	B1
	E/B	An horizon predominantly (more than 50%) of material like B horizon, but with tongues or other inclusions of material like an E horizon	A&B
Subordinate descriptors	a	Highly decomposed organic matter	—
	b	Buried soil horizon (used only for pedorelict horizons within paleosols: otherwise redundant)	b
	c	Concretions or nodules	cn
	e	Intermediately (between a and i) decomposed organic matter	—
	f	Frozen soil, with evidence of ice wedges, dikes or layers	f
	g	Evidence of strong gleying, such as pyrite or siderite nodules	g
	h	Illuvial accumulation of organic matter	h
	i	Slightly decomposed organic matter	—
	k	Accumulation of carbonates less than for K horizon	ca
	m	Evidence of strong original induration or cementation, such as avoidance by root traces from adjacent horizons	m
	n	Evidence of accumulated sodium, such as domed columnar peds or halite casts	sa
	o	Residual accumulation of sesquioxides	—
	p	Plowing or other comparable human disturbance	p
	q	Accumulation of silica	si
	r	Weathered or soft bedrock	ox
	s	Illuvial accumulation of sesquioxides	ir
	t	Accumulation of clay	t

Table 3.1 *Continued*

Category	New term	Description	Old term
	v	Plinthite (in place, pedogenic laterite)	—
	w	Colored or structural B horizon	—
	x	Fragipan (a layer originally cemented by silica or clay and avoided by roots)	x
	y	Accumulation of gypsum crystals or crystal casts	cs
	z	Accumulation of other salts or salt crystal casts	s

Note: This table has been adapted for use with paleosols from one by Guthrie & Witty (1982) showing proposed terminology of the new edition of the U.S.D.A. Soil Survey Manual compared to that of the 1951 edition. Some of the subordinate descriptors are thought to be more important than others so these letters (a,e,i,h,r,s,t,v,w) should all be written first after the master horizon if in combination with other letters and they should not be used in combination with each other. Master horizons can be subdivided by numbers (e.g., B1,B2,B3). If the parent material of a paleosol consists of interbedded shale and sandstone, these will show different kinds of alteration in the same profile. Such different layers separated by discontinuities are numbered from the top down without using the number 1 (e.g., A,E,E/B,Bt,2Bt,2BC,2C,3C). If you can form a clear mental picture of this profile, you are well on the way to mastering this pedological shorthand.

which are based on the nature and arrangement of horizons (Soil Survey Staff 1975).

Describing soil horizons

In order to characterize a paleosol in a way that is amenable to interpretation, horizon thicknesses, grainsize, color, reaction with acid and the nature of horizon boundaries must all be recorded in the field. Some observations will prove more important than others in ultimately understanding a paleosol, but it is difficult to anticipate which features these will be. Hence, it is useful to have a comprehensive standardized form of horizon description (Retallack 1988a). Especially useful is graphical field logging in a large, quad-ruled notebook, as is the custom in making geological sections of sedimentary sequences. Like the interpretative shorthand used to describe soil horizons (Table 3.1), these various symbols and style of representation may at first appear intimidatingly technical (see Figs. 2.5, 15.5, 16.3, 16.5, 18.5, 18.8, 19.2, 19.4, 19.5, 20.4, 20.9, 21.4, and 21.5). With familiarity, however, this becomes a rapid way of summarizing and comparing profile characteristics.

Information on grain size is needed for the classification and interpretation of many soils and paleosols. A profile showing mean grain size, separate from the column with lithological symbols, is a useful feature of graphical profiles of paleosols. Grain size can be estimated in the field using a grain-size comparison card or sediment samples. It can later be reassessed by laboratory studies of samples of the horizons, which should

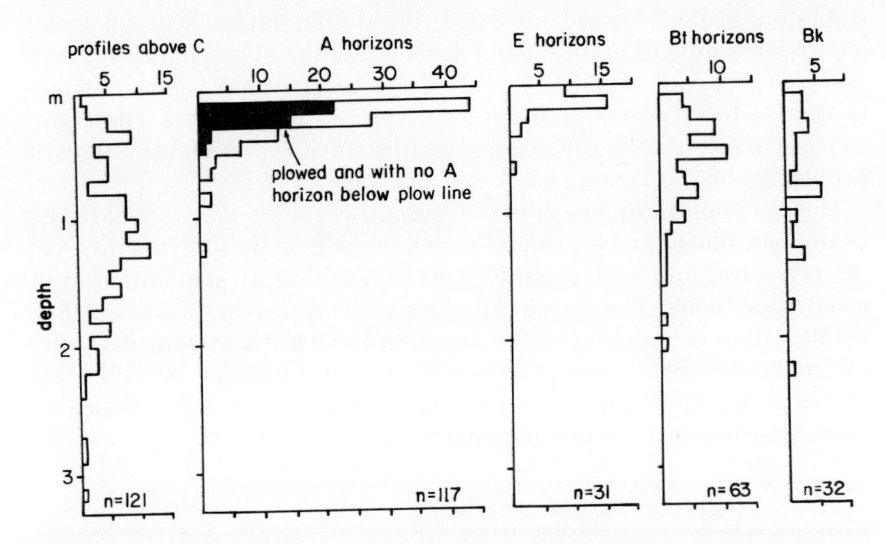

Figure 3.9 Typical depths for different kinds of soil horizons as revealed by their frequency distribution in North American soils (compiled from Marbut 1935).

be clearly labeled with an indication of their former orientation in the paleosol. The simplest way of marking orientation is to draw or scratch a large circle (1 in or 2 cm in diameter) on the upper surface of the sample or sample container in such a way that it is parallel to the ancient land surface. Such samples can be made into vertically oriented thin sections (Murphy 1986) and then point-counted under a petrographic microscope to determine more precisely the amounts of sand, silt, and clay. Common components can be determined with an accuracy of about 2% when 500 points are counted (Murphy 1983). Alternative methods of passing unconsolidated soil through sieves or through a column of water and then weighing each segregated size fraction are techniques used in soil science (Klute 1986). They are of limited usefulness in evaluating the grain size of lithified paleosols because there is no easy way of returning them to a disaggregated state that reflects their original grain-size distribution.

Color can be estimated in the field using a standard color chart such as that produced by Munsell Color (1975). This is arranged to reveal changes in the primary colors such as red and yellow (hues), in the degree of lightness of the color (values), and the degree of greyness of the color (chroma). It is best to consider only the hue initially in order to find the correct page of the charts. Then the color can be determined within a grid of value and chroma. Well indurated and metamorphosed paleosols may hold their color well, but little altered clayey paleosols of the kind widespread in scenic clayey badlands of Mesozoic and Cenozoic rocks in the western United States change color on exposure to air and on laboratory storage. Commonly they become paler (higher Munsell value) after a few hours of drying. After several months of laboratory storage, greenish gray parts of paleosol samples may become more yellow (warmer in Munsell hue) because of oxidation of iron-bearing minerals. Therefore, it is best to record color consistently on fresh rock within a few minutes of exposure.

The calcium carbonate content of paleosols can be determined in the laboratory, but should also be estimated in the field by applying drops of dilute hydrochloric acid (about 10% of standard 1 molar solution) from an eye-dropper bottle. The degree of reaction with acid can be divided into five readily observable stages (Table 3.2) proportional to the amount of calcium carbonate present. This property is used for classifying soils. An acid bottle also is useful for its more conventional geological use of distinguishing calcium carbonate from dolomite or chert.

Table 3.2 Scale of acid reaction to approximate carbonate content of paleosols (from Retallack 1988a)

Carbonate content	Reaction with dilute acid
Noncalcareous	Acid unreactive; often forms an inert bead
Very weakly calcareous	Only a little movement within the acid drop which could be flotation of dust particles as much as bubbles
Calcareous	Numerous bubbles, but not coalescing to form a froth
Strongly calcareous	Bubbles forming a white froth, but drop of acid not doming upwards
Very strongly calcareous	Drop vigorously frothing and doming upwards

Another feature of horizons which should be recorded in the field is the nature of their contacts with adjacent horizons. Two aspects of the contact are of interest: whether one horizon passes into another within a narrow or broad vertical distance, and whether the contact is laterally planar or somehow disrupted (Table 3.3). Abrupt or broken boundaries may lead to the suspicion that the profile contains erosional discontinuities. Diffuse planar contacts, on the other hand, may represent genetically related horizons of a single soil. The transition zones between diffuse horizon contacts may seem so thick as to warrant a separate horizon name. When assessing a profile with especially diffuse horizon boundaries, it is best to take an overview of the profile to decide which parts of the profile are most distinct and thus constitute the main horizons, and which are merely intergrades between them.

Horizons also may be characterized by laboratory analyses of hand specimens, but such studies are only as good as the fieldwork on which

Table 3.3 Sharpness and lateral continuity of boundaries of paleosol horizons (from Soil Survey Staff 1975)

Category	Class	Features
Sharpness	Abrupt	Transition from one horizon to another completed within 1 inch (2 cm)
	Clear	Transition completed within 1 to 2.5 inches (2-5 cm)
	Gradual	Transition spread over 2.5 to 5 inches (5-15 cm)
	Diffuse	One horizon grading into another over more than 5 inches (15 cm)
Lateral continuity	Smooth	Horizon boundary forms an even plane
	Wavy	Horizon boundary undulates, with pockets wider than deep
	Irregular	Horizon boundary undulates, with pockets deeper than wide
	Broken	Parts of the adjacent horizon are disconnected, e.g., by deep and laterally persistent clastic dikes in Vertisols

they are based. Chemical analyses for major and trace elemental composition may be useful. For lithified paleosols it is best to use whole rock chemical analyses by standard geological methods such as atomic absorption, X-ray fluorescence, inductively coupled plasma atomic emission spectrometry or neutron activation analysis. Full chemical analyses were once fashionable in the study of modern soils (Marbut 1935) and this important data base of analyses is occasionally supplemented with additional published analyses. Determination of materials extractable by solvents, such as iron by sodium dithionate, is now a more common approach of soil science (Page 1982). This readily soluble fraction of the soil is most susceptible to diagenetic change upon burial of a soil, and this limits the use of these analyses of paleosols. A useful adjunct to chemical analyses is to determine the density of samples by the standard method of weighing samples coated in paraffin, to prevent water infiltration, both in and out of water (Klute 1986). Using density measurements, chemical analysis results in weight percent can be converted to values in grams per cubic centimeter in order to gain an inventory of absolute chemical differences between horizons (Brewer 1976). Staining for feldspars (Houghton 1980), scanning electron microscopic observation (Mumpton & Ormsby 1976, Smart & Tovey 1981, Sudo *et al.* 1981), X-ray diffractometric studies of clay minerals (Klute 1986) and X-radiographs to reveal partly concealed structures (Bouma 1969) are just a few of a growing number of laboratory techniques for understanding particular aspects of paleosol horizons.

Soil structure

Soils may appear to be fragmented, featureless, or massive compared, for example, with cross-bedded dune sands or strongly cleaved slates. However, like these geological structures, soils have characteristic structures of their own. These structures are developed to different degrees and progressively overwhelm pre-existing structures of parent material such as sedimentary bedding, metamorphic foliation, or igneous crystal outlines. With experience, both the characteristic features of soil structure and their contrast with other structures of enclosing rocks enable paleosols to be readily recognized.

Structures of paleosols often can be discerned in geological descriptions by such terms as "massive," "structureless," "jointy," "slickensided," "veined," "mottled," or "nodular." Such language does not serve adequately to characterize soil or paleosol structures. On the other hand, the technical jargon for these features used by soil scientists can be intimidating. In this account, the technical terms used are based on the system outlined by Brewer (1976). Common words are suggested here as equivalent terms. In some cases these ordinary terms may prove inadequate, but they serve to introduce the subject.

Structural elements

Perhaps the most striking feature of soils is their intricate system of open cracks and hollows (Fig. 3.10). These open spaces may form an interconnected network (packing voids), small irregularly shaped pockets (vughs) or near-spherical holes (vesicles). In paleosols such open spaces are often crushed out of existence by compaction under overlying rocks. Fortunately, some indications of former cracks remain because of modification of soil material where water and air could circulate within the soil. These surfaces can be modified in various ways, e.g., by encrustation with washed-down clay or staining with iron oxide. Such irregularly planar features in soils and paleosols are called cutans. They could also be called clod skins because it is the cutans and the voids with which they are often associated that define the fundamental units of soil structure, the individual clods or peds. Peds may be of various sizes, ranging from large prisms occupying most of the thickness of the soil to small granules the size of sand grains. Large peds may be made up of smaller peds.

Burrows and root traces are two examples of an additional general class of tubular features found in soils: pedotubules or soil tubes. Such a general term is needed for an objective description of these features because their origin is not always obvious.

Another general class of soil structures is glaebules, or naturally hardened soil lumps. These are masses of material which have a distinctive

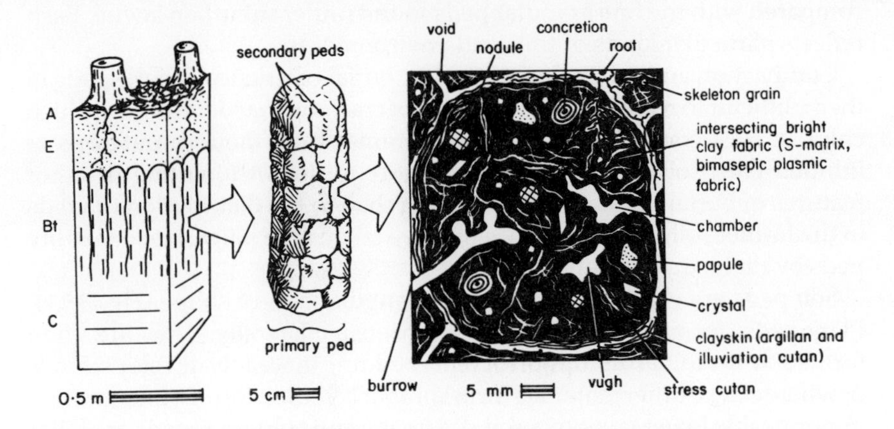

Figure 3.10 Structural units of soils in hand specimens and petrographic thin section.

mineralogical and chemical composition. Calcareous nodules of aridland soils are a good example. Crystals of minerals such as gypsum also are found within soils.

Pedotubules, glaebules, and crystals have a distinctive and often also simple mineralogical composition, unlike the clayey material making up most of the soil. On a microscopic scale this fine-grained soil matrix (plasma) consists of clay and amorphous iron and aluminum oxides which support small grains of rocks and minerals (skeleton grains) inherited from the parent material of the soil. The process of weathering tends to create plasma from skeleton grains. Especially prone to alteration are minerals such as plagioclase and hornblende. Quartz, on the other hand, is resistant and is a common skeleton grain in highly weathered soils and paleosols.

Much can be done to characterize soil structure in the field, but a complete description requires microscopic examination. Petrographic thin sections are useful for documenting differences between horizons of paleosols in addition to their microfabric (Bullock *et al.* 1985).

Peds

Peds are aggregates of soil: the clods of earth between cracks, roots, burrows, and other soil openings. Soil peds commonly can be crushed by hand and are not as indurated as rock. They are the persistent clods of the soil that tumble loose when one digs into it. Peds may initially seem so irregular and to be formed by such a random concatenation of processes as to be unworthy of careful scientific scrutiny. Even a little experience with soils is sufficient to dispel such doubts. Consider, for example, the very different domed columnar peds of soils around salt pans of desert regions compared with the fine granular peds found under suburban lawns. Each reflects particular kinds of soils and environments.

Compaction and other alteration after burial of paleosols make peds in them difficult to recognize. The pattern of cracking that determines which chips of rock come loose under the hammer from modern outcrops of lithified paleosols in most cases has more to do with jointing or other features of burial or modern weathering than to the distribution of voids in the former soil. Therefore, care must be taken with paleosols to identify peds by their surrounding cutans.

Soil peds are classified by their size, angularity and shape (Fig. 3.11). Platy peds, for example, are thin but extensive laterally. They are often formed by the initial disruption of relict bedding in weakly developed soils or where clay or other materials form laminar layers on top of a fine-grained impermeable lower layer of a soil. Prismatic and columnar peds are taller than they are wide and may extend through a considerable portion of a soil. They form in clayey soils by swelling and shrinking associated with wetting and drying. Some especially expandable clays such as sodium smectites form bulging tops to the peds. These bulges distinguish columnar from prismatic peds. Blocky peds (Fig. 3.12) are irregular in shape and usually not exactly equidimensional. They may have angular interlocking faces. Where there has been erosion or coating of ped margins they appear less sharp and subangular. A distinctive kind of blocky ped is formed in soils with highly smectitic clays in which diagonal stresses are associated with the swelling of clay as it takes in moisture after each rain. These lentil peds are separated by slickensides (Krishna & Perumal 1948). Granular and crumb peds are differentiated on the sharpness of their interlocking edges, as are angular and subangular blocky peds. Both granular and crumb peds are smaller and more equidimensional than

TYPE	PLATY	PRISMATIC	COLUMNAR	ANGULAR BLOCKY	SUBANGULAR BLOCKY	GRANULAR	CRUMB
SKETCH							
DESCRIPTION	tabular and horizontal to land surface	elongate with flat top and vertical to land surface	elongate with domed top and vertical to surface	equant with sharp interlocking edges	equant with dull interlocking edges	spheroidal with slightly interlocking edges	rounded and spheroidal but not interlocking
USUAL HORIZON	E,Bs,K,C	Bt	Bn	Bt	Bt	A	A
MAIN LIKELY CAUSES	initial disruption of relict bedding; accretion of cementing material	swelling and shrinking on wetting and drying	as for prismatic, but with greater erosion by percolating water, and greater swelling of clay	cracking around roots and burrows; swelling and shrinking on wetting and drying	as for angular blocky, but with more erosion and deposition of material in cracks	active bioturbation and coating of soil with films of clay, sesquioxides and organic matter	as for granular; including fecal pellets and relict soil clasts
SIZE CLASS	very thin < 1 mm	very fine < 1 cm	very fine < 1 cm	very fine < 0.5 cm	very fine < 0.5 cm	very fine < 1 mm	very fine < 1 mm
	thin 1 to 2 mm	fine 1 to 2 cm	fine 1 to 2 cm	fine 0.5 to 1 cm	fine 0.5 to 1 cm	fine 1 to 2 mm	fine 1 to 2 mm
	medium 2 to 5 mm	medium 2 to 5 cm	medium 2 to 5 cm	medium 1 to 2 cm	medium 1 to 2 cm	medium 2 to 5 mm	medium 2 to 5 mm
	thick 5 to 10 mm	coarse 5 to 10 cm	coarse 5 to 10 cm	coarse 2 to 5 cm	coarse 2 to 5 cm	coarse 5 to 10 mm	not found
	very thick > 10 mm	very coarse > 10 cm	very coarse > 10 cm	very coarse > 5 cm	very coarse > 5 cm	very coarse > 10 mm	not found

Figure 3.11 A classification of soil peds (data simplified from Soil Survey Staff 1975).

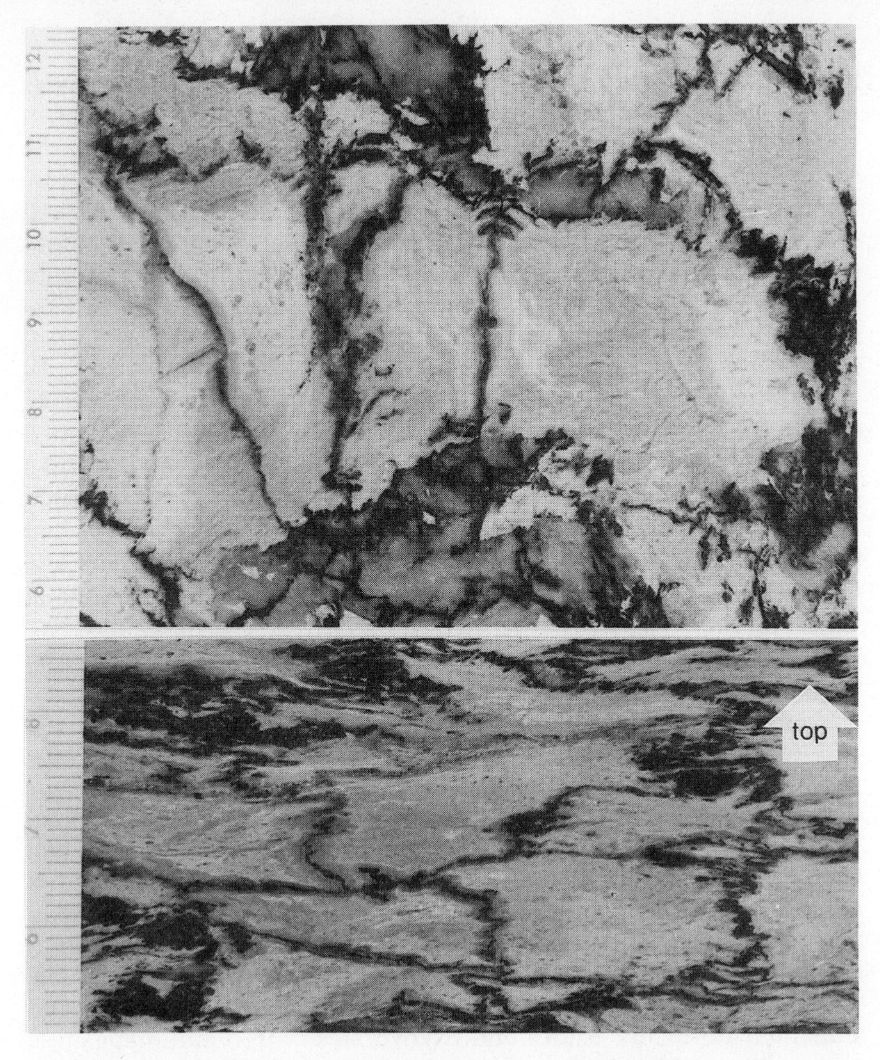

Figure 3.12 Angular blocky peds in polished slabs sawn horizontally (above) and vertically (below) from the subsurface (AC) horizon of the Waterval Onder clay paleosol, 2200 million years old, Transvaal, South Africa. Scale is graduated in centimeters and millimeters (from Retallack 1986a, reprinted with permisson of Elsevier Science Publishers).

blocky peds. They are found in upper portions of soils whereas blocky peds are more often encountered in subsurface (Bt) horizons. The gradation in structure from blocky to granular and crumb peds may be related to the increasing role of soil organisms in churning the soil and producing organic compounds such as polysaccharides that bind ped surfaces. Granular ped structures are common in the surface horizons of grassland soils, and so are crumb peds which may in part be fecal pellets of earthworms and other creatures. A distinctive kind of crumb ped called spherical micropeds are sand-sized spheres of deeply weathered and ferruginized clay abundant in some tropical soils.

Spherical micropeds and blocky and prismatic peds are superficially similar to clay clast breccias or conglomerate beds, to tectonically formed fitted breccias or tesselated pavements, and to mineralized vein networks or boxworks. These other geological structures commonly are associated with high-temperature minerals, marine fossils, or other indications that they did not form in a soil (Suppe 1985; Edwards & Atkinson 1986). The most diagnostic features of peds in paleosols are complexly altered bounding surfaces, peds within peds, traces of former surfaces annealed within peds, and cracks between peds that widen in a way that indicates the peds were loose enough to rotate slightly (Fig. 3.12).

Cutans

Cutans are modified surfaces of peds. They vary greatly in composition, which is the main basis for their classification. In the terminology proposed by Brewer (1976), various kinds of cutans are given latinate names indicating their composition, followed by the suffix "-an" to indicate that they are cutans. Argillans, for example, are clay skins. Ferrans are iron-stained surfaces. Skeletans are veins of skeleton grains or what a geologist would call a clastic dike. Other cutans may consist of oxides of iron and aluminum (sesquan), oxides and oxyhydrates of iron and manganese (mangans), soluble salts such as gypsum (soluans), calcite (calcans), silica in the form of opal or chalcedony (silans), or organic matter (organans). Similar cutanic features can form also during diagenesis and metamorphism of paleosols. Such later features are usually thicker, more coarsely crystalline and related to other distinctive metamorphic textures such as kink folding or schistosity. In contrast, the cutans of the original soil before metamorphism are irregular, thin, and fine grained. The chemical composition of cutans thought to be original can be important guides to former chemical conditions in paleosols (Fig. 3.13). Non-calcareous, nonclayey, ferruginous cutans (ferrans), for example, indicate acidic and highly oxidizing conditions as would be found in well drained, sandy soils of humid climates. This kind of information can be useful for interpreting and classifying paleosols.

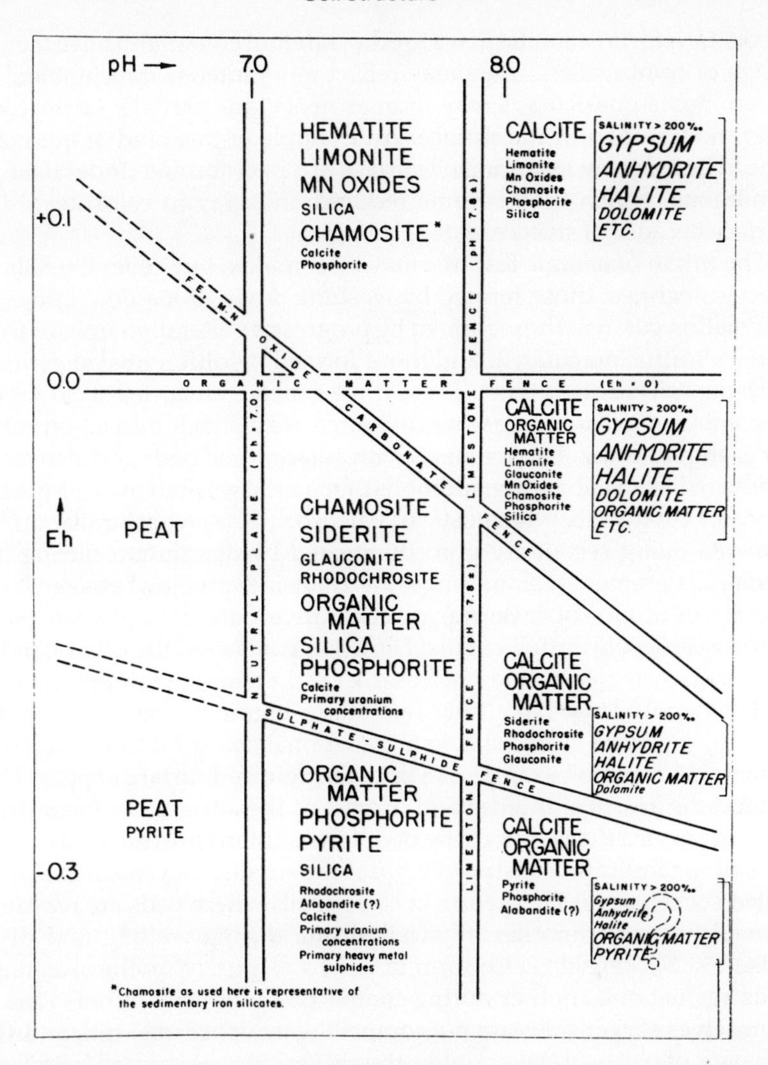

Figure 3.13 Stable mineral associations under different conditions of aeration (Eh) and acidity (pH) in sediments and soils. "Limonite" is a mixture of goethite and other ferric oxyhydrates (from Krumbein & Garrels 1952, reprinted with permission of University of Chicago Press).

Subcutanic features show a fixed relationship to a ped surface within the soil, but are not restricted to that surface. Two kinds are recognized. Neocutans (deep clod skins) are extraordinarily thick cutans which often lose intensity of development over a zone of centimeters away from the surface. Quasicutans (halo clod skins) also are thick and show a relation to a surface but are not right at that surface. They form a diffuse zone of alteration beneath the surface which faithfully follows its outline.

Neocutans form in similar ways to other kinds of cutans and have the same range of composition. They may reflect more intense development of a cutan. Some quasicutans may form as neocutans partially destroyed by later modification of the surface. An example of this kind of quasicutan would be a deeply ferruginized surface of a ped, formed under oxidizing conditions, which then became reduced and gray in color nearest the surface because of surface water gleying.

The origin of cutanic features may be complex, but generally falls into three categories: those formed by washing down of material into cracks (illuviation cutans), those formed by progressive alteration inward from a surface (diffusion cutans), and those formed by differential shear forces with the soil (stress cutans). These three genetic categories of cutans may be combined in confusing ways, but often are distinct. Illuviation cutans, for example, show sharp contacts with adjacent soil peds and also may be laminated from successive additions of material washed in. These are the kinds of cutans most diagnostic of paleosols. Somewhat similar are clay cements filling secondary porosity created by dissolution during deep burial. Clay cements seldom show the grainsize variation between layers, lateral continuity of layering, or correlative banding between broken pieces seen in illuviation cutans. Diffusion cutans, on the other hand, do not consist of material different from that of the peds but rather of modified ped material. They have only one sharp boundary on the outermost surface, the inner surface being a gradational change into unaltered soil material. A common example is a thick ferruginized surface of peds. These neoferrans are most intense and opaque at the surface, and less deeply stained toward the center of the ped. Stress cutans are less well defined. A common indication of stress cutans is the striated and smeared surfaces called slickensides. These form in clayey soils where peds are repeatedly heaved past one another by swell–shrink during wetting and drying episodes. Slickensides also form in paleosols purely by the crushing of peds against one another during compaction following burial. Thus, by themselves slickensides are not compelling evidence of shrink and swell behavior of paleosol clays. Unlike the slickensides associated with faults, those formed in soils or during compaction of paleosols are randomly arranged along diffuse zones rather than unidirectional and concentrated in narrow bands. In thin section, on either side of the highly birefringent, striated clay of the slickensided surface, the transition outward to unoriented clay is gradational with decreasingly abundant wisps of highly birefringent clay.

Glaebules

Glaebules are naturally segregated lumps of soil material (Brewer 1976) formed of the same wide variety of materials as cutans. Common examples are the calcareous nodules of desert soils. Glaebules range from highly irregular to almost spherical in shape and also vary in their distinctness and internal structure, but can be distinguished from cutans by their non-planar shapes and generally more distinct outlines.

Two common kinds of glaebules are nodules and concretions. These two terms are widely confused as synonyms, but for soil science are distinguished on the basis of their internal structure. A nodule is massive internally whereas a concretion contains concentric layers. This difference can be discerned by breaking them open or making a thin section through them (Fig. 3.14). Internal structure may have significance for the way in which they form. Continuous growth or recrystallization forms nodules, whereas discontinuous, often seasonal, growth forms concretions. Some nodules have a system of cracks radiating from their center. These septarian nodules appear like miniature turtle shells. They form when nodules change volume, e.g. by drying out irreversibly. This sometimes happens when siderite nodules, formed in stagnant water in lowland soils, are exposed to air. Both nodules and concretions are hard and have a sharp and distinct outer boundary. They contrast with another kind of glaebule, called mottles, which are diffuse patches of the same kinds of materials. Mottles may in some cases be early stages in the differentiation of nodules or concretions, but for the most part are irregular patches of discoloration.

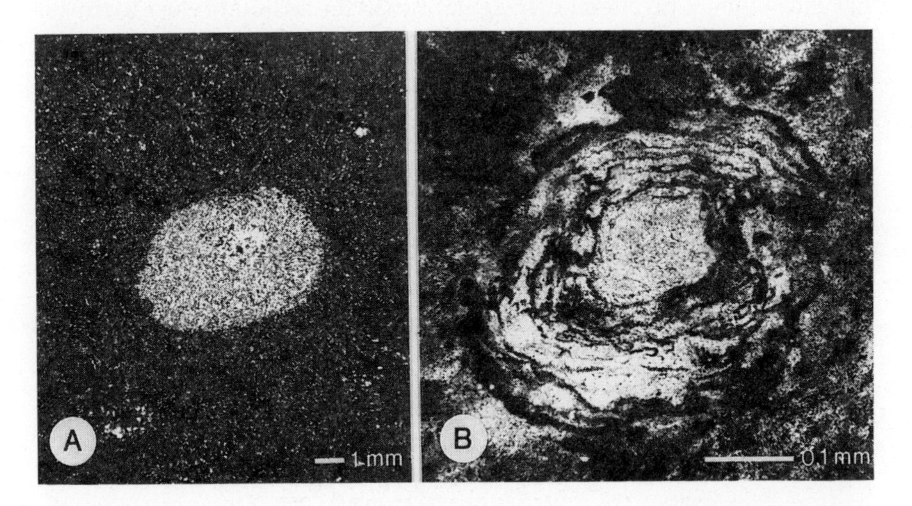

Figure 3.14 Thin sections of a calcareous nodule (A) and ferruginous concretion (B). The nodule is from horizon Bt of the type Lal clay paleosol (Oxic Haplustalf) from the Late Miocene (8 million years old) Dhok Pathan Formation near Kaulial village, northern Pakistan [Smithsonian specimen 353779; paleosol described by Retallack (1985)]. The concretion is from the Bt horizon of the uppermost Alfisol in bluffs of mid-Cretaceous (Cenomanian) Upper Dakota Formation along the Saline River near Russell, Kansas, USA [Retallack specimen R94; paleosol described by Retallack & Dilcher (1981a, b)]. Scale bars are 1 and 0.1 mm, respectively.

A final kind of glaebule is local aggregations of clay with sharp boundaries. In some cases it may be clear that these are fragments of older shale or claystone in the parent material of the soil (lithorelicts), fragments of other soils in the parent material (pedorelicts), fragments of clay-filled cavities in the soil (illuviation argillans), pockets of minerals locally weathered to clay (clay nodules), or feces of soil-ingesting animals (fecal pellets). For the cases where the origin of such features is not so clear that they can be assigned to one of these genetic categories, the terms papule and clay gall are convenient, descriptive terms.

Glaebules can be described by their composition, which may be sesquioxidic, manganiferous, calcareous, sideritic, pyritic or siliceous. These compositions reflect particular soil conditions (Fig. 3.13), provided they can be shown to have originated in the soil. Glaebules can be described as spherical, ellipsoidal, tuberose (irregular like a potato) or irregular (variable, but with rounded edges unlike a cutan). Mottles vary in the distinctness of their boundary with surrounding soil. These and other features of glaebules and mottles are best assessed in the field using classes set out by the US Soil Conservation Service (Table 3.4).

Glaebules are abundant and conspicuous features of many soils and paleosols, but are not diagnostic of them. Marine sedimentary rocks also contain glaebules. In some cases nodules in marine rocks are thought to have formed within chemical irregularities in the sediment close to the sea floor in a way analogous to their formation in soils, but they also form

Table 3.4 Size, abundance and contrast of mottles in paleosols (from Soil Survey Staff 1975)

Category	Class	Features
Contrast	faint	Indistinct mottles or glaebules visible only on close examination - both mottles and matrix have closely related hues and chromas
	distinct	Mottles are readily seen - with hue, value and chroma different from that of surrounding matrix
	prominent	Mottles are obvious and form one of the outstanding features of the horizon - their hue, value and chroma differing from that of the matrix by as much as several Munsell color units
Abundance	few	Mottles occupy less than 2% of the exposed surface
	common	Mottles occupy about 2 to 20% of the exposed surface
	many	Mottles occupy more than 20% of the exposed surface. This class can be subdivided according to whether (a) the mottles are set in a definite matrix or (b) the sample is almost equally two or more kinds of mottle
Size	fine	Mottles less than 5 mm diameter in greatest visible dimension
	medium	Mottles between 5 and 15 mm in greatest dimension
	coarse	Mottles greater than 15 mm in greatest dimension

during deep burial (Coleman 1985). Glaebules also form in cooling volcanic tuffs (thunder eggs; Staples 1965), on ocean floors (manganese nodules; Sorem & Fewkes 1979), in shallow lakes (lime balls; Kindle 1934), and around springs (tufa; Dunn 1953). Care must therefore be taken to determine whether glaebules were formed in the parent material of a soil, in the soil itself, or later during burial of a paleosol. Sometimes this can be established from their relationship with original features of the soil such as root traces and burrows, and from their general chemical compatibility with other original features of the paleosol. Abundant nodules in a rock may spark a suspicion that it was a soil, but considerable care must be exercised in interpreting them.

Crystals

Crystals occur in some kinds of soils and paleosols, especially in cracks and cavities such as root channels and burrows (Maglione 1981). Crystals can form tubular aggregates (crystal tubes), nodular masses (crystal chambers), spherical aggregates of radiating crystals (spherulites), or sheet-like aggregates (crystal sheets), or occur as single isolated crystals embedded in a matrix (intercalary crystals). Some distinctive kinds of crystal enclose matrix grains of the soil. Instead of appearing uniform, translucent, and sharp-edged like normal crystals, these look like a piece of sandstone or siltstone, but in the shape of a crystal. A common example of these is the so-called "sand crystals" of gypsum (MacFayden 1950). In very old saline paleosols, sabkhas, or playas in which original crystals have been strongly modified by diagenesis or metamorphism, only the impression or filled cavity of crystals may remain. The form of crystals in some cases may be sufficiently distinctive to allow identification. Small cubes after halite, often with concentric muddy fill and slightly bowed sides, are a common example of such crystal pseudomorphs (Haude 1970).

Crystals are mostly found in alkaline and salty soils. Like nodules, their composition may reflect a particular soil chemistry (Fig. 3.13). The most common kinds of crystals in soils are calcite and gypsum, which are found in soils of arid climates. One indication that crystals are an original part of a paleosol is displacive fabric, i.e., the way in which peds, nodules or other coherent parts of the soil have fallen or rotated into the crystal-filled region (Fig. 3.16C). Crystals also form during burial of a paleosol, e.g., in joints and veins or in cavities formed by dissolution of soluble parts of a paleosol. Crystals, like glaebules, can be common but are not diagnostic features of paleosols

Pedotubules

Burrows and root traces are prominent tubular features in soils and paleosols, but are not easy to distinguish in all cases from fulgurites, water escape structures or other tubular features. A convenient nongenetic term for all such tubular features of soils and paleosols is pedotubule (Brewer 1976). Within this system, pedotubules are classified according to their composition and contrast with surrounding material. Internal fabric is of four general kinds. Granular tubules (granotubules) are filled with mineral grains and little clay. Pelletoidal tubules (aggrotubules) are filled with pellet-like masses consisting of clay and clastic grains, usually the fecal pellets of soil invertebrates, but in some cases rounded soil peds. Clay tubules (isotubules) are filled with mixed clay and mineral grains without any preferred direction. Meniscate tubules (striotubules) contain clayey and granular layers transverse to the long axis of the tubule. In most striotubules the layers are semicircular in cross section, like a meniscus, but they can be V- or W-shaped. They form by means of the burrowing motion of soil organisms and are the same as the backfills (spreiten) in the terminology of trace fossils (Häntzschel 1975). Somewhat similar, very narrow V-shaped layering can be found in water or gas escape structures through bedded sediment (Neumann-Mahlkau 1976).

Tubules often penetrate several levels of a soil and can be classified according to their contrast with the surroundings (Brewer 1976). Compatible tubules (orthotubules), for example, are similar in fabric and composition to their matrix. They are recognized by slight differences in coloration or by distinctly stained margins. Mixing tubules (metatubules) have an internal composition dissimilar to their surrounding soil material but similar to other parts of the profile. In the long term, such tubules tend to homogenize the various horizons of soils. A common example in calcareous light-colored subsurface (Bk) horizons is burrows filled with dark-colored organic material from the surface (A) horizon. This kind of metatubule is a krotovina, a Russian term for a burrow filled by material washed in from above. A final kind is exotic tubules (paratubules), which consist of material unlike anything else in the profile. Pellet-filled tubules may be like this because pellets are seldom so abundant or well preserved in a soil matrix as in burrows. Exotic tubules also can form under special circumstances, such as volcanic ash filling burrows in soil, but dispersed or weathered near the surface of the profile. These various kinds of tubule can be compounded into descriptive categories such as exotic pelletoidal tubules (para-aggrotubules) or mixing granular tubules (metagranotubules).

Tubular features are common in soils but not diagnostic of them. Many burrows in soils are similar to those found in lake and ocean bottoms (Ratcliffe & Fagerstrom 1980). Burrows in paleosols may be valuable indicators of paleoenvironment, especially when sufficiently distinctive that their makers can be identified. Burrows of earthworms, millipedes, beetles, bees, and rodents can be distinguished when suitably preserved. None of these creatures can tolerate waterlogging and their depth of penetration into a paleosol may indicate the former level of the water table. As clearly original parts of a paleosol, the relationship of burrows to nodules and crystals may reveal their time of formation. The degree of deformation and crushing of burrows may be indicators of compaction in paleosols. It is surprising what can be learned from a careful study of burrows.

Microfabric

Some kinds of microscopic structure are characteristic of soils and paleosols (Brewer 1976, Fitzpatrick 1984). Especially distinctive is the appearance of the fine-grained part of the soil (plasmic fabric) in thin sections viewed under crossed nicols. Its appearance varies with the magnification used and the orientation and thickness of the section. Most plasmic fabrics are described from observations made at total magnifications of about 100–250 times natural size. For work in which it is important to distinguish paleosols from sedimentary rocks, it is best to cut the thin sections vertical to paleosol horizons and to the bedding planes of enclosing sedimentary rocks. In addition, thin sections of paleosols must be ground more carefully and thinner than usual for most rocks so that enough light can penetrate the clayey soil matrix, which usually is more opaque than mineral grains such as quartz (Murphy 1986).

The increased abundance of cutans obscuring original sedimentary, metamorphic, or igneous textures as a soil develops is expressed on a microscopic scale by the development of bright clay fabric [sepic plasmic fabric of Brewer (1976)]. Bright clay is highly oriented clay that has a high birefringence under crossed nicols (Fig. 3.15). In contrast, weakly oriented clay appears dull and randomly flecked (asepic). Different kinds of bright clay fabrics reflect increasingly extensive areas of oriented clay formed by washing into cracks, internal stresses in the soil, and surface weathering of soil peds. Bright clay may form isolated streaks longer than an individual clay crystal. This is a fabric called incipient bright clay fabric (insepic plasmic fabric). With further soil development, the bright clay streaks become more extensive and are partly adjoining, producing streaky bright clay fabric (mosepic plasmic fabric). Bright clay may also line cavities in the soil (cavity lining bright clay fabric or vosepic plasmic fabric). This would be crushed in paleosols to produce a symmetrical arrangement of bright clay around a midline where the cavity was elided. Bright clay also may coat grains to form grain-lining bright clay fabric (skelsepic plasmic fabric). With further development, bright clay may form the extensive criss-crossing networks of intersecting bright clay fabric (masepic plasmic fabric). Special cases of intersecting bright clay fabric include fabrics in which the angles of intersection are at right-angles (lattice-like bright clay or lattisepic plasmic fabric), or at other angles with two preferred directions

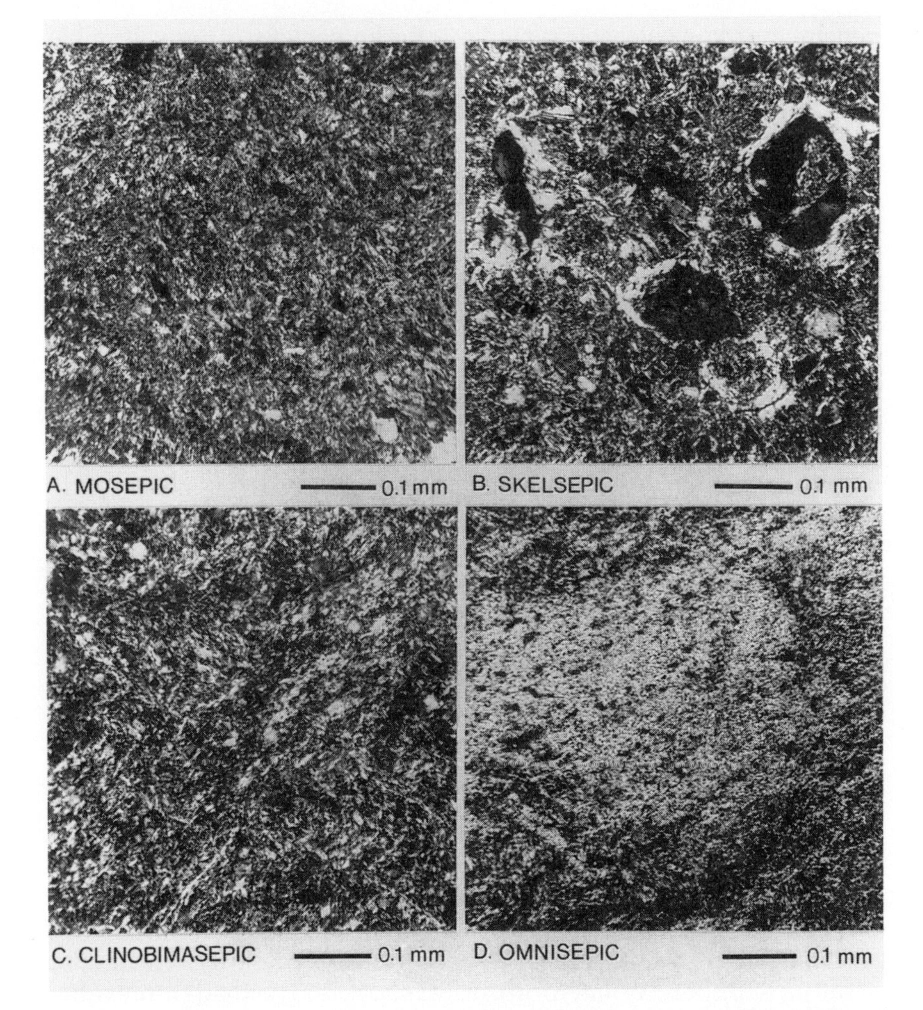

Figure 3.15 Sepic plasmic (bright clay) microfabrics. All under crossed polarized light and with bar scales of 0.1 mm: A, mosepic plasmic (streaky bright clay) fabric from clayey surface (A) horizon of a paleosol (Sulfihemist) in the mid-Cretaceous Dakota Formation near Russell, Kansas (Retallack & Dilcher 1981b; Retallack specimen R101); B, skelsepic plasmic (grain-lining bright clay) fabric from the subsurface (C) horizon of the Ogi silty clay loam light-colored variant paleosol (Fluvaquentic Eutrochrept) in the Oligocene Sharps Formation in Badlands National Park, South Dakota (Retallack 1983b; Indiana University specimen 15621); C, clinobimasepic plasmic (trellis-like bright clay) fabric in the subsurface (Cg) horizon of a paleosol (Aquent) in the mid-Cretaceous Dakota Formation near Fairbury, Nebraska (Retallack & Dilcher 1981b; Retallack specimen R110); D, omnisepic plasmic (woven bright clay) fabric in the clayey subsurface (Bt) horizon of a paleosol (Alfisol) in the mid-Cretaceous Dakota Formation near Russell, Kansas, USA (Retallack & Dilcher 1981b; Retallack specimen R94).

(trellis-like bright clay or clinobimasepic plasmic fabric), or with three preferred directions (net-like or clinotrimasepic). In very well developed and well drained soils or highly metamorphosed paleosols, all the clay may be bright and have a woven appearance (woven bright clay fabric or omnisepic plasmic fabric).

The degree of development of bright clay fabric is due in part to time available for soil formation and in part to the intensity of soil-forming processes such as stresses imposed by wetting and drying, filling and closing of cracks, and other alterations (Brewer & Sleeman 1969). Woven bright clay fabric (omnisepic) forms most rapidly in swelling clay soils of seasonal climates, but also forms eventually in very old soils (Holzhey *et al.* 1974). Lattice-like bright clay fabric (lattisepic) forms in the subsurface clayey (Bt) horizons of soils where clays are under moderate confining pressure (McCormack & Wilding 1974). Most kinds of bright clay fabric are characteristic of soils and paleosols. Important exceptions are cavity-filling and grain-lining bright clay which also are found in sediments (Scholle 1979). Coated grains may form by pressure of compaction of hard grains against a clayey matrix or by rolling or encrustation in a variety of sedimentary and soil environments. Cavity-lining bright clay can form during late diagenetic dissolution of grains or fossils as readily as within original cavities of the soil. A more highly oriented, but otherwise omnisepic plasmic fabric is found in paleosols metamorphosed to greenschist facies and beyond.

Other microfabrics (Fig. 3.16) are not diagnostic of soils, but can reveal much about the conditions of their formation. Clayey textures in which the only bright clays are isolated, randomly oriented, tiny crystals can be called flecked clay fabric (argillasepic plasmic fabric). Sandy textures lacking bright clay can be called flecked sand fabric (silasepic plasmic fabric). These are common textures of sediment and sedimentary rocks unaltered by soil formation. Usually sediments also show some microscopic indications of bedding or cleavage and can be regarded as laminated (strial fabrics). Laminated fabrics can be unidirectional (unistrial), as is usually the case for bedding, or bidirectional (bistrial), as in metamorphic rocks with both bedding and cleavage, or with two cleavage planes. Crystalline microfabric [or crystic fabric of Brewer (1976)] is characterized by discernable crystal outlines. Such fabrics are widespread in sediments, especially evaporites, limestones, and dolomites, and also in alkaline soils of dry climates (Scholle 1978). Pyrite and siderite of swampland soils are often crystalline. Opaque microfabric (one kind of isotic fabric) appears black in petrographic thin section. Opaque fabrics are produced by iron oxides and hydroxides which form in well drained soils. Isotropic microfabrics (another kind of isotic fabric) appear clear under ordinary light but black under crossed nicols. This kind of fabric is found in materials replaced by isotropic minerals such as opal or some zeolites which are

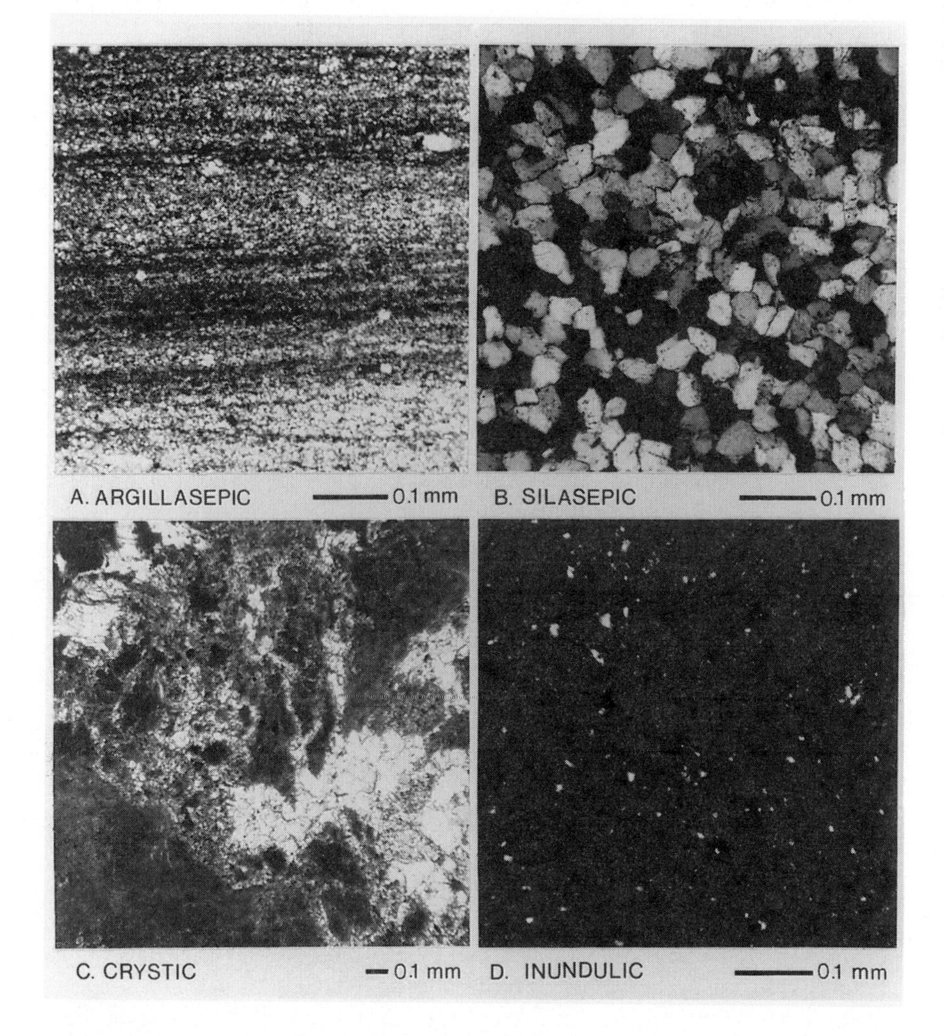

Figure 3.16 Asepic (dull clay) microfabrics. All under crossed polarized light and with bar scales of 0.1 mm: A, argillasepic (flecked clay) fabric in a lacustrine oil shale in the Middle Eocene, Green River Formation in Ulrich's Quarry near Kemmerer, Wyoming (Retallack specimen P6969B); B, silasepic (grainy) fabric from the near-surface (E) horizon of a paleosol (Aquept) in the Pennsylvanian, Fountain Formation near Manitou Springs, Colorado, USA (Suttner & Dutta 1986; Retallack specimen P6185); C, crystic (crystalline) fabric of fibrous and sparry calcite in the subsurface (Bk) horizon of a paleosol (Natrustoll) in the late Pliocene (3.0 million years), Denen Dora Member of the Hadar Formation at a locality ALX333 4 km northwest of the junction of the Kada Hadar and Awash rivers, Afar region, Ethiopia (Johanson *et al.* 1982; Taieb specimen G0); D, inundulic (cloudy) fabric from the surface (A) horizon of a paleosol (Fibrist) in the mid-Cretaceous Dakota Formation near Bunker Hill, Kansas (Retallack & Dilcher 1981b; Retallack specimen R108).

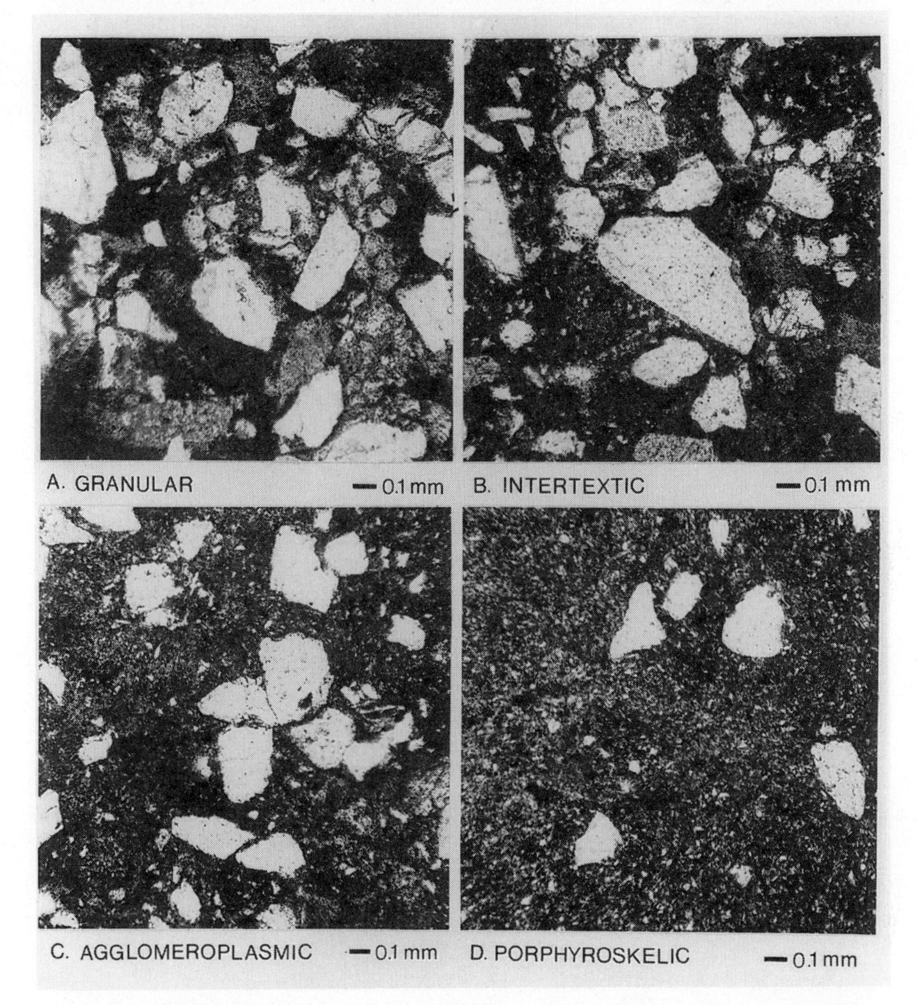

Figure 3.17 Grain fabrics. All under plane light and with a scale bar of 0.1 mm, from the type Naranji clay paleosol (Psammentic Haplustalf) in the Miocene (7.5 million years) Dhok Pathan Formation near Kaulial, Pakistan (Retallack 1985): A, granular fabric from subsurface (C) horizon (Smithsonian specimen 353837); B, intertextic fabric from subsurface (Bt) horizon (Smithsonian specimen 353831); C, agglomeroplasmic fabric from subsurface (Bt) horizon (Smithsonian specimen 353830); D, porphyroskelic fabric from surface (A) horizon (Smithsonian specimen 353828).

common in soils formed in volcaniclastic materials and in arid climates. Few soil fabrics are completely isotropic, but some are almost so. For these the terms somber fabric (undulic) and cloudy fabric (inundulic) may be more appropriate. These are distinguished by having a uniformly dark and featureless appearance (somber or undulic) or patches of varying darkness (cloudy or inundulic). Both of these dull fabrics may be the result of the drying of highly flocculated colloidal material such as mixtures of clay with organic matter or iron oxides and oxyhydrates. These dull fabrics are especially common in the clayey portions of swampland soils and their paleosol equivalents, the seat earths of coal seams. They also can be found in clayey matrices of mud flows and other deposits of weakly oriented clay.

Another aspect of soil microfabric (Fig. 3.17) is the proportion of original mineral grains and rock fragments (skeleton grains) to fine-grained material (plasma). Point counting to quantify these proportions can be useful, but so are the following four descriptive categories. Granular fabric is that in which the skeleton grains are touching, with little or no fine-grained material in the interstices. Intertextic fabric is similar but with occasional intergranular braces of fine-grained material between the grains. In agglomeroplasmic fabric, fine-grained material forms local pockets and an incomplete matrix to the larger grains. This is a common texture of a strongly burrowed soil. Porphyroskelic fabric has larger grains set within the fine matrix. This can be the original texture of the parent material or it can be the end result of soil formation in which many mineral grains have been broken down to clay by weathering.

In both microscopic and readily visible features, soils and paleosols have features inherited from their parent material, in addition to features acquired during soil formation. Only a few of the soil features are useful for recognizing paleosols in the rock record.

4 Soil-forming processes

Implicit among the features by which soils are recognized and classified are processes that formed them. Such processes are more readily observed in modern soils than in paleosols in which they ceased to act long ago. Nevertheless, processes often can be inferred from features of paleosols or from suites of paleosols. The reconstruction of processes gives meaning and connection to features and kinds of paleosols. One way to learn about soils is to memorize the main features of the principal kinds known. Real profiles do not always match such models exactly. Identification can be aided by understanding soil-forming processes which may have interacted to produce peculiarities of a particular profile.

Soil-forming processes can be reduced to physical, chemical, or biological components of weathering. That this distinction is an artificial one can be seen from the example of plant roots, which are biological structures but produce exudates with chemical effects and grow in girth to force the soil apart physically. Nevertheless, the large body of literature on soil biology, chemistry and physics is a testimony to the way in which these traditional scientific divisions have permeated soil science (Kühnelt 1976, Bohn *et al.* 1985, Hillel 1980). Important to the interpretation of modern soils is the measurement of physical, chemical and biological features of soils such as the linear expansion of clays, the pH of soil solutions and the distribution of soil organic matter. Few of these features can be measured directly from paleosols altered by burial. Other sources of information pertinent to these indices of soil formation must be sought, with due regard for common kinds of alteration after burial.

Indicators of physical weathering

A variety of physical processes play a role in forming soil from underlying rock or sediment. Granite and schist formed deep within the Earth under great confining pressure naturally expand and break into joints when unroofed by erosion and exposed at the surface. This initial loosening allows the circulation of gases and water, which may introduce ice in freezing weather or salts in dry climates. Both expand to force the cracks wider, as do expansion and contraction with daily and seasonal temperature fluctuations. Additional thermal expansion caused by forest fires may be sufficient to crack open large rocks. Some kinds of clay swell when moist and contract when dry. Effects of these physical processes can be observed and in some cases quantified in paleosols.

These physical weathering processes are difficult enough to study in modern soils (Hillel 1980), let alone paleosols, whose physical properties have been changed almost beyond recognition. Direct measurement of the physical properties of paleosols are therefore of little use unless they can be related to original conditions using some kind of correction factor. Other physical features of paleosols are best estimated by using empirical relationships or constants established by work on modern soils and calibrated by features that are measurable in paleosols. In other cases, physical weathering of paleosols can only be assessed in a very general way by analogy with such processes and their products in modern soils.

Loosening

Expansion of rocks and sediments accompanies the development of joints and soil cracks during weathering. This results in a lower bulk density

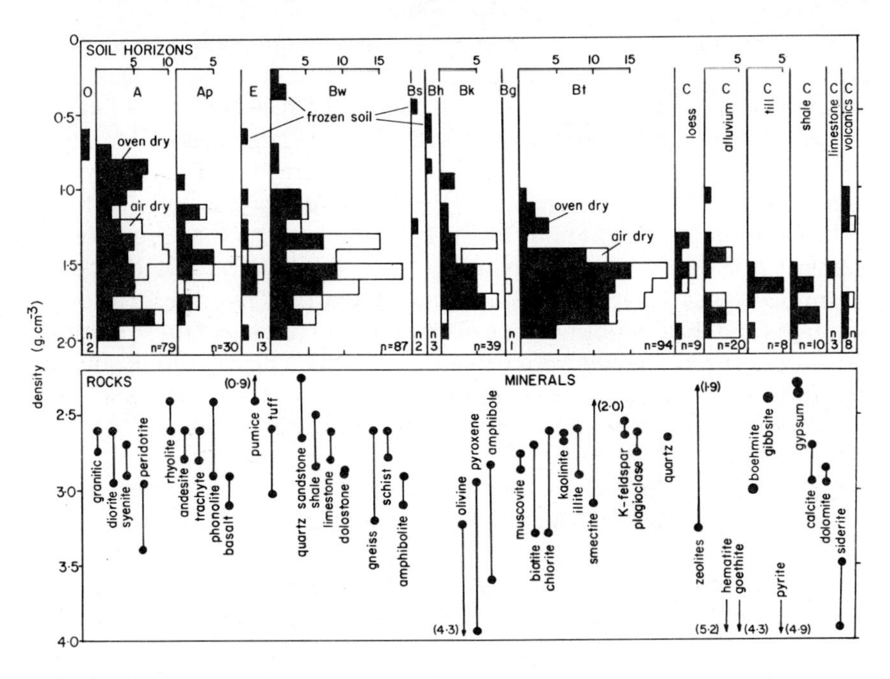

Figure 4.1 Density of soil horizons, rocks and minerals (compiled from Williams 1899, Deer *et al.* 1962–3 and Soil Survey Staff 1975).

measured in grams per cubic centimeter (Klute 1986). Most rocks have a density more than twice that of water. They range from high values for peridotite and basalt (3 or more) to lower values for granite (about 2.5). Soils have densities close to 1.0. They range from a low of 0.17 in organic horizons to a high of 1.85 in subsurface layers (Fig. 4.1). Paleosols vary over this entire range for rocks and soils, depending on the extent to which alteration during burial has changed them into rocks.

The original soil density of most paleosols is not determinable because of compaction, cementation, and other changes after burial. In some cases, however, compaction can be calculated from the folding of formerly straight and erect features of the soil such as skeletans or quartz veins. A compaction factor can be calculated from the present length of the vein by

Table 4.1 Estimation of original density, moisture equivalents and porosity of paleosols (from Birkeland 1984 and Bodman & Mahmud 1932)

$$C_f = \frac{T_f}{T_s} \qquad M_s \doteqdot 0.023P_a + 0.25P_i + 0.61P_c$$

$$D_f = \frac{W_f}{V_f} \qquad O_f = [1 - D_f/(x_1D_1 + x_2D_2 ...)] \times 100$$

$$D_s = C_f.D_f \qquad O_s = [1 - C_f.D_f/(x_1D_1 + x_2D_2 ...)] \times 100$$

C_f	=	diagenetic compaction ratio (fraction of the original thickness of the soil due to burial)
D_f	=	density of paleosol ($g.cm^{-3}$)
D_s	=	density of original soil ($g.cm^{-3}$)
D_1, D_2	=	standard density of a given mineral ($g.cm^{-3}$)
M_s	=	moisture equivalent of soil (%)
O_f	=	porosity of a paleosol (%)
O_s	=	porosity of soil (%)
P_a	=	sand in paleosol (%)
P_i	=	silt in paleosol (%)
P_c	=	clay in paleosol (%)
T_f	=	thickness of paleosol (cm)
T_s	=	thickness of soil (cm)
V_f	=	volume of paleosol (cm^3)
W_f	=	weight of paleosol (g)
x_1, x_2	=	proportion of given mineral in paleosol (fraction)

the shortest straight distance compared with the former length of the vein along its contortions (Table 4.1). Less reliable compaction factors can be estimated from comparison with the behavior of comparable materials at the likely depth of burial (see Figs 7.4–7.6).

Fluid flow

The amount of water and gases that a soil can hold is related to the open spaces between the grains of the soil (Hillel 1980). Of interest is the amount of water that can move through a soil in a given time (permeability), the amount of water that can be held in a soil (porosity), and the amount of water that can be retained by surface tension in the smallest cavities of the soil when the large pores have drained out (field capacity). None of these characteristics can be measured effectively in buried soils in which the cavities have been crushed or filled during burial.

Some idea of former field capacity can be gained from paleosols by studying their grain size distribution as established by point counting thin sections. Sandy soils are more porous and permeable, but have a lower

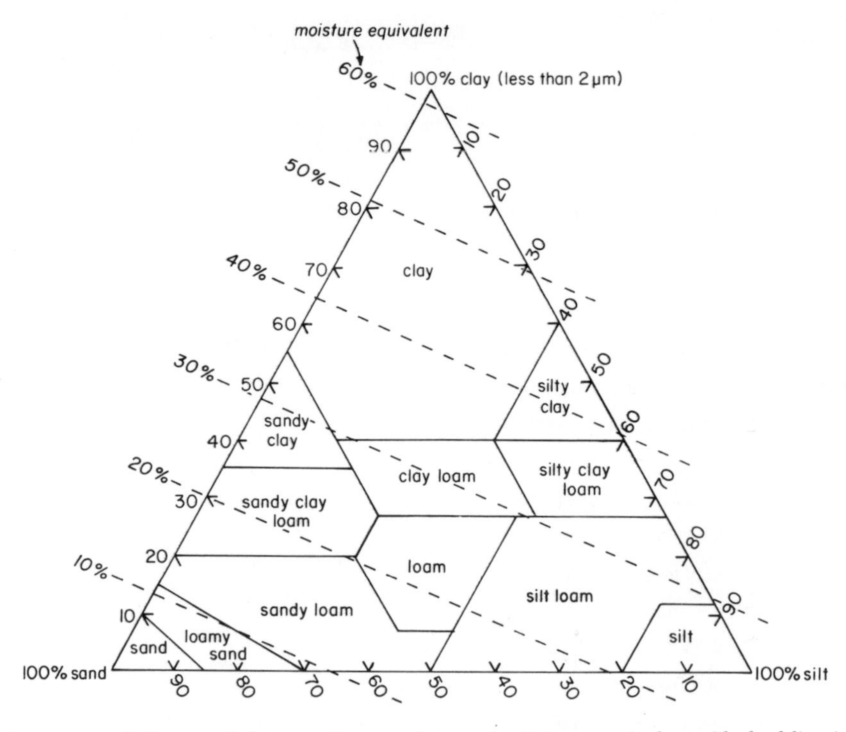

Figure 4.2 Soil textural classes with superimposed moisture equivalents (dashed lines) approximated by an empirical formula (Table 4.1) for soils low in organic matter (from Birkeland 1984, reprinted with permission of Oxford University Press).

field capacity than clayey soils. Field capacity can be approximated by moisture equivalent using an empirical equation based on the percentage of sand, silt, and clay (Fig. 4.2, Table 4.1).

An accurate estimate of former porosity can be obtained if the mineral composition and former density of the paleosol are known. The proportions of different minerals can be estimated by point counting thin sections. The original density of the paleosol can be estimated from its present density using compaction factors. Porosity is then calculated from the proportions of the various minerals, the density of the particular minerals from mineralogical reference works (Deer *et al.* 1962–3), and the former density of the paleosol (Table 4.1).

A more general idea of water flow and retention in soils may be gained by observations of root traces and soil structures. In general, the more copiously rooted and burrowed a paleosol, the more porous and permeable it may have been. If root traces and burrows sidle along rather than penetrate a particular layer, then it may have been an impermeable and hard barrier within the soil. Ferruginous concretions indicate oxygenated, free-flowing water. Siderite nodules, on the other hand, indicate stagnant, poorly oxygenated water (see Fig. 3.13).

Clay swelling

The amount of physical expansion due to wetting of clay can be measured by changes in length or density. Most clays expand to some extent when wet, but this behavior is most marked in smectite (28 ± 2% linear extensibility), less so in illite (20 ± 1%) and kaolinite (12 ± 2%) (Nettleton and Brasher 1983). Compaction, dewatering, and cementation of paleosols alters the swelling properties of clays. Their mineralogical composition also changes because of cations lost during dewatering. A well known diagenetic alteration of clay is the formation of illite from smectite by addition of K^+ and loss of Al^{3+} and Si^{4+} (Curtis 1985). Unraveling the amount and nature of such changes can be difficult, and this compromises the use of typical values of expansion for different modern clays to interpret paleosols.

Despite these difficulties, structural effects of clay heave may still be visible in paleosols. The most striking of these features is the hummock-and-swale microtopography called gilgai microrelief (Paton 1974).These natural undulations are thought to arise by swelling and upward buckling of the soil along deeply cracked hummocks. When dry soil falls into the cracks, this causes the hummocks to become even more prominent as they heave following the next rain. Material from the hummocks may be eroded and redeposited in swales. Thus begins a cycle of circulation of material that produces highly differentiated undulating layers of swale fill between the more massive ridge material. This subsurface differentiation is called

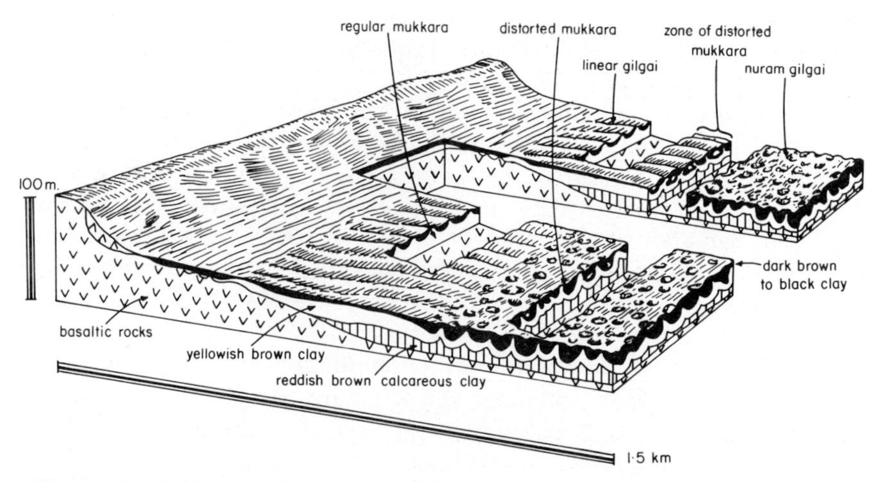

Figure 4.3 Mukkara structure in soils of the Darling Downs, southeastern Queensland, Australia (from Paton 1974, reprinted with permission of Elsevier Publishers).

mukkara structure (Fig. 4.3), and has been recognized in very ancient paleosols (see Fig. 9.9).

Smaller scale structures also may reflect former clay heave. Subangular blocky peds in heaving clay soils are often sheared by a conjugate system of slickensides into lentil peds (Krishna & Perumal 1948). Microfabric also may show a relationship to linear extensibility of clays which is greatest (up to 12%) for woven bright clay (omnisepic plasmic) fabrics but minimal (up to 3.8%) for flecked bright clay (insepic plasmic) fabrics (Holzhey *et al.* 1974).

Fire heating

Charcoal remains in the soil after a fire and is remarkably resistant to decay and chemical degradation (Retallack 1984a). Fossil charcoal has been found in paleosols and sediments as old as earliest Carboniferous (Cope & Chaloner 1985). In such old paleosols, it can be difficult to distinguish from unburnt wood which has been coalified during burial. There are, however, a number of differences between charcoal and coal fragments (Table 4.2; see also Fig. 4.6). The abundance of charcoal can be quantified in the same way as other paleosol constituents by point counting thin sections.

A second indicator of fire is the characteristic pattern of stone spalling produced by forest fires over stony soils. The heating of stones creates concentric fractures. Curved exfoliated shells of rock are thrown off the little-weathered, rounded core of the boulder. In the absence of fire, on the other hand, rocks are more deeply weathered chemically and biologically. They may maintain an irregular shape covered with lichens and a rind of

Table 4.2 Criteria for distinguishing between fossil charcoal and coalified wood fragments (from Harris 1981 and Cope & Chaloner 1985)

Fossil charcoal	Coalified wood
equant shape	usually elongate splinters
sharply broken or rounded ends	irregular or frayed ends
black and opaque	brown to black
broken surface fibrous	broken surface conchoidal
broken surface showing cell structure	broken surface structureless
no middle lamella between cell walls as seen in S.E.M.	middle lamella visible between crushed cells under S.E.M.
resistant to oxidation	easily oxidized
found in oxidized and gleyed paleosols	found in gleyed paleosols
glows on burning	burns with a bright flame

strongly weathered rock. These differences between fired and unfired rocks are used to assess fire frequency in bouldery soils of alpine regions (Birkeland 1984).

A final indicator of fire is the magnetic susceptibility of soil (Tite 1972). This is the ability of the magnetic minerals in the soil to enhance an applied magnetic field. Fires have the effect of reorienting and forming magnetic minerals, resulting in a significantly higher magnetic susceptibility. These kinds of measurements have been used by archeologists to detect ancient campfires, but have not yet found application to older paleosols.

Freezing

Just as extremes of heat from fire may leave traces in paleosols, so may extremes of cold in the form of periglacial structures. The most distinctive of these are ice and sand wedges. Sand wedges are sharply tapering, vertically bedded, sand-filled structures (Fig. 4.4). Ice wedge casts are similar in outline but filled with slumped and horizontally layered sediment. Both sand and ice wedges have a strong taper that is different from the deep, often contorted cracks of mukkara structure. Ice and sand wedge casts commonly are found in much coarser-grained material than the clayey soils usual for mukkara structure. Many periglacial structures form under particular climatic conditions (Table 9.2).

A variety of other periglacial features are known in soils and paleosols. String bogs, rock glaciers, and pingos are a few examples among many (Washburn 1980). Each forms in a particular modern climatic regime and this can be inferred for fossil examples.

Figure 4.4 Near-vertical sandstone wedge penetrating quartzite of the Cattle Grid Breccia, both of Late Proterozoic age (about 680 million years) in the Mt. Gunson Mine, South Australia (photograph courtesy of G. E. Williams).

Indicators of chemical weathering

Soil-forming chemical reactions are mostly of four main kinds: hydrolysis, oxidation, hydration, and dissolution. The reverse reactions of alkalization, reduction, dehydration, and precipitation also occur within soils and within paleosols during burial. Hydrolysis is usually the reaction of carbonic acid with a cation-rich mineral grain which produces clay and cations. Clay accumulates in the soil and cations are washed out or taken up by plants in solution. As an example, consider the hydrolysis of albite (Table 4.3). In essence, hydrolysis involves the displacement of cations by hydronium. It is the main chemical reaction resulting in the destruction of silicate minerals (Chesworth 1973).

Oxidation reactions are those in which an element suffers electron loss when forming a compound. The reverse reaction of reduction is more

Table 4.3 Common kinds of chemical reactions during weathering (from Garrels & Mackenzie 1971)

I. Hydrolysis

$$2NaAlSi_3O_8 + 2CO_2 + 11H_2O \rightarrow Al_2Si_2O_5(OH_4) + 2Na^+ + 2HCO^-_3 + 4H_4SiO_4$$

albite, carbon dioxide, water → kaolinite, sodium ions, bicarbonate ions, silicic acid

II. Oxidation

$$Fe^{2+} \rightarrow Fe^{3+} + e^- \quad \text{(partial reaction)}$$

ferrous ion → ferric ion, electron to other element

$$2Fe^{2+} + 4HCO^-_3 + \frac{1}{2}O_2 + 4H_2O \rightarrow Fe_2O_3 + 4CO_2 + 6H_2O$$

ferrous ions, bicarbonate ions, oxygen, water → hematite, carbon dioxide, water

III. Dehydration

$$2FeOOH \rightleftharpoons Fe_2O_3 + H_2O$$

goethite ⇌ hematite, water

$$CaSO_4.2H_2O \rightleftharpoons CaSO_4 + 2H_2O$$

gypsum ⇌ anhydrite, water

IV. Dissolution

$$CaCO_3 + CO_2 + H_2O \rightleftharpoons Ca^{2+} + 2HCO^-_3$$

calcite, carbon dioxide, water ⇌ calcium ion, bicarbonate ions

easily memorized from an odd phrase: "reduction is electron gain." Reduction occurs in waterlogged soils and in buried soils, but for well drained soils oxidation is the rule. Oxygen in the atmosphere is the most powerful naturally occurring electron scavenger and elements combining with oxygen surrender their electrons to its orbitals. Of the major elements making up most Earth materials, manganese and iron can readily spare an electron, and of these two elements iron is much more abundant. At its simplest level the oxidation of iron can be expressed by a partial reaction for iron (Table 4.3). Minerals containing iron in the ferrous state (Fe^{2+}), such as olivine or siderite, tend to be gray or green. Ferric iron (Fe^{3+}), on the other hand, is found in minerals, such as goethite and hematite, that are yellow, brown, or red. When liberated from silicate minerals by hydrolysis, ferrous cations are readily soluble in water and often are washed out of the profile. Ferric cations are relatively insoluble in water and so remain as oxides and oxyhydrates. The formation of hematite from olivine is a two-step process of hydrolysis followed by oxidation (Table 4.3).

Hydration or dehydration reactions involve the addition or loss, respectively, of water that is structurally part of a mineral. Some soil minerals such as goethite and gypsum are formed by hydration reactions and also can dehydrate to non-hydrous minerals (Table 4.3). Finally, dissolution is a reaction in which a compound disaggregates into its constituent ions in water like a cube of salt in water. Hydrolysis reactions are similar in that ions are formed in solution, but in hydrolysis reactions a new insoluble product, such as clay, is produced. A common dissolution reaction in soils is the dissolution of limestone in a weak solution of carbonic acid derived from atmospheric carbon dioxide and water.

The problem of recognizing chemical reactions in paleosols would be daunting were it not for such a limited array of common reactions. The problem is also simplified by the fact that most soil consists of only a few elements and minerals. Eight elements make up 98% by both weight and volume of average crustal rocks and soils. These are [with average weight percent in parentheses after Mason & Moore (1982)]: O (46.60), Si (27.72), Al (8.13), Fe (5.00), Mg (2.09), Ca (3.63), Na (2.83) and K (2.59). Trace elements may be important for interpreting many aspects of paleosols, but it is these major elements which define the net effect of chemical weathering. A working familiarity with common values for the weight percent of oxides of major elements and of ways of manipulating such analytical data can be useful for interpreting ancient chemical weathering from paleosols. Minerals also are present in limited variety in soils compared with the exotic crystals listed in mineralogical compendia (Deer *et al.* 1962–3). A few silicates (quartz, muscovite, microcline), clays (smectite, kaolinite, illite), carbonates (calcite, dolomite), sulfates (gypsum), oxides (hematite) and oxyhydrates (goethite) are the commonest of soil minerals. These minerals have characteristic optical properties and fields of chemical stability (see Fig. 3.13). Their textural relationships and degree of etching observed in thin sections can be guides to the nature and severity of chemical alteration in paleosols. They preserve in an arrested state the progress of chemical reactions because chemical alteration in soils is slow and seldom reaches thermodynamic equilibrium.

Acidification

A widely used average measure of the prevalence of hydrolysis reactions in modern soils is their acidity as measured by pH, i.e., the negative logarithm of the hydronium ion (H^+) activity. It is usually measured by a chemically treated electrode which develops an electrical potential with respect to a standard electrode when placed in a solution of equal amounts of soil and distilled water (Page 1982). This approach is not suitable for paleosols in which the soil has been so compacted and cemented that it is not possible to resurrect its originally reactive surfaces. Studies of Quaternary paleosols (Simonson 1941) have shown that pH changes substantially on burial of a soil or subsidence below the water table.

Fortunately, the pH of paleosols can be estimated within broad categories from the stability fields of minerals thought to have been original (see Fig. 3.13) and from biological and structural features (Buol *et al.* 1980). The extreme range of soil pH from 2.8 to 10 can be subdivided into six general categories (Baas-Becking *et al.* 1960).

Extremely acidic soils (pH 2.8–4.5) are noncalcareous and consist largely of peat, pyrite, quartz, or a deeply weathered clay, such as kaolinite. They show little evidence of burrowing or plant decay. Organic acids are produced by plants and microbes, carbonic acid by dissolution of high concentrations of soil carbon dioxide, and sulfuric acid by oxidation of sulfides. Extremely acidic soils contain a dearth of exchangeable cations such as Ca^{2+}, Mg^{2+}, Na^+ and K^+ and are dominated by hydronium (H^+) along with aluminum (Al^{3+}) through its reaction with water to produce hydroxyaluminum ions [$Al(OH)^{2+}$ and $Al(OH)_2^+$].

Moderately acidic soils (pH 4.5–6.5) are similarly noncalcareous and base poor, but may contain a wider range of easily hydrolyzed minerals such as feldspar and of base-rich clays such as illite or smectite. Limited biological activity may be evident from the presence of fecal pellets or burrows. Plant material in moderately acidic soils shows some signs of decay.

Near-neutral soils (pH 6.5–8) are weakly calcareous. Apatite, birnessite, and siderite also are stable under these conditions along with a variety of easily weatherable minerals. These soils are 70–90% saturated with bases (Ca^{2+}, Mg^{2+}, Na^+ and K^+) depending on the kind of clay present. They have no exchangeable cationic aluminum (Al^{3+}) and little exchangeable hydronium (H^+). Fecal pellets, burrows, and other evidence of biological activity may be abundant in near neutral soils.

Moderately alkaline soils (pH 8–8.5) are calcareous with free crystals and nodules of calcite or other carbonate minerals. Their clays are cation-rich (largely with Ca^{2+} and Mg^{2+}). Pelletoidal structure from the copious activity of soil invertebrates is common.

Alkaline soils (pH 8.5–9) are similarly calcareous, nodular and pelletoidal. Their clays tend to be smectitic with exchangeable cations including sodium and potassium (Na^+ and K^+). These are swelling clays that may create a mukkara structure (Paton 1974). Gypsum, anhydrite, dolomite, and zeolites also are characteristic of alkaline soils.

Extremely alkaline soils (pH 9–11) are often entirely sodium saturated and commonly contain halite, gypsum, zeolites, and other evaporite minerals. These soils have base-rich clays such as smectites. Domed columnar ped structures are characteristic (Northcote & Skene 1972). Root traces and other evidence of life in these soils are usually sparse because these soils form in very dry and saline desert basins.

These broad categories of pH can be refined by calculating mineral weathering ratios within a profile. The abundance of easily hydrolyzed minerals compared with hydrolysis-resistant minerals increases with pH, among other factors. Such ratios are more meaningful when compared with the ratio of the minerals in the parent material and when there is independent evidence of the time over which the profile developed. A commonly used mineral weathering ratio is quartz over feldspar. Microcline and plagioclase feldspar may be distinguishable from quartz by their twinning, but it may be necessary to stain a thin section with potassium dichromate (Houghton 1980) in order to count the proportions of orthoclase and quartz. Because feldspar is much more readily hydrolyzed than quartz, this ratio is higher in acidic than alkaline soils. Another mineral weathering ratio is that of zircon and tourmaline over pyroxene and amphibole. The latter two minerals are much less resistant to weathering than zircon and tourmaline. These minerals do not occur in sufficient abundance to be counted in thin sections. They should be concentrated from the soil for counting.

These mineral weathering ratios are based on the observed differential stabilities of minerals in soils (Fig. 4.5). Among mineral grains of sand and silt size, microcline (a potassium feldspar) is more resistant to weathering than albite (a sodium feldspar). Both feldspars and mafic minerals (olivine,

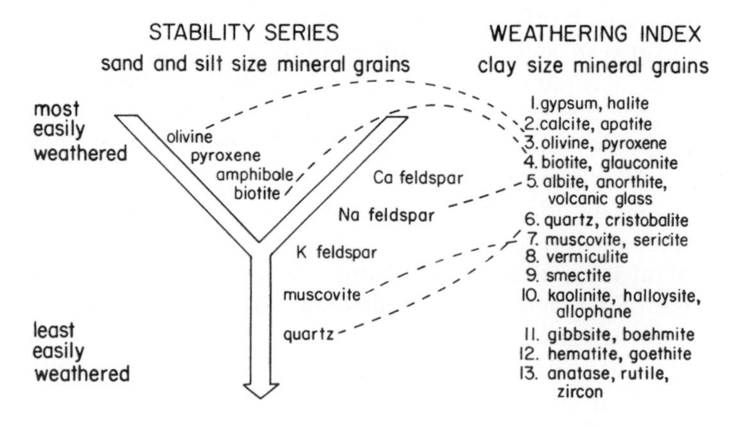

Figure 4.5 Relative stability of mineral grains of silt size or larger (left) and of clay-size minerals (right) under conditions of weathering in soils (data from Goldrich 1938, Jackson *et al.* 1948).

pyroxene, amphibole, and biotite) are more readily hydrolyzed than quartz (Goldich 1938). Quartz is not so weather resistant when fine-grained, however, because of its increased surface area (Jackson *et al.* 1948). From these general relationships, a variety of mineral weathering ratios can be devised to suit particular profiles.

For metamorphosed paleosols it may be worthwhile to calculate mineralogical composition from a chemical analysis. Various recipes and computer programs are available for calculating mineralogical modes of igneous rocks from a chemical analysis, but these are based on minerals and chemical reactions inappropriate for soils. For paleosols, once the likely minerals originally present have been established, their proportions can be calculated by "reacting" mole fractions of each major elemental

Table 4.4 Estimation of major kinds of chemical reactions in paleosols using molecular weathering ratios (examples of some of these ratios calculated for modern soils are given by Marbut 1935).

	I. Hydrolysis	
bases / alumina	=	$(CaO + MgO + K_2O + Na_2O) / Al_2O_3$
silica / alumina	=	SiO_2 / Al_2O_3
barium / strontium	=	Ba / Sr
	II. Oxidation	
ferric / ferrous iron	=	Fe_2O_3 / FeO
total iron / alumina	=	$(Fe_2O_3 + FeO) / Al_2O_3$
total iron & manganese / alumina	=	$(Fe_2O_3 + FeO + MnO) / Al_2O_3$
	III. Hydration	
silica / sesquioxides	=	$SiO_2 / (Al_2O_3 + Fe_2O_3)$
	IV. Salinization	
alkalies / alumina	=	$(K_2O + Na_2O) / Al_2O_3$
soda / potash	=	Na_2O / K_2O
soda / alumina	=	Na_2O / Al_2O_3

V. Atomic weights

Si	=	28.09	Ca	=	40.08
O	=	16.00	Mg	=	24.32
Al	=	26.98	K	=	39.10
Fe	=	55.85	Na	=	22.99
Ba	=	137.36	Sr	=	87.63

oxide on a large balance sheet (Garrels & Mackenzie 1971). This manual method of calculating modes is appropriate to the complexity of weathering reactions and deserves further attention and computerization.

An alternative approach is to use chemical analyses of paleosols directly by molecular weathering ratios. These are calculated by dividing the weight percent of each relevant oxide by its molecular weight and then adding or dividing as specified by the particular ratio. Molecular weathering ratios useful for assessing former extent of hydrolysis reactions are bases/alumina, alkaline earths/alumina, silica/alumina and barium/strontium (Table 4.4). A bases/alumina ratio of greater than unity reflects an alkaline or weakly developed soil. In very acidic or well developed soils this ratio is close to zero because of depletion of bases from weatherable minerals by hydrolysis reactions. In acidic soils the silica to alumina ratio may be more informative. Below a pH of about 4.5 there is dissolution of clay rich in alumina, but quartz is resistant. Marbut (1935) calculated ratios as high as 138 for leached (E) horizons of acidic soils (Ultisols) and 53 for comparable horizons of neutral soils (Alfisols). Values for alkaline soils (Inceptisols, Mollisols and Aridisols) are usually between 7 and 16. The ratio of barium to strontium is an indicator of leaching, degree of free drainage, and time of development. Strontium is the more soluble of these two otherwise chemically similar trace elements. The barium to strontium ratio ranges from near 10 in acidic, sandy soils (Spodosols) to near 2 in most rocks and soils (Vinogradov 1959).

A more rigorous method of analyzing chemical change is to normalize chemical compositions to a constituent thought to have remained stable in both the soil and its parent material (Table 4.5). This is a more complex procedure that relies on a variety of assumptions about parent materials, to be dealt with in a later chapter devoted to that topic.

Oxidation

An overall measure of the extent of oxidation and reduction reactions in modern soils is Eh, i.e., the electrode potential in volts induced by soil solutions. A low negative Eh is an indication that the soil can readily donate electrons and is reducing. A high positive Eh, on the other hand, is found in oxidizing soils with a high demand for electrons. The principal oxidizing agent in modern soils is free oxygen from the atmosphere. Its exclusion by waterlogging creates reducing conditions. Because the water content and pore structure of paleosols are so altered by burial, it is not possible to establish their former Eh by direct measurement. Like pH, Eh is altered substantially with burial or subsidence of a paleosol below the water table (Bohn *et al.* 1985).

The former Eh of paleosols can be reconstructed within broad limits from mineral assemblages thought to have been original (see Fig. 3.13). Because

Table 4.5 Formulae relating volume, thickness and chemical changes in soil and fossil soil horizons during original soil formation and subsequent burial and compaction (adapted from Brewer 1976).

$$C_s = \frac{V_s}{V_p} = \frac{T_s}{T_p} \qquad C_f = \frac{V_f}{V_s} = \frac{T_f}{T_s} \qquad D_s = C_f.D_f$$

$$K_s = \frac{X_s.R_p}{X_p.R_s} = \frac{P_s.D_s.R_p}{100.X_p.R_s} \qquad K_f = \frac{P_f.D_f.R_p}{100.X_p.R_f}$$

$$T_s = \frac{T_f}{C_f} \qquad T_p = \frac{T_s.R_s.D_s}{R_p.D_p} = \frac{T_f.R_f.D_f}{R_p.D_p}$$

$$V_p = \frac{V_s.R_s.D_s}{R_p.D_p} = \frac{V_f.R_f.D_f}{R_p.D_p} \qquad W_p = \frac{V_s.R_s.D_s}{R_p} = \frac{V_f.R_f.D_f}{R_p}$$

$$X_{gs} = \frac{V_s.D_s}{100}\left(\frac{P_p.R_s}{R_p} - P_s\right) \qquad X_{gf} = \frac{V_f.D_f}{100}\left(\frac{P_p.R_f}{R_p} - P_f\right)$$

$$X_{ns} = \frac{P_s.D_s.R_p}{100.R_s} \qquad X_{nf} = \frac{P_f.D_f.R_p}{100.R_f}$$

$$X_s = \frac{P_s.D_s}{100} \qquad X_f = \frac{P_f.D_f}{100}$$

where

C_s = pedogenic compaction ratio or fraction of original thickness due to soil formation

C_f = diagenetic compaction ratio or fraction of original thickness of soil due to burial

D_f = density of fossil soil ($g.cm^{-3}$)

D_p = density of parent material ($g.cm^{-3}$)

D_s = density of soil ($g.cm^{-3}$)

K_f = concentration ratio of a chemical constituent in fossil soil compared to parent material ($g.cm^{-3}$)

K_s = concentration ratio of a chemical constituent in soil compared to parent material ($g.cm^{-3}$)

P_f = amount of a chemical constituent in fossil soil (weight %)

P_p = amount of a chemical constituent in parent material (weight %)

P_s = amount of a chemical constituent in soil (weight %)

R_f = amount of a stable chemical constituent in fossil soil (weight %)

Table 4.5 *Continued*

R_p	=	amount of a stable chemical constituent in parent material (weight %)
R_s	=	amount of a stable chemical constituent in soil (weight %)
T_f	=	thickness of a fossil soil or horizon derived from a soil or soil horizon (cm)
T_p	=	thickness of parent material forming soil or soil horizon (cm)
T_s	=	thickness of soil or soil horizon derived from parent material (cm)
V_f	=	volume of fossil soil or horizon derived from soil or horizon (cm)
V_p	=	volume of parent material forming soil or horizon (cm^3)
V_s	=	volume of soil or horizon formed from parent material (cm^3)
W_f	=	weight of fossil soil formed from soil (g)
W_p	=	weight of parent material forming soil (g)
W_s	=	weight of soil formed from parent material (g)
X_f	=	weight of a chemical constituent in a given volume of fossil soil ($g.cm^{-3}$)
X_{gf}	=	weight of a chemical constituent in a fossil soil lost or gained with respect to a stable constituent remaining from parent material ($g.cm^{-3}$)
X_{gs}	=	weight of a chemical constituent in a soil lost or gained with respect to a stable constituent remaining from parent material ($g.cm^{-3}$)
X_{nf}	=	weight of a chemical constituent in a fossil soil normalized to a stable constituent remaining from parent material ($g.cm^{-3}$)
X_{ns}	=	weight of a chemical constituent in a soil normalized to a stable constituent remaining from parent material ($g.cm^{-3}$)
X_p	=	weight of a constituent in a given volume of parent material ($g.cm^{-3}$)
X_s	=	weight of a constituent in a given volume of soil ($g.cm^{-3}$)

chemical reactions of mineral transformation that define these categories also are pH dependent, it is not possible to give boundary values of Eh categories in millivolts. However, there are three recognizable categories of Eh: reducing, intermediate, and oxidizing, which correspond to waterlogged, wet, and well-drained soils, respectively (Baas-Becking *et al.* 1960). This may not always have been the case because atmospheric oxygen may have been lower in the distant geological past so that well drained soils at that time would have been moderately reducing. This may be a problem for interpreting the paleodrainage and former atmospheres of such ancient paleosols, but it does not affect their inferred redox status.

Reducing soils (strong negative Eh) are usually bluish or greenish gray. The iron in their minerals is in the ferrous state (Fe^{2+}), which is naturally drab colored. Additional grayness may be due to organic matter. These are the soils that accumulate peat which on burial and diagenesis is converted

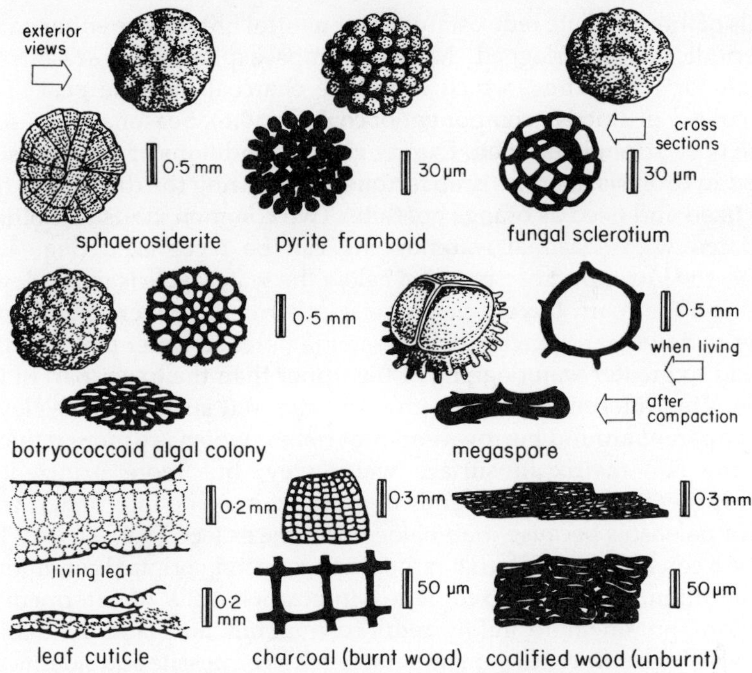

Figure 4.6 Some common constituents of coals and waterlogged paleosols. In some cases (botryoccoccoid algae, megaspores, cuticles, and coalified wood) specimens show marked compaction and in other cases (sphaerosiderite, pyrite, framboids, and sclerotinite) little change due to compaction.

to coal. One bright mineral associated with reducing soils is pyrite, often in the form of framboids (Fig. 4.6). In natural exposures of paleosols, pyrite may be destroyed by modern oxidation. It can still be detected as dustings of its weathering product, a light yellow powder of jarosite which has a characteristic smell of rotten eggs. Another distinctive mineral of reduced soils is vivianite, which is white as nodules or replaced arthropod carapaces when reduced, but turns a distinctive bright blue when mildly oxidized. Strongly reduced soils also contain well preserved plant remains because lack of oxygen curtails the activity of aerobic microbial decomposers. The proportions of amorphous compared with structurally preserved plant material, and of fossil roots compared with shoots, are useful indicators of the degree of former oxidation of a paleosol. The activity of large animals and plants in reducing soils is also limited. Strongly reduced soils contain few burrows and fecal pellets. Root traces do not penetrate them deeply, but spread in a tabular pattern (see Fig. 3.3). Soil structures also are weakly developed in reducing soils. Original sedimentary structures such as bedding and ripple marks may persist. On a microscopic scale this may be reflected in striated (unistrial plasmic) or in dull cloudy (undulic and inundulic) fabrics.

Soils of intermediate redox status (near neutral Eh) may be either partly or periodically waterlogged. Many swamps experience a season dry enough for forest fires, which introduce charcoal into the peat. This charcoal is a distinctive component of coal (Fig. 4.6). Seasonally dry soils also may be strongly mottled. During gleyed conditions of a wet season, limited hydrolysis liberates drab ferrous iron. During the dry season this is oxidized and fixed as orange goethite. Two common kinds of mottling associated with seasonal waterlogging can be recognized (Fig. 4.7). Usually the lower part of a paleosol below the water table is more gleyed than the upper part. In very clayey soils impermeable layers may form a barrier to downward percolation of water (a perched water table), which can lead to greater waterlogging of the upper than the lower part of the profile. The difference between groundwater and surface-water gley is most apparent around burrows and root traces, which are more reduced than the soil matrix in surface water gley, but more oxidized in groundwater gley. Different kinds of mottling must be interpreted with care for paleosols because drab-haloed root traces (see Figs. 3.6 and 3.7) may be a consequence of burial rather than reflecting original conditions.

In addition to these mixed redox indicators, soils of intermediate oxidation contain more mildly reduced minerals and lack the sulfide minerals of extreme reduction. Minerals such as birnessite and siderite are characteristic of soils of intermediate redox status. Sand-sized spherical aggregates of siderite crystals (sphaerosiderite) are common in paleosols of coal measures and may be products of original soil formation at

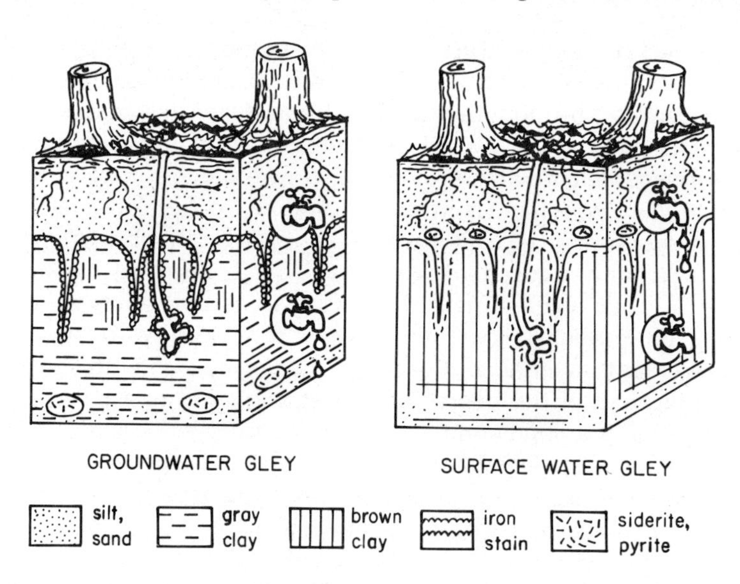

Figure 4.7 Schematic models for the formation of groundwater gley (left) and surface-water gley (right).

intermediate Eh or of gleization attendant on burial. Siderite is readily oxidized to goethite or hematite. All three minerals may be present in the same paleosol owing to an original soil regime of fluctuating oxidation or to oxidation of originally gleyed minerals in the modern outcrop.

Fluctuating oxidation in the original soil would be likely if the paleosol also showed a mixture of well decomposed and recognizable organic matter or patterns of root traces that include both deeply penetrating and tabular root systems. Swamps that are periodically dry have very low rates of peat accumulation. Paleosols formed under such conditions have little or no coal. Fecal pellets and burrows may be present and also some crude blocky and platy ped structures. Microfabrics of soils of intermediate redox status may include bright clay (sepic plasmic fabric).

Oxidized soils (high positive Eh) are warm colored with iron oxyhydrates such as brown ferrihydrite and yellowish brown goethite and oxides such as brick-red hematite. Hematite is especially common in paleosols, in part because of dehydration reactions accompanying burial (Walker 1967). Organic matter is not noticeable in oxidized soils except near the surface and around living roots. Dead organic matter in well drained soils is decomposed and recycled by a community of microbial decomposers. Although organic matter may be scarce, there may be abundant evidence of life in oxidized soils in the form of burrows, fecal pellets and copiously branching, deeply penetrating root systems. Well drained soils also have well developed soil structure including peds, illuviation clayskins, and sepic plasmic microfabrics.

Mineral weathering ratios are not especially useful for quantifying soil oxidation because iron oxide minerals are too fine grained and dispersed to count easily. The oxidation state of a paleosol can, however, be chemically characterized using molecular weathering ratios (see Table 4.4) or by normalizing the relevant chemical analyses to their abundance in parent material (Table 4.5). Ferric to ferrous iron ratios are the most effective proxy measure of oxidation state. In many methods of bulk chemical analysis the two oxidation states of iron are not distinguished. The weight percent value from such analyses may be cited as ferric (Fe_2O_3) or ferrous (FeO) iron and can be converted from one to the other using the ratio of their different molecular weights. An independent method of determining the abundance of one of the oxidation states is needed, for example, by potassium dichromate titration (Maxwell 1968). The abundance of the other oxidation state can be estimated by difference. When such additional analytical data are not available, the total iron or total iron plus manganese values may be useful guides to the degree of oxidation of a paleosol. The oxidized minerals of these elements are insoluble and tend to accumulate in soils, whereas their reduced cations tend to be lost from gleyed soils in groundwater flow. Total iron to alumina ratios calculated for a variety of North American soils are mostly less than 0.4 (Marbut 1935).

The ratio can reach 1.2 in ferruginous (spodic) horizons, 1.9 in clayey (argillic) horizons, and 1.5 in red and deeply weathered (oxic) horizons. Iron loss is best assessed by comparison with parent material by normalizing to a stable constituent, as outlined in a later chapter on parent material.

Hydration

Hydrated minerals common in soils include ferrihydrite, goethite, and gypsum. The degree of hydration of a soil can be approximated by identifying and quantifying the abundance of these minerals using Mössbauer spectroscopy for iron oxyhydrates (Dixon & Weed 1977) and X-ray diffraction for clays (Klute 1986). More direct estimates of hydration can be gained by weighing oven-dried soil after heating to temperatures that have been shown to drive off different volatile fractions. However, these volatile materials include organic matter and carbon dioxide in addition to structural water (Page 1982).

The degree of hydration of paleosols could be determined in the same way were it not for dehydration reactions following burial and compaction. Gypsum, for example, is dehydrated to anhydrite, ferrihydrite and goethite are converted to hematite, and smectite is altered to illite, then sericite, and ultimately muscovite. The transition from acicular low-density gypsum to orthorhombic high-density anhydrite may be obvious from the relict crystal outlines of the formerly larger crystals of the hydrated mineral. For fine-grained minerals such as hematite and illite, however, there is no such guide. Exactly how hydrated they may have been is difficult to estimate, but the maximum possible hydration is likely to be proportional to the abundance of these minerals.

Molecular weathering ratios may also be a crude guide to former hydration of minerals in a paleosol (see Table 4.4). The ratio of silica to sesquioxides is high in quartz-rich paleosols with few hydrated minerals. Conversely, the ratio is low in clayey ferruginous soils which may have been full of iron oxyhydrates and hydrated clays.

Salinization

The disappearance of minerals into solution and the opposite effect of precipitation of soluble salts are widespread in soils. In modern soils a useful indicator of dissolution is an analysis of ionic concentration in soil water. A good indicator of salinization is the saturation conductivity of the soil. This is measured using a wheatstone bridge and electrical conductivity cell on a vacuum-filtered extract of saturated soil paste (Klute 1986). The conductivity of the paste increases with increasing salt content.

Neither of these direct measurements is useful for paleosols, but both dissolution and salinization may leave observable effects. Etching and pitting of mineral grains can be observed in thin sections or scanning electron micrographs of paleosols, but it is difficult to distinguish the effects of dissolution during soil formation from former hydrolysis in the soil or from comparable diagenetic changes after burial. Dissolution is more usual in carbonate or sulfide minerals than in silicates and it may be especially obvious in calcareous or evaporitic paleosols. Karst topography is a striking geomorphic example of limestone dissolution. The irregular ferruginized dissolution planes in limestone called stylolites are another example on the scale of hand specimens (Bathurst 1975).

Salinization in paleosols is more amenable to analysis than dissolution. Salts common in soils of very dry climates include halite, gypsum, and mirabilite. Also characteristic of very saline soils are the phyllosilicates palygorskite and sepiolite and zeolites such as clinoptilolite, analcime, and laumontite. Point counting of thin sections for these minerals may provide a useful index of salinization. These values can then be combined in a mineral weathering ratio with other minerals that are stable under highly alkaline conditions. Most evaporite minerals are subject to alteration on burial of a soil by dissolution of salts and by dehydration of zeolites and hydrated phyllosilicates. However, the highly alkaline conditions under which they form in soils also promote dissolution of quartz and other forms of silica which may cement indurated horizons of a soil. Such aridland chert may be firm enough to preserve undeformed pseudomorphs of the original crystal outlines of soluble salts (Groves *et al.* 1981) and brittle enough to break into stratabound brecciated zones after their dissolution and removal (Bowles & Braddock 1963). In very ancient rocks such pseudomorphs and breccias constitute the main evidence for former evaporites.

Molecular weathering ratios are not especially useful for assessing dissolution. Salinization may be approximated by ratios of alkalies over alumina, soda over potash, or soda over alumina (see Table 4.4). Sodium is generally a more reliable indicator than potassium because it is more soluble and also less susceptible to diagenetic alteration (Curtis 1985).

Indicators of biological weathering

The influence of organisms in soil formation is so pervasive that it is difficult to draw the line between biological and other weathering processes. Plant roots are critical to the development of soil structure, but the associated microbial life of soil plays an even more important role in regulating chemical conditions and physical properties of soil. Their activities are difficult to comprehend because they are so diverse and far reaching. It is partly for this reason that soil physics and chemistry have remained so empirical in orientation. Much can be learned by applying thermodynamic chemical models or the theory of fluid mechanics to soils, but predictive physicochemical models continue to be elusive because of complications imposed by soil biology.

From an organism's point of view, including our own agricultural perspective, the most important biological processes are those affecting productivity of the soil. Three of these will be considered here. First, humification is a complex process by which organic matter from once living creatures is reincorporated into the soil for further use by organisms. Second, nutrient availability is the degree to which essential elements are recycled in soil ecosystems. Some bacteria live from the energy of electron release during the oxidation of iron. For animals such as ourselves, sources of reduced iron are needed as a vital constituent of molecules such as hemoglobin, which allows oxygen transport in blood. Magnesium is a vital constituent of chlorophyll which makes plants green and allows them to harness the energy of sunlight for the photosynthetic manufacture of organic matter. A variety of other elements are needed by organisms and their availability aids soil fertility. Third, bioturbation is the degree of churning by organisms. The activity of roots and burrowing animals varies widely in ecosystems of different biomass and has profound consequences for soil structure.

Humification

Progressive decay at the surface of modern soils can be followed from the time a leaf is shed from a tree to its attack by successive waves of microorganisms on the ground, and its final disappearance as amorphous dark organic matter. Chemically, this is an oxidation in which complex reduced organic compounds are converted to simpler compounds such as carbon dioxide. Physically, it is a comminution of large structures into fine amorphous organic matter. In practice, however, it is a series of life activities of successive waves of decomposing organisms. The soft watery cell contents are consumed first by a variety of soil microarthropods and worms. Then the decomposing remainder is broken down further by fungi and bacteria. Only certain kinds of microbes can break down cellulose cell walls and very few can destroy lignin, which may remain as a woody leaf skeleton after the rest of the leaf material has been destroyed. A good deal of the material of the leaf is metabolized by soil organisms and then by the organisms that feed on them. Some organic matter is released to the soil in a soluble (humic, fulvic, or other acids) or insoluble form (mainly polysaccharides). The process of humification in modern soils can be characterized by the degree of destruction of identifiable plant fragments and by analysis for the amount of humic or fulvic acids or amorphous organic carbon in the soil. Such analyses may involve treatment with an alkali to separate soluble and insoluble fractions of organic matter, followed by treatment of the soluble part with acid to separate humic from fulvic acids. Total organic matter can be determined by the loss in weight during combustion in a furnace of a soil sample already treated with acid to remove carbonate. An alternative method for estimating the content of organic matter is the Walkley–Black titration procedure in which ferrous ammonium sulfate neutralizes chromate utilized in the oxidation of organic matter within a solution of potassium dichromate in sulfuric and orthophosphoric acid (Page 1982).

Each of these techniques for estimating the degree of humification of modern soils also can be used for paleosols, although not always with meaningful results because of modification of organic matter during burial. Diagenetic alteration affects fine-grained organic matter more than recognizable plant fragments, so that the bigger pieces are an especially useful guide to the degree of humification experienced by a paleosol.

Three general stages of humification can be recognized in leaf litter and surface horizons of soils and their fossilized equivalents. This kind of fossil leaf locality is different from those formed within the fine layers of lacustrine shales or incorporated in river and delta channel deposits. In fossil litters, leaves are mixed with roots of various kinds that pass downward into the paleosol. The plant remains show varying stages of comminution, skeletonization, nibbling by insects, clumping because of the gluing effect of decompositional polysaccarides, and creasing from the activity of leaf miners and fungal hyphae. A wide range of such features distinguish a fossil moder humus of the kind found under broadleaf oak–hickory forests of moderately neutral to alkaline soils. A mor humus, on the other hand, shows little evidence of humification. This kind of humus is characterized by a high proportion of intact leaves and other plant remains. A typical mor humus is the carpet of pine needles found under conifer forest. The needles are protected from decay by the abundant phenols and resins within the needles and acidic soil conditions hostile to decomposing microbes. A mull humus, at the other extreme, contains little in the way of identifiable plant fragments. Most have been broken down into fine-grained organic matter intimately admixed with clay of the soil. Mull humus is formed by the activity of productive decomposer ecosystems, which may leave abundant evidence in the form of small ellipsoidal fecal pellets. These often dominate the microfabric of mull humus. The rich, dark surface horizons of grassland soils are a good example of a mull humus. These three kinds of humus found in well drained soils can be recognized in paleosols (Retallack 1976; Wright 1983) and quantified by point counting thin sections or estimating from fossil plant collections the proportions of amorphous organic matter, structureless plant fiber, and plant fragments with clearly recognizable cellular structure. Depending on which of these predominate, fossil humus can be classified as mull, moder, or mor.

A comparable division of the thick, peaty surface horizons of wetland soils also can be applied to the interpretation of humification in coal-bearing paleosols. Fibric peats are those formed in such stagnant or acidic bogs that more than a third of the plant fragments retain recognizable cellular structure. Sapric peats, on the other hand, contain less than a third of recognizable plant fragments and appear structureless and fine-grained in thin section because of extensive decay in periodically well drained or alkaline bogs. Humic peats contain intermediate amounts of partly decayed plant fiber. These three kinds of peat observed in modern peaty soils may be distinguished in coals of paleosols using techniques of coal petrography (Stach *et al.* 1975).

A terminology different from this approach of soil science traditionally has been used to describe the petrographic composition of coals. This also can be used to detail the degree of humification in coal-bearing paleosols. The six main rock types of coal in order of increasing abundance of unstructured organic matter are vitrain, cannel coal, torbanite, clarain, durain, and fusain. Vitrain (bright coal) is a brilliant, black, nonlaminated coal, clean to the touch and breaking with a conchoidal fracture. Durain (dull coal), on the other hand, is hard, compact, and dull. Clarain is an intermediate kind of coal with a silky luster and with bright vitrain-like and dull durain-like laminae alternating on a scale of 1 mm or less. Fusain is a soft, powdery, friable, soot-like coal which is dirty to the touch. Cannel coal has a dull, waxy luster, black color, little evidence of layering, and a conchoidal fracture. Cannel coal also lacks the abundant fine (1–4 mm side) cubic fractures (cleat) that are characteristic of other kinds of coal. Torbanite (boghead coal) is similar to cannel coal but more waxy, less brittle, and browner. These different rock types can be distinguished in the field with practice, but are better characterized microscopically.

Coal is very dark to opaque in thin section, so that greater detail can be seen on polished surfaces examined under a microscope by reflected light (Stach *et al.* 1975). Polished surfaces are also easier to make than thin sections because many coals are brittle and extensively fractured. Mineral grains are a minor component; most coal is made up of plant fragments called macerals. The vitrinite maceral group is derived from wood and is the main component of vitrain. It includes structureless, decayed wood (collinite) and wood fragments with some crushed cellular structure remaining (telinite). Vitrinite is gray and yellowish white with low relief under the reflected light microscope (Fig. 4.8) and light orange to dark red in transmitted light. The maceral group inertinite consists of thermally stable materials that moderate the combustibility and coking properties of coal. Inertinite is the main component of fusain. This maceral group includes fossilized charcoal (fusinite; Fig. 4.8), partly degraded charcoal

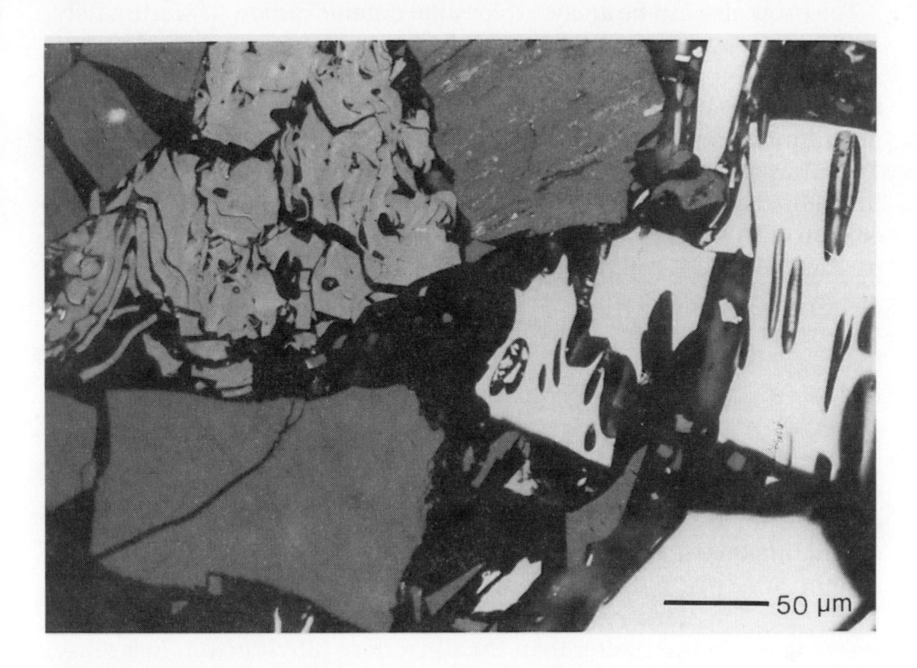

Figure 4.8 Broken fragments of semi-fusinite (light gray, upper left) flanked by vitrinite (gray) with fragments of fusinite (white, right) in a polished section viewed under reflected light of the upper Elkhorn #3 coal from Letcher County, Kentucky. Scale bar 50 μm (Photograph courtesy of J. Crelling).

(semifusinite), rounded woody objects with interlaced elongate hollows produced as resting stages of fungi (sclerotinite), and granular degraded material (micrinite). Inertinite is light gray to white in reflected light and nearly opaque in transmitted light, but not nearly so bright or opaque as pyrite. The maceral group exinite includes thermally unstable, oily, waxy, and resinous plant materials important to the coking properties of coals. Exinite is a partial component (with vitrinite) of clarain and (with inertinite) of durain. It is the main component of cannel coal. Torbanite is formed entirely of the exinite maceral alginite, which is usually the remains of the waxy coat of botryococcoid, unicellular, aquatic algae. Other exinite macerals include plant cuticles (cutinite), plant resins including amber and wax (resinite), and spore and pollen grains (sporinite). Exinite is generally dark gray to black in reflected light and yellow to yellowish brown in transmitted light. Each kind of exinite maceral has a distinctive shape that can be a clue to its identification (see Fig. 4.6). Point counting of thin sections or polished surfaces for decayed macerals (telinite, micrinite, and semifusinite) can yield quantitative data on the degree of humification of a coal.

Paleosols also can be analyzed for total organic carbon. Unfortunately, such analyses cannot automatically be assumed to reflect original organic matter distribution. Studies comparing modern surface soils with equivalent buried Quaternary paleosols have shown that the paleosols have only one fifth to one tenth of the organic matter originally present (Fig. 4.9). Two processes in particular can be blamed for this loss of organic matter. Burial of a soil may be preceded by stripping of its vegetation and some erosion of its surface. This and the leaf litter may be the most organic-rich part of the soil. Secondly, not all soils subside after burial into a completely reducing and sterile zone below the water table. Buried organic matter can be utilized by microbes deep within the overlying soil and in that way made available for the persisting ecosystem. Because of these losses of organic matter after burial, the shape of the organic matter profile rather than its magnitude may be more important for interpreting paleosols. In general, the abundance of organic matter declines with depth, but in some soils there is a subsurface zone (such as a Bh horizon) enriched in organic matter washed down cracks. Steadily declining organic matter with depth is the rule with grassland soils, but a subsurface accumulation may be found in woodland and forest soils (Stevenson 1969). In forests, the standing biomass of trees is a much more significant reservoir of organic matter than the subsurface (Bh) horizon. In a sense, then, these soils also have more organic matter near the surface, but it has been lifted out of the mineral part of the soil. The subsurface enrichment of organic matter in them is a minor irregularity in the decline of organic matter downward.

Many paleosols formed in well drained parts of the landscape or subsequently highly metamorphosed contain negligible amounts of organic carbon. Fossil root traces and burrows may provide evidence that it was once present despite very low analytical values. In these cases trace

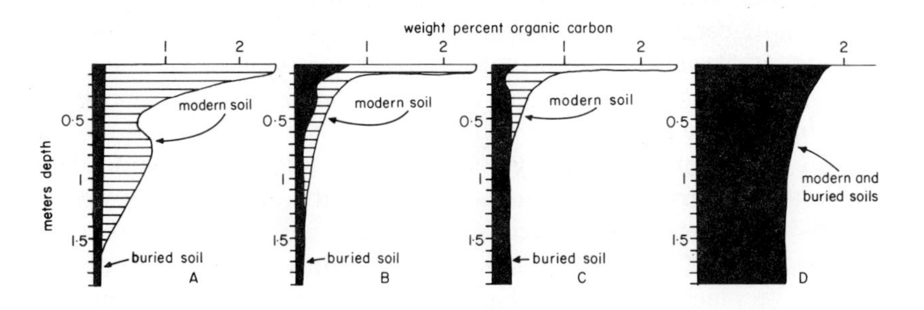

Figure 4.9 Comparison of the organic carbon contents of some paleosols with their supposed modern counterparts including Yarmouthian (1 million years) and Holocene Albaqualfs (A), Yarmouthian (1 million years) and modern Hapludalfs (B), Farmdalian (30 000 years) and modern Hapludalfs (C), and Wisconsinan (14 000 years) and Holocene Udifluvents (D) (from Stevenson 1969, with permission of Williams Wilkins).

elements characteristically complexed with organic matter in soils may provide a useful proxy indicator of the former profile of organic matter. Copper (Cu), chromium (Cr), nickel (Ni), and zinc (Zn) tend to follow organic matter and clay in soils. Their variation within a profile can be normalized to trace elements characteristically stable in soils such as lead (Pb) and zirconium (Zr). Phosphate (P_2O_5) also may show a distribution parallel to that of organic matter in soils (Smeck 1973) and a number of rare earth elements, such as yttrium (Y) and lanthanum (La), commonly follow phosphate. These various trace elements can be assayed by methods such as neutron activation analysis and inductively coupled plasma atomic emission spectrometry. A growing data base for trace elements in modern soils is increasing our understanding of their abundance and behavior during weathering (Aubert & Pinta 1977, Wedepohl 1969–78, Bowen 1979, Kabata-Pendias & Pendias 1984).

Nutrient availability

Biological productivity of soils is determined in part by the availability of nutrient elements needed for metabolism. For large plants the most needed nutrients (macronutrients) are hydrogen (H), carbon (C), nitrogen (N), oxygen (O), magnesium (Mg), phosphorus (P), sulfur (S), potassium (K), and calcium (Ca). Lesser amounts of the following elements (micronutrients) also are necessary for large plants: boron (B), chlorine (Cl), vanadium (V), manganese (Mn), iron (Fe), copper (Cu), zinc (Zn), and molybdenum (Mo). Hydrogen, carbon, and oxygen are readily available from air and soil as molecular oxygen, carbon dioxide, and water. There is also nitrogen as molecular nitrogen in the air, but large plants are unable to fix it in that form. Nitrogen, phosphorus, and sulfur are obtained in solution as cations such as nitrate, sulfate, and phosphate derived from other forms of these elements by soil microbes. The other plant nutrients are obtained by roots as ions in solution resulting from hydrolytic weathering.

The nutrient requirements of animals include the macronutrients of plants together with sodium (Na) and chlorine (Cl). Micronutrients of animals include also fluorine (F), silica (Si), chromium (Cr), nickel (Ni), cobalt (Co), arsenic (As), selenium (Se), tin (Sn), and iodine (I). Most of these are taken in as food but some (notably Na and Cl) may be obtained from salt licks and in some cases by eating soil (Jones & Hanson 1985).

Most of the macronutrients provided by soils form positively charged ions (cations) in solution. A good overall indicator of their abundance in soils is cation-exchange capacity (CEC). This has been traditionally determined by displacing cations with ammonium chloride and measuring its abundance by titration. Cation-exchange capacity includes freely available cationic bases, mostly Ca^{2+}, Mg^{2+}, K^+ and Na^+ (Page 1982). Two other cations abundant in soils (Al^{3+} and H^+) generate acids and these may be assayed by similar displacement techniques to measure the exchange acidity. Base saturation is another measure of soil nutrient availability. It is calculated as the percentage of the total cations that are basic. Available phosphorus, nitrogen, and sulfur can also be measured in soils using similar displacement techniques. Available phosphorus, for example, may be taken as that extracted with a 0.5 molar M solution of sodium bicarbonate.

These techniques for soils are not suitable for determining the base status of paleosols in which nutrients formerly available on the surfaces of weathered grains are the most likely to have been recombined during burial. However, cation-exchange capacity and base saturation are loosely linked to soil pH (Fig. 4.10) and can be approximated within broad classes of paleosol composition. Extremely acidic soils (pH 2–4.5) are noncalcareous and have a high exchange acidity due largely to Al^{3+}. Both their cation-exchange capacity and base saturation are low. Moderately acidic soils (pH 4.5–5.8) are similar but their exchange acidity is due largely to hydronium (H^+). Weakly acidic soils (pH 5.8–6.5) may have a high cation-exchange capacity and base saturation (70–90%), depending on the kind of clay present and its grain size. The cation-exchange capacity (in millequivalents per 100 g clay) of various kinds of clay [summarized by Grim (1968)] are as follows: kaolinite (2–10), illite (13–42), nontronite (57–64) and saponite (69–86), with the higher values in each range representing clay of finer grain size. Thus noncalcareous paleosols with appreciable amounts of original smectite are likely to have been highly base-saturated. Near-neutral soils (pH 6.5–8) are weakly calcareous with little remaining exchange acidity. Their cation-exchange capacity is moderate to high, depending on the kind of clay present, and their base saturation is close

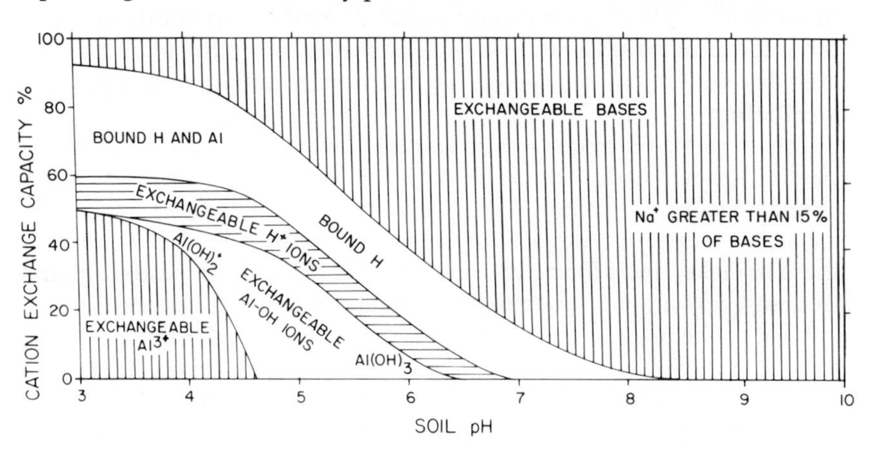

Figure 4.10 General relationship between soil acidity and cation-exchange capacity (from Birkeland 1984, reprinted with permission of Oxford University Press).

to 100%. This also is true of alkaline soils (pH 8–11). Excess calcium and magnesium may be obvious in the form of powder and nodules of carbonate in moderately alkaline to alkaline soils (pH 8–10). In extremely alkaline soils (pH 9–11), sodium dominates the exchange complex and an excess of this element may be detectable in traces of salts or soil structures such as domed columnar peds (Northcote & Skene 1972). Although nutrients are abundantly available in extremely alkaline soils, these are soils of dry, evaporitic climates in which water is so scarce as to limit growth.

The availability of other nutrients such as nitrogen, phosphorus, sulfur, and chlorine can seldom be assessed from mineralogical observations because they are usually only minor constituents of soils that are actively recycled by microorganisms (Stevenson 1986). Rock salt, gypsum, guano phosphates such as struvite, and nitrates such as sodaniter may be abundant in desert soils (Mueller 1968), but these soils are often too dry to be biologically productive. Similarly, sulfur in pyrite and phosphorus in vivianite are not easily available to plants because they are stable under conditions more anoxic than most organisms can tolerate. To some extent the availability of phosphorus in paleosols formed under more normal conditions (near-neutral pH and Eh) can be approximated by the abundance of apatite and bone fragments. More effective ways of characterizing phosphorus and sulfur reserves of paleosols are needed because in many ecosystems these are critical limiting nutrients.

Chemical methods for approximating cation-exchange capacity and base saturation include those already outlined as proxy indicators of former pH, such as the molecular weathering ratio of bases to alumina (see Table 4.4). Analyses for nitrogen, phosphorus, sulfur, and various trace elements also are useful. Phosphorus is commonly present in significant amounts (0.02–0.5%, average 0.05%; Dixon & Weed 1977) in the mineral fraction of soils. All three may be detectable in the organic fraction of paleosols. The various other trace elements also are amenable to study in paleosols using standard methods of chemical analysis, then correcting for compaction and normalizing to stable trace elements such as Pb and Zr.

Bioturbation

The amount of living material (biomass) and the fineness and degree of development of soil structure (tilth) are both related to the degree of reworking of the soil by organisms, or bioturbation. Different kinds of natural ecosystems vary considerably in biomass (Table 4.6). From their inception in colonizing barren soil, plant communities show a slow increase in biomass, usually building to a sustainable level (Kimmins 1987). These changes in plant succession are accompanied by changes in the nature of the soil. Roots, nematodes, and other soil organisms break soil

Table 4.6 Areas (x 10^6 km^2) and percentage of the Earth's surface occupied by major ecosystem types, their primary productivity (x 10^9 metric tons of carbon per year) and their productivity : area ratios (adapted from McLean 1978 and Lieth & Whittaker 1975)

Ecosystem type	Area (A) of Earth surface	Area %	Total net primary productivity (P)	Productivity %	P : A	P : A %
			Terrestrial ecosystems			
Tropical rain forest	17.0	3.4	15.3	22.13	0.90	12.18
Tropical seasonal forest	7.5	1.5	5.1	7.38	0.68	9.20
Temperate evergreen forest	5.0	1.0	2.9	4.19	0.58	7.85
Temperate deciduous forest	7.0	1.4	3.8	5.50	0.54	7.31
Boreal forest	12.0	2.4	4.3	6.22	0.36	4.87
Savanna	15.0	3.0	4.7	6.80	0.31	4.19
Woodland & shrubland	8.0	1.6	2.2	3.18	0.28	3.79
Temperate grassland	9.0	1.8	2.0	2.90	0.22	2.98
Swamp & marsh	2.0	0.4	2.2	3.18	1.10	14.88
Tundra & alpine meadow	8.0	1.6	0.5	0.72	0.06	0.81
Desert scrub	18.0	3.62	0.6	0.87	0.03	0.41
Rock, ice & sand	24.0	4.84	0.04	0.05	0.002	0.02
			Marine ecosystems			
Lake & stream	2.5	0.50	0.6	0.87	0.24	3.25
Continental shelf	26.6	5.4	4.3	6.22	0.16	2.16
Open ocean	332.0	66.9	18.9	27.34	0.06	0.81
Upwelling zones	0.4	0.08	0.1	0.14	0.25	3.38
Algal bed & reef	0.6	0.12	0.05	0.72	0.83	11.23
Estuaries	1.4	0.28	1.1	1.60	0.79	10.69

into smaller and smaller peds. These are then coated in clay and slimy polysaccharides so that their natural tendency to shrink and swell is subdued. The net effect of plant succession is to change a rocky or alluvial subsoil into an organic, well structured soil. In other words, plants promote what farmers call tilth. The epitome of tilth is the rich, dark soil formed under grasslands (see Fig. 3.8B). The building of biomass not only promotes tilth, but also preserves it from disturbance. Both soil erosion and redeposition are mitigated most effectively by the largest and most complex forested ecosystems. In contrast, desert soils show little tilth or biomass and are readily eroded.

The measurement of biomass is a difficult and time-consuming process for most natural ecosystems (Lieth & Whittaker 1975). Grasslands or agricultural crops can be mown or harvested and weighed. Because most of the biomass in such ecosystems is in their roots, these also must be excavated, washed free of soil, and weighed. Determining the biomass of

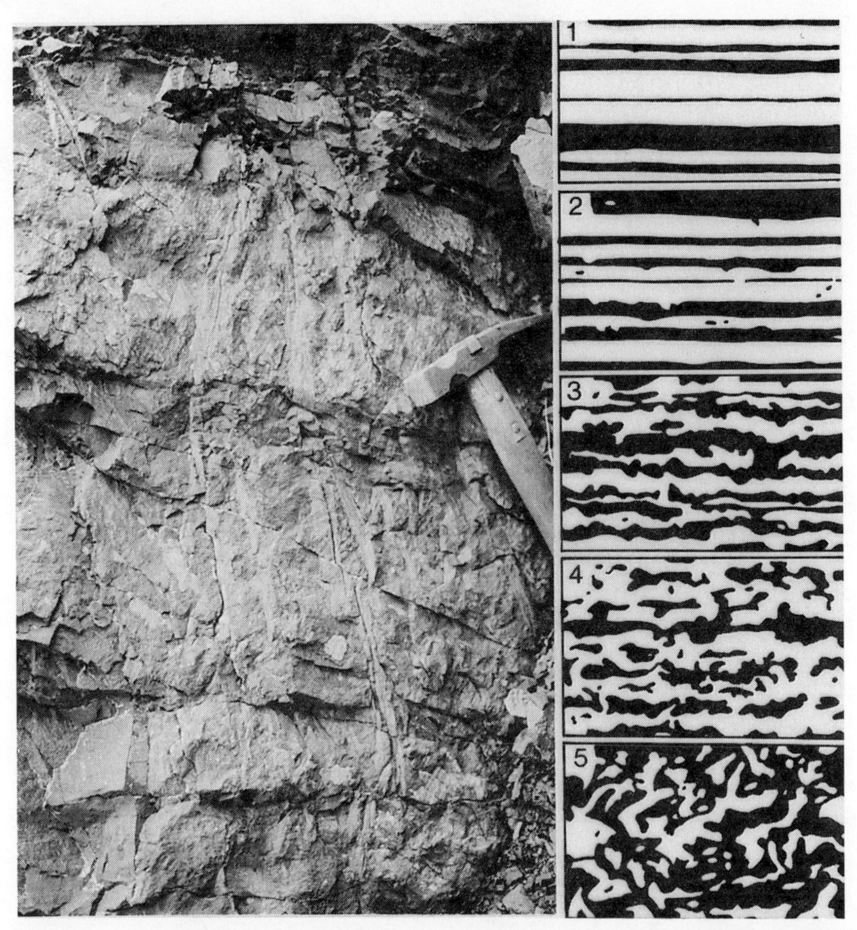

Figure 4.11 Fossil root traces emanating from the top (carbonaceous A horizon) of a paleosol (Aquept) and cutting across its relict bedding in the Middle Triassic, Tank Gully Coal Measures near Mt. Potts, New Zealand (left) (from Retallack 1979, with permission from the Royal Society of New Zealand), which is 3 in a schematic scale for the destruction of bedding by burrows (right) (from Droser & Bottjer 1986, with permission of the Society of Economic Paleontologists and Mineralogists).

forests is much more intimidating and has seldom been attempted directly. Usually it is estimated from empirical relationships to easily measured parameters such as diameter of tree trunks at breast height (about 1.5 m above the ground). A less precise but more feasible approach is to characterize the kind of vegetation in terms of general plant formations (Table 4.6) such as grassland, woodland, or forest.

For paleosols, quantitative estimation of biomass is not feasible and it is best to assess the general ecosystem type from features such as root traces, soil horizons and soil structures. Most modern kinds of plant formations can now be recognized from the fossil records of soils and plants, as outlined in a later chapter on the role of organisms in soil formation.

Compared with biomass as an indicator of bioturbation, soil tilth is easier to observe and measure. The degree of destruction of relict bedding may be a guide to the development of tilth in soils during early succession (Fig. 4.11, left). A comparative scale for the persistence of bedding developed for marine trace fossils (Fig. 4.11, right) can be used for the evaluation of paleosols. Another proxy indicator of tilth is the percentage of a horizontal line transect (1 m is sufficient in some paleosols) occupied by drab-haloed root traces formed by burial. As already argued, these may represent the last crop of large roots in a paleosol (see Fig. 3.7). This measure has proven especially useful for distinguishing paleosols formed under woodland from those of savanna (Retallack 1986d). Other root traces without drab haloes may also be present, but these represent roots that died and decayed before burial of the soil. Unfortunately, most animal burrows in paleosols do not have haloes or other evidence of their relative age. Estimates of animal activity from abundance of burrows in paleosols are compromised by the unknown degree of occupation of burrows and the unknown rate of destruction of abandoned burrows. Estimates from abundance of fecal pellets are compromised by similar considerations. Finally, quantification of soil structures and microfabrics can be achieved by point counting thin sections under the microscope or by measuring intersections along line transects. For example, the number of clayskins encountered along a 20-cm long transect in the subsurface (Bt) horizon of a paleosol is a useful measure of cutan development.

Generalized soil-forming regimes

The various weathering processes are interrelated in ways that tend toward a limited number of soil-forming regimes (Fig. 4.12). They are suites of processes thought to be important for understanding the main kinds of soils now found on Earth (Buol et al. 1980). Most classifications of soils implicitly acknowledge these generalized regimes, although they are rarely used explicitly in taxonomic nomenclature. Such theoretical underpinnings may provide meaning and direction in the often confusing business of soil classification.

Gleization

Waterlogged, mucky ground with a bluish or greenish gray color is called gley, a Russian agricultural term. Gleying or gleization is a general term used to characterize processes that produce these colors and other distinctive features of waterlogged soils. Stagnant water is exhausted of

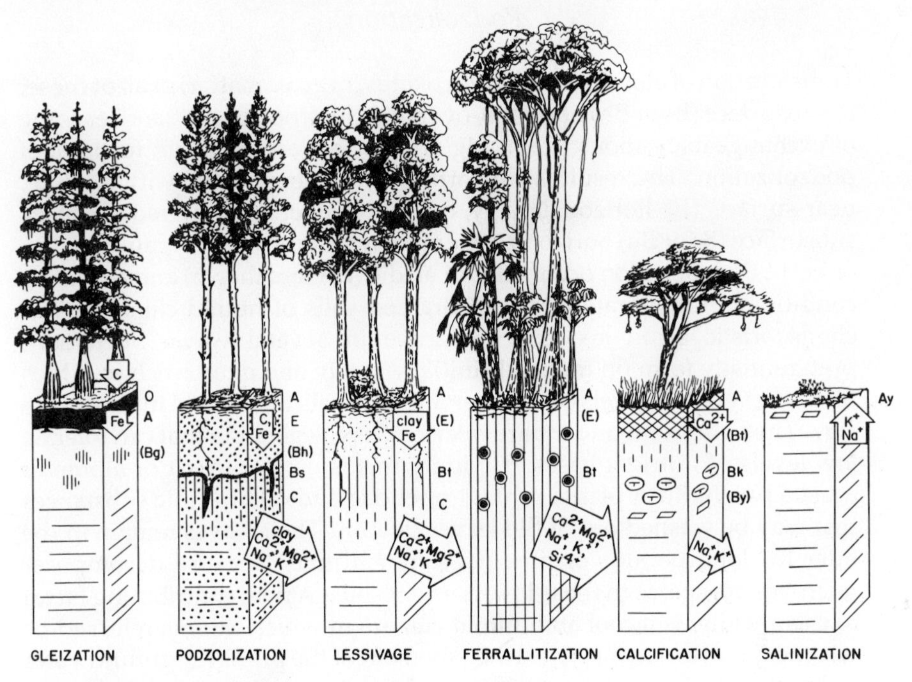

Figure 4.12 Common soil-forming regimes.

oxygen by microorganisms and under these anaerobic conditions other microorganisms reduce oxidized minerals. Microorganisms are almost always involved because thermodynamic constraints of the reduction of oxide and oxyhydrate minerals are such that these reactions proceed very slowly at surface temperatures and pressures (Stevenson 1986). Gleying results in the production of minerals containing iron in a reduced state (Fe^{2+}), such as drab-colored clays, siderite, and pyrite. Gleying can also be associated with the accumulation of organic matter because lack of oxygen discourages the activity of aerobic decomposer microbes. Also discouraged is the activity of large plants and animals which form distinctive low-diversity assemblages of swamps and marshes. Gleyed soils show little evidence of bioturbation and may have abundant relict structures of parent material. The root systems of trees within them tend to be planar and very near the surface because roots require oxygen to respire. Soil pH may be variable; it is often acidic owing to the organic acids, but can be made alkaline by abundant mafic minerals or nearby limestone bedrock. The process of gleying retards mineral weathering because water flow and removal of weathering products are limited. Therefore, easily weatherable minerals tend to persist in gleyed soils. Gleization reflects local waterlogging rather than wider effects of climate and vegetation evident from other soil-forming processes.

Podzolization

Translocation of aluminum and iron oxides, organic matter, or all of these, to subsurface (Bs or Bh) horizons, overall destruction of clay, and leaching of exchangeable cations (Ca^{2+}, Mg^{2+}, K^{+}, Na^{+}) are the main features of podzolization. The result is a colorful, contrasting profile with a white, near-surface (E) horizon largely of quartz over brown, red or black subsurface (Bs or Bh) horizons cemented by sesquioxides, organic matter, or both. Podzolization occurs under acidic (pH less than 5) and oxidizing conditions in moderately to well drained soils of humid climates. The characteristic acidic reaction can arise in several ways. Such soils preferentially form on materials initially sandy and quartz-rich in which there are very low levels of weatherable minerals that would hydrolyze to clay. They also form under particular kinds of vegetation that can tolerate low levels of nutrient bases. Typically this is conifer forest or alpine or coastal heath. Such plants contain phenolic and other acidic substances that may be washed out of the leaves by rain. Their leaves remain in the litter for long periods of time (in a mor humus) because decomposer microbes are sparse in these low-nutrient soils. A final contributing factor to their acidity is a cool and humid climate in which cations released by weathering are more likely to be flushed out of the profile by groundwater rather than taken up by organisms or clays or precipitated as salts.

Lessivage

Differentiation of a near-surface horizon which is leached and lighter in color (E) and a subsurface zone enriched in clay and darker in color (Bt) is the soil-forming process of lessivage. Unlike podzolization, lessivage results in an illuvial horizon that is clayey (Bt). It also may be reddish with sesquioxides or dark with organic matter, but these properties are judged incidental to clay enrichment. Evidence for clay illuviation may be sought in the form of clay skins. These are especially striking in thin sections of the subsurface (Bt) horizons (Holzhey et al. 1974). Those formed by lessivage tend to have bright clay (sepic plasmic) fabric and resistant grains floating in a clayey matrix (porphyroskelic). A podzolized subsurface (Bs or Bh) horizon, on the other hand, is usually grain-supported. Its grains are coated with a thin, opaque layer of sesquioxides or organic matter, often with fine desiccation cracks radiating from the grains (de Coninck et al. 1974). Soils formed by lessivage also are chemically distinct. They are neutral to moderately acidic. There is little exchange acidity and it tends to be dominated by hydronium (H^{+}) rather than aluminum ions (Al^{3+}), so that clays within the profile are more stable than in podzolized soils. Hydrolytic reactions, nevertheless, remove exchangeable cations (Ca^{2+}, Mg^{2+}, K^{+}, Na^{+}) from the weatherable minerals. These can be washed out of the profile and over time this can seriously reduce fertility. The process of lessivage is favored on parent materials such as basalt or shale which are rich in minerals that weather to clay. Lessivage is also favored by vegetation such as deciduous broadleaf woodland that is low in acid-generating phenolic compounds and resins. The fall of broadleaves encourages recycling of nutrients and abundant soil fauna that is a feature of soils formed by lessivage.

Ferrallitization

In well drained soils of wet tropical regions, weathering is intense and deep, and it may result in thick, uniform profiles depleted in exchangeable cations. As a result, the soil as a whole is enriched in clay and sesquioxides in the form of fine-grained, weather-resistant minerals such as kaolinite, gibbsite, and hematite. The process of ferrallitization is encouraged by a long period of soil formation and under ecosystems of great biomass and stability such as tropical rainforest. As in temperate woodland soils formed by lessivage, there may be some indication of a clayey subsurface (Bt) horizon, but it is usually thick (more than 1 m) and has diffuse boundaries. Soil reaction is buffered by abundant clay to near-neutral or mildly acidic. It is paradoxical that deeply weathered and infertile soils should be able to support such lush vegetation as rainforest. This is achieved through a nearly complete recycling of nutrients within a biologically active leaf litter (Sanford 1987). The roots of tropical trees in ferrallitized soils commonly are shallow and spreading. The soil itself may be deeply and copiously burrowed by ants and termites, which create sand-sized stable spherical micropeds that are characteristic of these soils (Stoops 1983).

Calcification

In well drained soils of semiarid to subhumid regions, only the surface of the soil may be moistened by rain before the water evaporates, is used by plants or animals, or soaks into organic matter and clay. Thus, rainwater may not flow through the soil as it does in more humid climates. In the moist and biologically active surface layers of the soil, exchangeable cations may be hydrolyzed out of weatherable minerals. In subsurface horizons, however, there may be sufficient water to remove some exchangeable cations (Na^+, K^+, Mg^{2+}) from the profile, but not calcium (Ca^{2+}), which accumulates as a distinctive subsurface (Bk) horizon near the depth of average wetting. Compared with ferallitization, lessivage, and podzolization, the process of calcification involves little loss of nutrient cations and the overall soil pH is alkaline. Soils of this kind can be fertile, supporting open grassland or grassy woodland vegetation with abundant soil fauna. In drier regions they support desert vegetation.

Salinization

In very arid climates, rainfall may be sufficient to hydrolyze soluble cations from weatherable minerals, but insufficient to remove them from the profile where they accumulate as surface crusts and crystals of salt such as halite, gypsum, and mirabilite. Salinization can also occur by precipitation of salts from the top of a temporarily high water table. Irrigation of aridland soils may exacerbate this process. Salinized soils are generally well drained for most of the year and extremely alkaline (pH 9–11). Their high salinity discourages the growth of plants other than small-leaved, evergreen, desert shrubs or cacti. The activity of soil fauna also is curtailed under such hostile conditions. At its extreme development in barren desert playas and coastal sabkhas, salinization may be regarded in part as a sedimentary process, but even there it involves modifications in place that are the hallmark of other soil-forming processes.

5 Soil classification

Many features of paleosols can be understood by comparison with modern soils in the same way as fossil bones are best interpreted by comparison with bones of similar modern organisms. For this reason, modern soil classifications are important for understanding paleosols.

Soil classifications date back to the latter half of the last century. From the beginning, these classifications had practical aims in planning for agricultural and other land use. Thaer's classification of 1857, for example, was based at the highest level on the kinds of crops that could be grown and on soil texture: wheat soils (clay), barley soils (sandy clay), oat soils (clayey sand), and rye soils (sand). Like other kinds of classification in natural history, soil classification provides basic units for extending and extrapolating research results. Most soil classifications have some theoretical underpinning in soil-forming processes and are designed to bring order to the multitude of combinations of different soil properties. Such organizational systems are more easily remembered than the raw data on which they are based. Systems of classification also become a way of thinking and communicating akin to a foreign language.

The optimal classification for paleopedological comparisons would be one based entirely on observable features of profiles that are relatively resistant to diagenetic alteration. The most suitable modern soil classification in this respect is the "factual key" of Northcote (1974) which has been applied to many Australian soils. Classifications of this kind may prove important to paleopedology in the future, but are not discussed further here for two reasons. First, the data base of soils classified by this scheme is small compared with that of other classifications. Second, the "names" of the soils are not memorable. They consist of descriptive strings of letters and numbers such as Uc2.36 and Dr2.73.

Three modern classifications command attention: the handbook of Australian soils of the Commonwealth Scientific and Industrial Organization (CSIRO) of Australia (Stace *et al.* 1968), the soil map of the Food and Agriculture Organization (FAO 1971–81) of the United Nations Economic and Scientific Organization (Unesco) and the soil taxonomy of the Soil Conservation Service (Soil Survey Staff 1975) of the United States

Department of Agriculture (USDA). The Australian handbook is based on soils of stable continental regions at mid-latitudes. The FAO classification is especially good for tropical soils. The US soil taxonomy, on the other hand, is based largely on soils of temperate climate in tectonically active, volcanic, and glaciated terranes.

Australian handbook

The handbook of Australian soils (Stace *et al.* 1968) was produced by the CSIRO on the occasion of the 9th International Congress of Soil Science in Adelaide in 1968. It is one of the few remaining classifications still in service that retains many of the classical names of soil science. Such old names as Podzol or Krasnozem, or common English names such as Calcareous Sand or Brown Earth, make it an approachable classification. Also useful are the numerous individual soil profiles described in the handbook and the detailed accounts of their micromorphology. The profiles also have been identified within an objective numerical system: the factual key of Northcote (1974). This kind of information is especially useful for unbiased comparison with paleosols.

The Australian classification is non-hierarchical in the sense that each of the names represents an independent kind of soil. Each soil type is based on actual field examples and represents a soil mapping unit widely recognized in Australia. Each is defined in objective terms on the basis of observable profile features and based on a "central concept." Some degree of variation is found within each named type and some of these variants are distinctive.

The typical vegetation, parent materials, and climate for each soil type are mentioned in the handbook, together with some consideration of how each type is thought to have formed. Another theoretical concession is the arrangement of soils into general categories reflecting a progressive increase in degree of profile development and leaching, from no profile differentiation to minimal profile development to mildly leached dark soils, and so on (Fig. 5.1).

The documentary basis of the classification is in some ways a weakness. New soils that do not correspond to the types described must be added to the existing list rather than interpolated between existing categories, as in other classifications. This is particularly a problem for soils not widely found in Australia such as arctic and alpine soils or soils of volcanic ash. The fine subdivision of desert soils is a mixed blessing. Many of these are relict paleosols that reflect more the particular geological history of the Australian outback than general features of soil formation applicable elsewhere.

Only soil cartoons (Fig. 5.1) can be offered here to provide an introductory impression of the classification. These cartoons cannot supplant careful consideration of the handbook itself for identifying paleosols.

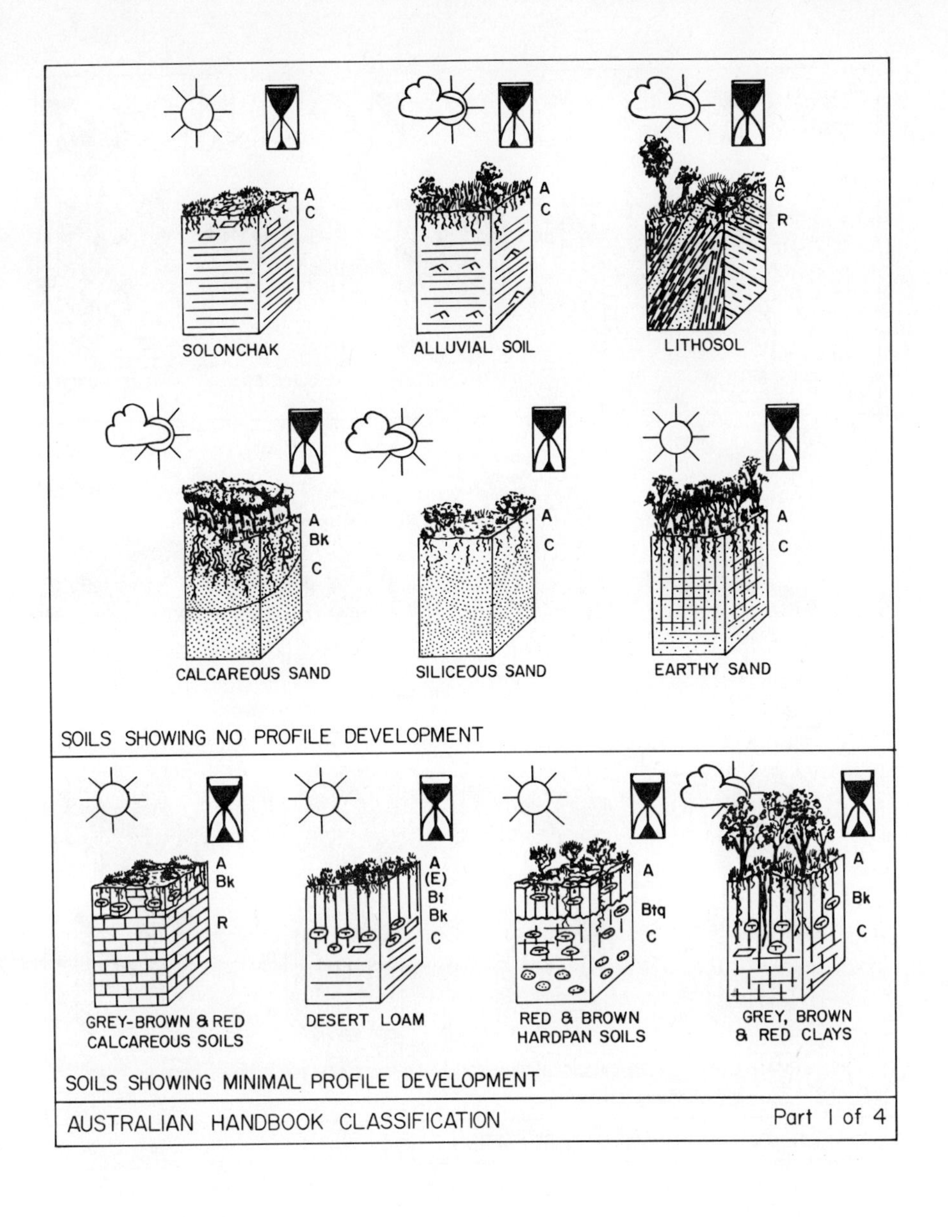

Figure 5.1 Cartoons of climate, vegetation, and soil profile form of the various kinds of soil recognized by the handbook of Australian soils (based on data from Stace *et al.* 1968). Sun or cloud and hourglass symbols reflect climate and relative time for formation.

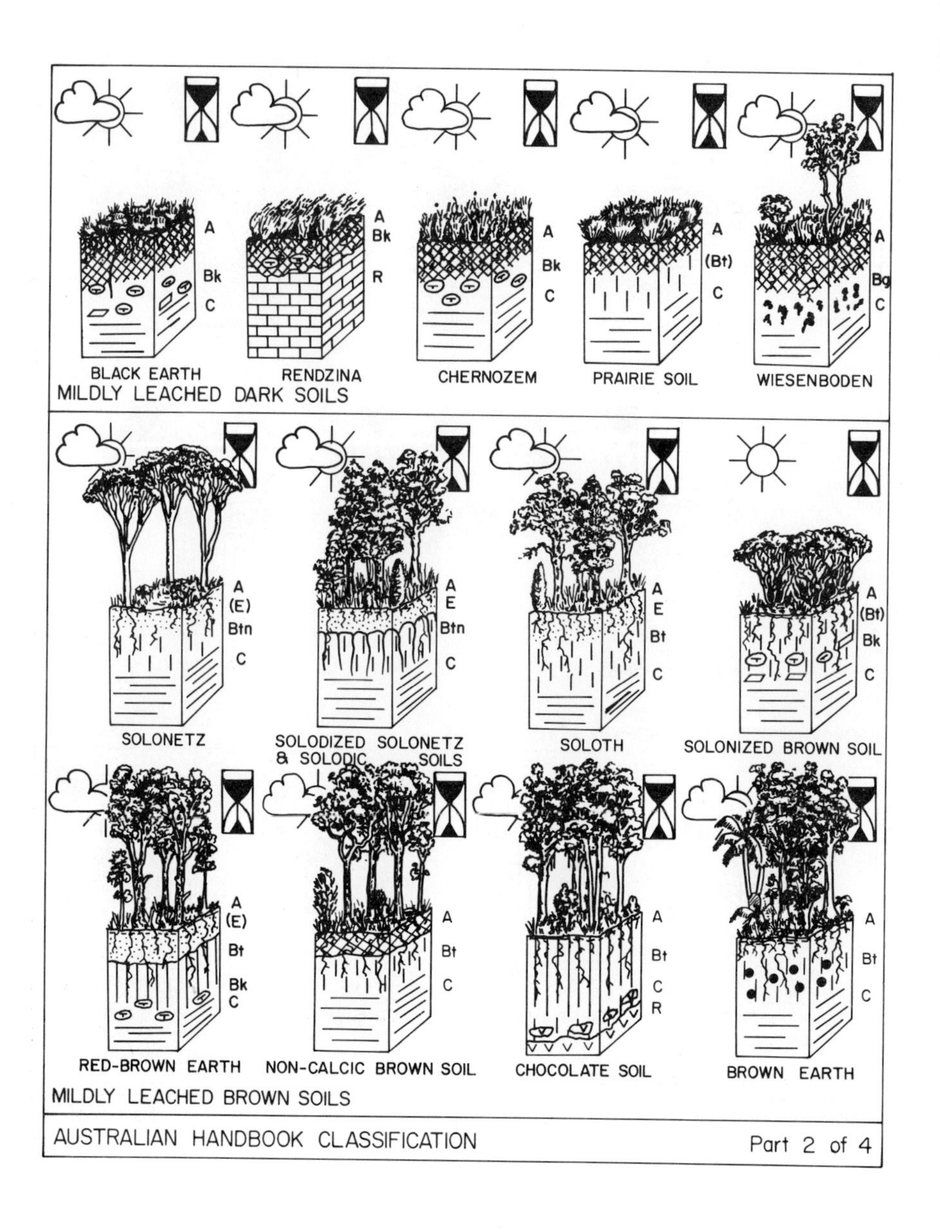

AUSTRALIAN HANDBOOK CLASSIFICATION Part 2 of 4

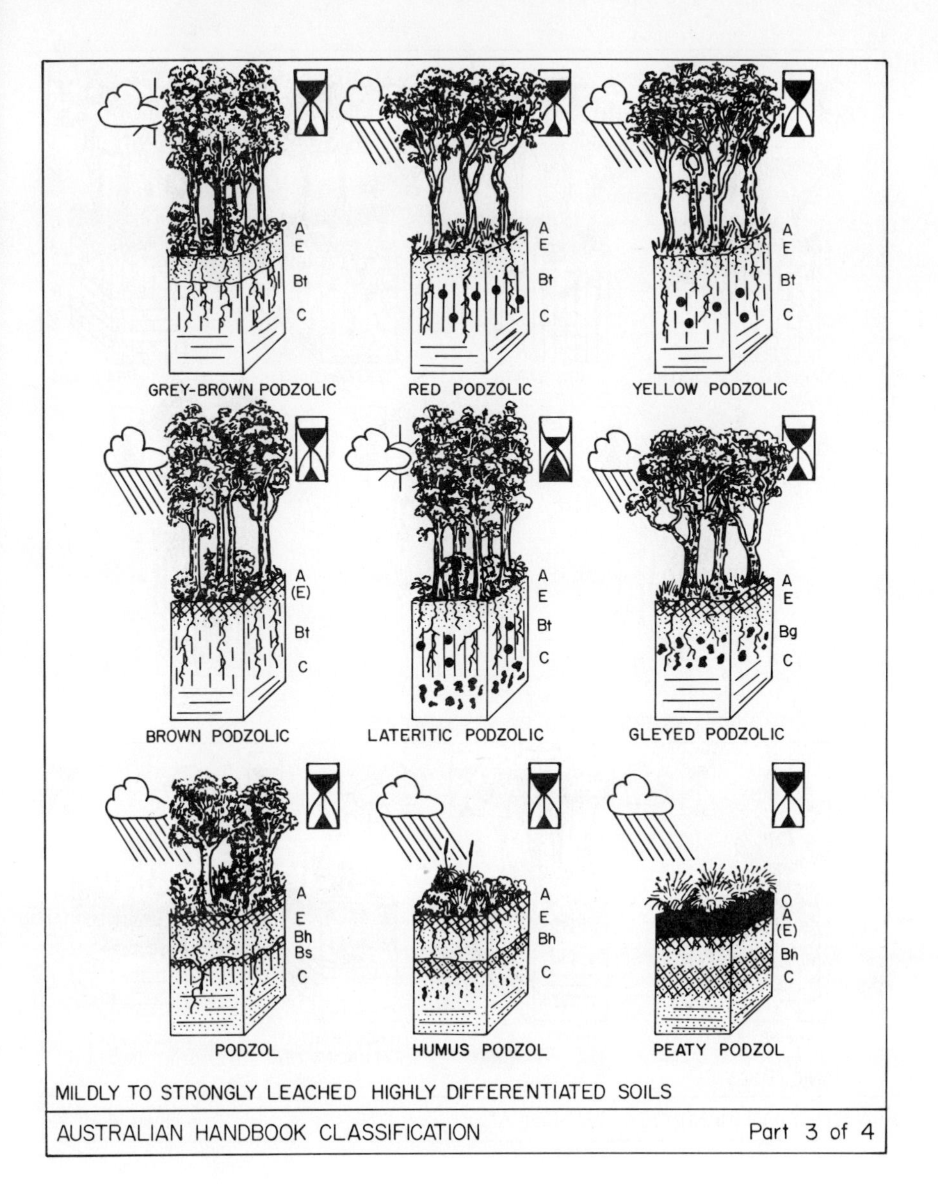
A
E
Bt
C
GREY-BROWN PODZOLIC
A
E
Bt
C
RED PODZOLIC
A
E
Bt
C
YELLOW PODZOLIC
A
(E)
Bt
C
BROWN PODZOLIC
A
E
Bt
C
LATERITIC PODZOLIC
A
E
Bg
C
GLEYED PODZOLIC
A
E
Bh
Bs
C
PODZOL
A
E
Bh
C
HUMUS PODZOL
O
A
(E)
Bh
C
PEATY PODZOL
MILDLY TO STRONGLY LEACHED HIGHLY DIFFERENTIATED SOILS
AUSTRALIAN HANDBOOK CLASSIFICATION
Part 3 of 4

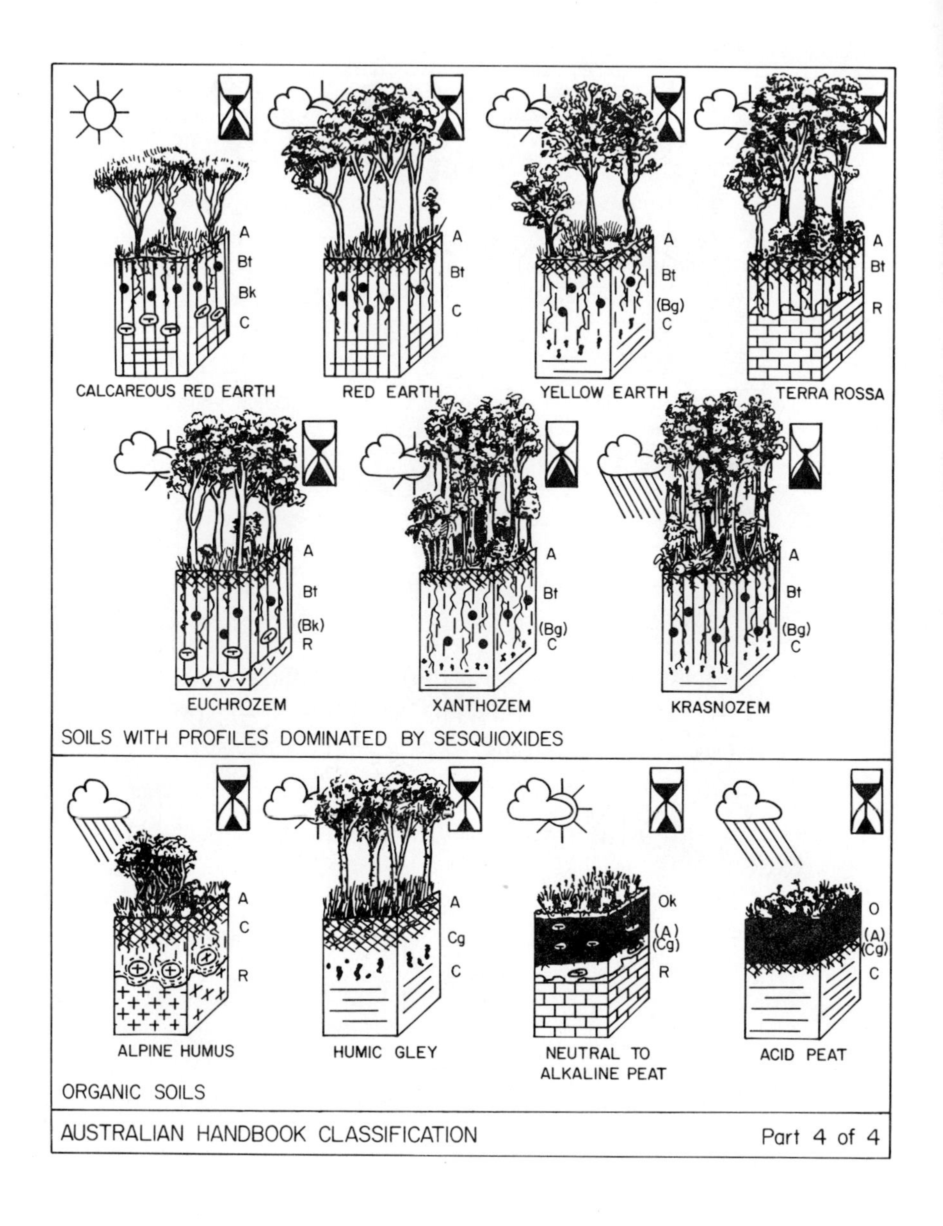

AUSTRALIAN HANDBOOK CLASSIFICATION Part 4 of 4

FAO world map

The classification of the FAO (Unesco) was begun as an international project in 1961. Explanatory volumes and soil maps began to appear in 1971 and the tenth (Volume V) was published in 1981. The maps are small scale (1 : 5 000 000), but the accompanying volumes make up for this lack of detail with extensive notes on climate, vegetation, geomorphological position, parent material and land use of the map units. Descriptions of individual soil profiles with analytical and site data are included as examples (FAO 1971–81). It is an unparalleled global data base for comparison with paleosols.

In concept, the FAO classification is a hybrid of traditional classifications like that of the Australian handbook and of more recent classifications such as that of the US Soil Taxonomy. It retains some traditional names such as Chernozem and Podzol, and adds new names such as Acrisol and Ferralsol. The classification is based on a system of diagnostic horizons and is hierarchical like the US Taxonomy. Instead of many hierarchical levels, the FAO world map has only two: the "major units," which have specific soil names such as Podzol and Acrisol, and "soil units," which are specified by adjectives, for example, Humic Podzol and Ferric Acrisol. Not all of the 26 major units are subdivided, but some include many subunits. The total number of soil units is 106. Such a hierarchical system implies that soils must first be classed within the 26 major units before being assigned to subsidiary soil units. This reflects a judgment that some features of soils, their diagnostic horizons, are more fundamental for their classification than are others. For example, Ferric Acrisols have a strongly ferruginized profile like Ferralsols, but they have less deeply weathered, base-depleted subsurface horizons than Ferralsols, and this feature of Acrisols is thought to be more important to their classification than ferruginization. Hierarchical systems thus involve a series of decisions. They are also easy to remember, at least at higher levels, where there are fewer categories.

A potential weakness of such a global system of classification is establishing correspondence between the regional schemes of classification and soil surveying from which much of the information was extracted. It is difficult enough to achieve consistent classification in a national soil survey of scientists speaking the same language and following the same philosophy of soil genesis developed at a small group of universities. This problem was addressed at length by numerous working parties on the correlation of different national soil classifications. However, not all scientists were happy with all aspects of the resulting classification for the world map and it was not adopted by the Soil Conservation Service of the US Department of Agriculture.

The original volumes accompanying the soil maps are too detailed to afford a concise introduction to the FAO classification, but an excellent and well illustrated summary has been published (Fitzpatrick 1980). The soil cartoons here (Fig. 5.2) are arranged in general order of increasing development, except that waterlogged soils are at the end. This order facilitates comparison with the other two classifications outlined.

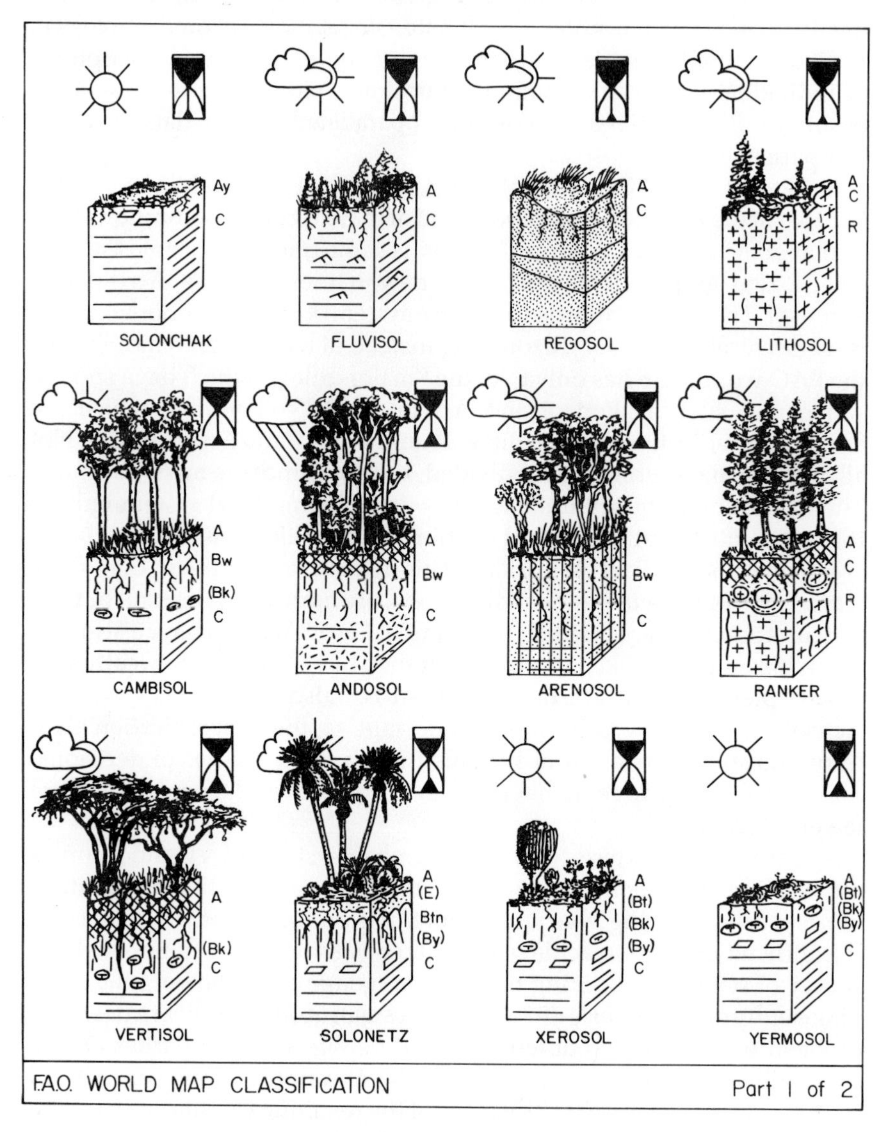

Figure 5.2 Cartoons of climate, vegetation, and soil profile form of the major soil units of the FAO soil map of the world (based on data from FAO 1971–81).

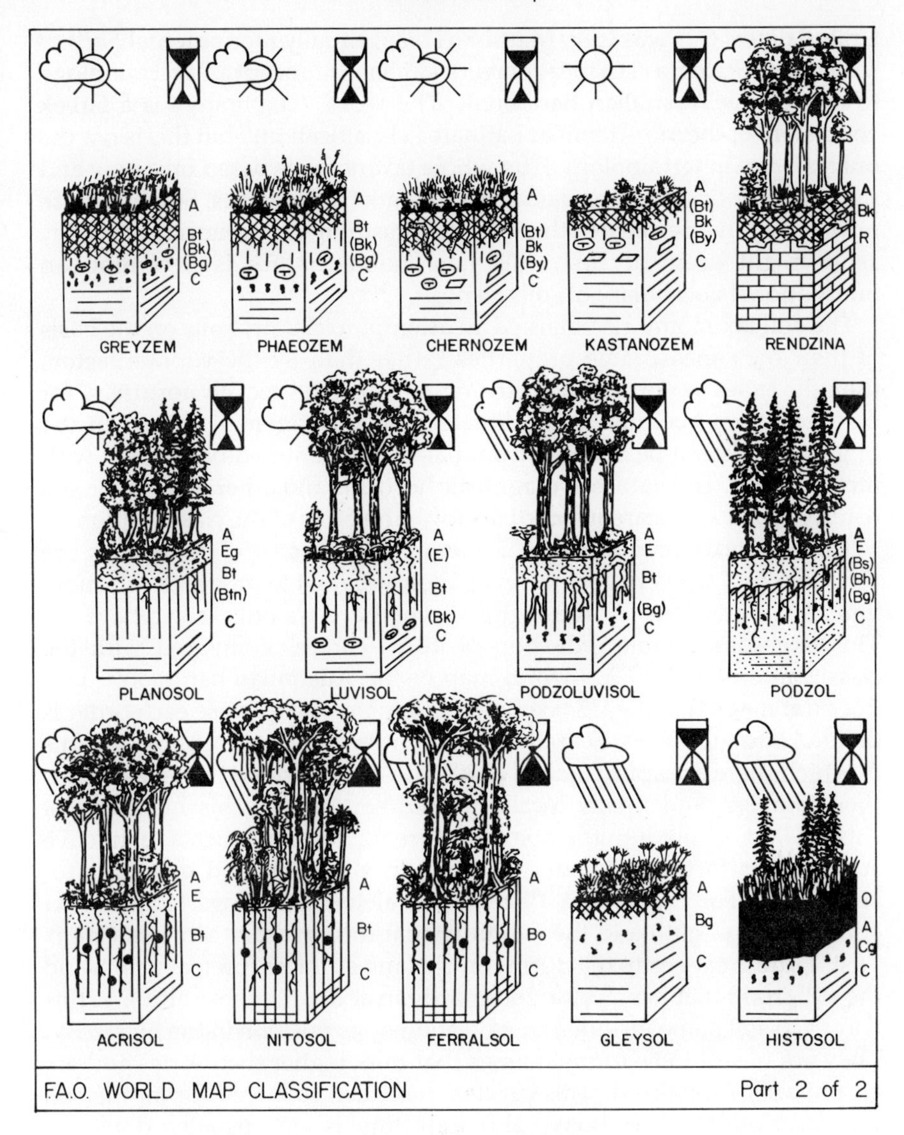

US soil taxonomy

The classification of the Soil Conservation Service of the US Department of Agriculture (Soil Survey Staff 1975) was to a considerable extent the work of Guy D. Smith. It was guided by him through a series of early drafts which were widely circulated for criticism and comment. The 7th Approximation was first published in 1960 and a detailed account appeared in 1975 under the distinctive title of *Soil taxonomy*.

This classification was meant to be a total break with the past. In the

United States this was C. F. Marbut's (1935) equally monumental *Soils of the United States,* a classification very similar in concept and terminology to that of the Australian handbook. The word "taxonomy" is a Greek equivalent of the more familiar Latinate "classification," but this is not the only change in terminology. The whole taxonomy is based on a new and rigorously defined nomenclature for soils and soil features. Some of these new terms have permeated the FAO classification. As foreign and complex as it might seem at first, this new nomenclature is becoming an international vocabulary of soil science.

The soil taxonomy also marked an attempt to classify soils on the basis of their own measurable properties rather than by the various factors thought to be important in their origin. This is especially fortunate for paleopedology, because paleoclimates and other genetic factors in the origin of soils can be inferred from paleosols identified objectively with modern soils. The nature of diagnostic horizons and other observable soil features such as calcareous nodules form the basis of the classification.

The soil taxonomy also has a hierarchical structure. Thus soils are classified first at a very general level, then assigned to progressively more limited subdivisions. At the highest level there are only ten soil orders. This is a considerable reduction of initial choices compared with the classifications of the FAO world map or the Australian handbook. This does not mean that the US taxonomy is less precise because each order is divided into suborders, 47 in all. Each suborder is divided into great groups of which there are approximately 230. These are divided into subgroups, then families, and finally locally named soil series. This hierarchy of subdivision is built into the nomenclature for soils in which a distinctive suffix refers to the soil order, a prefix to suborder, and qualifiers for subgroups. For example, a Petrocalcic Paleustalf belongs to the order Alfisol (suffix "alf") and the suborder that forms in dry climates (prefix "ust"), the great group that forms on old land surfaces (prefix "pale") and the subgroup that has a calcareous hardpan at depth (petrocalcic). In this way each soil name includes an indication of its position in the hierarchy. They are meant to be formal names that rank higher than series and are customarily capitalized. This practice has been followed for all the soil classifications in this book, although this is not usually done for classifications used in other countries.

The three main features of the soil taxonomy, emphasis on observable features, new nomenclature, and hierarchical organization, are also a source of some difficulty. The nomenclature is like a foreign language on first acquaintance, and it takes time and practice to become fluent. The emphasis on objective features in the soil taxonomy has led to much of it seeming like a legal document complete with specific stipulations, provisos, and caveats. Such legalistic strictures often obscure the main idea behind the particular soil type. The aim to get away from changeable genetic concepts that have influenced other soil classifications is admirable, but in this case it often is difficult to gain a definite idea of the kind of soil specified from a complex sequence of boundary criteria. On the other hand, there are some places where the aim to abandon genetic concepts was not carried far enough. Climatic conditions are the main criteria for distinguishing suborders of Vertisols. The particular information needed about how many days per year the cracks are open is difficult to find for modern soils and impossible for paleosols. Another problem is inherent in the hierarchical structure of the taxonomy which calls for selective weighting of different features of soils. The most diagnostic horizons and the degree of development at which they become diagnostic are clearly specified, but these same features may also be used at lower levels to define intergrades. This can be confusing until one learns to operate strictly within the hierarchy by choosing among alternatives at each level in the precise order specified. Choices must be made with care because a bad choice may lead one into an area of this large taxonomy distant from the correct choice. If a choice is difficult to make it may be wise to explore alternative pathways down through the hierarchy to see how they end. For paleosols it may be difficult to make decisions on diagnostic criteria at high levels and the identification should stop at that level. A final problem is that the soil taxonomy is not comprehensive. Deeply weathered, red tropical soils (the order Oxisols), for example, are rare in the United States and this part of the classification now is under revision.

Despite these drawbacks, the US Soil Taxonomy has had an undeniable influence on soil classification throughout the world. Although the FAO decided on its own classification, it incorporated basic ideas and terms from the US taxonomy for horizons and properties diagnostic for classification. Only an outline of the classification (Fig. 5.3) can be offered here. Well illustrated, regional descriptions of soils using this classification are now available (Aandahl 1982, Barker *et al.* 1983). Introductory accounts can be found in textbooks of soil science (Buol *et al.* 1980). For scientific work on soils and paleosols, however, there is no substitute for the thick green tome of *Soil taxonomy* and its amendments.

Diagnostic horizons and properties

Fundamental to the soil taxonomy are 17 diagnostic subsurface horizons and six diagnostic surface horizons. A surface horizon is called an epipedon (the plural is usually given as "epipedons"). This does not correspond exactly to the A horizon; B horizon material is included in some cases.

The mollic epipedon is a soft, dark, humus-rich, finely structured surface horizon found under grasslands including suburban lawns. High base saturation characteristic of this horizon may be implied for paleosols if they have dispersed or nodular carbonate, abundant burrows and root

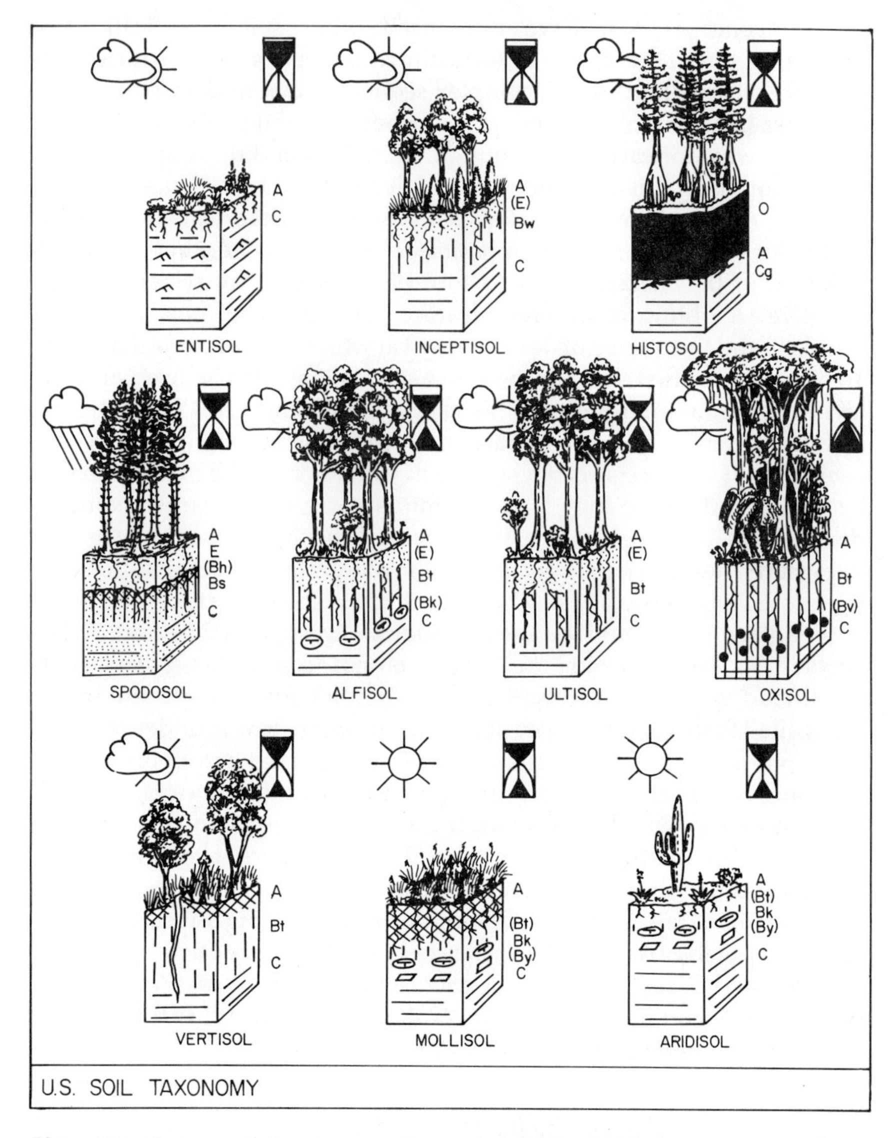

Figure 5.3 Cartoons of climate, vegetation and profile form of the various orders of soils defined by the US soil taxonomy (based on data from Soil Survey Staff 1975).

traces, or common easily weathered mineral grains such as feldspar. Characteristic granular or crumb ped structure may be preserved in paleosols, but organic matter is seldom preserved at original levels (Stevenson 1969). For this reason, care must be exercised in considering the requirement for a mollic epipedon to have more than 2.5% organic matter in the upper 18 cm or to have a dark Munsell color (value and chroma less than 3.5 moist and value less than 5.5 dry). The thickness requirements for a mollic epipedon are also difficult to apply to paleosols because of their likely compaction and erosion. These are more than 10 cm thick if developed on bedrock and more than 18 cm if there are other soil horizons within 75 cm of the surface or if the soil is loamy or clayey, and more than 25 cm if these other horizons are more than 75 cm down or if the soil is on sandy or gravelly materials. For paleosols adjustments to the amounts of organic matter and thickness of horizons need to be made based on estimates of burial diagenetic modification.

The anthropic epipedon may resemble a mollic epipedon, but has been long influenced by human use. It generally has a less finely developed structure than a mollic epipedon and may show signs of trampling or fertilization with bone scraps or shell fragments. A high content of phosphate is diagnostic. Evidence of artifacts, campfires, or hut foundations also can be used to recognize an anthropic epipedon. In theory this epipedon also includes other human works such as roadways and railway grades.

An umbric epipedon is difficult to distinguish by eye from a mollic epipedon. In contrast to the mollic epipedon, an umbric epipedon has base saturation of less than 50% and so is noncalcareous with low reserves of weatherable minerals. Some umbric horizons are more organic and less finely structured than mollic epipedons.

A histic epipedon is a peat layer. In paleosols, the histic epipedon is converted to a coal seam. The thickness of a peat required for a horizon to qualify as a histic epipedon is at least 20 cm. For a histic epipedon to qualify as a Histosol, peat on bedrock must be more than 10 cm thick. On other materials low-density peat of the kind formed under *Sphagnum* moss must be more than 60 cm thick and other kinds of peat more than 40 cm thick. The original thicknesses of coaly surface horizons of paleosols can be reconstructed from field evidence or general information on peat compaction (see Fig. 7.6). This is not a simple calculation because peat at the base of a thick, unburied histic epipedon is already compacted compared with peat at its surface. Compaction to as low as 0.02 times the original peat thickness is found in deeply buried coals of anthracite rank. By this ratio a seam of woody coal only 0.8 cm thick could qualify as a Histosol. Another diagnostic criterion for histic epipedons and Histosols is the amount of organic matter and clay. The percentage of organic carbon (y) relative to clay content (x) is given by the following conditions $y > 18$, and $y > 0.1x + 12$. Thus, it can have no more than 60% mineral matter and no less than 18% organic matter. In practice, this means that histic epipedons and Histosols are very dark to black with organic matter rather than gray with clay. A low mineral content and substantial thickness are also desirable qualities for mineable coal. Almost all economically mineable seams qualify as fossil histic epipedons and Histosols.

A plaggen epipedon is a soil surface created artificially by manuring and plowing. It is a dark organic horizon with a poor structure, but in other ways similar to an anthropic epipedon. Spade or plow marks may be visible, in addition to pieces of brick or pottery. Like the cultivated field in which they are formed, plaggen epipedons commonly occupy square or rectangular areas of ground on moderately level sites.

A final kind of surface horizon which accommodates most others is the ochric epipedon. This is too thin, too light colored, or not organic enough to qualify as one of the other kinds. An ochric epipedon may contain organic matter, but it is less intimately mixed with clay and may contain recognizable leaf litter. It also is more likely to be sandy, blocky, or otherwise more crudely structured than, for example, a mollic epipedon.

Among subsurface diagnostic horizons, the argillic horizon is one of subsurface clay enrichment. Exactly how the clay has been reorganized to achieve this enrichment is less important than demonstrating that clay moved during soil formation. Clayskins along the margins of soil peds and root channels are evidence of clay that was washed down the profile (Buol & Hole 1961). Clay rinds extending into fractured and partly hydrolyzed grains may be evidence of clay formation in place. Compactional effects may obscure these features in petrographic thin sections of paleosols. Compaction also alters the degree of clay enrichment and thickness of the clayey horizon in paleosols. The percentage clay (y) of the subsurface horizon compared with that of the surface horizon (x) required for an argillic horizon varies with the degree of clayeyness of the soil as a whole: if $0 < x < 15$, then $y > x + 3$; if $15 < x < 40$, then $y > 1.2x$; and if $40 < x < 100$, then $y > x + 8$. The requirement that an argillic horizon be at least one tenth as thick as all overlying horizons also must be applied to paleosols after considering compaction.

Another kind of clayey subsurface horizon is the agric horizon which forms after cultivation. Clay washes into the large cracks opened up by the plowed layer, and these wedge-shaped masses of clay may be layered with increments of washed-in topsoil. The agric horizon has a sharp lower boundary at the base of the plowed layer and may have associated human artifacts. The agric horizon is found only in Holocene soils and paleosols.

The natric horizon is a clayey subsurface horizon strongly base-saturated with sodium. This can be estimated for paleosols from soda to potash molecular ratios greater than unity and from a columnar or prismatic structure with a sharp upper boundary at the top of the horizon (Northcote & Skene 1972).

A sombric horizon is like a subsurface version of an umbric epipedon. The dark organic matter of a sombric horizon is not associated with iron and aluminum oxides as in a spodic horizon, nor is it base-saturated as in a natric horizon. Sombric horizons are not found under an albic horizon.

They are found mainly in moist soils of high plateaus and mountains in tropical to subtropical regions.

A spodic horizon is a sandy, usually quartz-rich, subsurface horizon cemented by amorphous iron and aluminum oxides, organic matter, or different layers and combinations of these. To meet the criteria for a spodic horizon this dark, cemented horizon must be laterally continuous and at least 2.5 cm thick. There are also requirements for the amount of iron and aluminum as extracted by pyrophosphate and dithionite–citrate. These requirements cannot easily be determined for well lithified paleosols. The spodic horizon can be identified in thin section by complete grain coatings of opaque, amorphous sesquioxides and organic matter which commonly have numerous radial cracks as if they had shrunk around the grains on drying (de Coninck *et al.* 1974). Spodic horizons form in humid, acidic soils in which clay is destroyed and thus have very little clay. Sesquioxidic or organic clayey subsurface horizons are better regarded as argillic, oxic, or sombric than spodic.

The placic horizon is a black to dark-reddish hardpan cemented by amorphous iron and manganese oxides or by an iron and organic matter complex. They are mostly thin (2–10 mm), but are variable in thickness (1–40 mm). They are brittle and break into angular segments. Often they appear wavy. In some soils two or more of them may bifurcate and anastomose more or less parallel with the ground surface. In thin section they show a massive opaque cement enclosing the clastic grains. Also, unlike spodic horizons, they are found in clayey soils and at various depths in the profile other than below a sandy eluvial (E) horizon.

The cambic horizon is a mildly weathered, slightly clayey or oxidized subsurface horizon more altered than underlying material, but lacking the distinctive properties and degree of development required for other kinds of subsurface horizons. This mild weathering is best judged by comparison with underlying parent material or saprolite. The cambic horizon may appear more massive and structured, show less sedimentary or other relict features, seem colored more yellow, brown or red due to oxidation, or be less calcareous or salty.

Oxic horizons are so highly weathered that they have few exchangeable cations remaining (cation-exchange capacity less than 16 millequivalents per 100 g clay). This is reflected in the abundance of kaolinite and other 1 : 1 lattice clays, trace amounts of weatherable minerals such as feldspar, and molecular ratios of bases to alumina close to zero. Oxic horizons also are red or brown with oxides of iron and aluminum. Although very clayey, there is usually little evidence of clayskins. Unlike argillic horizons, which show a subsurface peak of clay enrichment, oxic horizons are normally deep (at least 30 cm) and show fairly constant amounts of clay with depth or a very diffuse zone of subsurface clay enrichment. Stable, sand-sized, spherical micropeds are characteristic of oxic horizons (Stoops 1983).

A final distinctive subsurface horizon is the albic horizon. This is a light-colored, white, and sandy layer from which clay and oxides of iron and aluminum have been leached, leaving naturally light-colored minerals such as quartz and feldspar (Dumanski & St. Arnaud 1966). Clay and sesquioxides washed out of the albic horizon may be just below in an argillic or spodic horizon. The albic horizon may be at the surface and commonly is near the surface just below a thin, organic, and rooted horizon. The light color of an albic horizon is often conspicuous by contrast with overlying and underlying materials. These are the pastel shades in the upper left-hand corners of the Munsell color charts (value generally greater than 4 and chroma less than 3, with some exceptions noted by Soil Survey Staff 1975).

In addition to these common kinds of subsurface horizons, there are several specially hardened or cemented subsurface materials. These may appear indurated in a modern soil compared with the friable enclosing material. In a paleosol entirely lithified, these former hardpans are best recognised by the way in which root traces avoid them. A duripan is a hardpan cemented by silica. These vary in appearance with the materials they cement because silica cement is partly transparent.

A fragipan is a dense subsurface hardpan of clay. Commonly it is mottled and has a prismatic structure. Groundwater perched on top of a fragipan forms reduced and drab-colored surfaces on the hardpan and its prismatic peds. The origin of fragipans is still in doubt. They could be buried soil horizons or zones altered by permafrost. They are found in areas of humid climate and under forest vegetation (Birkeland 1984).

A petroferric contact is a strongly ferruginized upper surface to bedrock at the base of a soil profile. Its close relationship with bedrock contacts and small amounts of organic matter distinguish it from spodic and placic horizons. Specimens from petroferric contacts are heavy, dark red, and rich in iron (often more than 30 wt. % Fe_2O_3). They are extensive in forested or once-forested soils of tropical and subtropical regions.

Plinthite is a new term for a particularly distinctive kind of laterite. Plinthite is a horizon of a soil (not the whole soil) formed in place (not redeposited) and has the unusual property of drying irreversibly on exposure to air. It is a material rich in iron with scattered red mottles of hematite and goethite in a matrix of highly weathered, light-colored clay (usually kaolinite). Hardened, vesicular, pisolitic or brecciated laterites are not included in this more restrictive definition of plinthite, though these other lateritic soil materials can be derived from plinthite by drying or redeposition. Plinthite is thought to form deep within forested soils in humid, tropical to subtropical climates.

A distinctive kind of subsurface horizon found in marine-influenced waterlogged soils is the sulfuric horizon. This is either flecked with bright golden specks from pyrite or is a dull yellow color from jarosite formed by the oxidation of pyrite. Sulfides are fixed bacterially in these soils from sulfate. The sulfuric horizon is especially common in soils of mangal and salt marshes.

Subsurface horizons can also become cemented with calcium carbonate. These are called calcic horizons when the carbonate is in the form of powder or isolated nodules and petrocalcic horizons when extensively cemented to form a continuous brittle layer within the soil. These soil carbonates are generally micritic and are petrographically more complex than carbonate cements formed during burial. Calcic and petrocalcic horizons may have complex dissolution and cavity-filling structures. The remaining clastic grains characteristically have nibbled edges where replaced by carbonate. These horizons are found in aridland soils in which carbonate is not effectively leached by available soil water.

Gypsic and petrogypsic horizons are similar to calcic and petrocalcic horizons, but their cementing material is gypsum. Salic horizons are cemented with salts more soluble than gypsum, including mirabilite and halite. These form in even more arid climates such as the margins of desert playa lakes.

Entisol

The main feature of this order of soils is a very slight degree of soil formation, either because of a short time available or because of exceedingly unfavorable conditions. Entisols may be penetrated by roots and show some mineral weathering and surface accumulation of organic matter, but the original crystalline, metamorphic, or sedimentary features of their parent materials remain little altered by soil formation. Entisols are thus as variable as their parent materials, which range from fresh alluvium, till, and sand dunes to a variety of rocks. Their topographic setting also is variable. Most are found on young geomorphic surfaces such as floodplains and on steep slopes where erosion removes soil material as it is formed. Their climate also is varied. Those forming in humid, warm climates where soil formation is rapid are younger than those formed in dry or cold climates. Early successional vegetation of grasses and other herbs and shrubs is characteristic of Entisols. Some of these soils on steep rocky slopes and along streams support trees. For paleosols the presence of root traces is diagnostic of Entisols because in other respects they are little altered from their parent material. Some Entisols are too stony, infertile, or poorly drained for cultivation. However, large areas of Entisols in alluvial bottomlands are cultivated for a variety of grain and vegetable crops, and grassed over for pasture.

Inceptisol

These soils represent a stage in soil formation beyond that of Entisols but still short of the degree of development found in other soil orders. They may have some accumulation of clay in a subsurface horizon, but it is not sufficient to qualify as an argillic horizon which is diagnostic for Alfisols and Ultisols. Similarly, they may have organic matter at the surface but it is not so thick or peaty as in Histosols. Although varied as precursors of other kinds of soils, a typical Inceptisol can be imagined as having a light-colored surface horizon (ochric epipedon) over a moderately weathered subsurface horizon (cambic horizon). These soils have developed to the extent that some relict features from their parent material may be difficult to detect within the profile. These primary igneous, metamorphic, and sedimentary structures normally take some time to be obliterated entirely. In humid to subhumid climates this may be only a few thousand years, and in drier climates tens of thousands of years. Inceptisols form in low, rolling parts of the landscape in and around steep mountain fronts. In sequences of alluvial terraces they form at intermediate positions between Entisols nearest the stream and other better developed kinds of soils farther away from the stream. The parent material of Inceptisols is as variable as that of Entisols. Soils formed on volcanic ash with at least 60% recognizable pyroclastic fragments also are included within Inceptisols. Ash has a high internal surface area so that in climates of even moderate rainfall, weathering and clay formation may proceed rapidly. The climates of Inceptisols also are varied and their vegetation ranges from forest to tundra. The shrubby woodlands of "pole trees" that form during recolonization of disturbed ground by forest are especially characteristic, as are open woodland and wooded grassland. Many Inceptisols offer excellent natural grazing and they can be cultivated to improve pasture and grow a variety of vegetables and grain crops.

Histosol

These are organic-rich soils with thick peaty horizons (histic epipedon) that form in low-lying, permanently waterlogged parts of the landscape. The main process in their formation is the accumulation of peat, which means that organic matter is produced by growth of vegetation faster than it is decomposed in the soil. The breakdown of organic matter is related to waterlogging because the most effective microbial decomposers require oxygen and this is used up by microbes in stagnant groundwater. Sediment or rock (R, C, or Cg) underlying the peat is usually little altered by weathering. There may be some leaching or formation of gley minerals such as pyrite or siderite, but most of the weatherable minerals and structures of the parent material remain. This meager mineral weathering is due in part to a short time of formation. A typical rate of peat accumulation for woody peat is 0.5–1 mm/yr (Falini 1965). Rates are much lower in swamps that are drained for a part of the year so that there is some seasonal decomposition of peat. Rates are much faster under herbaceous vegetation such as marsh grasses, *Periphyton* algae, or mosses. The plant species of Histosols are usually low in diversity and restricted to such waterlogged sites. Histosols support bog, swamp, and marsh. These soils can be drained for cultivation, but are best left alone for specialty timber cutting or rough seasonal grazing.

Vertisol

These uniform, thick (at least 50 cm), clayey profiles have deep, wide cracks for a part of the year. This cracking may produce a hummock-and-swale topography (gilgai microrelief) and its subsurface expression of a disrupted, festoon-shaped surface horizon (mukkara structure). Pavements, fences, and trees may be unbalanced by the strong shrinking and swelling action of the smectitic clays in these soils. Other kinds of clay also are found, although less commonly. Most Vertisols are found on parent materials of intermediate to basaltic composition. They may form in only a few hundred years on claystones, shales, or marls of smectitic composition. It may take longer for them to form on limestone, volcaniclastic sandstone, or basalt. Vertisols are found mainly in flat terrane at the foot of gentle slopes. Their climate and vegetation are dry and sparse enough that alkaline reaction and good reserves of exchangeable cations can be maintained. These are subhumid to semi-arid climates (180–1520 mm/yr) with a pronounced dry season. Their vegetation ranges from grassland to open woodland. Wooded grassland is characteristic. These soils offer excellent natural grazing and with irrigation they can be made to produce rice, cotton, and sorghum.

Mollisol

These soils have a well developed, base-rich, well structured (granular or crumb) surface horizon of intimately admixed clay and organic matter (mollic epipedon). Subsurface clayey (argillic or Bt), calcareous (calcic or Bk), or gypsiferous (gypsic or By) horizons also may be present, but are not definitive of the order. The characteristic surface horizon is created by fine root systems of grassy vegetation and by burrowing activity of diverse populations of soil invertebrates. Mollisols are found under grassland vegetation in subhumid to semi-arid climates. Most are found in low, rolling or flat country, but there are also Mollisols above the snowline in alpine regions. They form under a wide range of temperatures from the equator to the poles and in lowlands to high mountain meadows. They also form on a variety of parent materials although favored especially by base-rich sediments and rocks such as clay, marl, and basalt. In drier regions these soils are used mostly for open-range grazing. In wetter regions they are widely cultivated for wheat and maize, and also can produce a variety of vegetables.

Aridisol

These are soils of arid to semi-arid regions. Rainfall in such regions is often insufficient to leach soluble salts, so these soils commonly have shallow calcareous (calcic, petrocalcic, or Bk), gypsiferous (gypsic, petrogypsic, or By) or salty (salic or Bz) horizons. These cementing materials are present in large nodules or continuous layers rather than dispersed as in Inceptisols. The surface horizons of Aridisols are light-colored, soft, and often vesicular. Subsurface horizons that are not cemented with salts, carbonates, or sulfates may be similarly friable and silty, but many Aridisols have clayey (argillic or natric) subsurface horizons. Both clay and carbonate in these soils are thought to be derived from weathering and flushing down the profile of extremely fine-grained dust of feldspar and other easily weatherable minerals, rather than the complex processes of weathering found in forested soils. Aridisols are mostly found in low-lying areas because steep slopes in arid regions tend to be eroded back to bedrock and Entisols. The parent material of Aridisols is varied. Unconsolidated alluvium, loess, and till are common parent materials. Vegetation on these soils is sparse and includes various prickly shrubs and cacti. Aridisols can be irrigated for cultivation, but at the risk of salinization. They are best left for sparse native grazing.

Spodosol

The diagnostic feature of these soils is a subsurface horizon enriched with iron and aluminum oxides or organic matter (spodic or Bs horizon). Commonly this underlies a bleached, sandy, near-surface (albic or E) horizon, although this is not essential for Spodosols. This has been a source of confusion because the broadly equivalent Podzols and Podzolic soils in the traditional sense have been recognized from their albic rather than spodic horizons and albic horizons are found in both Alfisols and Ultisols in addition to Spodosols. Spodosols form in only a few hundred years on quartz-rich sands, but also can form by deep weathering of materials of less felsic composition. They form on hilly bedrock or low, rolling quartz-rich sediments. Spodosols are found principally in humid climates in which clay and soluble salts are dissolved and washed out of the profile. Though most common in temperate regions, they also are found in the tropics and near the poles. Coniferous forest is their most characteristic vegetation, but they also support other kinds of evergreen woody vegetation that can tolerate low-nutrient levels. Spodosols are naturally infertile and most used for softwood timber production.

Alfisol

These base-rich forested soils have a light-colored surface horizon (ochric epipedon or A) over a clayey subsurface (argillic or Bt) horizon that is rich in exchangeable cations (base saturation greater than 35%). Such base saturation can be assumed for paleosols when they contain nodules of carbonate in a horizon (Bk) deep within the profile. If these are lacking, fossil Alfisols may be distinguished from otherwise superficially similar base-poor soils (Ultisols) by the abundance of base-rich clays, such as smectites, of easily weatherable minerals such as feldspar (more than 10% in the 20–200 μm size fraction) or by a molecular weathering ratio of bases to alumina of greater than unity (see Table 4.4). The prefix "alf-" of Alfisols is derived from a traditional division of soils into calcareous (pedocals) and noncalcareous soils (pedalfers). This can be a source of confusion because some Alfisols are calcareous. Alfisols form on parent materials in climates and under vegetation that allow maintenance of reserves of mineral nutrients. In general this means sediments and rocks of intermediate to basaltic composition, rainfall ranging from subhumid to semi-arid and vegetation ranging from wooded grassland to open forest. Their topographic setting and temperatures are extremely varied. Alfisols are naturally fertile soils used for forestry and grazing. When cleared they can be cultivated for a variety of fruit, vegetable, and grain crops.

Ultisol

These base-poor forested soils are similar to Alfisols in overall profile form that includes a well-developed clayey subsurface (argillic or Bt) horizon. Unlike Alfisols, Ultisols are more deeply weathered of mineral nutrients. They do not include calcareous material anywhere within the profile, have low reserves of weatherable minerals (less than 10% in the 20–200 μm fraction), and have molecular weathering ratios of bases to alumina of less than unity (see Table 4.4). Base-poor clays such as kaolinite and highly weathered aluminous minerals such as gibbsite are common in these soils. Their low base status is commonly related to a long time of formation (tens to hundreds of thousands of years). Over such a period they can form on a wide variety of parent materials. They form mostly in older parts of the landscape such as rolling hills or bedrock, high alluvial terraces, and plateau tops (Cady & Daniels 1968). Ultisols form most readily in humid, warm climates. There are some examples in polar and desert regions, but these are thought to be relicts of former climates more favorable to deep weathering. Their natural vegetation is coniferous or hardwood forest. Some also support wooded grassland, which can sometimes be traced to human deterioration of more luxuriant former vegetation. Most of these soils are used for forestry. In tropical regions some produce pineapples and sugar cane. Some also are cultivated for vegetables and grain, but only after extensive fertilization.

Oxisol

These are deeply weathered soils with uniform profiles, no more than trace amounts of easily weathered minerals, and dominated by kaolinitic clays (oxic horizon). Their molecular weathering ratios of bases to alumina are close to zero (see Table 4.4). Deeply weathered mottled horizons (plinthite) may also be found in these soils. Most Oxisols are a striking brick-red color, but some also are yellow or gray. A stable microstructure of sand-sized, spherical micropeds of clay is characteristic (Stoops 1983). The advanced weathering of these soils is due in part to their great age, often amounting to tens of millions of years. They are found mostly on stable continental locations on gentle slopes of plateaus, terraces, and plains. Their development is especially favored by tropical humid climates where weathering is most intense. They are known also in arid and cool climates where they are sometimes found to be relict soils or to have formed on highly weathered sediments. Their natural vegetation is rainforest. Large areas of these soils on Precambrian rock in tropical regions are now covered with wooded grassland, but were initiated as forested soils in the distant geological past (at least Miocene). Oxisols can be used for rough grazing and for forestry. Some produce tree crops such as cocoa, coffee, sugarcane, and tropical fruits.

A word of caution

These brief outlines and cartoons of the different kinds of soils recognized by the Australian handbook, FAO world map and US soil taxonomy may prove useful as an initial guide to these classifications. Features of these soils discernible in paleosols especially have been stressed. However, such a brief outline does not do justice to any of these classifications. For effective identification of a paleosol within a classification of soils, there is no substitute for carefully reviewing the original accounts and then comparing the paleosol in detail with one of the representative described modern soil profiles. For the Australian classification, identification is a matter of finding the closest match. For hierarchical systems such as the US soil taxonomy, decisions at high levels must be made carefully. It may prove useful to explore the ultimate ramifications of several alternatives because it sometimes happens that specific profiles at lower levels of the classification are easier to match with a paleosol than abstractions at high levels.

Even with excellent field and laboratory data for a paleosol, it may not prove possible to identify it within a modern soil classification. It may only be possible to identify some paleosols to the level of suborder in the US taxonomy, whereas others can be identified to subgroup level. For example, the only Alfisols with a continuous calcareous horizon at depth are Petrocalcic Paleustalfs. Fossil Vertisols, on the other hand, can seldom be identified beyond the order level. There are other reasons also for not straining a paleosol to fit a modern soil classification. Some soil types may be extinct. Such anomalous paleosols should be recorded carefully until the known sample of them reaches a size that their true nature becomes understood as more than just a local peculiarity.

At first appearance these classifications and their terminology may seem intimidating. The widely different systems in use may also be a source of despair. Memorizing the essential features of the ten soil orders of the US soil taxonomy is a good way to approach the subject. With use they will become familiar and their similarities with other classifications become more striking than differences. The taxonomic languages of soil science convey a wealth of observations and ideas about soils.

6 Mapping and naming paleosols

Paleopedology is a field science. Its objects of study are too bulky to be brought back into the laboratory in their entirety, and must be characterized and sampled outdoors. Field work commonly involves identifying different kinds of paleosols and establishing their relationships with each other and the enclosing sediments or rocks. These dual activities of mapping and naming are a necessary first step in the study of paleosols. Such field observations will determine how a paleosol is sampled and later analyzed. More sophisticated indoor studies may refine or test hypotheses developed in the field, but are unlikely to supplant them. Therefore, it pays to have a logical plan for mapping and naming paleosols, and this plan will vary according to the aims of the study. Three common aims are considered in this chapter: paleoenvironmental interpretation of paleosols, use of paleosols as stratigraphic marker horizons, and portrayal of paleosols on geological maps.

In the face of a burgeoning scientific nomenclature for different kinds of fossils, rock layers, and soils, it may fairly be asked whether there is a need to map and name paleosols. The formal mapping and naming of fossils, rocks, and soils often becomes an end in itself. Their nomenclature is regulated by such long-winded and legalistic documents as the International Codes of Botanical (Voss *et al.* 1983) and Zoological Nomenclature (Ride *et al.* 1985), the International Stratigraphic Guide (Hedberg 1976), numerous local stratigraphic guides (such as that of the North American Commission on Stratigraphic Nomenclature 1982), and the USDA Soil Survey Manual (Soil Survey Staff 1951, 1962). The named entities are redefined, amended, and debated. If this is the result of naming and mapping, should not paleosols continue to be studied on an informal basis? I think not, for the simple reason that informally described paleosols have a tendency to slip into scientific oblivion. Similar problems were encountered in the early days of the study of trace fossils and fossil pollen. It is artificial to bestow biological binomial Latin names on such things as fossil tracks and trails, and on the acid-resistant walls of plant spores and pollen. Yet experience with the study of these materials has shown that scientists feel free to ignore prior work using informal or common names, acronyms, or descriptive alphanumeric terms. A name that is concise, constructed along a pre-established set of rules, and based on real and well illustrated examples is needed. There should be a clear sense that the material is a recognizable object. In this way, a large data base that is internally consistent can be built up over the years by different investigators.

Mapping and naming of paleosols have not yet been legalized to the extent of fossils, rocks, and soils, but there are signs that this is happening. The most recent version of the North American Stratigraphic Code (North American Commission on Stratigraphic Nomenclature 1982) includes provision for formal naming of paleosols as stratigraphic marker horizons. Perhaps there will follow a national registry for such names, as is maintained by the US Geological Survey for other stratigraphic names (Luttrell *et al.* 1986). Other approaches to paleosol mapping are accumulating in scientific publications. These other names at present must be regarded as informal. The distinction between formal and informal names can be indicated by capitalization. The Sangamon Geosol, for example, is a formal name, whereas the type Gleska clay paleosol is not. Until there is an officially endorsed code for naming paleosols there is scope for a number of individual approaches to mapping and naming them.

Paleoenvironmental studies

My own interest in paleosols has been concerned with interpreting ancient environments of soil formation. Each different kind of paleosol represents a different kind of paleoenvironment. There are no special difficulties in naming a paleosol in this kind of study if it involves a single profile at a geological unconformity. If work is extended to characterize variation in paleosols along an unconformity or in numerous paleosols interbedded with fluvial deposits, then many distinct names may be needed in a small area. In a study of paleosols in Oligocene alluvium of Badlands National Park, South Dakota, 87 successive paleosols of ten different kinds were recognized in 143 m of stratigraphic section (Retallack 1983b). For paleoenvironmental studies of such sequences a separate name is not needed for each paleosol, only for each different kind of paleosol insofar as it represents a discrete subenvironment of past landscapes.

The nature of such studies is similar to that of modern soil mapping, except that the soils of many superimposed landscapes must be named and mapped rather than those of just one soilscape. While modern soil mapping units were not designed with paleosols in mind, they are serviceable for paleopedological studies. The soil units defined by the US Department of Agriculture (Soil Survey Staff 1951,1962) are used for naming and mapping variations in soil across the landscape (Fig. 6.1).

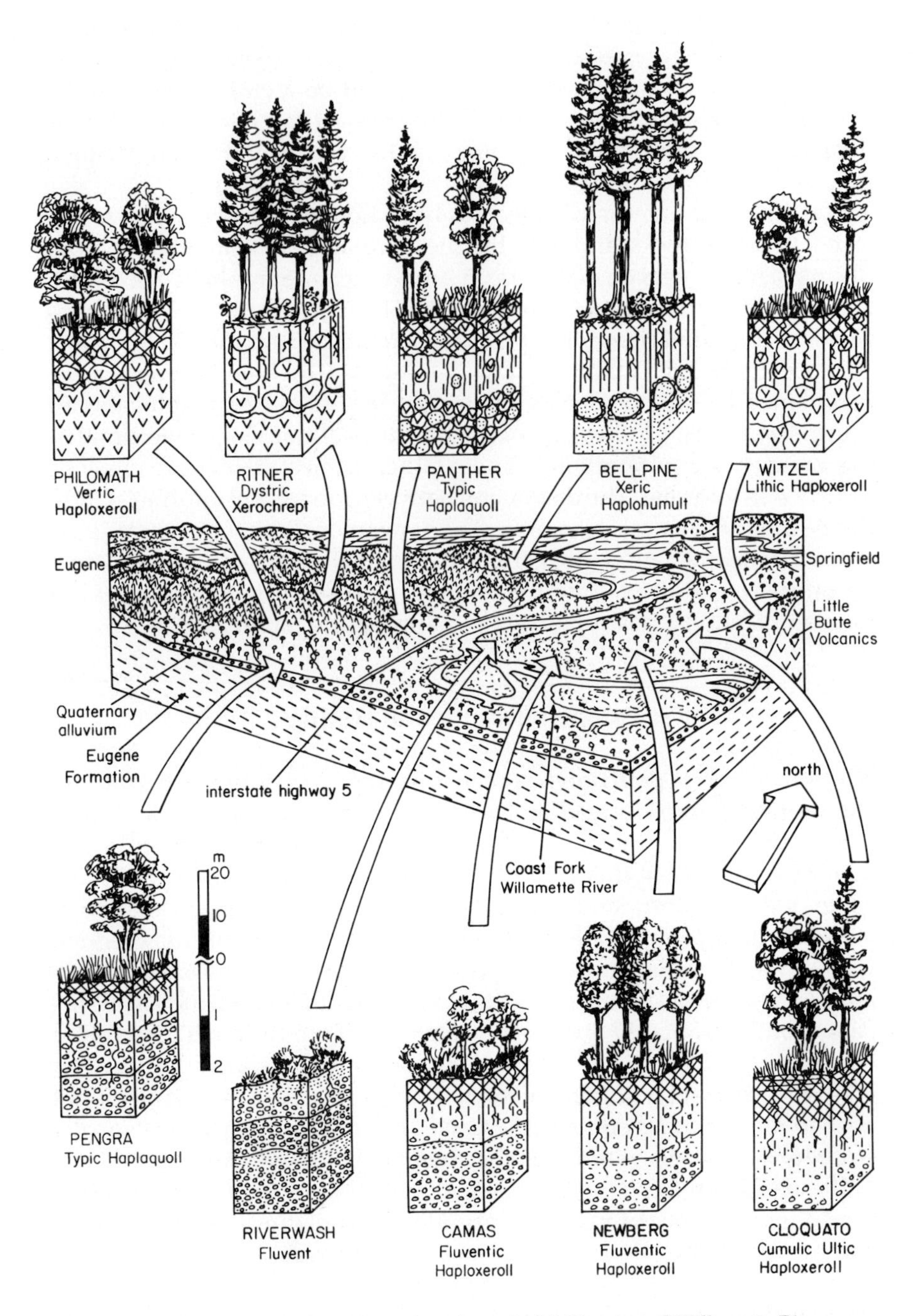

Figure 6.1 Soil series and their vegetation along the McKenzie and Willamette Rivers near Eugene in the southern Willamette Valley of Oregon, USA (based on data from Vokes *et al.* 1951, Patching 1987).

These soilscapes are made up of a number of irregularly shaped polypedons which interlock like a large mosaic. A pedon is the smallest volume of soil that can be recognized and identified. It may be a substantial area of soil (10 m^2) if soil horizons are laterally variable in thickness. Usually a pedon is the profile visible in a small pit (ca. 1 m^2 or less in area). The task of the soil mapper is to compile information from a number of soil pits so that areas of substantially similar soil may be delimited from areas with different soils. The basic unit of mapping is a soil series which includes one or more polypedons scattered over the landscape. A soil series is based on field and laboratory characterization of a particular pedon and named after a nearby geographic locality, e.g., the Bellpine Series (Fig. 6.1). An individual pedon can be named for the texture of its surface horizon, e.g., the Bellpine silty clay loam. This name may seem uninformative, but with local experience it conjures up an image of a thick, red, clayey soil (a Xeric Haplohumult) that forms on moderately sloping, well drained hills of marine volcaniclastic sandstones under mixed conifer forest dominated by Douglas fir (*Pseudotsuga menziesii*). It is a matter of judgment what amount of variation is permitted within a soil series. Minor features of the soil such as small ferruginous nodules may be used to define soil variants, such as the Bellpine silty clay loam nodular variant. Slight variations in thickness of horizons may be used to define phases, such as the Bellpine silty clay loam eroded phase. Soil series also can be grouped into larger units or associations. The Bellpine Series, for example, is part of an association of soil series formed on moderately sloping sedimentary bedrock. These are different from alluvial soils of basaltic gravel in the nearby valley bottoms.

For paleosols the problem of mapping and naming is similar. Paleosols may be characterized as they are encountered in a stratigraphic sequence either by examining cliff outcrops, excavating badlands, or logging drillcore. Each recognizable paleosol profile can be assigned to a limited number of generalized kinds. Soil series are ideal for such reconnaissance mapping of paleosol sequences (Fig. 6.2). For each series a type profile should be sampled and described in detail as a reference standard. With laboratory studies to determine the texture at the ancient surface, it may be possible to assign a particular soil name to a paleosol, e.g., the type Gleska clay paleosol. As in soil mapping, it is best if the type profile of a paleosol series is distinct and typical for the series rather than a composite profile or an intergrade between different paleosol types. This in the end is less critical than its careful and unambiguous characterization, so that later investigators can find and assess it for themselves.

There are several advantages to adopting soil mapping units for the naming and mapping of paleosols. The system has been tried and tested by soil scientists. The system is hierarchical, thus permitting various levels of generalization and interpretation. These distinctive names also prevent

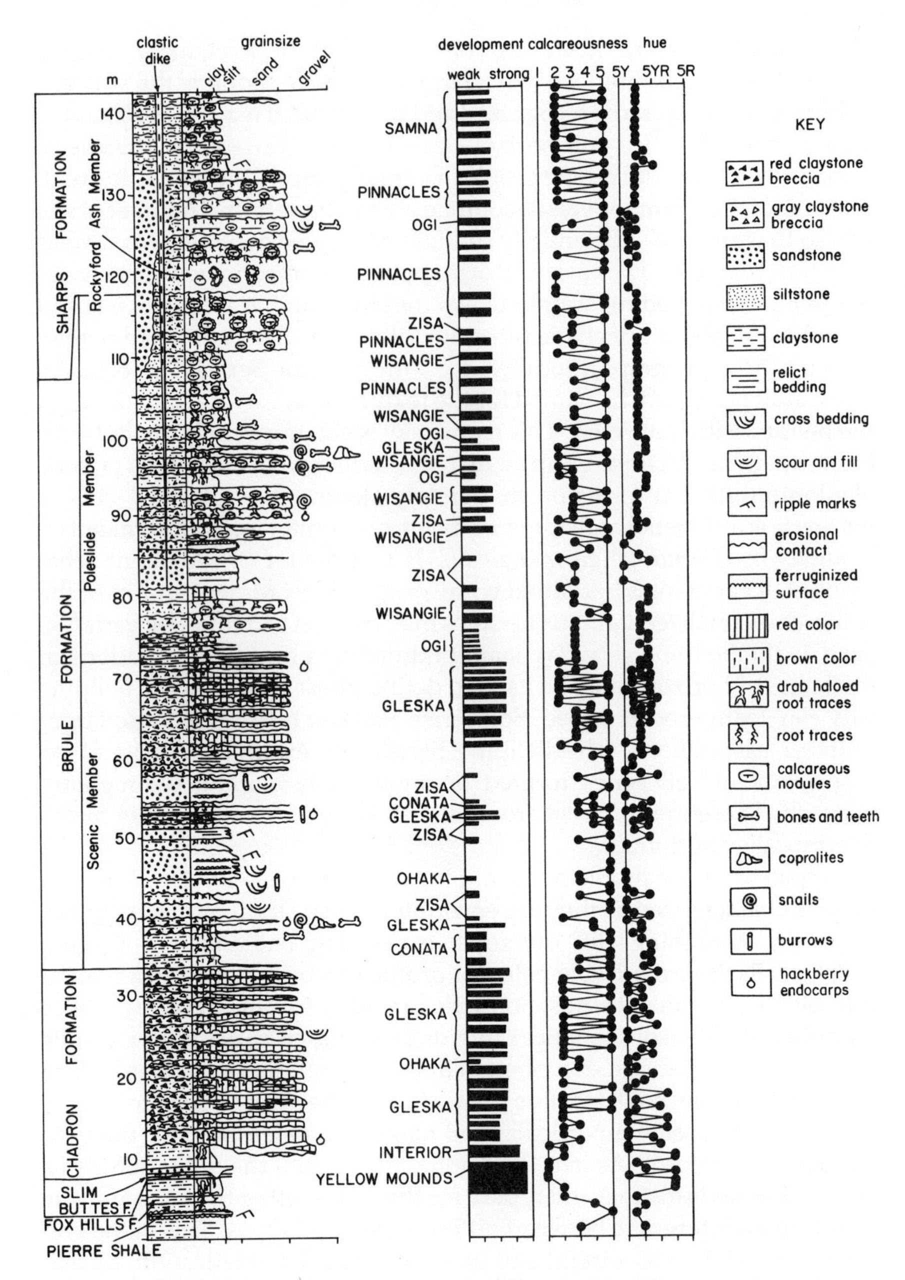

Figure 6.2 Paleosol series recognized in a measured stratigraphic section through mid-Tertiary alluvial rocks in the Pinnacles area of Badlands National Park, South Dakota, USA, also illustrated in Fig. 2.6. Position of paleosols marked by boxes, the width of which corresponds to degree of development (Table 13.1), calcareousness by reaction with dilute acid (Table 3.2), and hue from a Munsell color chart (modified from Retallack 1983b, reprinted with permission of the Geological Society of America).

confusion with paleosols of other areas. Finally, the names are simply mapping units, independent of how the paleosols are ultimately interpreted within a soil classification or in terms of the paleoenvironment that they represent.

This kind of mapping and naming has been applied to sequences of paleosols as measured in the field or from drill cores. The lateral distribution of different kinds of soils across an ancient landscape can be reconstructed from a vertical section by using a common kind of geological inference often called Walther's Facies Law. This states simply that different kinds of sediment deposited side by side in nature will be preserved on top of one another in a sedimentary sequence. Stream channels, for example, migrate laterally across their floodplains as they erode the outside of their meanders. Their levees and point bars inside the meander follow this lateral migration, so that point bar deposits become overlain by levee and then floodplain deposits in a characteristic fining-upward sequence. The relationship of paleosols in a sequence to these different point bar, levee, and floodplain sediments can be a guide to their former distance from streams and distribution across alluvial landscapes (Fig. 6.3). Provided observations are restricted to a single genetic package of sediments and do not extend across a major unconformity, this facies method of reconstructing past landscapes may be reliable (Hallam 1981). Major unconformities can be discerned by the presence of paleosols that are especially thick or well developed, or so closely superimposed that they overlap in a confusing manner.

These techniques for mapping and naming soils can be applied also to the study of lateral variation of paleosols across a single buried land surface. This may seem a straightforward way of reconstructing soilscapes, but it is complicated by difficulties in determining relative ages of different parts of the landscape. Consider, for example, the modern rock-bound alluvial landscape of the Willamette River in central western Oregon (Fig. 6.1). The bedrock unconformity may have been covered by a soilscape when it was first excavated, but progressive infilling with alluvium has resulted in continued modification of soils high in the landscape long after the lowest soils were covered by river deposits. Thus, the buried soils on the unconformity between modern alluvium and bedrock represent conditions of the distant past whereas higher soils represent a combination of past and present soil-forming conditions. On the present landsurface also, the soils are of very different age. Those closest to the stream are the youngest and in some cases are forming on flood deposits only a few years old. Soils of higher terraces are older and soils on hilly bedrock are the most ancient. Because landscapes are renovated in this patchwork fashion, lateral mapping of paleosols is not necessarily a good guide to soilscapes of the past. There are rare cases where time planes can be identified using paleomagnetic reversals

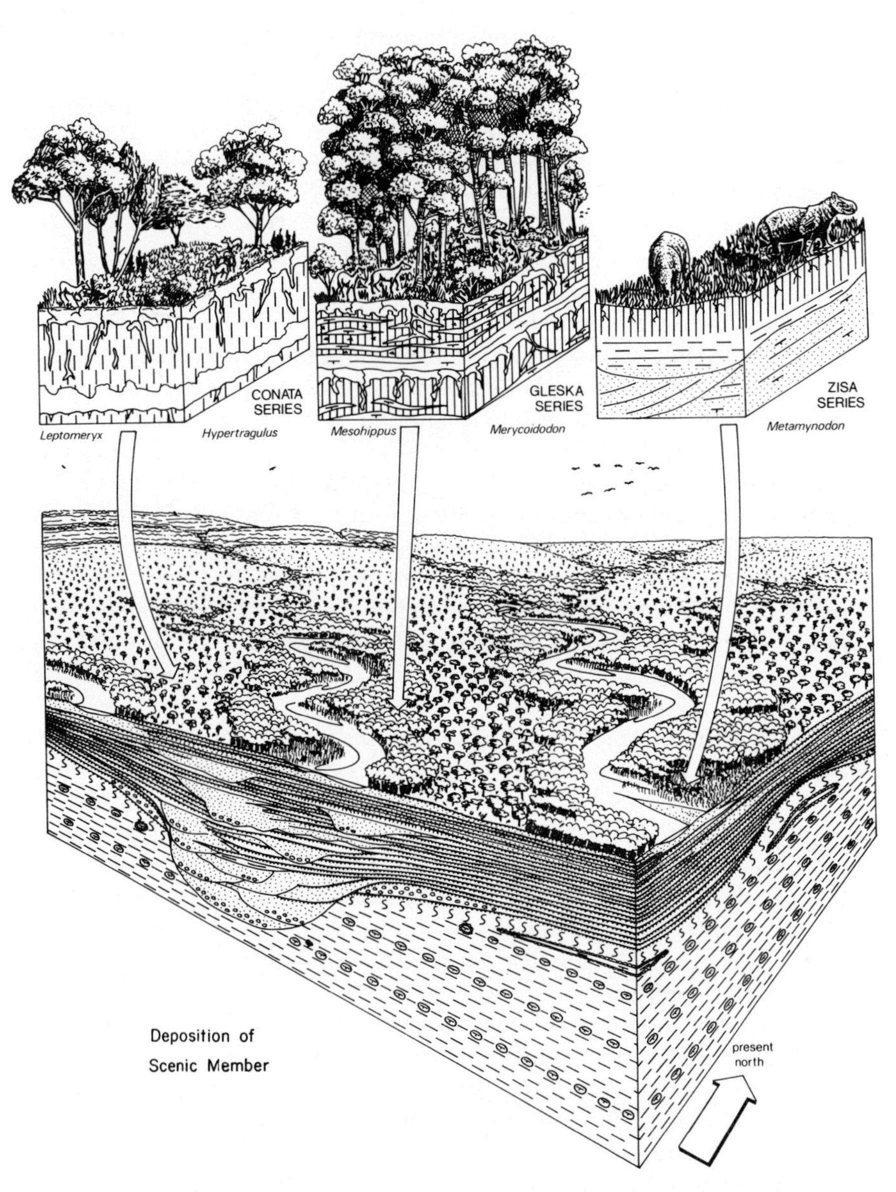

Figure 6.3 A reconstruction of soil series and vegetation during mid-Oligocene time in the Pinnacles area of Badlands National Park, South Dakota, USA (from Retallack 1983b, reprinted with permission of the Geological Society of America).

(Behrensmeyer & Tauxe 1982) or volcanic ash (Burggraf *et al.* 1981). These allow the reconstruction of an ancient soilscape in detail.

Other difficulties in applying soil mapping units to paleosols are fundamental to classification and naming in general. The kinds of feature useful for distinguishing one soil from another are based on experience and may seem subjective to a beginner or outsider to the field. What to one investigator may seem a diagnostic differentiating feature may to another seem a trivial variation in only one kind of soil. The kinds of features of soils stressed by various soil classifications represent a consensus of those aspects of soils generally held to be important. Subjectivity in classification may be mitigated by adopting a strictly hierarchical set of differentiating criteria as in cladistic taxonomy (Cracraft & Eldredge 1979) or by statistically analyzing a number of soil properties as in numerical taxonomy of fossils (Sokal & Sneath 1963).

Another problem in naming paleosols is finding a name that is distinctive and appropriate, and that is not already used for a stratigraphic unit, modern soil series, geomorphic surface, distinctive fossil faunule, or other natural feature. In remote regions there may be few place names of any kind that can be used. Even in well populated regions the diversity of paleosols within small outcrops may quickly exhaust available local names. In my own studies I have resorted to using descriptive terms from politically unrecognized native languages such as Lakota Sioux (American Indian) and Dholuo (Kenyan). Even with a distinctive name, a paleosol series should be clearly labeled a paleosol in order to prevent confusion with modern soil series, e.g., the Bellpine silt loam vs. the Gleska clay paleosol.

The term "soil facies" has been used in a similar sense [by Birkeland (1984)] to paleosol series, as outlined here. Soil facies and paleosol series have a similar relationship to soil stratigraphic units as do sedimentary facies to sedimentary formations. Unlike sediments, however, soil facies and paleosol series represent conditions during non-deposition rather than during sediment accumulation. These distinctions are blurred by a recent proposal to recognize "pedofacies" (Bown & Kraus 1987, Kraus 1987). These are simply a kind of sedimentary facies, typically floodplain deposits whose main features are pedogenic rather than sedimentary (Fig. 6.4). Pedofacies are thus distinctive sedimentary rock units containing one or more paleosols. Pedofacies have been named after localities such as the Sand Creek and Elk Creek pedofacies of the Early Eocene Willwood Formation of northwest Wyoming, USA. They are better labeled informally as is usual for sedimentary facies. "Red striped" and "orange nodular facies," respectively, would be more appropriate for the Sand Creek and Elk Creek facies. Pedofacies mapping is a convenient way to map broadly different groups of paleosols over large areas without the effort of digging out and characterizing individual profiles.

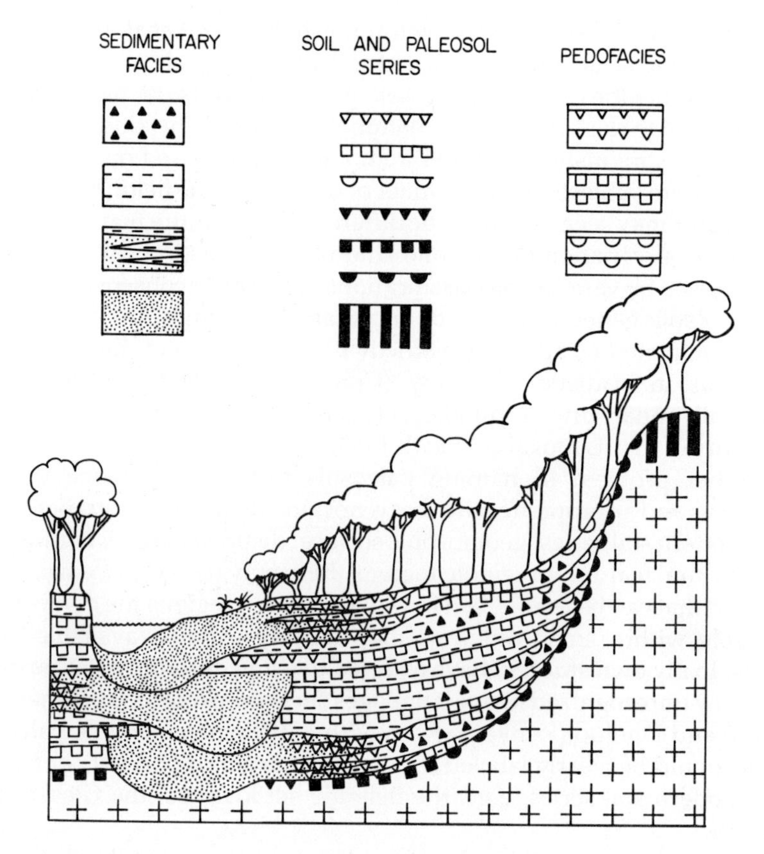

Figure 6.4 The relationship between sedimentary facies, pedofacies, and soil and paleosol series.

Stratigraphic studies

An additional use for paleosols is as stratigraphic marker horizons, i.e., distinctive horizons that can be traced in order to establish the relationship in time and space between different sedimentary units. Paleosols have proven especially useful for establishing relative age within Quaternary sediments that are too old for radiocarbon dating and too young or poorly fossiliferous for establishing their age paleontologically (Fig. 6.5). Quaternary paleosols may be distinctive, often reddish and clayey zones. They form natural divisions between deposits of alluvium, loess, and till that may look so similar as to be difficult to distinguish from one another. Paleosols also have been used as stratigraphic markers in older rocks such as the Early Triassic Buntsandstein in West Germany (Ortlam 1971).

The formal term now recommended for such soil stratigraphic units in the most recent North American Stratigraphic Code (North American

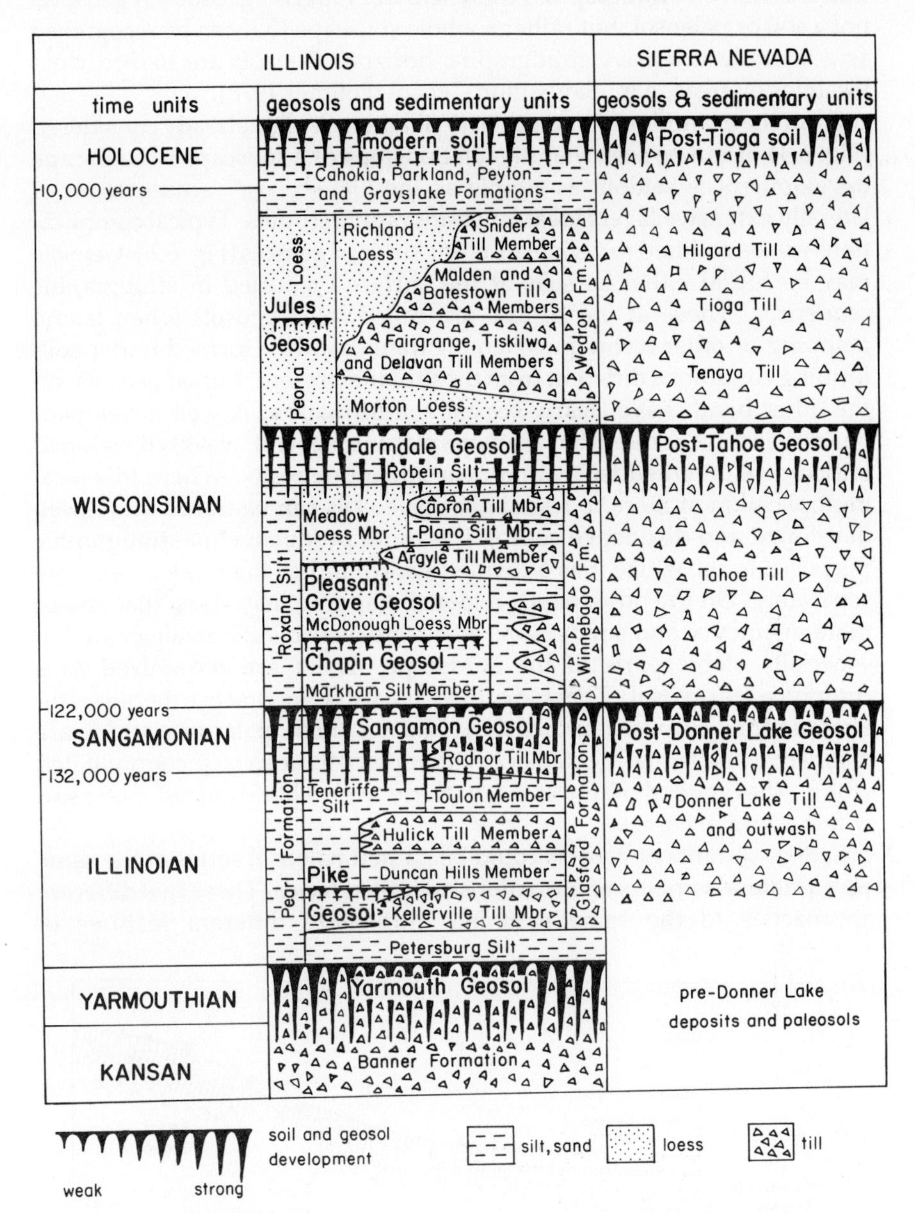

Figure 6.5 A correlaton of Quaternary soil stratigraphic units (geosols) and sedimentary formations between central Illinois and the western Sierra Nevada, California, USA (data from Birkeland 1984, Follmer *et al.* 1979).

Commission on Stratigraphic Nomenclature 1982) is "geosol." A geosol is not a soil or paleosol, but rather a whole soilscape that can be recognized as a laterally extensive stratigraphic horizon. Geosols are named from localities or areas, e.g., Sangamon Geosol (Follmer 1978).

From the example of the southern Willamette Valley already considered (Fig. 6.1), it can be seen that the relationships of a soilscape to landscape development is seldom as simple as the "layer cake" arrangement of laterally extensive layers ideal for stratigraphic studies. Typical complications can be codified to define different kinds of geosols (Fig. 6.6). Geosols may be found either as soils at the surface or buried in stratigraphic sequences. Those at the surface are called relict geosols when lateral mapping or other techniques demonstrate that they formed under soil-forming conditions different from those of the present. Buried geosols, on the other hand, have clearly ceased soil formation. A well developed, buried geosol of a stream terrace may splay into several, weakly developed buried soils interbedded within near-stream deposits. Where this local multiplication of more or less similar paleosols can be traced into a well developed and distinctive geosol, it is more productive for stratigraphic purposes to recognize it as a divided geosol rather than name each one separately. On the other hand, regional mapping may show that one or more distinctive and stratigraphically useful geosols are amalgamated in especially stable parts of the landscape. These are recognized as a compounded geosol. These distinctions work in practice because for stratigraphic purposes emphasis is laid on those paleosols which are distinctive and laterally extensive. Other paleosols can be accommodated within the general framework of these well characterized paleosols without formal names.

Lateral variants of geosols could be named paleosol series in the same way as lateral variants of formations are called facies. These two different approaches to the same materials emphasize different features of

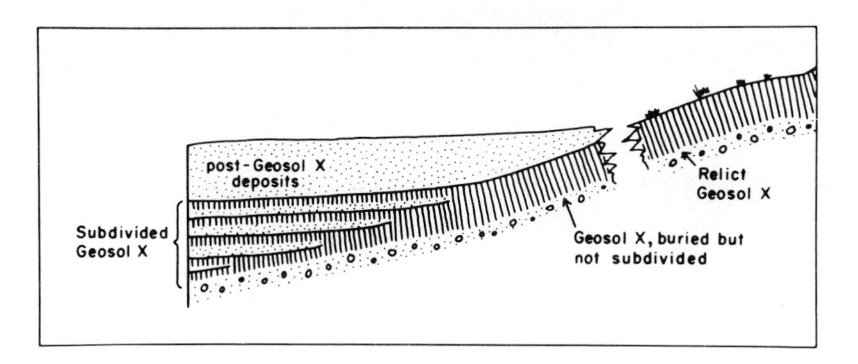

Figure 6.6 Concepts and terminology for the recognition of geosols (from Morrison 1978, reprinted with permission of Geoabstracts Inc.).

paleosols. Geosols are recognized from features thought to reflect time. Within sequences deposited by successive advance and retreat of glaciers this includes the degree of development of a paleosol insofar as it reflects the time between glacial advances. It also includes the degree of calcareousness, clayeyness, or red hue as this is related to regional climatic fluctuations. In contrast, paleosol series are recognized by a wider variety of features that reflect soil-forming processes specific to a particular site. These include patterns of root traces insofar as they relate to local variations in former vegetation or drab hues and peat accumulation as indications of locally waterlogged parts of the landscape.

The use of paleosols in mapping Quaternary sediments is a well established scientific tradition (Chamberlain 1895), especially in the continental United States and Europe (Fig. 6.5). The term geosol was proposed by Morrison (1967). In previous versions of the North American Stratigraphic Code, they were called "soils." This term has such varied meaning to soil scientists, engineers, farmers, and geologists that it was deemed inappropriate for this special use. A geosol really is not a soil in the sense of a pedon, a construction material, an agricultural resource, or a zone of weathering, but is rather an assemblage of paleosols representing an ancient landscape. The need for a new term was independently recognized by Brewer, *et al.* (1970) who proposed the term "pedoderm" for much the same reasons. Pedoderm is incorporated into the Australian Code of Stratigraphic Nomenclature (Stratigraphic Nomenclature Committee 1973) and is now the term of choice of the International Quaternary Association (Birkeland 1984). Time and further usage will reveal whether the term geosol or pedoderm prevails internationally.

The main theoretical difficulty with the use of geosols for stratigraphic mapping is the assumption that they are of about the same age everywhere. Morrison (1978) has argued for this by proposing that geosols represent geologically brief "soil-forming intervals," when climate was especially conducive to soil formation. These soil-forming intervals were envisaged to punctuate longer periods of time during which the climate was too cold or dry for soil formation. The opposite view that soil formation proceeds at a constant rate unless interrupted by sedimentation or erosion is equally unjust. The true position probably lies somewhere between: the rate of soil formation varies considerably with climatic conditions, degree of prior development, and other factors (Birkeland 1984). On present evidence, however, geosols cannot be seen as markers of time planes that are as ideal as beds of volcanic ash. They are imperfect, yet useful stratigraphic markers.

Practical problems arise in the lateral correlation of geosols. It is easier to correlate long sequences of geosols at different localities than individual geosols. The degree of development of each geosol in a sequence is a pattern that may be matched over wide areas.

Many regions have a shortage of names unused for other natural features that might be appropriate for geosols. This problem is not so serious as it can be for naming more numerous paleosol series because weakly developed or laterally impersistent paleosols or the individual profiles of a divided geosol are seldom so stratigraphically significant that they need to be named. Geosols are named only when they are mappable over a wide area. The Sangamon Geosol, for example, has been traced through much of the midcontinental United States from Ohio to Texas. Some soil scientists who find the proliferation of geosol names confusing prefer to adapt pre-existing stratigraphic names. The paleosols on top of the Tahoe Till (Fig. 6.5), for example, could be called the post-Tahoe Geosol and those underneath it the pre-Tahoe Geosol. A similar practice has been adopted informally for naming Precambrian paleosols at major geological unconformities (Grandstaff *et al.* 1986).

Deeply weathered rocks

Some zones of deep weathering are so thick and distinctive that they cover larger areas of ground than geological formations (Fig. 6.7). They are not the same as geological formations because their lower boundary is diffuse and because several different thick weathered zones may overlap substantially. Nor are they the same as geosols which consist of soil solum. Deep weathering profiles, in contrast, are mainly saprolite and thick duricrusts: laterites, bauxites, calcretes and silcretes (Goudie 1973). The soil solum seldom is preserved in these deep weathering profiles that can reach hundreds of meters into their parent materials.

A way of portraying these deeply weathered zones on geological maps has been suggested by Senior and Mabbutt (1979). They proposed the use of a locality name together with the simple and succinct term "profile." It may also be desirable to use a term descriptive of the profile, e.g., the Curalle silcrete profile of southwestern Queensland, Australia (Fig. 6.7). Terms other than the locality name should not be capitalized because at present these names are informal. Nevertheless, they should be defined according to standards appropriate to other stratigraphic units. Several features of a profile should be characterized for adequate definition: the locality of a reference section, a description of the lithologies and thickness of the parent and altered rocks within the reference profile, an estimate of the thickness and other variations in the profile elsewhere, and indications of its age, distribution, and relationships with enclosing rocks.

Advantages of this system lie in its adherence to general principles for establishing other stratigraphic units. This approach clarifies the only record of much geological time in some continental regions. With such

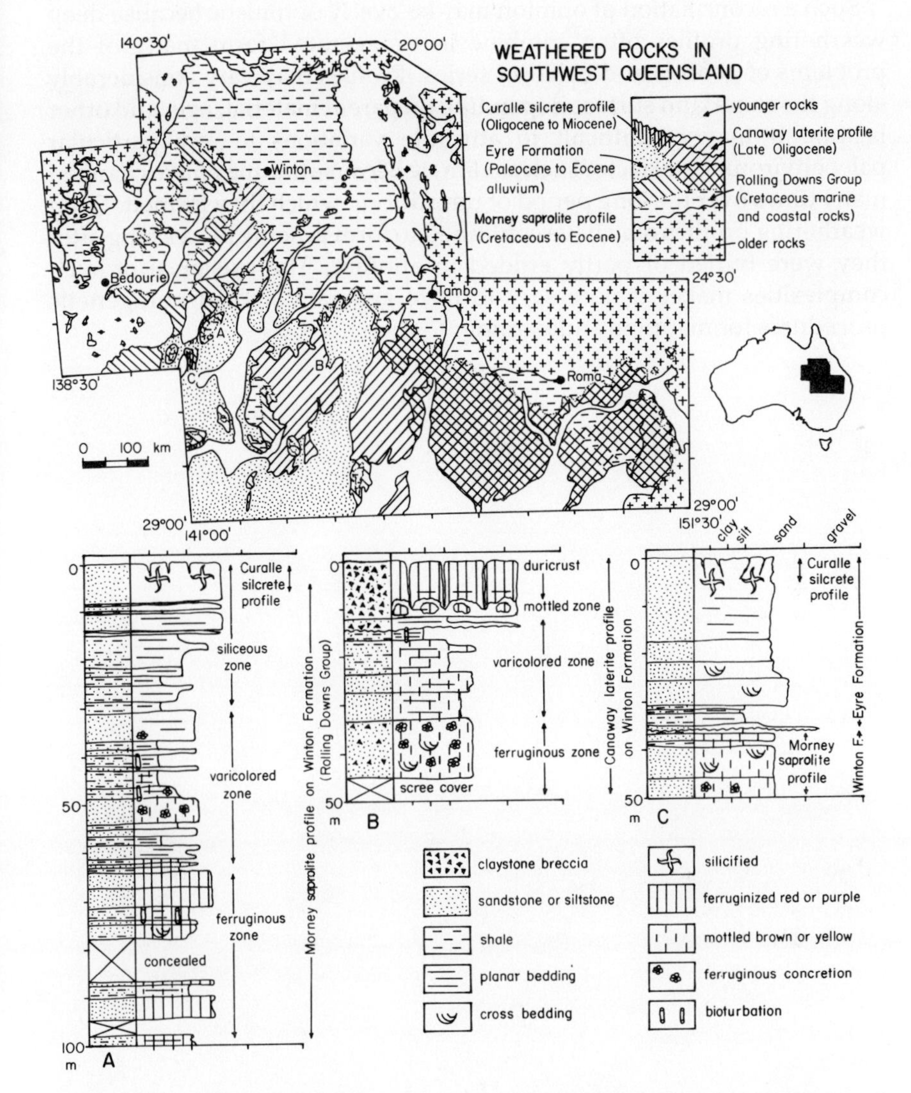

Figure 6.7 Weathering profiles, their stratigraphic relationship, and distribution in south-western Queensland, Australia (based on data of Senior & Mabbutt 1979).

difficult and controversial rock materials it is important to have agreed names for examples independent of theories of their origin.

Such a reconciliation of opinion may be overly optimistic because deep weathering profiles often combine in exaggerated form many of the problems of geosols and paleosol series. Profiles may vary considerably along the same land surface depending on parent material, age, and other factors. It may be difficult to attribute variations to any particular paleoenvironmental factor such as climate because this may have changed many times over the long period of time during which they formed. Deep weathering profiles may have continued to form in some places long after they were buried or partly eroded in another. The existence of such complexities makes it necessary to have generally agreed non-genetic procedures for mapping and naming.

7 Alteration of paleosols after burial

Part of the problem with unraveling the alteration of paleosols after burial is clarifying the terminology for these changes. Diagenesis is the term used in studies of sedimentary petrology to describe alteration after deposition. Diagenesis thus includes soil formation, which on Earth occurs at 1 bar pressure and temperatures from 84 to −88°C (Kimmins 1987). Additional changes after burial pass into what would be called metamorphic alteration at temperatures of 200°C or above (Fig. 7.1). Because of the natural gradient of increased temperature downward in the Earth's crust, there is a practical pressure limit to diagenetic alteration of less than 7 kbar lithostatic pressure, which is found at depths of about 25 km within the crust. Metamorphic alteration proceeds at more extreme temperatures and pressures where new mineral assemblages form at the expense of the old ones and rock structure is reconstituted by the growth of new crystals. Destruction of primary features of soils and imposition of metamorphic foliation, schistosity, and crystalline texture limit the interpretation of paleosols. However, many diagnostic features of paleosols such as root traces, soil horizons, and soil structure survive such changes. These structures have been found in paleosols metamorphosed to zeolite, prehnite–pumpellyite, and lower greenschist facies (Retallack, 1979, 1985). With greater metamorphic alteration, the only remaining indication of a paleosol may be a highly aluminous bulk composition and a mineralogy dominated by kyanite, sillimanite, garnet, or corundum (Reimer 1986, Barrientos & Selverstone 1987). With only this compositional and mineralogical information it may be difficult to distinguish between paleosols and zones of hydrothermal alteration.

Although diagenesis is defined as alteration of sediments after deposition, it sometimes is taken to mean only alteration after burial. Mere documentation of dissolution of grains and their coating with clay may provide evidence of diagenesis, but does not in itself demonstrate whether it occurred in the soil or after its burial. Much information about soil formation of the past is masquerading under the guise of diagenetic studies of nonmarine rocks. The distinction between alteration occurring during soil formation and that attendant on burial can be made by establishing the

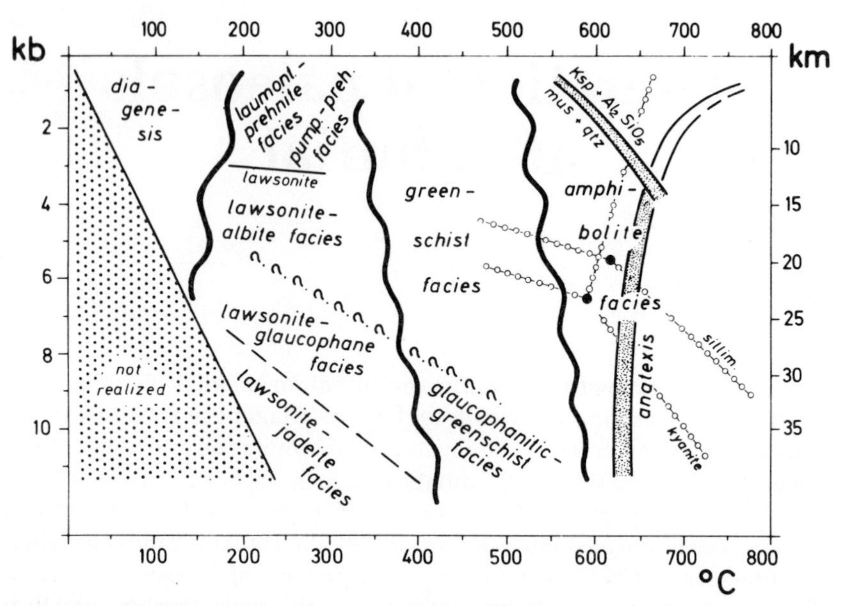

Figure 7.1 Pressure and temperature conditions of diagenesis and various metamorphic mineral assemblages including some high-temperature aluminosilicates common in metamorphosed paleosols (from Winkler 1976, reprinted with permission of Springer).

relationship of mineral grains and other features with those parts of the soil thought to have been original, such as root traces.

Field relationships provide the most striking and unequivocal illustrations of the relative time for formation of soil and burial diagenetic features. Sand-filled burrows approaching, then sidling around, siderite nodules have been found (Fig. 7.2), unequivocally indicating that the nodules or some precursor of them were there when the burrows were excavated. Nodules of this kind also are formed in marine and lacustrine rocks at some depth below the bottom, and they have been assumed previously to be features of burial diagenesis. This is still a common explanation for sideritic, calcareous, and ferruginous nodules in sedimentary rocks (Coleman 1985). The following additional features of pedogenic nodules may prove useful in distinguishing them from late diagenetic nodules. Fossil root traces may sidle around them, but if they penetrate the nodule they are better preserved and more fully inflated than in the surrounding matrix. Other fossils such as skulls, turtle shells, and hollow bones also may be less compacted within nodules than outside them. This is evidence that the nodules formed before there was any appreciable compaction and that they subsequently resisted crushing better than the surrounding matrix. Original pedogenic nodules may also contain more easily weatherable minerals such as pyroxene and olivine than are found in the surrounding matrix. This indicates that nodules formed early during soil development and protected these minerals from weathering. Nodules of paleosols may form horizons at a fixed depth below the ancient surface. Finally, nodules of the same kind within paleosols and as pebbles in associated fluvial paleochannels provide evidence that the nodules were formed in the soil and eroded out of the banks of streams.

Cross-cutting relationships in petrographic thin sections also provide evidence for the sequence of mineral development in paleosols. Cross-cutting relationships may be difficult to see within clay-sized minerals, but often are clear between larger crystals. Diagenetically formed crystals may be large enough to cut across large structures such as burrows. If, on the other hand, the burrow wall truncates or has rotated crystals, then the burrow can be assumed to be of about the same age as or younger than the crystal and the crystal to be an original part of the soil. Long sequences of events can be reconstructed by stringing together such observations. In the search for order, however, it should not be overlooked that a particular mineral may be of different age in different parts of the paleosol. Care and patience are needed to assemble a representative paragenetic sequence.

Other kinds of evidence also may be useful in unravelling the sequence of diagenetic modifications. For example, the strontium isotopic composition of clay coatings to grains of Miocene to Oligocene sandstones from the Great Plains of North America provides clues to their age (Stanley & Faure

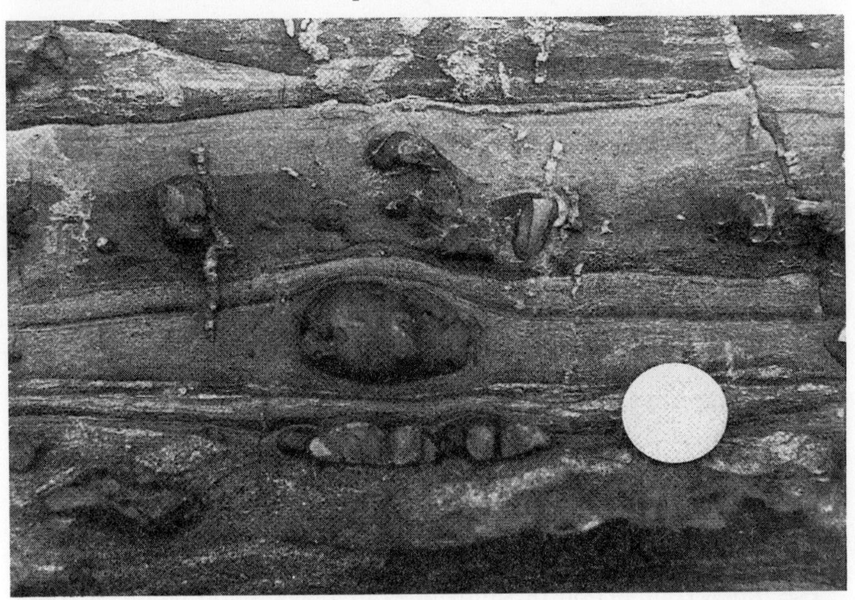

Figure 7.2 Burrows (metagranotubules) of cicada-like insects sidling around siderite nodules in a very weakly developed paleosol (Aquent of Warriewood series of Retallack 1977) from the Early Triassic, Gosford Formation near Avoca, New South Wales, Australia. The coin is 28 mm in diameter (from McDonnell 1974, reprinted with permission of the Geological Society of Australia).

1979). The main source of the light isotope (^{86}Sr) is the radioactive decay of rubidium (^{86}Rb). The ratio of ^{86}Sr to ^{87}Sr can be used for radiometric dating, but clay coats are not suitably free of contamination from groundwater for accurate dating. Nevertheless, strontium isotopic ratios demonstrated younger geological age of samples collected higher in the sequence. This is consistent with their origin during or shortly after deposition rather than by equilibration with groundwater at a much younger date. A variety of geochemical approaches to dating paleosol features remain to be more fully exploited.

Another method for establishing the age of alteration seen in paleosols is by considering how consistent a feature is with known kinds of alteration after burial. Various features of paleosols can be compared to assess whether their conditions of formation, such as Eh and pH (see Fig. 3.13), are internally consistent. Well known kinds of alteration after burial also can be considered and the remainder of this chapter constitutes an exploration of these.

Compaction

Because of compaction of paleosols by overburden, their thickness and the relative proportions of their constituent horizons can be distorted compared with those of modern soils. Many physical properties are changed during compaction and it is difficult to predict changes in thickness of paleosols unless there is some clear evidence from the paleosol itself. Perhaps the best evidence is the ptygmatic folding of vertical cracks filled with distinctive soil material, for example, cracks within a paleosol filled with sand from an overlying paleochannel (Retallack 1986b). Comparison of the length of these clastic dikes around the convolutions with their length in a straight line can yield a compaction ratio that is useful for reconstructing former soil thickness, density, and chemical composition (see Table 4.5). Root traces, burrows, and other near vertical features also may be useful, as is deformation of the soil matrix around compaction-resistant parts of a paleosol such as calcareous or sideritic nodules or permineralized tree trunks (Ryer & Langer 1980). If there are other indications that the nodules are original features of the soil, then relict bedding within nodules can be compared with compacted bedding outside them. The crushing of skulls, turtle shells, eggs, burrows, and logs in paleosols may be additional guides to degree of compaction.

The initial compaction of sandstones involves deformation of softer matrix, cement, and deformable grains such as claystone. With further load and filling of the pores between the grains with deformable materials, the hard grains become crushed and then sutured together. When grains are first pressed together they touch without deformation. As the grains

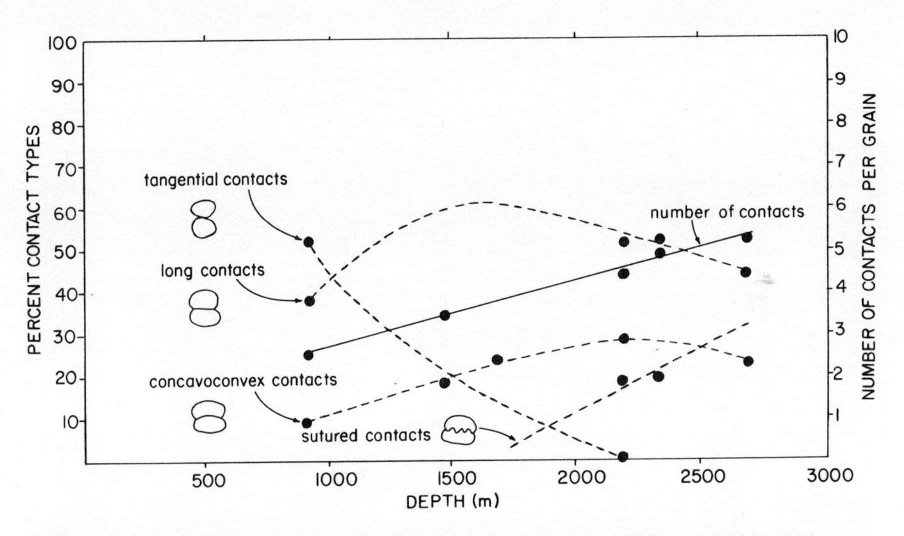

Figure 7.3 Relationship between depth of burial and types of grain contacts in Mesozoic sandstones of Wyoming (modified from Taylor 1950, with permission of the American Association of Petroleum Geologists).

accommodate greater compactional pressures they form long, touching surfaces, then concavo-convex, and finally sutured contacts. With progressive compaction, the number of contacts and the proportion of contacts of various kinds change (Fig. 7.3). From this sequence of events it can be seen that the compaction of sandstones depends largely on the ability of individual grains to resist compaction. Quartz-rich sandstones are less easily compacted than are those with grains of claystone or schist (Fig. 7.4). Reduction of porosity to less than half its original value (from 40 to 20%) is typical after burial to about 5 km, which represents a decrease to about 80% of the original thickness. These figures vary widely, but the degree of compaction observed for sandstones is not so extreme as that for claystones and coals (Chilingarian & Wolf 1976).

The compaction of clays or clayey sandstones is greatest if they are extremely fine-grained, have small amounts of silt and sand grains, are smectitic rather than kaolinitic in composition, are rich in sodium, magnesium, or iron rather than calcium or potassium, or are non-calcareous, non-pyritic, or rich in organic matter (Rieke & Chilingarian 1974). Local differences in clay composition or crystallinity and their behavior during compaction are especially apparent from radially slickensided patches in many ancient shales ("*Guilielmites*" of Byrnes *et al.* 1977). Clay compaction also is limited by the amount of enclosed fluids and the adjacent system of porous rock layers that can relieve overpressuring of contained water and gases. Another complication is the stability of clay minerals under different metamorphic regimes. Conversion of smectite to

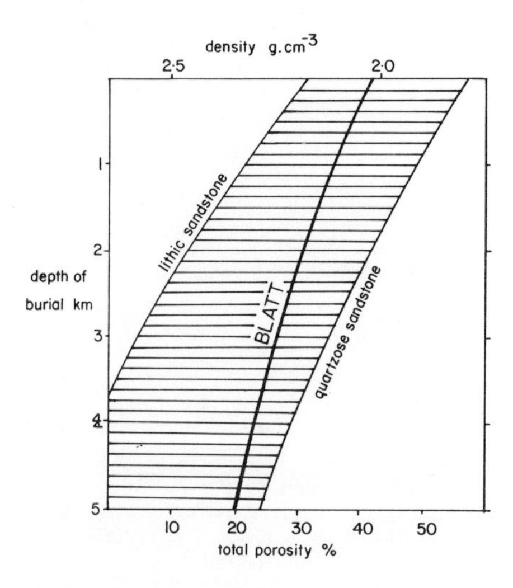

Figure 7.4 Variation in porosity and density of sandy sediments with increasing depth of burial (data from Maxwell 1964, Galloway 1974, and Blatt 1979).

illite may be accompanied by considerable volume reduction. Because of these complications, the degree of compaction of claystones varies within wide limits (Fig. 7.5). The generalized compaction curves of Weller (1959) and Vassoevich (1960) differ to some extent, but can be taken as representative. Both show that initial reduction of porosity and compaction to about 80% of the original thickness requires little loading (less than 1 km of overburden) as clay minerals become oriented parallel to bedding and fluids are expelled. From this point to a porosity and compaction approaching 11% of what they were originally, further reductions in thickness are more difficult unless there are temperature-related changes in clay mineralogy or exceptional losses of fluid pressure. Further reduction in thickness beyond this point is slow and effected by proportionally much greater loading, deformation, and growth of metamorphic mineral assemblages. Highly metamorphosed paleosols may not be much more compacted than deeply buried, little-metamorphosed paleosols.

Complications considered for the compaction of clayey rocks also apply to coals which commonly contain at least some clay and also undergo complex chemical changes in the coalification process. The initial steps in the formation of coal from peat involve flattening of organic matter and loss of moisture under the weight of overburden. This can occur to an advanced stage at the bottom of thick peat bogs (Elliott 1985). Contrary to older ideas concerning coalification, it is not so much pressure as temperature that is important in the formation of bituminous coal and anthracite (Stach *et al.* 1975). Overpressuring can even inhibit the

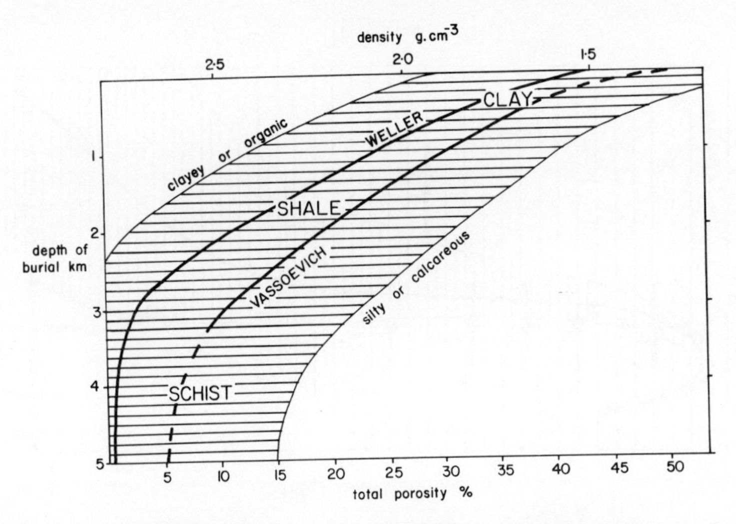

Figure 7.5 Variation in porosity and density of argillaceous sediments with increasing depth of burial showing the typical curves proposed by Weller (1959) and Vassoevich (1960) and outside limits from various sources cited by Rieke & Chilingarian (1974).

development of higher grades of coal because this involves chemical reactions in which volatiles must be removed. As volatile oxygen and hydrogen are lost, the porosity of coals reaches a minimum at a point where it contains about 89% carbon when analyzed dry and ash-free (Fig. 7.6). Coal density generally varies little between about 1.5 and 1.2 g/cm^3. It reaches a minimum value when the coal has 89% carbon and then increases with the formation of massive graphite which may reach a density of 2.2 g/cm^3. Changes in density and porosity can be used as indicators of thickness change due to compaction and they are generally in accord with other estimates based on the thickness change of laterally associated sediments, on crushing of woody material observed in thin and polished sections, and on the degree of compaction observed around compaction-resistant inclusions such as tree stumps and coal balls. Estimates from such inclusions seem to be the most reliable indices of compaction.

Coal balls are a distinctive kind of calcareous nodule found in coals of the midwestern United States, Europe, and Ukraine. They are known to have formed very early in the peat. They contain exquisitely preserved, uncompacted plant tissue, often passing out directly into coalified fragments beyond the nodule. Coal balls have been found at the base of thick coals to have less compacted plant remains than at the base of comparably thick modern peats. Coal balls also have been found as clasts within paleochannel sandstones overlying and eroding into coals with coal balls. Coal balls thus preserve a small area of uncompacted coal within a coal seam (Scott & Rex 1985). The varied data from porosity and density

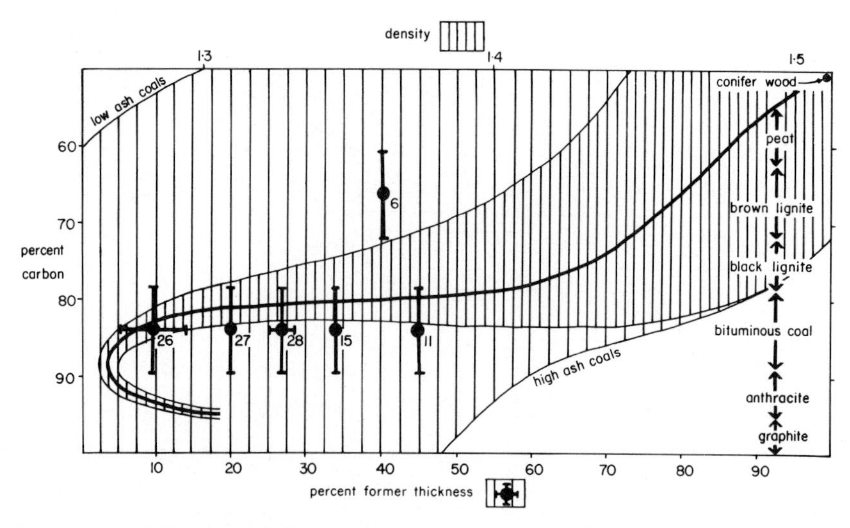

Figure 7.6 Variation in thickness and density of coal with increased rank as indicated by percentage of carbon on a dry and ash-free basis (data from Stach *et al.* 1975) The black dots and error bars are coal/peat compaction percentages based on observations of hard inclusions in coals (numbered in the order of entries in Table 1 of Ryer and Langer 1980).

changes and from inclusions such as coal balls indicate a wide range of possible compaction. Reduction of thickness of at least 50% is likely for most lignites and bituminous coals. There is further reduction in thickness with progressive coalification. Information on porosity, density, inclusions, and coal rank can all be important to estimating former thickness of the peat that formed a coal seam.

Cementation

Much of the transition from soft, friable soil or sediment to hard indurated rock is due to precipitation of cement. Cement fills cracks between the grains, often forming successive coatings to the wall of the crack. Cements precipitate from groundwater passing through the pores at various times during formation and burial of a sediment or soil. Cementation during soil formation results in the formation of hardpans and nodules (Flach *et al.* 1969). Hence, cementation is not solely a phenomenon of deep burial.

The most widespread cements are silica and calcite, but hematite, clay, and siderite are locally abundant cements. The chemical conditions favoring the formation of each kind of cement are different (see Fig. 3.13) and may provide clues to their origin. Calcite forms cements in soils at a pH generally greater than 8.3 and silica at a pH less than 9. The Fe to Mn ratio of calcite increases with increased oxidation, and this crude measure of Eh can rapidly be observed by the degree of cathodoluminescence of carbonate cements in thin sections (Frank *et al.* 1982). The Eh of silica cements may be apparent from included black organic matter or red hematite. This information can be a useful supplement to studies of grain relationships in thin section for determining the relative age of cements. In general, deep burial environments are reducing because of the scavenging of oxygen in pore waters by mafic minerals. Low-grade metamorphism creates acidic conditions because of carbon dioxide generated during the thermal breakdown of buried organic matter and carbonates.

A more time-consuming, but powerful, technique for assessing the timing of cementation during burial is isotopic analysis. The oxygen isotopic ratio ($^{18}O/^{16}O$) of carbonate may be appreciably lightened during burial (Dickson & Coleman 1980, Beeunas & Knauth 1985). This difference is also related to temperature, pore-water reactions, and hydrocarbon generation in ways that are becoming better understood and quantified.

Neomorphism

Neomorphism is a general term for recrystallization, which now has a more precise meaning (Bathurst 1975). Neomorphism includes a variety of processes in which the crystal shape or size of minerals changes during diagenesis. It includes replacement of a mineral by its polymorph, such as aragonite by calcite (inversion), the replacement of a deformed crystal by an undeformed crystal (strain recrystallization), and replacement of an undeformed crystal by another undeformed crystal (recrystallization). Thus the size and shape of crystals in paleosols may have little relation to what they were originally.

The widest variety of crystalline textures are found in calcium carbonate, but some comparable textures are found in silicate rocks. For example, calcite cement deposited in optical continuity around a carbonate grain is called a syntaxial cement and is similar to quartz overgrowths of quartz grains. Similarly, fibrous calcite corresponds to chalcedony in crystal habit, as does spar to megaquartz and micrite to microquartz (also known as chert). A variety of textural features can be used to determine whether these crystal habits are neomorphic or were original cements (Folk 1965). These rely on indications of the relative age of the crystals compared with original grains of the sediment or paleosol. Isotopic and cathodoluminescence studies of the kind used to evaluate cements (Dickson & Coleman 1980, Frank *et al.* 1982) can also be applied toward resolving problems of neomorphism.

Despite a large amount of published information on neomorphism, few studies have been explicitly concerned with paleosols. A variety of neomorphic fabrics have been recorded from Late Carboniferous coal balls (Rao 1985) which I regard as a kind of soil caliche of peaty soils (Histosols). I also have seen coalescive neomorphic spar in caliche nodules of Ordovician and Silurian age from Pennsylvania (Retallack 1985).

Authigenesis

The formation of new minerals in place is called authigenesis. Unlike neomorphism, in which new crystals form, authigenesis involves the appearance of minerals of new chemical composition as well as crystal form. Authigenic minerals are distinguishable from cement by many of the same criteria used for neomorphic minerals. Usually it is more of a problem to distinguish authigenic crystals from detrital grains. Compared with rounded and worn detrital grains, authigenic crystals are euhedral with sharp angular outlines when viewed in petrographic thin sections. Authigenic crystals also may truncate pre-existing grain boundaries and may be conspicuous as unusually large grains in a fine-grained matrix.

It may not always be easy to determine whether authigenic minerals formed in the original soil or during subsequent burial. Feldspars have not been found to form in soils, but do form over a wide range of conditions of temperature and pressure (Baskin 1956). Metamorphic minerals such as prehnite, pumpellyite, sericite, lawsonite, glaucophane and jadeite form under specific conditions of elevated temperature and pressure (Winkler 1976).

Replacement

The process of one mineral changing into another bit by bit is called replacement. This term does not strictly include the dissolution of a grain to leave an open cavity that is later filled by another mineral. Sometimes, however, the term "replacement" is used loosely to cover both cases, which can be difficult to distinguish.

The pyritization of shells in estuarine soils is an example of replacement (Clark & Lutz 1980). Oxidation of bacterially created pyrite produces sulfuric acid, and respiration of plant roots and intertidal invertebrates produces carbonic acid. Aragonite of molluscan shells is unstable in such an acidic environment and with reducing conditions in underlying anaerobic parts of these waterlogged soils pyrite can be precipitated locally as aragonite is dissolved. Similar pyritization of calcareous or phosphatic shells also may occur after burial of marine invertebrates in the dark, organic muds of poorly oxygenated seas well away from intertidal soils. Pyritization of wood and other plant material, on the other hand, seldom occurs by replacement. The pyrite usually fills cavities defined by cell walls once the cell cytoplasm has decayed away (Matten 1973). Although pyritic petrifactions of wood may appear to be a case of cellular replacement, they represent solution and fill.

Dissolution and fill is most easily distinguished from replacement in cases where the substitution of one mineral for another has been incomplete. This is especially clear when corroded relicts of the original mineral remain with a rind of the replacing mineral or when the original mineral grains have an irregular or "nibbled" appearance. Crystal orientations cutting across older mineral boundaries may also be a clue.

Dissolution

Some materials may be removed from soils and rocks by simple dissolution and transport in pore water. This can be by removal of cement (decementation), of grains or of entire layers of grains. Cavities of dissolved grains filled with another material are called pseudomorphs. Clay cube pseudomorphs of halite crystals are a familiar example (Haude 1970). Brecciated pipes and beds may be the only record remaining of some evaporite beds (Bowles & Braddock 1963).

Dissolution of other minerals depends on the pH or Eh of pore water, the size of crystals dissolving, and the pressure of overburden and of pore water. In deeply buried sandstones there is preferential dissolution and suturing of grains where they touch one another. Material released by pressure solution may form cements nearby, and this is one likely mechanism for the formation of quartz overgrowths on grains (Renton *et al.* 1969). Dissolution of quartz also proceeds more rapidly if the pore waters are highly alkaline (pH more than 9) and if the quartz is very fine-grained. On the other hand, slightly acidic conditions are needed for the dissolution of limestone and other carbonates. Rainwater is acidic enough (about pH 5.7) to dissolve limestone outcrops and produce pitted and irregular topography called karst. Dissolution of limestone under pressure forms stylolites stained with residual iron and aluminum. Acidic solutions produced by thermal maturation of hydrocarbons at depth also are capable of selectively dissolving or hydrolyzing grains of sandstones to create secondary porosity (Schmidt & McDonald 1979). Stylolites, sutured grain contacts, and secondary porosity are forms of dissolution characteristic of deep burial environments (Bathurst 1975), but other forms of dissolution such as empty grain coats may appear similar in soil and late diagenetic alteration.

Dehydration

Chemical reactions involving both addition of water to mineral structure (hydration) and its subtraction (dehydration) are common in soils. With compaction on burial and dewatering, however, dehydration reactions prevail. These reactions may result in a considerable change in volume and crystal habit. Elongate, rhombic gypsum is stable at the surface, but inverts to a smaller volume of orthorhombic anhydrite at depths of only 35 m (Murray 1964). Anhydrite can be converted back to gypsum if it is re-exposed at the surface in a porous outcrop in which water is available.

A more widespread dehydration reaction in paleosols is the diagenetic change of reddish brown, microcrystalline–spherical ferrihydrite to yellowish brown, microcrystalline–acicular goethite and ultimately to brick-red, microcrystalline–hexagonal–platy hematite (Walker 1967). An increasingly red to purple color may also be due to increasingly coarse (up to 2 μm) grain size of iron oxide crystallites (Blodgett 1988). These dehydration steps can occur during soil formation and also after its burial. In midwestern North America (Ruhe 1969), soils less than 10,000 years old are brown (10yr–2.5yr) whereas similar buried soils 100,000 years (or more) old are red (5yr–7.5yr). The effects of diagenetic dehydration can be seen more strikingly over longer time spans. Paleozoic paleosols are almost invariably hematite-rich with little or any original ferrihydrite or goethite. Their distinctive brick-red or purple color may bear little resemblance to their former color (Fig. 7.7). These changes may be important for interpretation of paleosols. Hematite-rich paleosols may not necessarily have formed under tropical climates where the most hematite-rich modern soils are found (Birkeland 1984). Interpretations based on redness of hue of paleosols may still be useful for comparison of paleosols with a common burial history, but in general should be mistrusted compared with other lines of evidence for paleoenvironmental conditions.

Reduction

Chemical reactions involving electron gain or reduction are prominent in waterlogged soils. They also may have dominated the formation of well-drained soils during Precambrian time when atmospheric oxygen levels are thought to have been much lower than at present. Most late diagenetic and metamorphic environments also are reducing. The opposite kind of reaction (oxidation) is mainly found in well drained soils or in coarse-grained, shallowly buried materials through which oxygenated water may circulate.

One of the most important reduction reactions for paleosols, because it controls their color, is the conversion of red, oxidized, ferric (Fe^{3+})

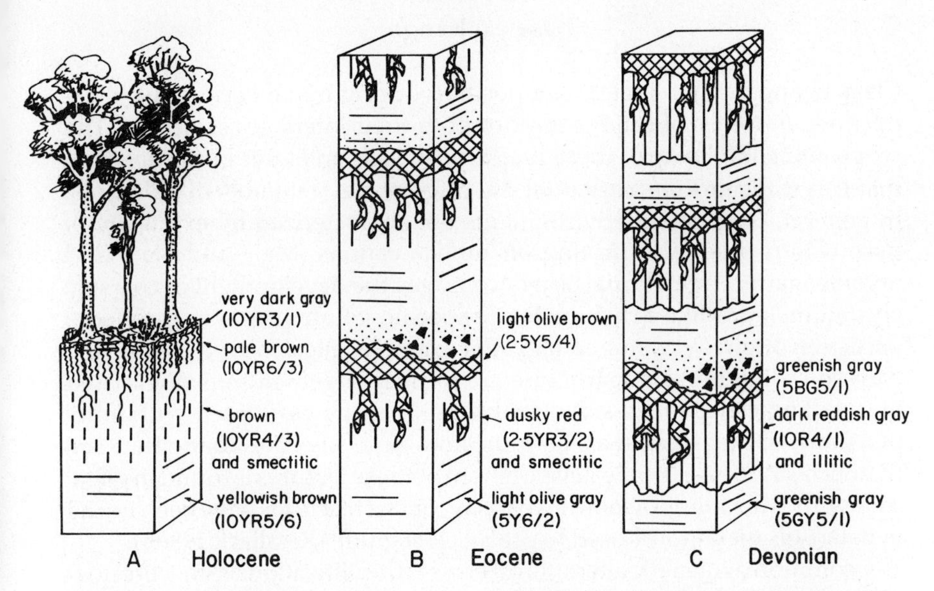

Figure 7.7 Hypothetical modification of a woodland soil by burial gley of organic matter, dehydration reddening of ferric oxyhydrates, compaction, and illitization of smectite. The profiles illustrated from left to right are modern soil profile 7 of Appendix I of Buol *et al.* (1980), a paleosol in the Early Eocene, "Sand Creek facies" of Bown & Kraus (1981a) and the Late Devonian, Peas Eddy clay of Retallack (1985).

minerals, such as hematite, to reduced, gray, ferrous (Fe^{2+}) minerals such as siderite. This reduction is greatly aided by anaerobic bacteria, which mediate the reaction as a source of electrons that they use to ferment or oxidize organic matter for energy. Thus roots, fragments of organic matter or litter horizons of soil can be strongly reduced upon burial and subsidence into a biologically active water table. Dispersal of remnant organic matter and development of drab colors in formerly organic parts of otherwise reddish paleosols is common (Fig. 3.7). Minerals formed by reduction include pyrite, siderite, and a variety of green and gray clays. The development of these features after burial can be called burial gleying. The contrast between burial gley features and gley due to original waterlogging may be difficult to disentangle unless there is other evidence of former drainage such as patterns of root traces. In contrast to reduction in waterlogged soils and shallowly buried soils in which bacteria probably play a role (Allen 1986b), the reduction of hematite during deep burial is remarkably sluggish. Hematite persists in paleosols into the greenschist metamorphic facies (Retallack 1985). Only by the amphibolite facies is it appreciably converted to magnetite (Thompson 1972).

Base exchange

Clays change mineralogical composition during burial because of introductions or losses of cations important to their interlayer structure. The scope and complexity of these reactions in soils and after burial are such that it is difficult to be certain of the original clay minerals of paleosols. In general, deep burial environments are characterized by expulsion of pore water, often also driving off mobile cations (Mg^{2+} and Na^{+}) and organic matter. Deep burial also encourages the development of coarsely crystalline, mica-like structure. Some specific soil structures may reflect a particular original clay mineralogy. For example, gilgai microrelief and the corresponding mukkara structure are formed largely in soils dominated by swelling clays. Steep-sided mukkara structure associated with lentil peds and slickensides may be indicators of a smectite-dominated soil (Paton 1974). Less extreme development of these features is found in gilgai soils with other kinds of more stable clay. Such structures have been found in paleosols metamorphosed to greenschist grade (Retallack 1986b).

A common diagenetic alteration of clays is the illitization of smectite now documented in deep boreholes in many parts of the world. Some of these studies have shown that the increase of potassium-rich clay (illite) is at the expense of potassium-rich minerals (such as potash feldspar) and potassium-poor clay (smectite). In these cases illite appears to be forming from smectite by addition of K^{+} from the dissolution of mineral grains of K-feldspar (Fig. 7.8). The liberation of K^{+} may be effected by hydrolysis

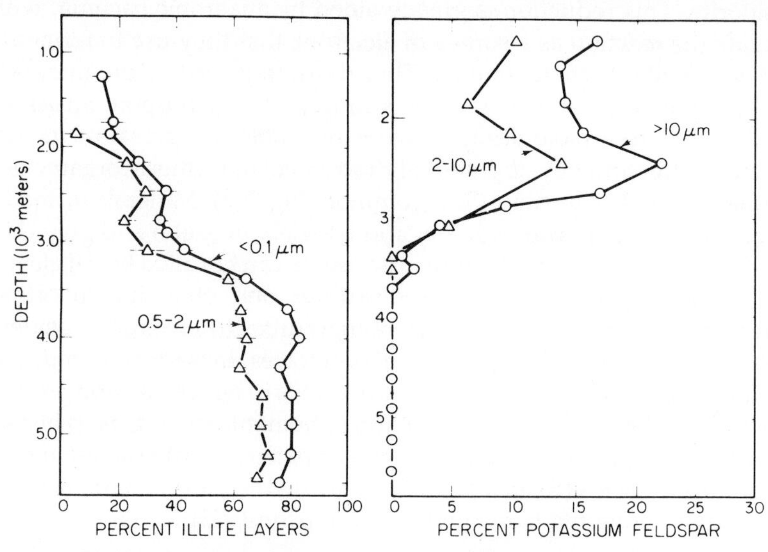

Figure 7.8 Changes in the percentage of illite and potash feldspar with depth of burial in Oligocene–Miocene sediments in a borehole (CWRU#6) in Harris County, coastal Texas (from Hower *et al.* 1976, reprinted with permission of the Geological Society of America).

reactions due to acids from organic matter or carbon dioxide in groundwater. The illitization of smectite is promoted by removal of other common cations (Na^+, Ca^{2+}, Mg^{2+}) and of silica in pore water. These changes have been documented at moderate depths (2000–3000 m) and temperatures (60–180°C) (Curtis 1985), but theoretically such alteration could proceed at much lower temperatures over very long periods of time (Bethke & Altaner 1986). Progress of the reaction may be limited by the availability of potassic minerals (such as potash feldspar and muscovite) and of local porosity allowing circulation and removal of pore water (Morton 1985). Because the reactions seldom go to completion, those rare paleosols and enclosing rocks lacking K-feldspar are unlikely to be illitized (Retallack 1986b). Illitization can be gauged from major element chemical analysis of paleosols, which may show molecular ratios (K_2O:Al_2O_3) too potassic for known clay minerals of soils, or which demonstrate surficial enrichment of potash (K_2O) but depletion of soda (Na_2O), these alkalis tending to track

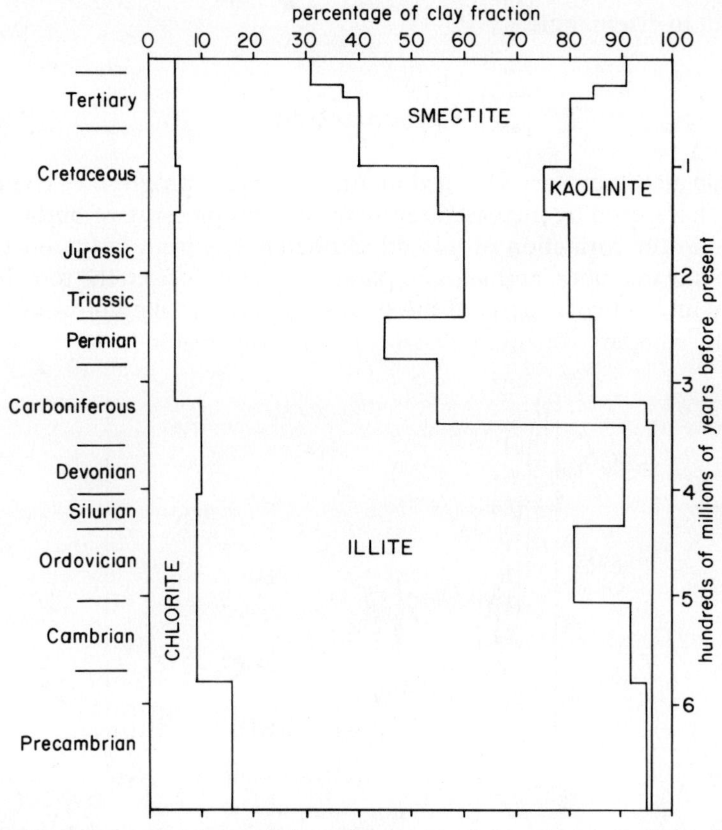

Figure 7.9 Relative abundance of the major groups of clay minerals in shales through geological time [based on compilation of Weaver (1967) of 40 000 analyses from North America, reprinted with permisson of Springer].

each other in modern soils (Feakes & Retallack 1988). These trends and their difference from unaltered soils can be made more obvious by plotting them on triangular diagrams with selected major elements at the apices (Nesbitt & Young 1989). Another approach is to estimate the degree of illite crystallinity from X-ray diffractogram traces as an index of degree of metamorphic alteration (Frey 1987).

Illitic soils are less deeply weathered than smectitic soils and the abundance of illite in mid-Paleozoic and older mudrocks has been taken as evidence for less severe weathering before the advent of forests to stabilize the landscape (Fig. 7.9). This idea is supported by studies of Paleozoic sandstones (Basu 1981), but because most rocks of this age have been buried deeply, the clay data also can be explained by illitization during deep burial. The whole question of trends of clay mineral formation over geological time therefore remains unsettled. Original illitic composition and diagenetic illitization are both likely for some Precambrian and early Paleozoic paleosols, but the relative contributions of each are difficult to disentangle.

Carbonization

Organic matter preserved in sediments undergoes progressive chemical changes induced by increased temperature and pressure of burial. This results in the formation of oil and ultimately gas from fatty and waxy remains of microbes, animals and plants. Coal forms from the remains of woody and cuticular parts of the plants. Diagenetic alterations of these initially complex organic substances generally result in an increased

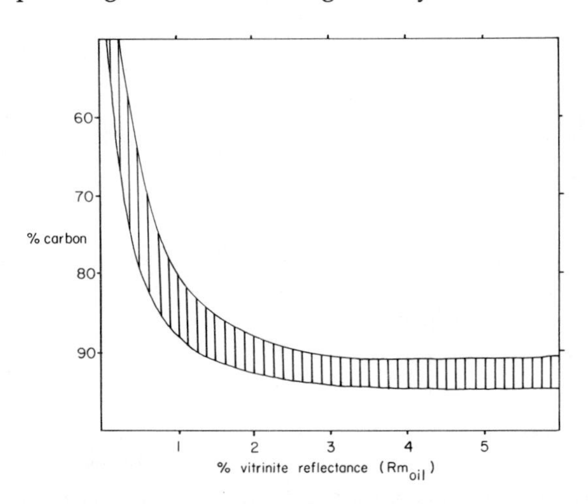

Figure 7.10 Correlation between the percentage of carbon on a dry, ash-free basis and the reflectance of the coal maceral vitrinite (based on data of Stach *et al.* 1975).

concentration of carbon and a loss of volatile elements such as hydrogen, oxygen, phosphorus, and nitrogen (Stach *et al.* 1975).

Each kind of organic matter alters differently, but some substances have proven very useful indicators of degree of alteration. Wood, for example, is altered to vitrinite. This is a common constituent (maceral) of coal and also can be found isolated (as a phytoclast) in shale or sandstone. With increased coalification vitrinite becomes increasingly shiny. The measurement of its reflectance under a reflecting microscope provides a fairly accurate measure of coalification and severity of diagenesis of surrounding sediments up to medium grades of metamorphism (Fig. 7.10). The pressure and temperature gradient and burial history of each sedimentary basin is unique so that a local trend in vitrinite reflectance must be independently calibrated in order to yield estimates of burial depth and temperature. This kind of information can be very useful for assessing the degree of burial alteration of a paleosol.

PART TWO

FACTORS IN SOIL FORMATION

8 Models of soil formation

Different definitions of soils reflect different approaches to them. Soils present a different kind of challenge to farmers and to engineers. Soil scientists have a different view again, or rather several different views (Johnson & Watson-Stegner 1987). These varied views of soils are in effect different theoretical bases for their study. So far, in describing various features and kinds of soils, they have been considered as if they were natural history specimens. Like the trilobites, cowries, or narwhal tusks in a gentleman's cabinet in Victorian England, the variety of soils found in nature can be described and classified. Fundamental aspects of soil formation can emerge from such comparative studies. The quest for a "natural classification" aims to reveal distinct processes and products of soil formation. Mapping soils as objects of natural history is also useful for guiding effective human use of the landscape. In contrast, soils also can be viewed as open systems, as energy transformers, and as environmental products (Fig. 8.1). Each of these different views of soils, as detailed below, elicits different kinds of research questions, experiments, and observations.

Soils can be envisaged as open systems to the extent that they represent a boundary between earth and air through which materials move and are changed. Four basic kinds of fluxes can be imagined: additions, subtractions, transfers, and transformations (Simonson 1978). Additions include mineral grains brought in with airborne dust. Organic matter also is added to the soil in the form of leaf litter. Subtractions include surface erosion of mineral and organic matter. Mineral matter is lost in solution in groundwater, and organic matter is lost through cropping by organisms, which then migrate elsewhere. Transfers include movements of material within a soil profile. An example is the downward leaching of clay to produce a clayey subsurface (Bt) horizon in Alfisols and Ultisols. Relocation of soil material in burrows is another example. Transformations involve changes in composition and form of soil materials. Organic matter, for example, is recycled through a variety of forms ranging from decayed plant material to parts of living soil organisms. The formation of clays from easily weathered minerals such as plagioclase is another example. Flux

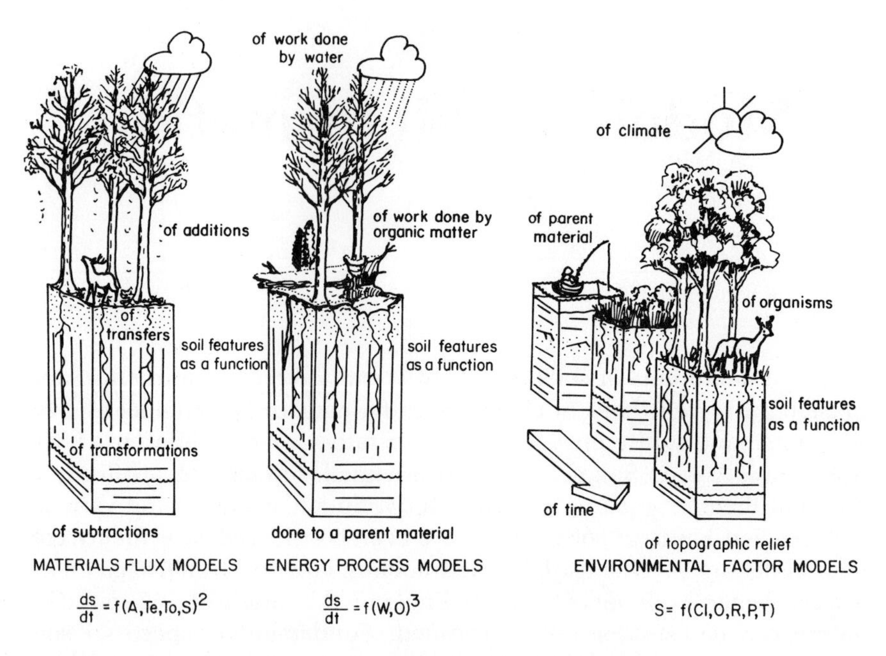

Figure 8.1 Three different mathematical models of soil formation.

models aim to quantify movements of material within and around a soil, such as rates of accumulation of calcium carbonate in desert soils (Machette 1985).

Soils also can be viewed as energy transformers, that is, as a body of material changed by the continuing efforts of natural processes (Runge 1973). The pattern of energy flow through soils can be as complex as the flow of material. The primary source of energy in soils is heat from the sun, which far surpasses (by about 7500 times: Eagleman 1980) heat flow from deep within the crust generated by decay of radioactive elements. Energy also can be gained, in a sense, from deposition of sediment on the soil, or from the addition of groundwater or rain, because these new minerals and water are capable of altering pre-existing materials, or doing work within the soil. Less apparent, but no less important, is energy gained by exothermic mineral alterations, transformation of organic matter, biological growth, friction, and wetting and thawing of soil. Solar energy is lost by radiation, reflection and evaporation. Other energy sinks include erosion, drainage, drying, or freezing. Energy also can be transformed within a profile, by conduction, convection, condensation, evaporation, percolation, and chemical reaction. In addition to energy flow within a soil, there is also impedance of energy flow, such as a rate-limiting factor for a chemical reaction or restricted groundwater movement because of low permeability of the soil. All of these processes and the amount of energy available to drive them are what create a soil profile. Energy models aim to quantify patterns of energy transformation in soils and their results. For example, a mathematical model for the development of calcic horizons in desert soils should be based on estimates of fluxes of materials together with equations for their rate of chemical reaction under different physical conditions (McFadden & Tinsley 1985).

Soils also can be viewed as an environmental product, molded over time from whatever material was available, by climate, organisms and geomorphic processes. It would be ideal if there was a one-to-one correspondence between features of soils, such as clay content, and a particular environmental factor, such as rainfall. Unfortunately, clay content, like many other features of soils, is also dependent on the temperature of the soil, the nature of minerals available to be weathered and the time available for weathering. Soil clay content is thus less like an automated rain gauge than a poorly calibrated "synthetograph" which averages out a number of overlapping modifications. The aim of environmental factor models is to tease apart these separate influences on the soil. The multitude of specific influences on soil formation can be reduced to five main factors: climate, organisms, topographic relief, parent material, and time (Jenny 1941). These five classical factors are most easily remembered by the acronym "CLORPT." This is a useful set of categories for mentally considering all aspects of the formation of a soil or paleosol in the field. More importantly, CLORPT provides a theoretical framework for devising natural experiments to investigate processes of soil formation.

Because soil formation is a multivariate process, then in order to study any one of the factors in isolation it is necessary to consider only those cases where all the other factors are constant, or at least nearly so. In order to study climate, for example, what is needed is a number of soils formed in different climates, but as similar as possible in their ecosystem, topographic setting, parent material, and time over which the soil formed. Such a group of soils is called a climosequence, and mathematical relationships relating soil features to climate are called climofunctions. A well known example is the simple formula relating depth of the calcareous horizon to mean annual rainfall in postglacial grassland soils of the midwestern United States (Jenny 1941). Biofunctions, topofunctions, lithofunctions, and chronofunctions are useful for quantifying the effects of organisms, topography, parent material and time available for soil formation, respectively. The approach assumes that these factors are independent, but it may be difficult to find situations where vegetation, for example, is uniform over regions of different climate. There is also some question as to whether time can really be considered an active factor in the same way as rainfall or root growth. Nevertheless, each of the factors does have recognizable effects, and with such a strategy of dividing to conquer, it is possible to make some sense out of the complex multivariate process that is soil formation (Yaalon 1975).

Each of these approaches to the study of modern soils provides information valuable for the interpretation of paleosols, even though it is not possible to study paleosols in exactly the same way because the fluxes of materials, flow of energy, and environmental factors controlling them ceased to act long ago. From material flux models comes information about rates of development of soil features. From energy process models comes an understanding of the relationships between variables in soil formation. From factor function models comes information about the role of selected environmental influences in soil formation. More and better information about the past will be gleaned from paleosols as such studies of Quaternary soils gain in scope and precision.

Although each approach is of value, this book emphasizes the environmental factor approach. Field studies of environmental factors have been widely reported, and there is a tremendous amount of information about soil formation already available in this form (Birkeland 1984). The five classical factors of soil formation are of special value in structuring paleoenvironmental interpretations of paleosols (Retallack 1983a). In addition, changes in particular factors over geological time can be studied by choosing paleosols for which other factors were similar (Retallack 1985).

9 Climate

During the founding period of soil science, climate was considered one of the most important factors in soil formation. There are widespread acidic sandy Spodosols in temperate regions, but in the tropics red clayey Oxisols are common. Such observations encouraged the concept of zonal soils, which are soil types restricted to particular climatic zones. Other soils, e.g., Entisols, are azonal because of limiting local circumstances, such as a short time of development, so that climatic effects are not easily discernible. Some of the generalizations upon which the zonal concept of soils was founded are still regarded as valid. However, a more detailed understanding of soil geography has also revealed numerous exceptions.

The tendency to incorporate climatic information within soil classifications has been countered by efforts to base soil classification on observable features of soils. Each soil order of the US soil taxonomy does have a climatic range, but in most cases it is broad. Finer subdivisions of hierarchical soil classifications may have a more restricted climatic range. In the US taxonomy, units as broad as suborders of Vertisols are defined by climatic criteria. Hence, the zonal concept of soils has not been entirely abandoned. This is unfortunate, because paleoclimatic interpretations cannot be made by identifying paleosols using paleoclimatic data. Paleosols accurately classified using other criteria may be compared with modern examples to gain some idea of paleoclimate. This method of paleoclimatic reconstruction is rather like using a particular species of fossil leaf to infer paleoclimate from the climatic range of a related living species. Another way of interpreting paleoclimate is to examine the size, shape, and marginal outline of fossil leaves as physiological indicators of paleoclimate, independent of what particular leaf species are involved. It is this kind of interpretation of features of paleosols, independent of their classification that will be attempted in this chapter. Numerous published studies of climatically sensitive features of soils provide a data base for the interpretation of paleoclimate from paleosols. Most of these studies have striven to isolate the effects of a particular climatic variable by keeping constant other climatic variables, in addition to other factors.

The terms climate and weather sometimes seem interchangeable in common usage, but have distinct scientific meanings. Weather is the record of rainfall, temperature, and humidity as reported daily in newspapers and on television. Climate, on the other hand, is the average of data compiled from weather reports, usually summarizing records from at least 30 years of observations. Climatic data are based on a particular weather station, but these stations are generally chosen to reflect conditions of the surrounding region. Local frost hollows and exposed high ridges are to be avoided for regional weather stations because they have their own microclimate. These local climatic deviations are significantly different from regional climate and can be especially important for small animals and plants.

Soil climate is a special kind of microclimate. It refers to the conditions of moisture, temperature, and other climatic indices within soil pores. In well drained soils, the soil climate is a muted version of regional climate, with extremes of variation damped out. In waterlogged soils, however, soil climate is unrelated to regional climate. The temperature and oxygenation of waterlogged soils depend more on local rates and pathways of groundwater flow than on atmospheric conditions. Waterlogged soils can be found in desert oases in addition to wet forests. Such differences between soil climate and regional climate can be a source of confusion for interpreting paleosols. For example, the climatic drying often inferred for the transition between Carboniferous coal measures, with their drab paleosols, and Permian red beds, with their variegated red and orange paleosols, may reflect improved drainage of paleosols rather than regional climatic change. The distinctive soil climate of waterlogged soils has long been appreciated. Estimates of other kinds of soil climate are now finding their way into soil classifications, models for soil formation and studies of soil biology. In few cases, however, have observations of soil climate continued for periods of 30 years or longer. Because of this and because the conditions of human life and agricultural production are more easily related to regional climate, the climate above ground is most often used in studies of the relationship between soil features and climate.

Some classifications of climate

Some climatically sensitive features of soils can be related clearly to particular climatic variables, but even the best of these lack precision. Soils are not as sensitive for recording climatic conditions as are meteorological instruments, but climate can be interpreted from them within fairly broad categories. Suitably broad categories are provided by a number of classifications of climate, a few of which are reviewed here.

One of the most influential large-scale classifications of climate was devised in 1918 and modified over the next two decades by the German

Table 9.1 Köppen's classification of climates (modified from Trewartha 1982)

I. Tropical wet climates (***A***)

These are hot rainforest climates with temperature of the coolest month above 18°C (64.4°F). With monthly temperatures lower than 18°C certain sensitive tropical plants do not thrive. Within the ***A*** group of climates two main types are recognized: one has adequate precipitation throughout the year (***Af***) while the other includes a distinct dry season which affects vegetation adversely (***Aw*** and ***Am***).

Af = Tropical wet climate; rainfall of the driest month is at least 60 mm. Within this climate there is a minimum of seasonal variation in temperature and precipitation, both remaining high throughout the year.

Aw = Tropical wet-and-dry climate; distinct dry season in low-sun period or winter. A marked seasonal rhythm of rainfall characterizes ***Aw*** climates; at least 1 month must have less than 60 mm. Temperature is similar to that in ***Af***.

Am = Monsoon; short dry season, but with total rainfall so great that ground remains sufficiently wet throughout the year to support rainforest. ***Am*** is intermediate between ***Af*** and ***Aw*** resembling ***Af*** in amount of precipitation and ***Aw*** in seasonal distribution.

II. Dry climates (***B***)

These are dry climates in which there is an excess of evaporation over precipitation. No surplus of water remains, therefore, to maintain a constant groundwater level so that permanent streams cannot *originate* within ***B*** climates. There are two main subdivisions of ***B*** climates; the arid or desert, type ***BW*** (***W*** from the German word *Wüste* meaning "desert") and the semiarid or steppe type ***BS*** (***S*** from the word *Steppe* meaning "dry grassland").

BW = Arid or desert climate

BS = Semiarid or steppe climate.

Other small letters used with B climates are as follows:

h (heiss) = average annual temperature over 18°C (64.4°F). ***BWh*** and ***BSh*** therefore are low-latitude, tropical, deserts or steppes, respectively.

k (kalt) = average annual temperature under 18°C (64.4°F). ***BWk*** and ***BSk*** therefore are middle-latitude, cold, deserts or steppes, respectively.

III. Warm temperate to subtropical climates (***C***)

These are temperate forest climates; average temperature of coldest month below 18°C (64.4°F) but above -3°C (26.6°F); average temperature of warmest month over 10°C (50°F). The average monthly temperature of -3°C (26.6°F) for the coldest month supposedly roughly coincides with the equatorward limit of frozen ground and a snow cover lasting for a month or more. Within the ***C*** group of climates three contrasting rainfall regimes are the basis for recognition of three principal climatic types: the ***Cf*** type with no dry season, the ***Cw*** type with a dry winter and the ***Cs*** type with a dry summer.

Cf = No distinct dry season; difference between the rainiest and driest months is less than for ***Cw*** and ***Cs*** and the driest month of summer receives more than 30 mm (1.2 in).

Cw = Winter dry; at least 10 times as much rain in the wettest month of summer as in the driest month of winter. Alternative definition: 70% or more of the average annual rainfall is received in the warmer 6 months. This type of climate has two characteristic locations: (1) elevated sites in the low latitudes where altitude reduces the temperature of the ***Aw*** climates which prevail in the adjacent lowlands and (2) mild middle-latitude monsoon lands of latitudinally oriented elongate continental land masses, such as northern India and southern China.

Table 9.1 *Continued*

Cs = Summer dry; at least three times as much rain in the wettest month of winter as in the driest month of summer and the driest month of summer receives less than 30 mm. Alternative definition: 70% or more of the average annual rainfall is received in the winter 6 months.

IV. Cool temperate climates (***D***)

These are cold snow-forest climates; average temperature of coldest month below -3°C (26.6°F), average temperature of warmest month above 10°C (50°F). The average temperature of 10°C for the warmest month approximately coincides with the poleward limits of forest. *D* climates are characterized by frozen ground and a snow cover of several months duration. Three principal subdivisions of the ***D*** group are recognized: (1) ***Df*** with no dry season, (2) ***Dw*** with dry season in winter and (3) ***Ds*** dry in summer (Mediterranean).

Df = Cold climate with humid winters.

Dw = Cold climate with dry winters; characteristic of northeastern Asia where the winter anticyclone is well developed.

Ds = Summer dry; at least three times as much rain in the wettest month of winter as in the driest month of summer and the driest month of summer receives less than 30 mm or 70% or more of the average rainfall is received in the winter 6 months.

V. Frigid climates (***E***)

These are polar climates; average temperature of the warmest month below 10°C (50°F). In the high latitudes, once the temperatures are well below freezing and the ground frozen it makes little difference to plant life how cold it gets. Rather, it is the intensity and duration of a season of warmth which are critical. For this reason a warm month isotherm is employed as the boundary of ***E*** climates. Two climatic subdivisions are recognized: (1) ***ET*** in which there is a brief growing season and a meager vegetation cover and (2) ***EF*** in which there is perpetual frost and no vegetation.

ET = Tundra climate; average temperature of warmest month below 10°C (50°F) but above 0°C (32°F).

EF = Perpetual frost; average temperature of all months below 0°C (32°F). Such climates persist only over the permanent icecaps.

meteorologist Vladimir Köppen. He recognized five main groups of climates, corresponding to the main kinds of terrestrial vegetation. Each kind of climate is designated by letters, as a kind of meteorological shorthand. Capital letters are used for the main climatic groups and lower-case letters for subsidiary climatic features (Table 9.1). One of the large climatic groups includes dry climates. Humid climates are divided into climatic groups based on temperature. On a global basis the main categories of climate form distinct latitudinal belts. Subdivisions of the main climatic groups are based on a variety of features, mainly temperature in the case of dry and frigid climates, and seasonality for the other categories. Mediterranean climates (Cs), for example, have cool to warm temperatures but very low summer precipitation. Lack of moisture in the growing season means that these climates are more difficult for plants and less encouraging for soil formation than other climates with comparable mean annual temperature and precipitation.

There have been many attempts to improve, modify, and replace Köppen's classification of climates. Some of its criteria seem arbitrary and unnecessarily complex. From the point of view of soil science, it is perhaps not the best classification because it subdivides extensively on the basis of temperature and seasonality, which are less significant for soils than is precipitation. From this perspective, the classification of Holdridge (1947) is useful because it relates particular kinds of vegetation to the amount of mean annual precipitation, potential evapotranspiration ratio, and mean annual biotemperature in a straightforward manner (Fig. 9.1). Mean annual biotemperature is a climatic index based on temperature records, adjusted for the observation that large vascular plants become physiologically dormant at temperatures of less than 0°C or more than 30°C. Mean annual biotemperature can be calculated from hourly or daily temperature records by substituting 0°C for all temperatures lower than that, and 30°C for all higher temperatures. Biotemperature is reflected in changes in plant formations with altitude and latitude. The frost line or critical temperature line is a notable discontinuity, around which vegetation and human use of the landscape are often markedly different. The potential evapotranspiration ratio is a measure of moisture available to plants, which depends as much on losses of moisture by evaporation from the ground and from streams, and transpiration from leaves, as on gains of moisture by precipitation. Mean annual potential evapotranspiration ratios can be calculated from an empirical formula, by multiplying the mean annual biotemperature (in degrees Celsius) by 58.93 and dividing that figure by the mean annual precipitation (in millimeters). Holdridge's classification was devised and works best for tropical and subtropical vegetation (Holdridge *et al.* 1971). The typical vegetation for each climatic type is a useful concrete image of that climate, but it can be misleading. Vegetation of an unusual soil, of waterlogged sites, or of a distinct microclimate may reflect regional climate less faithfully. Another disadvantage of this classification for paleopedological studies is that few of these plant formations have a long geological history. Even if Ordovician soils formed in the climate of the wet forest life zone, they could not have supported trees because these had not yet evolved.

Some of the problems with climatic classifications stem from their attempt to place limits on a complexly moving atmospheric system interacting with obstacles in the form of continents and mountain ranges (Eagleman 1980). Large-scale paleoclimatic patterns can be reconstructed for a variety of continental configurations of the geological past by using atmospheric circulation models. Such reconstructions can be useful for understanding the wider paleogeographic context of paleosols. The simplest circulation pattern that can be imagined is one in which the Earth is a flat, featureless sphere, either all water or all land. In this case, there would be a zone of westerly wind flow in middle latitudes, consisting of a

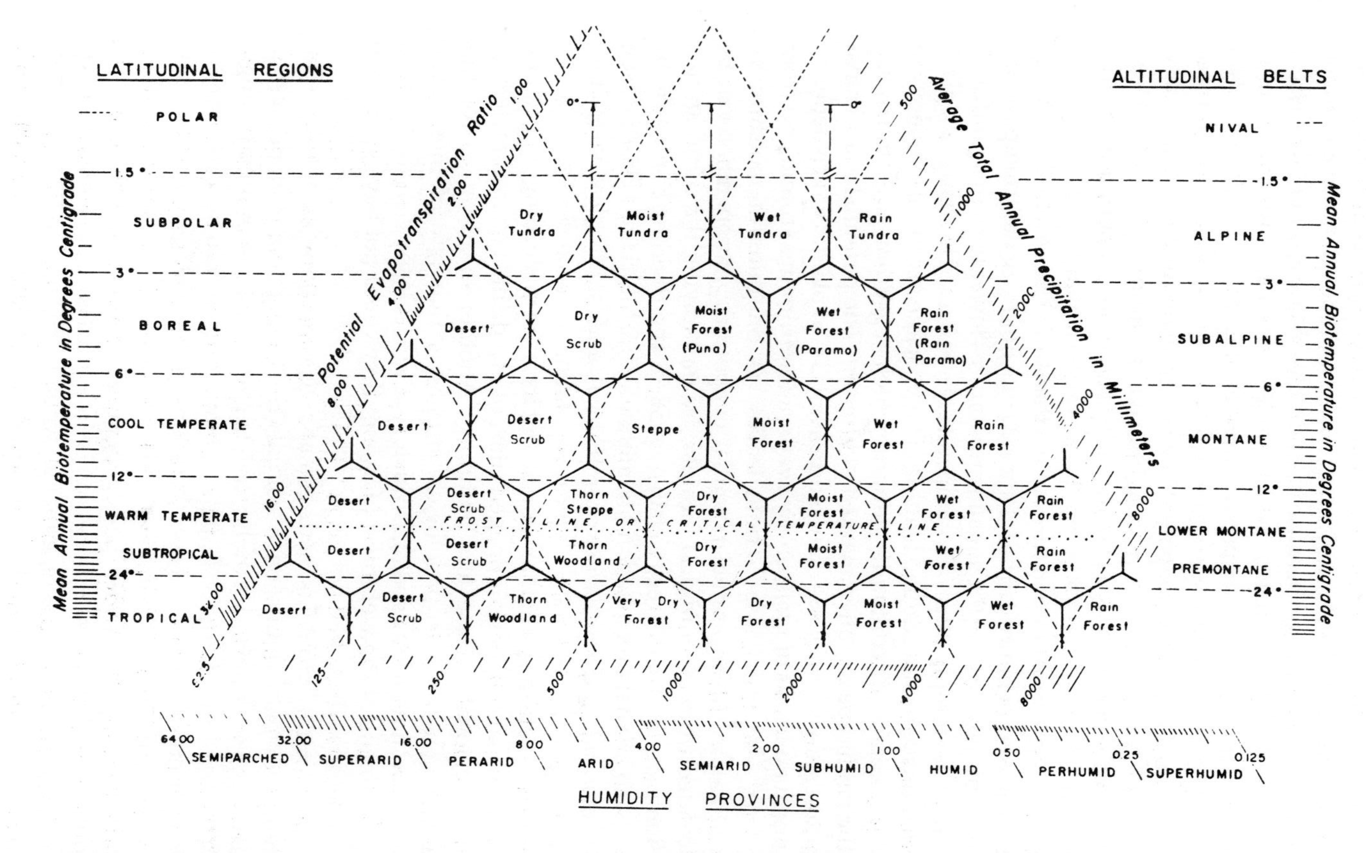

Figure 9.1 Holdridge's climatic classification of world life zones (from Holdridge *et al.* 1971, reprinted by permission of Pergamon Press).

succession of cyclones (low-pressure centers, with clockwise circulation in the southern hemisphere) and anticyclones (high-pressure centers, with clockwise circulation in the northern hemisphere). Wind directions would converge to run easterly near the equator, and they would also be easterly near the poles. This general circulation pattern is modified by intervening continental masses. Meridionally oriented continents with marginal mountain ranges astride the westerly circulation gather clouds and precipitation on the windward side, leaving large areas of dry climate, the so-called "rain shadow," downwind (Fig. 9.2). These are found east of the Rocky Mountains of North America, the Andes of South America, and even the Southern Alps of the much smaller land mass of New Zealand.

Other climatic effects are created by large continents oriented latitudinally. Large land masses cool more in winter and warm more in summer because they are remote from heat circulated by ocean currents. The large landmass of Eurasia, for example, develops a high-pressure cell during the winter because its air is generally colder than that of the Indian Ocean, and this deflects the intertropical convergence zone southward. In summer, however, its high-temperature (or low-pressure) cell attracts the intertropical convergence zone northward over India and southern China, bringing a season of onshore winds and torrential rains known as the monsoon. Intertropical regions of the Americas and northern Australia also have monsoonal circulation, and presumably so did other large latitudinally oriented land masses in the geological past. Summer-dry climates, as in the Mediterranean region, in the Pacific Northwest of the United States, in southern South Africa, and in southeastern South Australia, are a special case of a similar phenomenon. These are produced

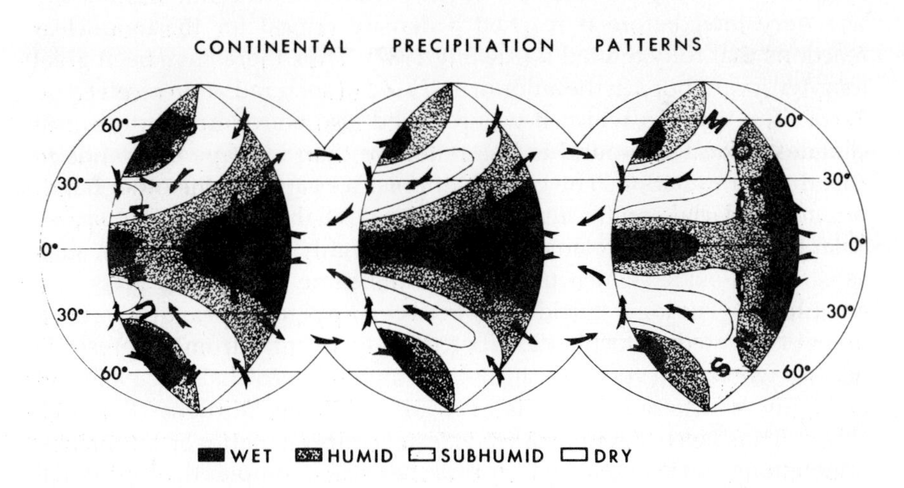

Figure 9.2 Idealized models of continental precipitation patterns (from Ziegler *et al.* 1979, reprinted by permission from Annual Reviews Inc.).

in the downwind (usually eastern) end of summer high-pressure (low-temperature) cells developing over a cool ocean.

Each of these classifications and general circulation models is based on climates that have been observed in historical times. Over geological times, there is evidence for considerable variation in such fundamental constraints as the amount of solar energy received on Earth and the amount radiated back into space. Some of this paleoclimatic variation can be related to wobbles in the angle of the Earth's rotational axis with regard to its plane of orbit around the Sun, which produce the precession of the equinoxes. Precession results in cyclical changes of orientation over 23 000 years or so, and variation in obliquity of the axis of rotation is cyclical over 42 000 years. There are changes in the eccentricity of the Earth's orbit also on periods of 100,000 of years (Hays *et al.* 1976). Although these are small orbital changes, their effects in changing the amount and distribution of solar radiation at the Earth's surface are magnified by surface processes to make big differences in climate: the difference between Chicago's present cool temperate climate and a full glacial climate in which Chicago is locked in continental ice. Pleistocene climatic fluctuations have been well documented from sequences of paleosols (Kukla 1977). The best known Ice Age is that proceeding at present (over the past 2 million years), but there also were earlier Ice Ages (Frakes 1979) during the Permo-Carboniferous (about 290 million years ago), Ordovico-Silurian (about 440 million years), Late Precambrian (900–600 million years), and Early Precambrian (about 2300 million years). These may reflect longer-term astronomical cycles, perhaps the passage of the solar system through the dusty spiral arms of the galaxy. Even earlier in Earth history, as the solar system formed, there would have been a period when the Sun's luminosity was very low, before it reached a density critical for thermonuclear reactions that fully ignited it (Kasting 1987). Thus there have been great long-term variations in the amount and kind of solar radiation received on Earth. The ways in which it was received also were important to past climates. Forested ground is less reflective than ice caps and tends to absorb solar radiation. The growth of polar ice caps, on the other hand, promoted glacial conditions by reflecting sunlight back out into space. Solar heat also can be retained on Earth by highly absorptive gases, such as carbon dioxide, which trap heat in the same way as the glass of a greenhouse. The long "torrid ages" between Ice Ages appear to have been times of elevated atmospheric carbon dioxide, to judge from the presence of fossil wood with very wide growth rings (Creber & Chaloner 1985) and of highly weathered paleosols even at high paleolatitudes (Retallack 1986c). Thus paleoclimatic change was not simply a matter of continental orientations, but of a series of historical changes so complex that additional information from paleosols is welcome.

Indicators of rainfall

Many soil-forming chemical reactions occur in dilute solutions, within pore spaces intermittently supplied with water by rain. Even when soil pores are full, these reactions differ from those in open water, such as rivers and lakes. The wetting and drying of soil change the volume of small water films available for reactions, removing or concentrating reacted products. Thus, freely drained soils show more profound chemical weathering than waterlogged soils, and freely drained soils of humid climates are more altered than those of dry climates.

Numerous chemical and mineralogical features of soils can be related to mean annual rainfall, but only a few of the most characteristic soil features useful for interpreting paleosols are discussed here. Some features of modern soils, such as organic matter and clay content, show strong correlation with rainfall (Jenny 1941). These are not mentioned further here. Organic matter seldom is preserved at original levels in well drained paleosols (Stevenson 1969). Soils of more humid climates tend to be more clayey and more red and have fewer easily weathered minerals, but this also is true of soils in hotter climates and soils of greater age (Birkeland 1984). These competing factors may prove difficult to disentangle from the effect of rainfall on these aspects of paleosols.

Presence of calcium carbonate

Calcium is a common component of rocks, for example in the minerals feldspar, apatite, and amphibole. These minerals are among the most easily weathered, and as a result calcium is one of the most easily mobilized elements in soils. In soils of humid climates, these calcium-bearing minerals are hydrolyzed, and calcium disappears as cations in groundwater. In soils of dry regions, in contrast, there may not be sufficient moisture completely to remove calcium liberated by hydrolytic weathering. Evaporation of soil water at a wetting front can result in the precipitation of low-magnesium calcite in a subsurface (Bk) horizon. Dolomite can form in soils in the same way, although it is uncommon (Dixon & Weed 1977).

Soils have been divided into two main kinds: those with free carbonates [the pedocals of Marbut (1935)] and those lacking free carbonates (Marbut's pedalfers). Discounting mountain soils, most soils west of the prairie–forest transition zone near Lincoln, Nebraska, in the midwestern United States, are pedocals, and most soils east of that line are pedalfers. In the midcontinent, the dividing line is at about 500 mm mean annual rainfall in the cool climate (5–6°C mean annual temperature) near Red Lake Falls in northwestern Minnesota, and about 600 mm in the warmer climate (22°C mean annual temperature) near San Antonio in southern Texas. In the southwestern United States from the humid mountain slopes of the Sierra Nevada down into the intermontane deserts of Nevada, carbonate first appears in soils near Reno, Nevada, which has a mean annual precipitation of 180 mm and a mean annual temperature of 9.5°C (Birkeland 1984). This variation in the critical rainfall separating calcareous and non-calcareous soils can be attributed to the greater amount of water lost to evaporation in warmer climates. Rainfall is more effective for weathering when it falls mainly during winter in the Mediterranean climate of Nevada, compared to the summer wet climate of the mid-continental United States. Although calcareous paleosols are a reliable indicator of dry climates, and non-calcareous paleosols of wet climates, it is not possible to specify the dividing line accurately without information on other aspects of paleoclimate.

Depth to calcic horizon

Soils with free calcium carbonate usually have a distinct calcareous layer or calcic (Bk) horizon. The depth of this horizon below the surface reflects the depth of wetting of the soil by available water. In drier areas of the Great Plains of midcontinental United States, the calcic horizon is closer to the surface than in wetter eastern areas (Fig. 9.3). In this region the parent materials of the soils are fairly uniform loess, with little topographic relief. The vegetation is mostly grasses, and the age of the soils is largely postglacial (younger than 13 000 years). There are some variations in these other factors (Ruhe 1984), but this remains a reasonable climosequence. Another study of the depth of the calcic horizon in soils of the Mojave Desert (Arkley 1963) has shown a similar general relationship, but with slightly different values. The relationship between depth to the top of the calcic horizon (D cm) and mean annual rainfall (P cm) in the Great Plains of North America was found to be $D = 2.5(P-30.5)$. In the desert soils of California and Nevada, a slightly different relationship was found: $D = 1.63(P-1.14)$. These differences can be related to cooler temperatures during the winter rainy season in California, compared with warmer temperatures and high evaporation during the summer rains in the Great Plains. Other factors may include grain size and drainage of parent materials, influx of calcareous dust and amount of limestone in source regions (Marion *et al.* 1985). Similar relationships between carbonate leaching and rainfall have also been noted for tropical soils in Tanzania (de Wit 1978) and India (Sehgal *et al.* 1968).

Difficulties arise in applying this information to the interpretation of former rainfall from paleosols. For example, there may be erosion of a paleosol before its burial by later deposits. This can be assessed by features such as the nature of root traces and soil structure in the surface of the paleosol. Significant erosion may be suspected if a paleosol is overlain by

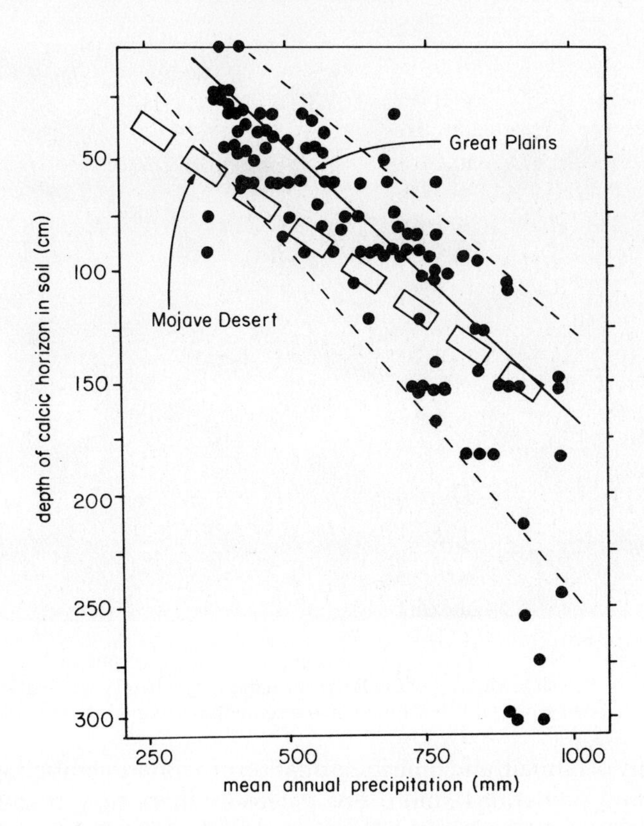

Figure 9.3 The depth of calcic horizon in soils formed under different regimes of mean annual rainfall in postglacial loess soils of the Great Plains of North America (after Jenny, 1941) and of the Mojave Desert, California (after Arkley, 1963).

cross-bedded sandstone of a paleochannel rather than by silts or clay that settled from ponded floodwaters. A second problem is compaction of paleosols after deep burial. This is more significant for clayey than for sandy paleosols (see Figs. 7.4 and 7.5). For these reasons, the depth to calcic horizons of paleosols and the inferred mean annual rainfall are usually minimum values. Such interpretations gain strength from examination of many paleosols rather than just one (Fig. 9.4). In sequences of paleosols they can be reliable indicators of relative change in rainfall (Fig. 9.5).

Clay mineralogy

The kinds of clays formed in soils represent the products of hydrolysis of weatherable minerals and can be related to the amount of rainfall available to soils. Grain size and mineralogy of parent materials, temperature,

Figure 9.4 Calcareous nodules (white) forming calcic (Bk) horizons in Late Oligocene paleosols of the Poleslide Member of the Brule Formation in Badlands National Park, South Dakota. The eroded pipe of the benchmark in the center foreground is 4 cm in diameter.

seasonality of rainfall, and time for formation of a soil are competing factors determining clay composition. For paleosols there may be additional problems determining which clays were products of soil formation, as opposed to inherited from a parent material or altered after burial. The bright clay fabric and finer grain size of soil clays compared with alluvial clays may be obscured by diagenetic alteration (Retallack 1986b).

With these cautionary issues in mind, clays of wetter climates generally have a 1 : 1 rather than 2 : 1 layer structure, are less rich in cations, and lower in the general weathering sequence of clay-sized minerals (see Fig. 4.5). In basaltic soils of tropical, continuously wet parts of Hawaii, for example, smectite is the dominant soil mineral below about 1000 mm annual precipitation, kaolinite between 1000 and 2000 mm, and iron oxide and alumina above 2000 mm (Fig. 9.6). Similar results have been reported for tropical, alternating wet and dry climates (Sherman 1952). Progressively more base-poor clays in wetter climates have also been observed in Californian soils under a warm temperate Mediterranean climate (mean annual temperature 10–15.6°C). Some of the differences observed can be related to coarser-grained parent materials of felsic composition, such as granite, compared with finer grained rocks of mafic composition, such as diabase (Barshad 1966). In both, however, smectite is a major component of soils that receive less than 500 mm mean annual rainfall, and kaolinite

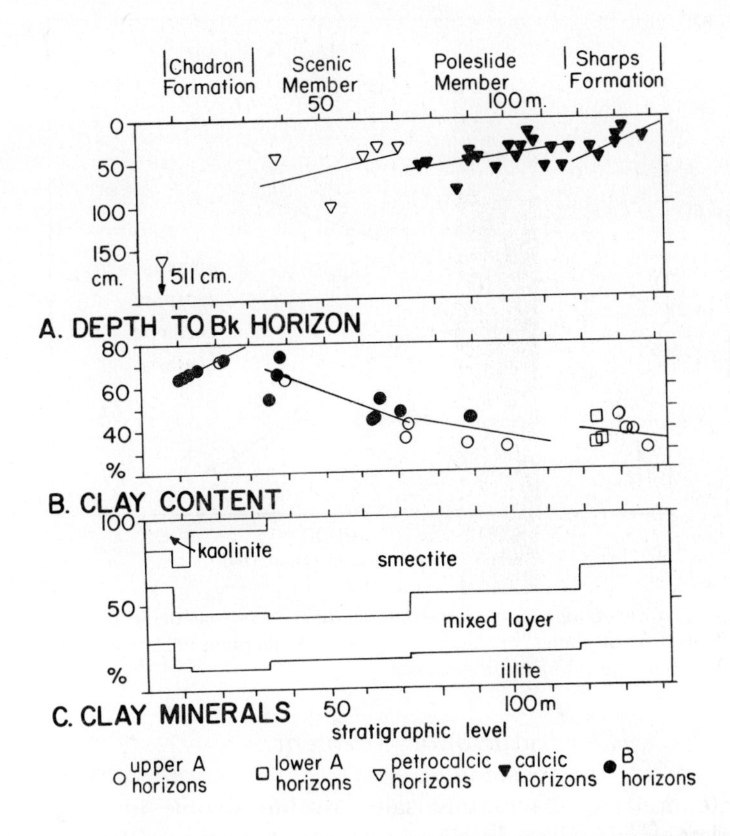

Figure 9.5 Climatic drying over geological time (approximated by stratigraphic level on the horizontal axis) is indicated by decreasing depth of calcareous nodules and layers, by declining amount of clay (determined by point counting) in upper A horizons, lower A horizons, and B horizons, and by more base-rich clay minerals, in a sequence of Late Eocene to Oligocene paleosols in Badlands National Park, South Dakota (from Retallack 1986d, reprinted with permission of Society of Economic Paleontologists and Mineralogists).

and halloysite dominate soils of wetter climates. Iron and aluminum oxides predominate in climates with more than 2000 mm mean annual rainfall. Similar changes in amount and kind of clay in ancient sequences of paleosols (Fig. 9.5) can be regarded as evidence of paleoclimatic change.

The fibrous clay minerals palygorskite and sepiolite are found in soils of very arid regions (Singer & Galan 1984). Palygorskite is common in the clay fraction of soils that receive less than 400 mm mean annual rainfall. These minerals are not very stable in paleosols although they have been recognized in examples as old as Permian (Watts 1976). Palygorskite may become hydrated to form sepiolite. Both minerals are readily leached by rain or groundwater.

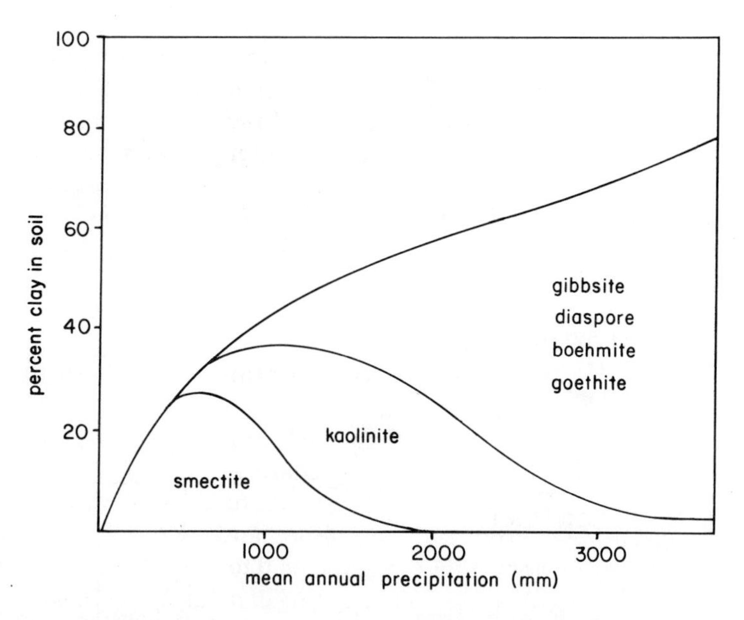

Figure 9.6 Clay mineralogy of soils formed under different regimes of mean annual rainfall in the equable, tropical climate of Hawaii (adapted from Sherman 1952, with permission of the American Institute of Mining Engineers).

Evaporite content

Evaporite minerals are mainly salts, readily soluble in water, that accumulate in soils of very dry climate. The most common of these minerals in soils is gypsum, usually found in the crystal habit of needles or laths (selenite). Other common evaporite minerals are cubic halite and fine needles of silvite or mirabilite. Zeolites such as analcime and clinoptilolite are also found in alkaline and salty soils. Evaporite minerals form where there is a source of salts, such as on ashy volcanic materials or in areas affected by sea spray. The most important constraint on their formation, however, is a climate in which evapotranspiration exceeds rainfall. How dry a climate is needed has been addressed in several studies (Birkeland 1984). Gypsum is abundant in soils receiving less than 175 mm mean annual rainfall in the hot semiarid climate of Tunisia, less than 165 mm in the cold semiarid desert rangelands of the Bighorn intermontane basin of northwestern Wyoming and less than 300 mm in the winter-wet, warm temperate mountains and plateaus of northern Iraq. In warm temperate, winter-wet Israel, gypsum is found in soils receiving up to 290 mm mean annual precipitation, but in such wet climates it is leached to depths of 2 m (Dan & Yaalon 1982). Like calcic horizons in soils, the gypsic horizon becomes shallower in drier climates. In Israel it is within 20 cm of the surface in soils receiving only 100 mm mean annual rainfall, and within 2 cm of the surface in those receiving less than 50 mm of rainfall. This relationship is more difficult to use in interpreting paleosols than the comparable relationship with calcic horizons. In addition to problems with surface erosion of paleosols and compaction after burial, salts are readily dissolved by groundwater. All that remains of many ancient evaporite layers are pseudomorphs of crystals (Groves *et al.* 1981) or zones of breccia where country rock has collapsed into the dissolved salt layer (Bowles & Braddock 1963).

Indicators of temperature

Although water is the medium for most soil-forming processes, the rate of these reactions also is controlled by temperature. According to Van t'Hoff's temperature rule, for every 10°C rise in temperature the rate of a chemical reaction increases by a factor of two to three. Since surface temperatures over which water remains liquid range from 0 to 84°C in soils (Kimmins 1987), there is the potential for several orders of magnitude differences in reaction rates between frigid and tropical soils. This tallies well with the observation that most tropical soils are much more deeply weathered than polar soils.

Despite these dramatic differences, in many cases it is difficult to disentangle the complementary effects of rainfall and duration of soil formation from those of temperature in promoting weathering. This particularly applies to the amount of clay, redness of hue, plinthite formation, depth of weathering, and depletion of weatherable minerals in soils. Each of these features of soils is more pronounced in soils of warmer climates (Birkeland 1984). Tropical regions, however, were less disrupted by glaciation of the past 2 million years and also were continuously unfrozen, unlike many soils of cool temperate climates. The competing effects of rainfall and time for formation compromise the paleoclimatic interpretation of these features of soils.

Soil microstructure

A distinctive feature of some tropical soils is the abundance of spherical micropeds in thin section (Stoops 1983). These are small (10 µm–1 mm) spherical aggregates of claystone (Fig. 9.7). Usually the clay is kaolinitic and strongly stained red and opaque with hematite or pink with gibbsite. There may also be small grains of quartz within the peds, but few other minerals are found. The micropeds are both physically tough and chemically inert. Although made of clay, they bestow a light, porous texture to the soil, so that it seems sandy to the touch. With compaction in

Figure 9.7 Spherical micropeds (dark) and associated burrows (light, sand-filled) in petrographic thin section, viewed under plain light, of the surface (A) horizon of the type Lal Clay paleosol, an Hapludalf from the Miocene (8.3 million years) Dhok Pathan Formation near Kaulial, Pakistan. Scale bar 1 mm (from Retallack 1985, reprinted with permission of the Royal Society of London).

a paleosol, the micropeds may not be obvious in hand specimens, but they can be revealed in thin sections. The micropeds are thought to originate in part as fecal and oral pellets of soil invertebrates.

Karst

Sculpted surfaces of limestone can form both as an erosional landscape (karst) and as a subterranean phenomenon of localized dissolution at the base of a soil or between geological formations (cryptokarst). Buried karst landscapes are covered by alluvium or colluvium, whereas cryptokarst is filled only with soil, residues of dissolution, or collapse breccias. Ancient examples of both kinds (paleokarsts) are widespread well back into the geological past (James & Choquette 1987).

The shape of karst topography is related to climate (Jennings 1985). Steep holes or dolines form along joint planes or by collapse of an underlying cave system. Dolines tend to be so abundant, closely spaced, deep, and strikingly fluted in tropical regions that the whole landscape consists of depressions (cockpit karst) or isolated spines (tower karst). The giant karst of Sichuan, in southeastern China, is a well known example of tower karst,

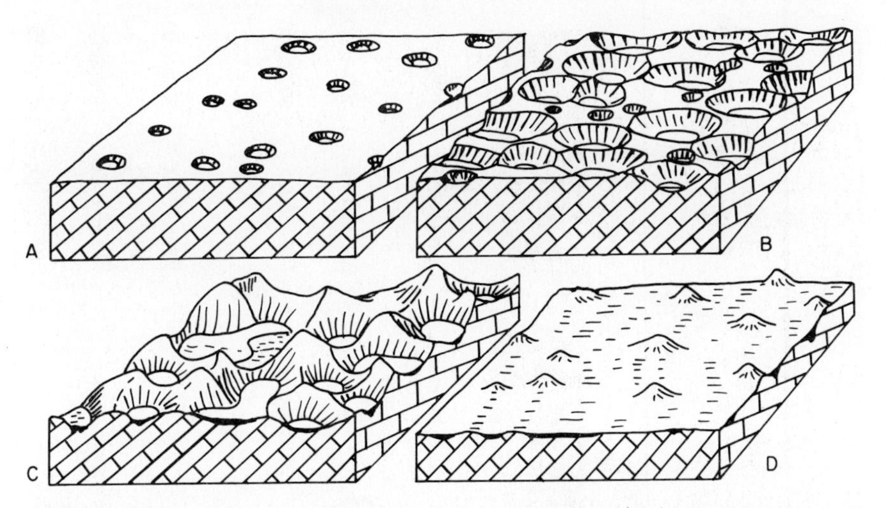

Figure 9.8 Different karst landforms: doline karst (A), uvala karst (B), cockpit karst (C), and corrosion plain (D), that form in increasingly warm and wet climates and over increasingly long intervals of time (from Jennings 1985, reprinted by permission of Blackwell Scientific Publishing).

celebrated in classical Chinese watercolors. Karst towers 10–300 m high, with diameter to altitude ratios of 1.5–8, are found in tropical karst. In cool regions, karst topography is more subdued, often preserving the original bed surface of the limestone as a plateau, and there are fewer narrow dolines per unit area (Fig. 9.8). Intermediate between these extremes are karsts with a big depression formed by overlapping dolines (uvala) or with a flat floor due to stream sedimentation (polje).

Strict climatic interpretation of karst topography may be compromised by variation in age of the surface, in its relation to subterranean cave systems, and in the geological structure of the limestone. However, the greatest impediment to interpreting paleoclimate from paleokarst has been the difficulty of finding outcrops large enough to reveal the whole topography of ancient karst. A spectacular exception is paleokarst from Illinois, USA, similar to a degraded cockpit karst (Leary 1981).

An unusually grotesque form of spongy and jagged karst, called black phytokarst, is attributed to the activity of endolithic algae. It appears to be restricted to humid tropical climates (Folk *et al.* 1973). A possible ancient example has been recognized from the upper novaculite of the Caballos Formation in western Texas (Folk & McBride 1976).

Several kinds of cryptokarst features also appear to be correlated with climate (Jennings 1985). Dense concentrations of tunnels running in all directions, or cavernous subsoil weathering, are mainly found in tropical humid regions. Fossilized examples of this form of cryptokarst have also been found (Wright 1981).

Table 9.2 Climatic thresholds and climate-diagnostic values of periglacial features (after Williams 1986)

Periglacial feature	Mean annual air temperature (°C)	Mean air temperature of coldest month (°C)	Mean air temperature of warmest month (°C)	Mean annual precipitation (mm)	Other climatic indication	Freezing index (°C.days/year)	Thawing index (°C.days/year)
Ice-wedge polygons	<- 4 to <- 8	<-25 to <-40	+10 to +20	> 50 to 500	rapid temperature drops in early winter with thin snow cover	2600->7000	100-1000
Sand-wedge polygons	<-12 to <-20	<-35	<+ 4	< 100	rapid temperature drops		
Seasonal frost crack polygons	< 0 to <- 4	<- 8			rapid temperature drops, moderately continental	1000->7000	1000-2000
Tundra hummocks (high latitude)	<-10					2300->7000	100-1500
Earth hummocks, thufur	<+ 3				gentle cooling	700- 3800	500-3000
Seasonal frost mounds	<- 1 to <- 3						
Microhummocks	<+ 1						
Mudpits, nonsorted circles	<- 2			> 400-800		1000->7000	200-1000
Palsas	< 0 to <- 3			thin snow cover	continental	1000->7000	300-2200
Closed-system pingos	<- 5					2700- 7000	< 100-1500

Table 9.2 *Continued*

Periglacial feature	Mean annual air temperature (°C)	Mean air temperature of coldest month (°C)	Mean air temperature of warmest month (°C)	Mean annual precipitation (mm)	Other climatic indication	Freezing index (°C.days/year)	Thawing index (°C.days/year)
Open-system pingos	<- 1					2300->7000	250-1700
String bogs	< 0 to - 1				continental	700- 5000	1000-3000
Rock glaciers	<+ 2 to 0			<1200	continental	1000- 5000	300-1000
Features resulting from thawing of ice-rich permafrost	<- 1					800->7000	< 100-3700
Sorted circles and stripes (>1 m)	<- 4					1500->7000	200-1500
Sorted circles and stripes (<1 m)	<+ 3						
Miniature sorted forms	<+ 1						
Microforms of gelisolifluction	<- 2						

Permafrost

Soils of frigid climates (mean annual temperature less than 0°C) show a variety of structures created by the formation of ice in the soil (Washburn 1980), and many of these may be preserved in paleosols. The most characteristic of these are sand-wedge casts. These strongly tapering vertical fissures are periodically filled with sand as ice melts (see Fig. 4.4). Other permafrost features include brecciated zones, where tabular bodies of ice have melted, allowing frozen soil to fall into its place. There are a variety of striped and polygonal arrangements of soil material produced by freezing and thawing. Many of these features in paleosols can be used to infer particular climatic limits (Table 9.2).

Indicators of seasonality

The seasonal availability of water and temperature may have an appreciable effect on the development of soil features. Seasonality of rainfall, of heat, of dust influx, and of other agencies of soil formation are important to the clayeyness, redness, and base saturation of soils. Each of these features, however, is controlled by other factors that make it difficult to tease out the contribution of seasonality. The nature and degree of seasonality are better revealed by other features of soils, only the most diagnostic of which are reviewed here.

Mukkara structure

Gilgai microrelief and mukkara subsurface structure of swelling clay soils (see Fig. 4.3) are indicative of a climate with pronounced dry and wet seasons (Paton 1974). They are not found in extremely arid or humid climates. The mean annual rainfall of gilgai soils in Australia ranges from 180–1520 mm. Some moisture is needed to promote clay formation and provide a contrasting wet season. In drier climates these soils form on parent materials already clayey and smectitic. On the other hand, clays become more deeply weathered and lose their swelling properties in humid climates. This is due in part to the hydrolytic loss of cations, but is also related to the conditioning and stabilization of soil with organic matter and roots of forest vegetation in wetter climates. These paleoclimatic generalizations may not have held for mukkara structure of the distant geological past (Fig. 9.9), before the advent of substantial land vegetation or before the development of the present oxygen-rich atmosphere during mid-Precambrian times (some 2000 million years ago).

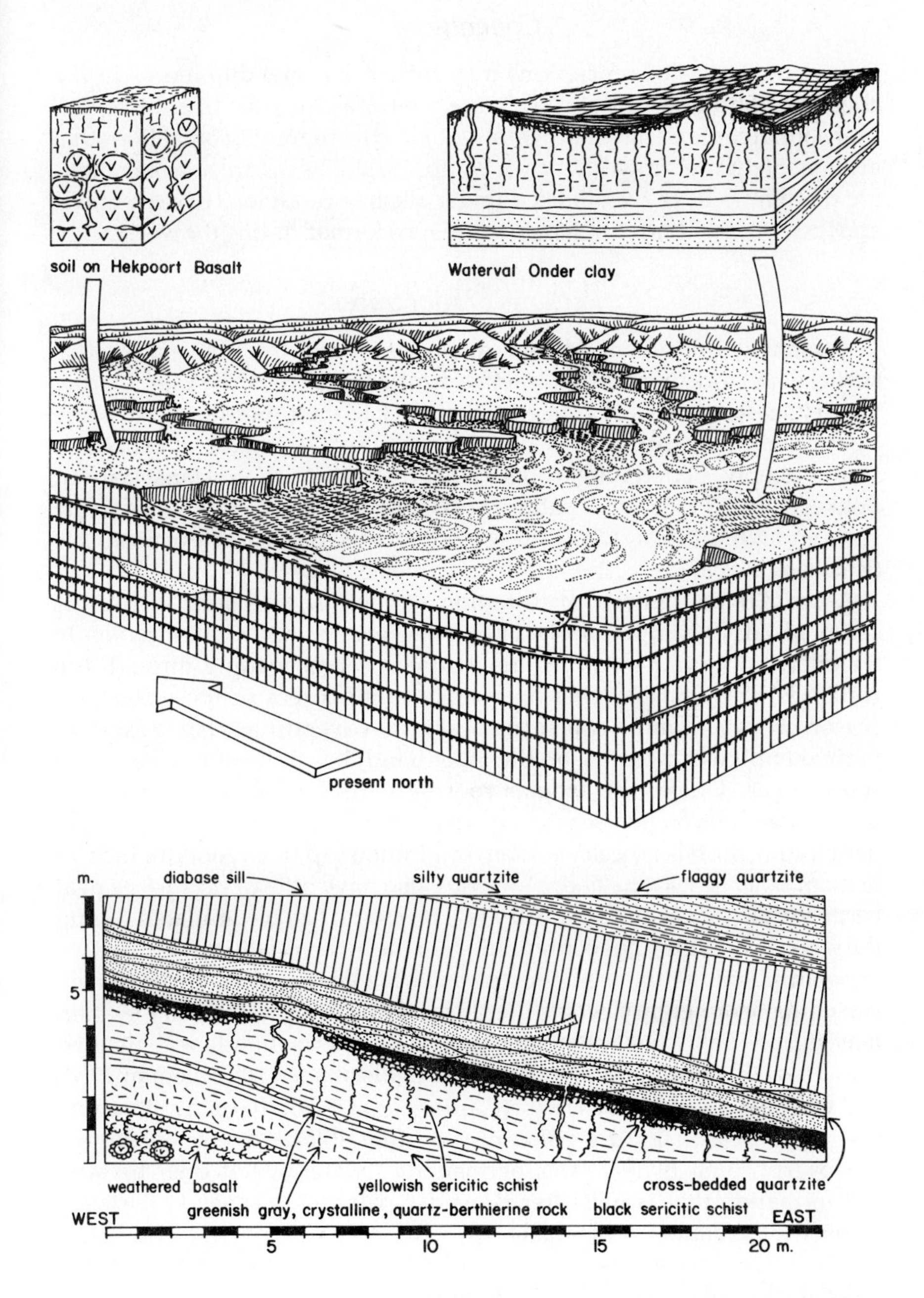

Figure 9.9 Reconstructed paleosol of early Proterozoic age (2200 million years old) in the lowest Dwaal Heuvel Formation near Waterval Onder, Natal, South Africa, with an annotated field sketch (below) showing the interpretation of its mukkara structure (from Retallack 1986b, with permission of Elsevier Publishing Co.).

Concretions

Concentric layers of concretions may reflect seasonal differences in the chemical condition of the soil. Some well drained soils of subtropical, monsoonal climates have both calcareous and ferruginous concretions that are intimately intergrown (Sehgal & Stoops 1972). The calcareous concretions are thought to have accumulated under alkaline conditions of dry periods and then corroded as ferruginous concretions formed during the wet season.

Patterns of root traces

In seasonally dry wooded grasslands, grasses and trees have a copious surficial network of roots that is active during wet parts of the year. During the dry season, leaves of grasses wither back to their root stock, but some trees obtain moisture from deep within the soil by means of especially stout, deeply penetrating roots called "sinkers" (Van Donselaar-ten Bokkel Huinink 1966). This pattern of a dense, near-surface network of fine roots with a few deeply penetrating stout roots is recognizable in paleosols (Retallack 1983a).

A comparable bimodal distribution of roots is also seen in soils of seasonally dry swamps (see Fig. 11.5). These are distinct from permanently waterlogged swamps even to the casual observer wading through the wetlands of Louisiana. One only sinks to the knees in the water of a seasonally dry swamp because the peat is thin over firmer clay or sand. In permanently wet swamps, on the other hand, it is difficult to walk in the soupy peat. Unlike the tabular root systems typical of permanently waterlogged swamps, seasonally dry ones may have some deeply penetrating roots for the dry season, in addition to tabular roots for the wet season. Soils of seasonally dry swamps also have little or no surface peat because this decays during the dry season after it has accumulated during the wet season. A serious complication for the interpretation of seasonality in swamp soils is the tendency of such lowland regions to be slowly subsiding sedimentary basins. Because of this, a seasonally dry swamp may subside below the water table to become a permanently waterlogged soil. This is a common phenomenon in Louisiana today (Coleman 1988) and was also common in the past (Gardner *et al.* 1988). In such a situation, the earlier seasonally dry paleosol can often still be discerned despite the overprinting by the later permanently waterlogged one, because soil formation that would alter the older paleosol is severely curtailed below the permanent water table.

Charcoal

Fossil charcoal is a common component of paleosols and can be evidence of fire in woody vegetation. Occasional fires rage in a wide range of climates, exacerbated by human carelessness. Occasional fires, however, do not produce charcoal with such regularity that it accumulates in soils faster than it decays. In nature, fire is most common in soils and vegetation of seasonally dry climates. The scrubby chaparral vegetation of California hillsides and the comparable maquis of southwestern Europe and mallee scrub of southeastern Australia are renewed every few decades by burning during the almost rainless summers they endure (di Castri *et al.* 1981). Even swamps, such as the Florida Everglades, burn extensively after short electrical storms during their dry season. Many seasonally dry grasslands also burn during the dry season, but fires through grass have neither the intensity nor the abundance of woody material that is needed to make charcoal. Not all fires are recorded in paleosols, but if fossil charcoal is common, then fire and dry seasons probably were frequent.

Frost cracking

Thick masses of ice, such as ice wedges or permafrost layers, are indications of climatic seasonality, in addition to cold. They form within soils from water supplied during the summer, and their growth during the early winter is maintained from year to year by incomplete melting during the summer (Washburn 1980). Without such seasonal freezing and thawing a glacier would form rather than a periglacial soil.

Frost cracking and patterned ground can also be taken as indicators of a moderate summer followed by contraction of the soil in subfreezing temperatures of a very cold winter. These structures are defined by oriented soil material often forming a polygonal pattern. Such polygonal ground is most pronounced in stony soils. It can be difficult to distinguish in clayey or silty paleosols unless these were distinctively layered (Retallack 1983c).

Frost-cracked or charcoal-rich paleosols represent very general and extreme climatic regimes. In most cases, only broad paleoclimatic limits can be untangled from other past influences evident in paleosols. Nevertheless, such broad constraints can be useful and even decisive additions to isotopic, paleobotanical, and other lines of evidence for past climates.

10 Organisms

Climate and vegetation are so closely linked in the modern world that these two controls on the degree and kind of soil formation may be difficult to tease apart. Indeed, some geographers refer to "forest climates" or "grassland climates." Consideration of vegetation as a factor in soil formation has been criticized as not a clearly independent variable from others, such as climate and topographic position. The concept of a "grassland climate" may create a more vivid impression than tables of climatic data, but is not an accurate characterization of climate. Open grasslands are found in a variety of climatic settings and show some climatic overlap with other plant formations. It is in these marginal situations that vegetation can be considered an independent variable, and its unique contribution to soil formation can be appreciated.

The role of vegetation in soil formation also can be understood from the example of artificial growth chambers, such as the climatron of the Missouri Botanical Garden in St. Louis, USA. This enormous growth chamber was built with heating and plumbing (a microclimate), soil material (parent material), and landscaping (topography). The planting of seeds and seedlings markedly affected the quality of its soils, independent of these other factors. Its various kinds of plantations have become lush, bulky, and resistant to perturbation. This experiment in recreating natural habitats has counterparts in nature in the form of plant succession on surfaces disturbed by fire, flood, landslides, volcanic eruptions, or human activities (Kimmins 1987). There are close parallels between the successional colonization of a surface by organisms and soil development, as was seen during recolonization of the volcanically devastated Indonesian island of Krakatau after 1883 (Fig. 10.1). There was a buildup of biomass, accumulation of reserves of organic matter, maintenance of stability, and mitigation of diurnal and seasonal fluctuations in water availability and in temperature. These changes went hand in hand with weathering of the coarse-grained ash to clay and its stabilization with organic matter and sesquioxides. On geological time scales, the colonization of land surfaces by vascular land plant communities during Paleozoic time provides another example of vegetation as an independent variable in soil formation.

These successional and evolutionary changes in vegetation and soils can be contrasted with plant zonation and migration. Zones of different kinds of vegetation develop along environmental gradients, such as the successively better drained and more windblown habitats on the remnant peaks of Krakatau (Fig. 10.1). In these cases the effects of vegetation on soil formation are mixed with those of topography and climate. Zonation of plants over larger regions controlled by regional climatic differences, or the migration of plant communities as climate changes, are also situations in which it is difficult to disentangle the effects of vegetation as an independent variable. In assessing the effects of vegetation alone, the critical question is what the soil would have been like under different plants.

Large plants are only the most obvious part of the organisms that contribute to soil formation. Many other creatures also play a role. Fungi, for example, in intimate association with the roots of many vascular plants, make available nutrients such as nitrogen and phosphorus in forms that can be utilized by their host plants. Fungi also play a role in decay of leaf litter, returning organic matter to the soil. Earthworms aid in the general process of mixing soil minerals with organic matter, thus improving its fertility and stability for other organisms. Many creatures condition the soil in a general way, but some kinds of burrows and fecal pellets are distinctive of particular organisms. Well preserved earthworm burrows, for example, can be recognized even in very ancient paleosols (Retallack 1976).

Not only particular organisms but also different kinds of ecosystems can be interpreted from features of paleosols. Different plant communities create distinctive patterns of root traces, soil structures and overall profile form. These features, reflecting different kinds of

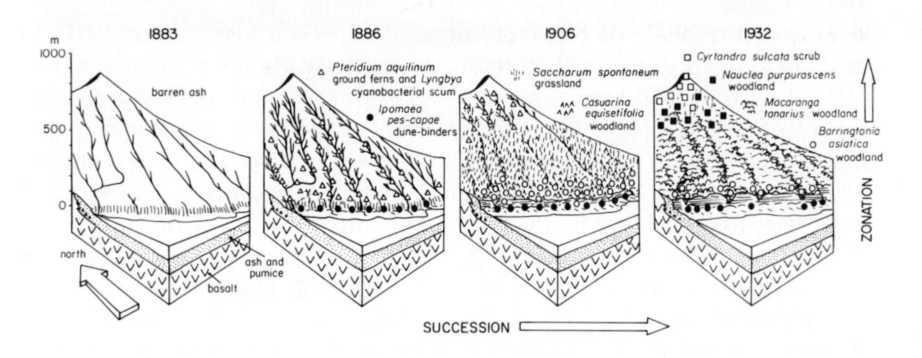

Figure 10.1 Plant succession and zonation following the 1883 devastation of the Indonesian volcano Krakatau (data from Simkin & Fiske 1983).

ecosystems, provide a useful overall assessment of the influence of organisms on a particular paleosol. Quantification of such general effects of organisms is difficult because their degree of expression is as much related to the ecosystem responsible as to the time available for soil formation. Emphasis is therefore laid here on the nature rather than degree of soil formation that can be attributed to general ecosystem types. Within such wide interpretive limits, fossil soils can be viewed as grounds for interpreting the evolution of terrestrial ecosystems, independent of the fossil record of plants and animals.

Much is already known about land organisms of the geological past from the study of spores and pollen (Traverse 1988), of fossil plant remains (Stewart 1983), and of bones and teeth (Carroll 1988). These various kinds of fossils are found in deposits of lakes and streams but also are preserved in fossil soils. Only a small fraction of the potential kinds of fossils that once lived in a paleosol are actually preserved in it (Retallack 1984a). Fossil soils are evidence of preservational environments in addition to living environments, and consideration of their fossils reveals the incompleteness of the fossil record. Both paleosols and fossils may provide supportive evidence for a particular kind of biota. If these two lines of evidence are not in concert, however, it may not be because one is in error. It could instead reflect preservational biases in the kinds of information recorded in fossils and paleosols. Alternatively, it could be that particular kinds of soils in the past supported biota that seem anomalous by modern standards. Even when the fossils found in paleosols are well understood, more can be learned of their preservation, ecology, and evolution from the paleosols on which they lived.

Traces of organisms

A footprint in shale or a burrow in sandstone is a fossilized record of past life, but one completely different from a body fossil, such as a bone or leaf. Footprints and burrows are not part of an organism, but rather a record of an organism's life activities. Whereas a bone or a leaf may be distinctive enough to be identified as a particular species, it is seldom possible to identify the makers of a footprint or burrow with such precision. On the other hand, trace fossils are found in the place where the animal lived. They are not transported, mixed, or selectively preserved in a burial environment, like many body fossils.

Trace fossils have been given Latin binomial names, but this is more a matter of convenience than of taxonomic consistency. A system for naming the great array of trace fossils now known was needed to communicate between interested scientists. Informal naming systems of letters or numbers tended to be ignored without the rules of priority and other archival procedures demanded by the International Code of Zoological Nomenclature (Ride *et al.* 1985). It is recognized, of course, that trace fossils are not actual fossils of organisms, and that in most cases their makers may never be discovered. Some admission of this is implicit in the proposal that generic names of trace fossils uniformly be distinguished by the suffix "-ichnus," from the Greek word for track (Basan 1979).

Classifications of trace fossils have been attempted, based on preservational types or behavioral groups, rather than the evolutionary relationships inferred by classification of other kinds of fossils. Many trace fossils are preserved best in, and on, thin beds of sandstone interbedded with shale. Sandstone is a durable casting medium. When weathered along the contact with shale, it reveals trace fossils more clearly than those entombed only by sandstone or by shale. Preservational types of trace fossils can be classified according to whether they are on the top, bottom, within, or beyond the casting sandy bed (Häntzschel 1975). Such distinct beds with trace fossils along zones of lithological contrast are known in paleosols (Retallack 1976), but generally are rare in them and so of limited value in characterizing terrestrial trace fossils. Another way of classifying trace fossils is on the kind of activity they represent, such as feeding burrows, dwelling burrows and crawling marks. This system is of limited value for trace fossils in paleosols, in part because so much more is known about the behavior of terrestrial animals. The complex burrow systems of bees, for example, are used for egg laying, feeding larvae, and dwelling. More appropriate for trace fossils in paleosols are common English terms used in studies of behavioral entomology, such as burrow, hive, cell, gallery, and chimney. The naming of trace fossils in paleosols will continue to rely on procedures established in the study of marine trace fossils, often using the same ichnogenera. Detailed interpretation, however, is better based on the enormous amounts of information gathered by soil biologists, as summarized in the ensuing paragraphs.

Microbes

This commonly used abbreviation for "microbiological organism" is a suitably vague term for a variety of organisms that seldom leave distinctive traces in soils but nevertheless have profound effects on soil formation. Microbes include a host of unicellular creatures, such as bacteria, algae, and protoctistans. Their effects on soils are best considered under general roles and nutritional groups (Table 10.1) rather than along taxonomic lines.

Basic requirements of microbes include a source of energy, which can

Table 10.1 Kinds of microbes, their metabolic requirements and role in soils (from Pelczar *et al.* 1986). Initials for trophic groups are autotrophic *(A)*, chemotrophic *(C)*, heterotrophic *(H)*, lithotrophic *(L)*, organotrophic *(O)*, phototrophic *(P)*.

General role	Kind of organism	Example genus	Energy source	Electron donor	Carbon source	Oxygen relations	Comments
Carbon cyclers	Algae	*Chlamydomonas*	Sunlight *(P)*	H_2O *(L)*	CO_2 *(A)*	Aerobic	Primary producer in normal to wet soils
	Cyanobacteria	*Nostoc*	Sunlight *(P)*	H_2O *(L)*	CO_2 *(A)*	Aerobic	Primary producer in normal to wet soils
	Purple non-sulfur bacteria	*Rhodospirillum*	Sunlight *(P)* sometimes organic compounds *(C)*	Organic compounds *(O)*, sometimes H_2S *(L)*	Organic compounds *(H)*, sometimes CO_2 *(A)*	Amphiaerobic	Primary producer in swamps
	Methanogenic bacteria	*Methanobacterium*	H_2 *(C)*	H_2 *(L)*	CO_2 *(A)*, sometimes formate *(H)*	Anaerobic	Creators of "swamp gas" (CH_4)
	Aerobic spore-forming bacteria	*Bacillus*	Organic compounds *(C)*	Organic compounds *(O)*	Organic compounds *(H)*	Aerobic	Decomposer of organic compounds in normal to wet soils
	Fermenting bacteria	*Clostridium*	Organic compounds *(C)*	Organic compounds *(O)*	Organic compounds *(H)*	Anaerobic	Decomposer of organic compounds in swamp soils
	Protoctistans	*Amoeba*	Organic compounds *(C)*	Organic compounds *(C)*	Organic compounds *(H)*	Aerobic	Predator in normal and wet soils

Table 10.1 *Continued*

General role	Kind of organism	Example genus	Energy source	Electron donor	Carbon source	Oxygen relations	Comments
Nitrogen cyclers	Nitrogen-fixing bacteria	*Azotobacter*	Organic compounds *(C)*	N_2 *(L)*	Organic compounds *(H)*	Aerobic	Makes N available as ammonium (NH_4^+) in normal and wet soils
	Root nodule bacterioids	*Rhizobium*	Organic compounds *(C)*	N_2 *(L)*	Organic compounds *(H)*	Aerobic	Makes N available as ammonium (NH_4^+) to host plant
	Denitrifying bacteria	*Pseudomonas*	NO_2 *(C)*, sometimes organic compounds *(C)*	N_2 *(L)*, sometimes organic compounds *(O)*	CO_2 *(A)*, sometimes organic compounds *(H)*	Anaerobic	Releases nitrogen gas (N_2) to atmosphere from poorly oxygenated soils
Sulfur cyclers	Sulfate-reducing bacteria	*Desulfovibrio*	Organic compounds *(C)*	SO_4^{2-} *(L)*	Organic compounds *(H)*	Anaerobic	Creates "rotten egg gas" (H_2S) and encourages formation of pyrite in swamp soils
	Sulfur bacteria	*Chromatium*	Sunlight *(P)*	H_2S, or S *(L)*, sometimes organic compounds *(O)*	CO_2 *(A)*, sometimes organic compounds *(H)*	Amphiaerobic	remobilizes sulfur in poorly oxygenated soils
	Sulfur metabolizing bacteria	*Thiobacillus*	S, FeS *(C)*	S, SO_4^{2-}, Fe_2O_3 *(L)*	CO_2 *(A)*	Aerobic	Creates rust and rock varnish and remobilizes sulfur in wet and normal soils

be either from the sun (phototrophs) or chemical compounds (chemotrophs), and a source of carbon, which can be from carbon dioxide (autotrophs) or other organic compounds (heterotrophs). These metabolic reactions can be driven by the reducing power of electrons from organic matter (organotrophs) or from minerals in the environment (lithotrophs). Because reduction and oxidation reactions are central to many metabolic processes, organisms also can be classified according to their relations with oxygen. Those which must use oxygen as the terminal electron acceptor for energy conversion (aerobes) can be distinguished from those which cannot use oxygen in this way (anaerobes) and those which can use either aerobic or anaerobic metabolic pathways (amphiaerobic). From this it can be appreciated that microbes influence the pH, Eh and nutrient availability of soils as a consequence of their livelihood (Baas-Becking *et al.* 1960). This would not be especially noteworthy if microbes were merely responding to predetermined physicochemical conditions, but this is not entirely the case. Chemical and physical equilibrium in soils is constantly thwarted by the metabolic activities of microbes. Depletion of oxygen in stagnant water within soils, for example, occurs more rapidly than would be expected by simple diffusion under a given temperature and pressure because of the oxygen demand of aerobic microbes. Once oxygen is depleted, these microbes die or migrate elsewhere and then anaerobic microbes will determine further changes in the chemistry of the soil water. To a certain extent, then, the nature of weathering is determined by microbes.

Another general effect of microbes is an increase in soil stability. Their binding effect is not as obvious as that of the roots of large plants but is nevertheless measurable. Experimental washing of barren soil slopes with water consistently produces greater erosion than hosing down comparable slopes armored with microbial scums (Booth 1941). Individual peds also are more stable in soils inoculated with microbes than those left alone (Metting 1987). Microbes can resist soil erosion in several ways (Griffiths 1965). Some microbes promote the hydrolytic comminution of sand and silt grains of the soil to colloidal materials, such as clay, calcium carbonate, and iron oxyhydrates, and these can act as cements. Mucopolysaccharides are produced by many microbes to aid motility, and these substances also act as a glue to bind soil particles together (Foster 1981). It is these materials that impart much of the distinctive fine structure of productive soils, or tilth.

Nodular and concretionary structures in soils also are produced by microbes. Laminated surficial limestone crusts and ministromatolites are formed by cyanobacteria in the same way as marine and intertidal stromatolites (Krumbein & Giele 1979, Reams 1989). Photosynthetic microbes in these crusts promote the precipitation of carbonate by using carbon dioxide dissolved in water, thus neutralizing its acidity. They also become encrusted in lime when the soil dries out after rain. Subsurface filamentous structures in caliche nodules also are formed around microbes (Klappa 1979a).

Another product of microbial metabolism is the distinctive sand-sized aggregates of tiny balls of pyrite known as framboids (see Fig. 4.6). These are produced in waterlogged soils, by reduction of ferric iron, sulphate, and sulfur, coupled with the breakdown of organic matter, by microbes such as *Desulfovibrio* (Altschuler *et al.* 1983).

A final distinctive chemical segregation widely attributed to microbes is rock varnish. These are thin (usually 20 μm but can be 2–500 μm), dark coatings of amorphous iron and manganese oxides (Dorn & Oberlander 1982). In polished sections, the thin coats of varnish commonly are lumpy (Fig. 10.2B), in part reflecting the shape and position of nearby microbes (Perry & Adams 1978). They also may have fine internal layering like miniature stromatolites (Fig. 10.2A). Varnish characteristically is developed on the tops but not the bottoms of pebbles and soil peds, a distribution also seen in very ancient examples (Retallack 1986b). Opinions differ on the way in which the iron and manganese are introduced and the exact nature of the microbes involved (Krumbein & Jens 1981, Staley *et al.* 1982, Margulis *et al.* 1983). There is, however, general agreement that these distinctive structures are biogenic in origin.

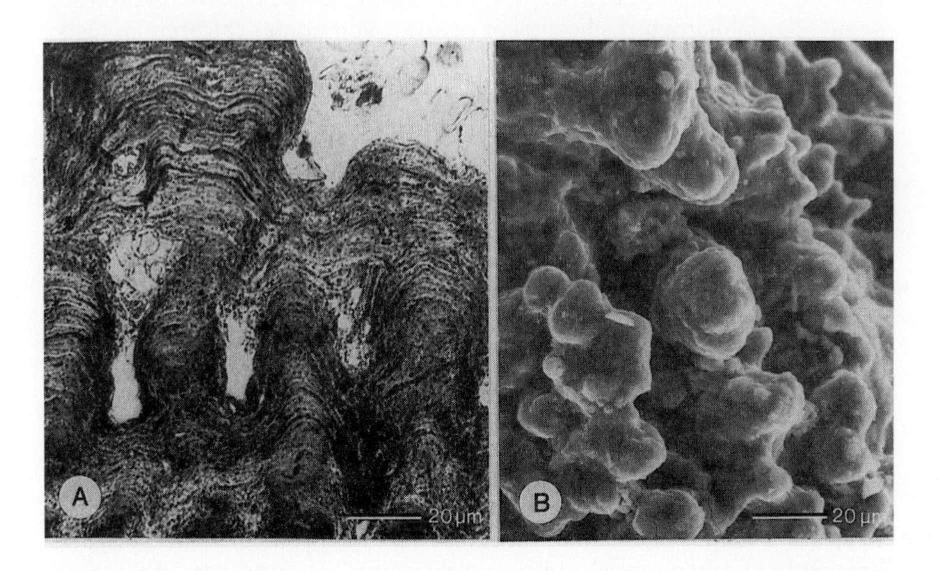

Figure 10.2 Micrographs of a petrographic thin section (A) and from a scanning electron microscope (B) of rock varnish from near Phoenix, Arizona. Scale bar 20 μm (photographs courtesy of John B. Adams).

The activity of microbes also can be inferred from the condition of fossils of larger organisms in paleosols. Large plants and animals are so dependent on microbes that one can reasonably infer that they must have been present wherever there are larger fossils. Trees, for example, depend upon a variety of bacteria within the mucigel surrounding their root hairs in order to obtain essential micronutrient elements. Ruminant animals have microbial communities within their guts that permit the breakdown of chemically inert portions of plant food. From the presence of large fossil root traces and the cake-shaped coprolites (fossil feces) of ruminants, one could infer that these various kinds of microbes were present. The preservational style of fossils also provides clues to the activity of microbes. Fossil leaves found in paleosols commonly show a range in decomposition comparable to that found in the leaf litter of modern soils (Retallack 1976). Some leaves are coriaceous, dry, and curled, whereas others are wilted and threadbare. The veins of lignin in a leaf commonly are the last to resist microbial decay, and these may be fossilized as a kind of leaf skeleton. The lack of fossil flesh attached to bones, except under unusual circumstances of salinity, freezing, or anoxia, is also a testimony to the effectiveness of microbial decomposition in soils.

Perhaps the most direct possible evidence of microbes in soils is the fossil hardparts of those microbes which have them. These are rare in paleosols and difficult to interpret. Soil-dwelling thecamoebans produce small (20 μm diameter) siliceous or organic, sometimes calcareous, tests, often bedecked with sand grains and other debris (Kühnelt 1976). These creatures also live in lakes and the sea, so it is difficult to discern whether their tests, like those of diatoms and foraminifera, might not have blown in from elsewhere, or have been part of the aquatic parent material of a paleosol.

Fungi

Many kinds of fungi can be regarded as microbes, but they are singled out here because of their distinctive effect on soils. Fungi have long tubular cells (hyphae) that are larger (2–10 μm in diameter and reaching lengths of several centimeters) than bacteria (0.5 to 2 um diameter and 10 μm long) but smaller than unicellular algae (10–40 μm) and protoctists (30–50 μm). Fungal hyphae form a loose network (mycelium) that thoroughly penetrates soil and leaf litter, in addition to forming solid fungal bodies recognized as mushrooms, toadstools, and brackets. Hyphae, mushrooms, and spores of fungi are preserved especially well in coal balls and silicified peats and have a long fossil record (Stubblefield & Taylor 1988).

Fossil root traces of vascular plants may have associated evidence of fungi. Needle-fiber cement associated with root traces in calcareous paleosols (Fig. 10.3) may have been precipitated by fungi during fungal decay of the root, or fungal activity associated with the rhizosphere (Wright 1986b). Many fungi, particularly those of the groups including yeast (ascomycetes) and mushrooms (basidiomycetes), enter into a mutualistic relationship with plant roots known as a mycorrhiza. These can form an external sheath (ectomycorrhiza) to roots with some hyphae penetrating the outer cells of the host plant. A more intimate relationship is found where some of the living cells of the host are invaded by hyphae (endomycorrhiza). Fungi benefit from such mutualistic relationships by gaining carbohydrates and other foodstuffs from the host plant. The plant benefits because fungi are capable of mobilizing nutrients that are important but difficult to obtain, such as phosphorus, nitrogen and potassium. Fossil endomycorrhizae have been found in silicified plants (Stubblefield & Taylor 1988). Mycorrhizal roots are stouter and more abruptly ending, and branch at shorter intervals and at higher angles than ordinary roots (Richards 1987). These megascopic features can also be observed in fossil roots (Pitt *et al.* 1961).

Some resting stages of fungi also are distinctive. Small woody balls with vermiform hollows common in peats and coals (see Fig. 4.6) are similar to

Figure 10.3 Needle fiber calcite in a tubular void remaining from a root trace, forming a septum separating septal filling calcite from an area of random needle fiber calcite, from the early Carboniferous, Heatherslade Geosol, on the Gully Oolite at Three Cliffs Bay, South Wales. Scale bar 0.1 mm (photograph courtesy of V. P. Wright).

sclerotia, the resting stages of some fungi. Enigmatic small balls of radiating calcite crystals found in modern and ancient caliche also may be casts of fruiting or resting stages of soil fungi. These have a long fossil record under the generic name *Microcodium* (Klappa 1978).

Effects of fungi can also be recognized in fossils associated with paleosols. Fine networks of creases on fossil leaves may represent hyphae, and masses of leaves closely adpressed together in fossil leaf litters are similar to those bound by fungi on modern soils (Retallack 1976). Some forms of fungal decay of wood, such as dry rot or pocket rot, are also distinctive, and are known from studies of permineralized wood to be of great antiquity (Stubblefield & Taylor 1988).

Effects of fungi on mineral weathering are also useful for their detection in paleosols. Endolithic fungi generate acids that etch deep impressions of hyphae into the surface of grains of carbonate and volcanic glass (Ross & Fisher 1986). Fungi are part of the microbial communities on rock varnish (Staley *et al.* 1982), but it is not yet clear whether they play an active role in its formation or are merely bystanders. Fungi loosen soil grains and encourage the hydrolytic formation of thin films of smectitic clay around them. There also are grounds for suspecting that fungi induce a qualitatively distinctive kind of weathering compared with that of other soil organisms. Studies of nutrient cycling in Douglas fir (*Pseudotsuga menziesii*) forest (Fogel & Hunt 1983) have demonstrated that the order of abundance of uptake of major cations by soil fungi ($N > Ca > Mg > P > K$) is distinct from that of mycorrhizae ($N > K > Ca = P > Mg$) or the trees as a whole ($N > Ca > K > P > Mg$). Other studies have shown that fungi preferentially adsorb rare heavy metals (Cooke 1979).

Lichens

These hardy organisms are plant-like in their sessile, somewhat leafy bodies, but are symbiotic associations of fungi and unicellar photosynthetic microbes. In most lichens the photosynthetic organism is a green alga, such as *Trebouxia* or *Pseudotrebouxia*, or a cyanobacterium, such as *Nostoc*. The fungal component in most cases is an ascomycete.

Lichens grow on bare rock or tree trunks and have been regarded as pioneers in the early successional colonization of new land surfaces. They are, however, very slow growing. Studies of long-term lichen growth on boulders abandoned by retreating glaciers have shown that lichen colonies grow in diameter only 0.1–0.03 mm per year (Calkin & Ellis 1984). Hence lichens differ from early successional plants, which grow and reproduce quickly. Lichens are better regarded as organisms able to withstand conditions hostile to most other forms of life (Grime 1979).

Traces of lichens may be preserved in calcareous soils and paleosols (Klappa 1979b). Individual thalli of lichens are patchy in distribution, and the acids generated by lichens form solution pits in limestone bedrock or around soil nodules exposed at the surface. This miniature karst topography may show outlines of the thallus. Lichens also may be encrusted to form a banded structure called a lichen stromatolite. Their laminae are irregular and less laterally extensive than those of stromatolites formed by mats of cyanobacteria. The banding of lichen stromatolites may be interrupted by small channels created by rhizines (rootlike bundles of hyphae). They also enclose calcareous spherulites that may be casts of fungal fruiting bodies.

Weathering effects of lichens on non-calcareous parent materials are difficult to distinguish from those of mosses, liverworts, algae, and cyanobacteria. Endolithic lichens produce distinctive leached zones under a surficial silicified crust that is prone to flaking (Friedmann & Weed 1987), thus encouraging the growth of larger surface lichens. The weathering rinds found underneath lichens on an 80-year-old lava flow in Hawaii were found to have a mean thickness of 0.142 mm, whereas the weathering rind on bare lava was only 0.002 mm (Jackson & Keller 1970). The chemical style of weathering under lichens on this tropical humid island corresponds to ferrallitization, with a strong concentration of iron, some concentration of aluminum and a loss of silica, titanium and calcium. Similar weathering of basalt under lichens has been found in Scotland (Jones *et al.* 1980). On gneiss and granites in Spain the lichens produced rinds of kaolinite, halloysite, amorphous silica, and less common goethite (Ascaso *et al.* 1976). Studies of the chemical composition of lichens (Syers & Iskander 1973) have shown that they are able to concentrate macronutrients such as P, S, Mg, Ca, K, and Fe, and also have high concentrations of Zn, Cd, Pb, and Sn.

Liverworts and mosses

These nonvascular plants are both bryophytes, but are distinct from each other (Fig. 10.4). Mosses have slender axes with numerous small, helically arranged leaves. Liverworts, on the other hand, are flat thallose plants. Both mosses and liverworts have a rudimentary system of cells with thickened walls (hydroids) for water transport, unicellular hairs (rhizoids) that function as roots, and thin patches of cells that allow gas exchange. They lack the water-conducting tracheids, nutrient-gathering true roots, and gas-exchanging stomates of vascular land plants. They live by aerobic photosynthesis like most plants.

Although some species can tolerate the dry conditions of sand dunes and deserts, mosses and liverworts are most abundant, diverse and obvious in bog soils of high latitudes and altitudes. *Sphagnum* moss, for example, forms thick accumulations of peat. This kind of peat has a very low bulk density and a high water retentivity compared with woody peat,

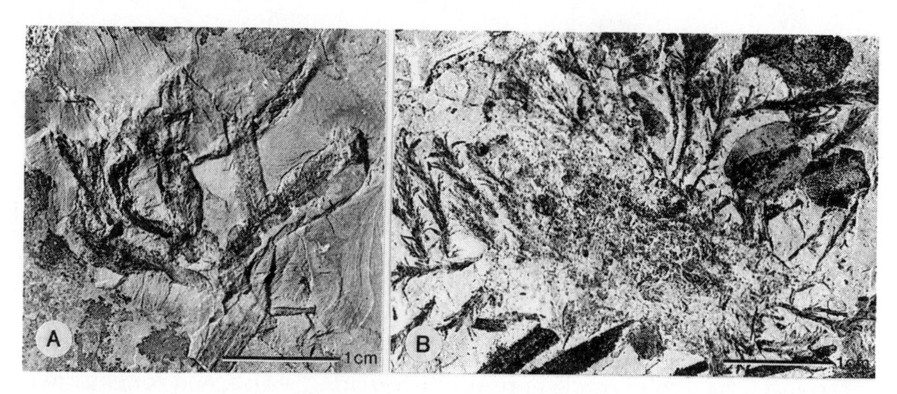

Figure 10.4 A fossil liverwort *Marchantites tennantii* (A) and moss *Muscites guelescinii* (B) from the Late Triassic, Molteno Formation near Birds River siding, South Africa. Scale bars 1 cm (A is BP/2/4658a and B is BP/2/4722) (photographs courtesy of H. M. Anderson).

and can accumulate even over rock outcrops or moderately sloping ground. The strongly acidic pH under *Sphagnum* bogs can promote podzolization in moderately drained sites. More obvious evidence of bryophytes is preserved in the peat itself. Carbonaceous shales and permineralized peats containing fossil mosses are known well back into geological time (Smoot & Taylor 1986).

The general effects of bryophytes on weathering are similar in many ways to those of lichens and fungi. Liverworts form stromatolites and solution cavities in limestone or caliche. Mosses have been found encrusted with tufa (Richardson 1981). The pattern of nutrient uptake of mosses and liverworts is more like that of fungi and lichens than that of vascular plants in that they accumulate a variety of heavy metals toxic to vascular plants. These elements, such as Cu, Fe, and Ni, accumulate in the soil following the death of the plants (Rinne & Barclay-Estrup 1980). These toxic elements allow bryophytes to resist herbivory. The accumulation of macronutrients derived from mineral weathering (Mg, P, S, K, and Ca) is lower by about a half under bryophytes compared with vascular plants, but mobilization of these elements can still be an asset to vascular plants colonizing soil previously occupied by bryophytes. Of these elements, potassium is anomalous in that it is present in bryophytes in proportions much lower than in vascular plants (Shacklette 1965). Unlike other nutrient elements, it appears to decline in abundance with the age of some mosses (Ruhling & Tyler 1970).

Vascular plants

These are plants with conducting tissue of elongate cells whose walls bear distinctive helical or annual thickenings. These tracheids form the veins of leaves and the wood of trees. Vascular plants include most plants of common human experience on Earth today, ranging from tiny filmy ferns to giant redwoods. Like mosses and liverworts, vascular plants photosynthetically reduce carbon dioxide to organic matter, using chlorophyll as a catalyst and releasing oxygen to the air. Their roots are water- and nutrient-gathering structures, which may respire and release carbon dioxide into soil solutions to create carbonic acid. These and other chemical effects of roots have already been discussed because they dominate the weathering of most modern soils. Only those features of soils and paleosols pertinent to interpreting the nature and distribution of individual plants are addressed here.

A dramatic illustration of the role of an individual tree in soil formation is the locally overthickened eluvial horizon called a basket podzol, found in some Spodosols (Retallack 1981b). The term "podzol" is meant here in its original Russian sense of "under ash," for the white, quartz-rich eluvial horizon. Studies of the chemical variation in soils under individual trees (Zinke 1962) and experimental work establishing the rapidity and mechanisms of podzolization (Fisher & Yam 1984) are indications that eluvial horizons of basket podzols are leached of clay, iron, and other materials through the action of a variety of chemicals produced by plants, particularly phenolic compounds washed out of leaves by rain. Some cryptokarst features, such as rounded solution hollows, may be a comparable phenomenon. These basin-like depressions are corroded into limestone bedrock beneath an individual plant because of the acid generated by respiration of its roots and associated microflora (Jennings 1985). Like fossilized stumps, basket podzols and cryptokarst solution hollows can be evidence of the spacing of trees in ancient forests (Retallack 1976, Decourten & Bovee 1986).

Another structure that can be an indicator of forest density is the disturbance of soils resulting from windthrows. Most trees have a radiating array of surficial roots in addition to some deeply penetrating ones. When upearthed by wind, a hemispherical mass of soil is left supported by roots vertically in the air. The hole in the soil surface created in this way is called a cradle knoll. These have a sloping, curved side away from the fallen trunk and a steep, straight side that is capped by a mound of fallen earth once the partly attached stump has rotted away (Johnson *et al.* 1987). Irregular surface horizons of many paleosols could be a record of windthrows. Similar features could be produced by the catastrophic clearing of vegetation by a mudflow or flood deposits, but the holes remaining in the soil after such violent uprooting are more symmetrical than in cradle knolls and also are prone to additional scouring before the energy that uprooted the trees is spent.

Decay of stumps in place can be associated with microbial precipitation of carbonate or other minerals. Some large subcylindrical nodules at the surface of paleosols may represent former stumps (Kraus 1988).

As can be seen from the example of basket podzols, some vascular plants have a marked chemical effect on soils. In humid climates the effect is acidifying and tends to result in the accumulation of Al, Fe, and Si, and marked loss of Ca, Mg, Na, and K. Other elements needed in only trace amounts are B, Cl, V, Mn, Fe, Cu, Zn, and Mo, which commonly show a distribution in soils under vascular land plants similar to the distribution of organic carbon (Stevenson 1969) and phosphorus (Smeck 1973). This is a pattern of enrichment near the surface, depletion below the surface and then a restoration to original levels in the unaltered parent materials. The uppermost zone of enrichment may be within strongly organic surface horizons, as in Mollisols of grasslands or Histosols of swamps. Alternatively, the uppermost zone of enrichment may be leaf litter and standing biomass in soils lacking a strongly organic surface horizon, such as Aridisols of deserts and Alfisols of woodlands. This pattern of deep reorganization of organic matter, phosphorus, and some micronutrient metals could be taken as evidence of a paleosol formed under vascular plants, because these communities have a greater nutrient demand and standing biomass than those of non-vascular plants. Unfortunately, a number of competing diagenetic effects (see Figs. 4.9, 7.7 and 7.8) make it difficult to interpret the nature of past vegetation from chemical criteria alone.

Nematodes

These worm-like creatures are extremely abundant in soils and include predators, microbial feeders, omnivores, herbivores, and plant parasites. Most nematodes are less than a few millimeters long. They circulate organic matter within the soil. Their movement on thin films of water tends to plaster over the surfaces of peds with fine clay and mucopolysaccharides, and so stabilizes soil structure. Even the large nematodes tend to move by wriggling through cracks in the soil rather than by creating burrows. Nematodes form distinctive sinuous surface trackways that have been fossilized in lacustrine muds (Moussa 1970). Galls of nematodes formed on leaves, twigs, and roots of vascular plants have also been fossilized (Berry 1923, Straus 1977). Nematode galls on plants are distinguished by their "giant cells" (Jones 1981). Damage to fossil arthropod cuticle also has been blamed on nematodes. Occasionally fossil nematodes themselves are found in amber (Poinar 1983).

Molluscs

Brackish adapted molluscs, including oysters (Plaziat 1970) and mussels (Retallack & Dilcher 1981b), have been found in mangrove paleosols formed within the intertidal zone. On land the principal soil molluscs are snails and slugs. Snail shells in paleosols can be very useful paleoenvironmental indicators (Evans 1972), and are known even in very ancient paleosols (Solem & Yochelson 1979). The prominent calcareous shell of snails is lacking in most slugs, which can negotiate narrower soil cracks and tolerate more acidic soils than snails. Some slugs have a small shell which is fossilizable (Runham & Hunter 1970). Also fossilizable is the distinctive slime-bordered, flat, undulating trail of snails and slugs. Snail trails have been reported in marine sediments (*Aulichnites*, *Scolicia*, and allied ichnogenera; Häntzschel 1975) and could also be recognized in weakly developed paleosols. The scraped-out feeding trails made by snail radulae during feeding on endolithic lichens in limestone bedrock (Shachak *et al.* 1987) also could be preserved on paleokarst. Slugs and snails eat other animals, fungi, algae, and leaves. Fossil leaves with nibble marks partly repaired with callus by the plant (*Phagophytichnus*; Häntzschel 1975) could equally be produced by herbivorous insects and their larvae as by snails and slugs.

Annelids

Several groups of segmented worms (Annelida) are common in soils but have a sparse fossil record (Morris *et al.* 1982). Best known are common earthworms (Lumbricidae) and potworms (Enchytraeidae). Leeches (Hirundinea) are found within leaf litter, but most leeches are aquatic.

The brown-to-pink lumbricid worms of calcareous grassland and garden soils commonly are 2–5 mm in diameter, but some, such as *Megascolides australis* of Australia, attain diameters of 2 cm and lengths of 1.4 m (Lee 1985). Earthworms feed on decomposing leaf litter and other organic detritus in and on the soil. They may gather surficial mounds of leaves that they particularly favor. They also ingest soil as they excavate their burrows, and this is intimately mixed with organic matter in their ellipsoidal fecal pellets. The granular and crumb structure of grassland soils is thought to be largely a product of earthworm activity. Fecal casts at the surface of the soil may form tall (up to 15 cm high by 3.75 cm diameter), erect pipes or slurry-like masses. Their burrows are also distinctive. Fresh burrows are nearly circular in outline, with scattered ellipsoidal fecal pellets at the bottom (Barley 1959). Once abandoned, the burrows are flattened and deformed into a digitate structure (Fig. 10.5). The walls may be lined with clay, organic matter, calcite, or iron oxyhydrates (Jeanson 1967). Both collapsed and inflated burrows have been found in paleosols (Meyer 1961, Retallack 1976), and several ichnogenera have been proposed (*Oligichnus*, *Edaphichnium*; Bown & Kraus 1983) to accommodate earthworm burrows.

Enchytraeids (potworms) are generally smaller (1 mm–5 cm long) than earthworms, but have similar life habits. Whereas few species of earthworms tolerate a soil pH lower than 4, potworms are often abundant

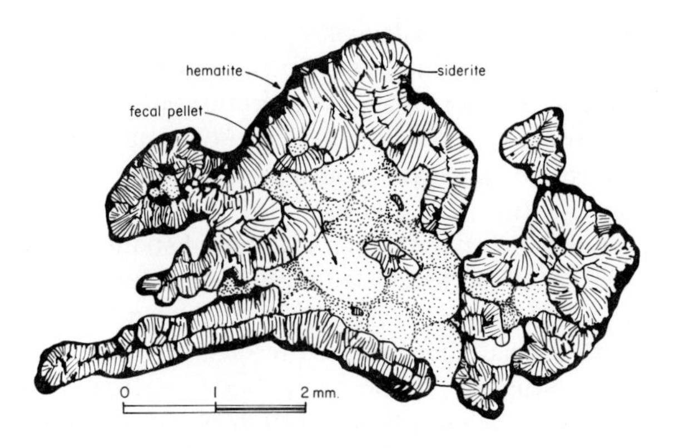

Figure 10.5 Sketch of petrographic thin section of a fossil earthworm burrow, including fecal pellets. The specimen is from the surface (A) horizon of the Turrimetta clay slightly eroded variant paleosol, a Fibrist, in the Early Triassic basal Newport Formation south of Bilgola, New South Wales (from Retallack 1976, reprinted by permission of the Geological Society of Australia).

in acidic woodland litter and peat (O'Connor 1967). They do not leave well defined burrows but use existing soil cracks. Their cocoons are covered with sand grains and organic debris cemented by mucus, and so are potentially fossilizable. Potworms contribute to the enormous numbers of very small fecal pellets found in some soils, but their feces cannot easily be distinguished from those of mites and springtails.

Velvet worms

These creatures are now a minor component of soil faunas and have been placed in their own phylum, the Onychophora. Their long fossil record (Thompson & Jones 1980, Whittington 1985) encourages the view that they may have been more abundant in soils of the past. Unfortunately, their activities leave little trace in soils. They are predators, capturing and then externally digesting insects and other creatures by means of slime strands. They have soft, flexible bodies, susceptible to desiccation, and so are found mainly in leaf litter or mosses of woodland soils in humid regions. Some species retreat to burrows during the heat of the day or a dry season, but these are simple cracks in the soil or burrows abandoned by other creatures (Endrody-Younga & Beck 1983).

Water bears

Tardigrades are commonly called "water bears" because of their rotund shape and paired appendages. They are about the same size as a sand grain (only 50 μm–1.2 mm long). Tardigrades pierce plant cells, rotifers, and nematodes with their stylets and suck out their body fluid as food. They have remarkable powers of resistance to extremes of desiccation, of temperature (withstanding a range from 151° to −270°C), and of X-rays (surviving 24-h exposure to 570 000 roentgens). They can become dormant for as long as 100 years in a little modified barrel-shaped form (Crowe & Cooper 1971). These tuns, as they are called, are potentially fossilizable in organic soils but do not seem to have been noticed. Adult tardigrades have been found fossilized in amber (Cooper 1964).

Crustaceans

Barnacles, lobster, and shrimp are abundant and diverse in oceans and lakes but also common in mangal and swamp soils. The considerable diversity of crustaceans found in mangal soils (Powers & Bliss 1983) is reflected in the great variety of burrows and other traces found in them. Barnacles may be in evidence from the segmented circular scars that they leave on roots, stems, and rocks. Ghost shrimp and crabs excavate complex branching systems of burrows (ichnogenus *Thalassinoides*), in some cases distinguished by finely striated walls (*Spongeliomorpha*) or walls lined with small balls of clay (*Ophiomorpha*, in Frey *et al.* 1978). Crabs and shrimp also excavate inclined tubes with ovoid basal living chambers (ichnogenera *Psilonichnus, Pholeus*, and *Macanopsis*) and tubular helical burrows (*Gyrolithes*; Häntzschel 1975). The fecal pellets of crabs and shrimp also may be distinctive in their regular arrangement of internal cavities (*Favreina* and several comparable coprolite ichnogenera; Häntzschel 1975).

Crustaceans found in freshwater, waterlogged and seasonally flooded soils include cladocerans (Cladocera), copepods (Copepoda), ostracods (Ostracoda), sand-hoppers (Amphipoda), crabs and cray-fish (both Decapoda). Extinct lipostracan branchiopods (*Lepidocaris*; Fig 10.6d) have been found in permineralized peaty paleosols (Histosols). In alluvial soils of the Mississippi River of North America as far north as the Canadian border, tunnels of crayfish (such as *Procambarus gracilis* and *Orconectes immunis*) extend as deep as 5 m into the permanent water table (Thorp 1949). The entrance to these burrows is marked by a short chimney formed of balls of clay. Similar structures have been reported from paleosols (Bown 1982, Bown & Kraus 1983).

Woodlice (Oniscoidea: Isopoda) also are crustaceans but live on dry land. Like other crustaceans, they are prone to desiccation once out of water. Many of them roll into a ball or cluster together to block their pseudotracheae during dry conditions (Warburg 1968). Because of this sensitivity to dry conditions, woodlice are most common in organic horizons of grassland soils or the leaf litter of deciduous woodlands. They are omnivorous, feeding on dead wood, fungi, leaves, and carcasses of

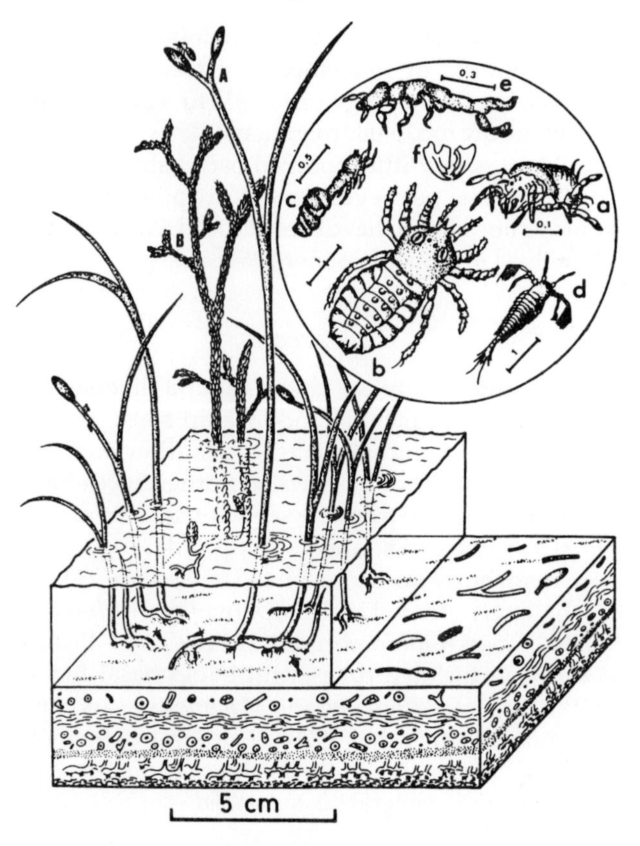

Figure 10.6 Reconstruction of semiaquatic and terrestrial arthropods from the Early Devonian Rhynie Chert of Scotland. The plants include an extinct non-vascular "cooksonoid" (*Aglaophyton major*) (A) and a vascular zosterophyll (*Asteroxylon mackei*) (B) and the animals include a pachygnathid mite (*Protacarus crani*) (a), trigonotarbid spider (*Palaeocharinus* sp.) (b), liphistid spider (*Palaeocteniza crassipes*) (c), branchiopod crustacean (*Lepidocaris rhyniensis*) (d), collembolan (*Rhyniella praecursor*) (e), and a possible thysanuran (*Rhyniognatha hirsti*) (f) (from Kühne & Schlüter 1985, reprinted with permission of *Entomologia Generalis*).

insects and mammals. Fossil trails attributed to terrestrial isopods have been found (Brady 1947). The exoskeleton of woodlice is hardened by biogenic accumulations of calcium carbonate like that of other crustaceans, rather than just the chitinophosphatic material of most other arthropods. They have been found as fossils in paleosols (Morris 1979) and are persistent as bleached skeletons on modern soils.

Millipedes and centipedes

These two kinds of elongate arthropods with many segments are distinct in their ecology and effects on soils. Millipedes (Diplopoda) have two pairs of legs for every segment and are herbivores and detritivores. The pill millipedes (Glomerida) are ecologically like woodlice, which they resemble. The extinct spiny pill millipedes (Amynilyspedida; Fig. 10.7) may have been similar. Flat-back millipedes (Platydesmida, Polydesmida, and Chordeumida) also are restricted to moist leaf litter and organic soils, as was likely for extinct spiny millipedes (Euphoberiida; Fig. 10.7) of Euramerican coal measures. One group of flat-backed millipedes (Polydesmida) produce distinctive small (2.5 mm diameter) disc-shaped feces of clay and sand, with a deep impression down the middle from their anal flaps (Kühnelt 1976). These are sometimes lined with clay from drying out of the semifluid feces (Romell 1935). None of these flat-back millipedes are efficient burrowers. If they burrow at all, they tend to use existing spaces in the soil.

In contrast, many round-back millipedes (Spirobolida, Stemmiulida, Polyzoniida, Siphonophorida, Iulida, Spirostreptida; Fig. 10.7) are active burrowers. Their burrows are round in cross-section, like their bodies, and

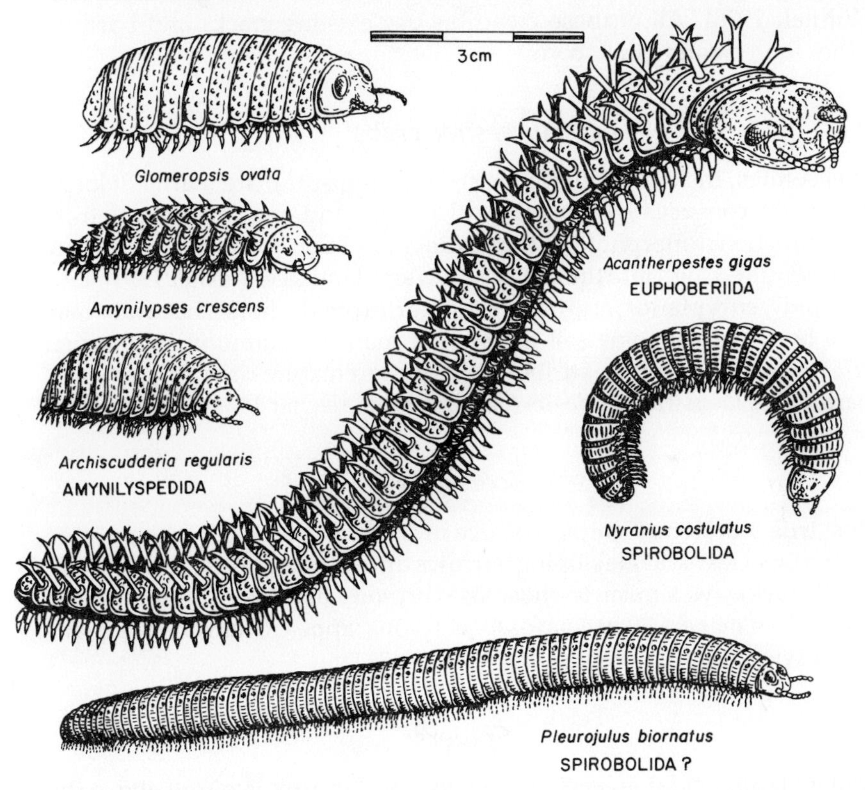

Figure 10.7 Reconstructions of fossil millipedes from the Late Carboniferous Gaskohle Series, near Nyřany, western Czechoslovakia, showing extinct spiny pill millipedes (Amynilyspedida), spiny millipedes (Euphoberiida) and round-back millipedes (Spirobolida) (based on data of Fritsch 1899 and Hoffman 1969).

lined with a thin rim of pressed clay (Paulusse & Jeanson 1977). The fecal pellets of some round-back millipedes (Iulida) are cylindrical and have concave and convex ends where cut off by the anal flaps (Kühnelt 1976). Similar burrows have been found in very ancient paleosols (Retallack & Feakes 1987). Fossilized millipede skeletons are also found in paleosols (Jell 1982, Leakey 1952), presumably because of unusually strong mineralization. The exoskeleton of some modern species is up to 55% by weight of calcium carbonate and calcium phosphate, and only 33.8% chitin and 10.1% protein (Neville 1975).

Centipedes (Chilopoda) have only one pair of legs per segment and are fast-moving predators. They live in leaf litter and bark of trees, and also in the soil. An allied group are "glasshouse centipedes" (Symphyla). These small, white myriapods feed on living roots, decaying plant matter, and microorganisms. The pauropods (Pauropoda) are another centipede-like group of animals with branched antennae and only 9–10 pairs of legs. They eat decaying plant matter and are found in loose, loamy grassland soils (Kühnelt 1976). All of these creatures use existing cracks and burrows rather than creating their own.

Horseshoe crabs

Chelicerates, including horseshoe crabs (Xiphosura), are marine. Horseshoe crabs congregate on beaches to spawn, and one could argue from their functional morphology that extinct kinds of horseshoe crabs may have ventured high into the canopy of swampland trees (Fisher 1979). Like aglaspids, eurypterids, and other extinct arthropods, horseshoe crabs may have been part of some salt marsh and mangal communities (Størmer 1977). Fossil trackways attributed to these creatures are known from marine and lacustrine deposits (Goldring & Seilacher 1971).

Scorpions

Most true scorpions (Scorpiones) live in cracks in the soil, but some large tropical species excavate sloping burrows up to 80 cm long (Kühnelt 1976). Fossil trackways similar to those of scorpions have been found (Brady 1947). Their makers, like most fossil scorpions, appear to have been aquatic (Kjellesvig-Waering 1986).

Spiders

Spiders (Arachnida) are active predators and mostly live well above the soil, but some spiders excavate burrows in the soil. The burrows of the Australian trapdoor spider (*Anidiops villosus*) have round lids and are lined with silk (Main 1985). Other trapdoor burrows have stout inner doors and chambers. Silk is unlikely to be preserved in most paleosols, although fossil cocoons and spider webs are known from amber (Gerhard & Rietschel 1968, Schlee 1980).

Mites

Most mites (Acarini) are microscopic (Fig. 10.6a), but these are the most abundant and diverse group of soil animals. Different kinds of mites parasitize and eat live prey, feed on plants, fungi, bacteria, algae, and lichens, and decompose dead organisms. They leave few obvious traces in the soil other than small ellipsoidal fecal pellets, and these are difficult to distinguish from pellets of other small soil creatures (Wright 1983).

Springtails and wingless insects

A variety of silverfish-like hexapods are found in soils (Thysanura, Diplura, Protura, Collembola). Different species eat live prey, fungi, algae, and decaying vegetation. They live in crevices in the soil, leaving small cylindrical fecal pellets but no distinct burrows. The springtails (Collembola) are so named because of the large spine on their tails that can be flexed to effect a quick escape. They are small (mostly less than 2 mm long) and notable for their ability to tolerate climatic extremes ranging into arid and polar soils, where few other soil organisms are found. Collembola also are the most ancient fossil hexapods known (Fig. 10.6e).

Butterflies and moths

Lepidopteran galls and mines in leaves and bark fragments have been found as fossils (Crane & Jarzembowski 1980). Fossil twig-covered cases of the larvae of case-worm moths are known from amber (Schlee 1980) and egg cases of noctuid moths in clay (Gall & Tiffney 1983).

Caddis flies

Most of these insects (Trichoptera) have aquatic larvae that construct larval and pupal cases from pebbles, shells, or whatever is at hand. Fossil cases of this kind are common in lake deposits (Lewis 1972). A few modern species of caddis fly larvae are completely terrestrial and live within leaf litter. Some of the fossil cases made completely of leaves [ichnogenus *Folindusia* of Sukacheva (1980)] could have been of this kind.

Crickets

Most crickets, grasshoppers, and related insects (Orthoptera) are herbivores that leave little trace in the soil. Two groups, the mole crickets

(Gryllotalpidae) and pygmy mole crickets (Tridactylidae), burrow in moist soil near lakes and streams (Ratcliffe & Fagerstrom 1980). Burrows of pygmy mole crickets have simple tear-shaped living chambers either at the end of vertical passages (like the ichnogenus *Macanopsis*; Häntzschel 1975) or terminating horizontal galleries, similar to those made by bugs, cicadas, beetles, and wasps. More distinctive are the burrows of mole crickets. Their spacious (1 cm diameter) galleries branch and join to create an irregular network just below and parallel to the ground (similar to the ichnogenus *Protopalaeodictyon*; Häntzschel 1975).

Bugs

Among the bugs (Order Hemiptera), there are a few kinds that burrow: shore bugs (Saldidae), toad bugs (Gelastocoridae), and burrower bugs (Cydnidae). Most of these burrows are simple inclined shafts with a slightly larger living chamber at the bottom (ichnogenus *Macanopsis*). Burrower bugs do not dig access shafts to the surface. Their burrows are just under the surface of the soil, under stones or logs, in sand, or in mold near the roots of tussock grass. Often the roof of the burrow is cracked through to the surface and partly collapsed after their passing (Ratcliffe & Fagerstrom 1980).

Cicadas (Suborder Homoptera, superfamily Cicadoidea) are locally abundant soil burrowers. Mature cicada nymphs excavate long, straight burrows directly up toward the surface from their root-feeding areas at depths of 25–50 cm into the soil. The larvae and nymphs spend many years (up to 17 in some species) at this level within the soil. During the spring of some years, they emerge in such great numbers that the soil is riddled with simple vertical burrows (ichnogenus *Skolithus*; Häntzschel 1975). Similar behavior may be of great antiquity among insects (Retallack 1976).

Scale insects (Coccoidea) are bugs, and their characteristic waxy patches may be preserved on fossil leaves (Zeuner 1938). Aphids (Aphidoidea) also are bugs. Their galls on leaves and twigs may be fossilized (Brooks 1955).

Beetles

Almost anything that is edible and some materials, such as dung, that might not seem so from a human perspective are eaten by beetles (Order Coleoptera). These are the most diverse group of insects associated with soil. Internal casts of ellipsoidal cocoons, such as those made by beetles, are locally abundant in some paleosols (Freytet & Plaziat 1982, Pickford 1986b). Only a few kinds of beetles leave distinctive burrows in the soil.

Adult tiger beetles (Cicindelidae) are predators. They excavate simple shallow burrows in order to retreat from nightly or winter cold or from rainy or hot weather. Their larvae also are predatory and excavate simple burrows that may be vertical, and have right-angled bends to sub-horizontal galleries (Ratcliffe & Fagerstrom 1980). They range in depth from a few centimeters to 125 cm deep. A distinctive feature of these burrows is their meniscate backfills ("spreiten") of clay and silt (as in the ichnogenera *Ancorichnus* and *Scoyenia*; Frey *et al.* 1984). Another distinctive feature of tiger beetle burrows is the occasional ellipsoidal pupation chambers located along the burrows.

Rove beetles (Staphylinidae) feed on algae, decaying vegetation, and other insects. Algal-feeding species are especially conspicuous in sandy soils of beaches and streamsides. Their burrows may be very complex, with multiple entrances and chambers situated above and doubled back over their entrance burrows.

Ground beetles (Carabidae) are a large group of predatory burrowing beetles that make simple vertical and horizontal burrows. They also make a distinctive kind of burrow, with radiating and branching subterranean galleries (like the ichnogenus *Megagrapton*; Häntzschel 1975).

Both larvae and adults of variegated mud-loving beetles (Heteroceridae) excavate horizontal galleries that meander and branch just below the surface of soils. These herbivorous beetles are especially common along the shores of streams and lakes.

Some of the most elaborate burrows made by beetles are excavated by scarabs (Scarabaeidae; Halffter & Edmonds 1982). Many of these beetles construct simple vertical or inclined burrows (like the ichnogenus *Macanopsis*) in which to spend a night or to over-winter. Burrows made in order to provision larvae are more elaborate. Fresh leaves, decaying leaf litter, dung, and humus are buried by various species, each with specific dietary preferences. These food reserves support the development of larvae which hatch from eggs buried alongside or within the food. Pupation also occurs underground, and the young adults emerge at the surface to feed and then provision future generations. The behavior of scarabs can be considered on a scale of increasingly elaborate parental care. Little care is invested in future generations by small dung beetles such as *Aphodius*. Their life cycle is short and egg production copious enough that a few are likely to survive to adulthood in a large dung pat. Other species, such as *Dichotomius carolinus* (Fig. 10.8A), pack an egg, food supply of dung, and partition of soil successively, end to end, in simple near-vertical burrows. More elaborate still are burrows made by dung beetles such as *Geotrupes stercorarius* (Fig. 10.8B). This species excavates a long vertical burrow beneath a large pat of dung. Working back toward the surface, successive lateral chambers are constructed. Each one is supplied with an egg and a large wad of dung before being sealed with earth. Trace fossil burrows of this kind have been referred to the ichnogenus *Pallichnus* (Retallack 1984b). Other species, such as *Phanaeus palliatus*, not only bury dung and eggs in subterranean chambers but sculpt a thick clayey rind

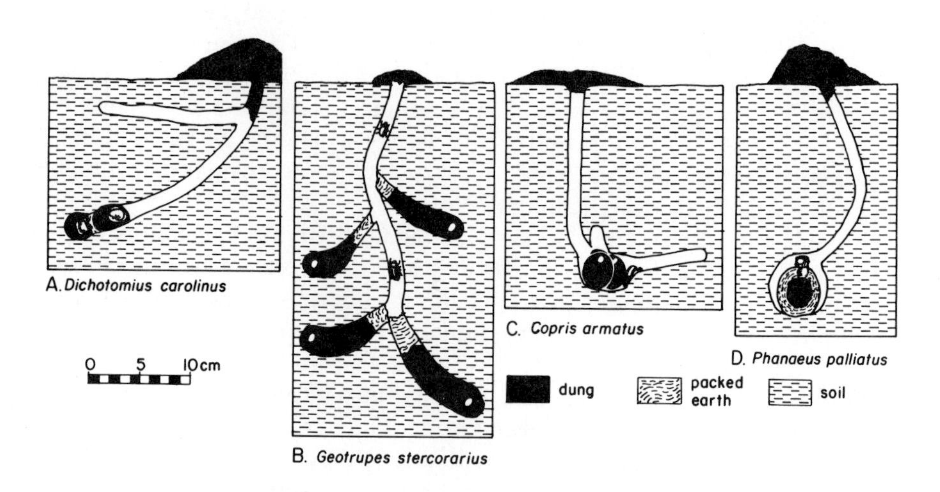

Figure 10.8 Brood burrows of dung beetles (Coleoptera, Scarabaeidae) below the dung they use to provision their larvae: A, *Dichotomius carolinus*, from Cuernavaca, Mexico; B, *Geotrupes stercorarius* from near Saalfeld, East Germany; C. *Copris armatus* from Salazar, Mexico; D, *Phanaeus palliatus* from Ocoyoacac, Mexico (A, C, D adapted from Halffter & Matthews 1966, B from Von Lengerken 1954).

around both (Fig. 10.8D) in order to deter predation and parasites. Trace fossils of this kind have been referred to the ichnogenus *Coprinsphaera*. The most elaborate nesting behavior found in dung beetles is the cutting of spheres of dung from a large dung pat, which are then rolled away for burial elsewhere. This behavior of *Scarabaeus sacer* was seen by ancient Egyptians as a metaphor for the rising and setting of the sun, and a sign of good luck that the universe was still in operating order. These dung-rolling beetles bury their dung less deeply than other beetles, and are more difficult to detect in the fossil record.

The activity of beetles also can be inferred from other fossils associated with paleosols. Many kinds of beetles (Scolytidae, Buprestidae, and Cerambycidae) bore into wood. Similar borings and fecal pellets have been found permineralized in fossil wood (ichnogenera *Palaeobuprestus*, *Palaeoipidus*, and *Palaeoscolytus*; Häntzschel 1975). The engraver beetles (Scolytidae) make a long central gallery occupied by the female beetle. Radiating galleries are excavated by numerous larvae as they feed outward from the central gallery. This pattern of burrowing is especially obvious when confined to bark, because this breaks open in some fossils to reveal the entire pattern clearly (Radwanski, 1977). Carrion beetles (Dermestidae) are effective in defleshing large carcasses, and also bore into bones. This type of damage has been recognized in fossil bones (Kitching 1980).

Termites

Large organized societies of termites (Isoptera) have a profound effect on the decomposition of wood, humus, lichens, and fungi in soils of tropical to warm temperate climates. Earthen termite mounds up to 10 m high dot the landscape of many tropical grasslands. The internal structure of these termitaria can be complex, including a centrally located system of chambers for the queen and larvae, fungus gardens, and extensive passageways for ventilation (Wilson 1971). These kinds of structures can be striking when preserved in paleosols (Bown 1982, Leakey & Harris 1987).

In addition to organized structures of termitaria, soils and paleosols also may be riddled by a network of galleries leading deep into the soil to water, or ramifying in all directions to sources of organic matter. The small spherical micropeds that dominate many tropical soils and paleosols (see Fig. 9.7) are thought to represent in part the degraded pellets of termites. Mound-building termites fashion small (125–750 μm diameter) spherical to ovoidal pellets from earth and saliva, and also expel them as feces (Mermut *et al.* 1984). In contrast, well-preserved fecal pellets of dry-wood termites are short rods, with a hexagonal cross-section (Rohr *et al.* 1986).

Termite nests are large structures that alter the physical properties of soils. Their presence may be discernible long after invaded and torn apart by aardvarks or reworked by burrows of other soil animals. The aerating effect of termite galleries can result in an appreciably more alkaline soil pH within a termitarium, compared with the soil beyond (Watson 1967). In some tropical countries, caliche from the center of termite mounds is an important local source of agricultural lime fertilizer (Thorp 1949).

Ants

This group of social insects (Hymenoptera, Formicidae) also produce complex nests. Harvester ants (such as *Pogonomyrmex maricopa*) create large mounds, up to 76 cm high. The mounds consist of particles of sand and granule size and are coarser-grained than the surrounding soil (Cole 1968). Extensive networks of burrows attributed to ants (Bown 1982) and colonies of fossil ants associated with fossil leaves (Wilson & Taylor 1964) have been found in paleosols. Like those of termites, the nests of ants aid in mixing of organic matter, aeration, circulation of water, and raising the pH of soil (Kühnelt 1976).

Wasps

Many of these predatory social insects (Hymenoptera, Superfamilies Pompiloidea, Vespoidea, and Sphecoidea) excavate burrows for their larvae. They bury their eggs with an anesthetized insect or spider, on which the hatchling larvae proceed to feed until pupation. The larval cells of wasp burrows are generally tear-shaped. They terminate a simple inclined entrance burrow or are arranged laterally along the entrance burrow (Bown 1982). More complex and copiously branching nests are made by more highly social species (Evans & Eberhard 1970).

Many wasps construct nests above ground, and under certain circumstances these also can be preserved in paleosols. Leaf galls known from fossil leaves (Brooks 1955) are created by wasps, flies, and aphids. Clay nests like those of potter wasps (*Eumenes pedunculatus*) also have been found fossilized (Handlirsch 1910). A complex multicellular nest in a siderite nodule has been thought to represent a paper nest of the kind constructed by highly social wasps (Brown 1941a), but not without some controversy (Brown 1941b, Bequaert & Carpenter 1941).

Bees

These common insects (Hymenoptera, Apoidea) can be regarded as specialized hairy wasps that feed their larvae with pollen and nectar collected from flowers or with other glandular secretions of plants. Burrows of solitary bees are similar in overall plan to those of wasps but can be distinguished by two fossilizable features: a coated or highly polished wall to the larval cells and clayey closures with concentric or helical grooving (Fig. 10.9). These also are features of multicellular nests of bees, such as the Miocene (probably not Cretaceous; Retallack 1984b) fossil nest *Uruguayichnus* (Schlüter 1984), which is similar to the nests of living tropical sweat bees such as *Augochloropsis sparsilis* (Michener 1974). These look like a crudely constructed piece of honeycomb, made from clay, and are fashioned within a larger chamber along the main galleries of the burrow systems. Simple shallow burrows are made by solitary bees, and more complex structures by social species, but there are many exceptions to this general rule (Sakagami & Michener 1962).

Other evidence of bees associated with paleosols includes fossil leaves with neatly cut circular holes of the kind made by leaf cutter bees (Brooks 1955), and shallow borings in wood (similar to ichnogenus *Trypanites*; Häntzschel 1975), like those of xylocopid bees (Schenk 1937). Fossilized wax honeycomb of the kind made by social honey bees has also been found in cave earth (Stauffer 1979).

Fish

A variety of fish fry are found within mangal soils. They persist in small shallow pools and burrows as the tide goes out. Some fish, such as the Australian mudskipper (*Periopthalmus gracilis*), climb out of the water and on to the exposed mudflat and mangrove roots (Nursall 1981).

Figure 10.9 Fossil larval cells (*Celliforma spirifer*) of bees (Hymenoptera, Apoidea) showing spiral closures, from the Middle Eocene Bridger Formation, near Mountainview, Wyoming. Scale at bottom is in millimeters (Retallack photograph of the type specimens of Brown 1934, Smithsonian Institution number 340878).

Fish may also be important in swamp and marsh soils. In the black water (varzea) swamps of Brazil, fish eat copious quantities of large seeds and aid dispersal of these plants (Gottsberger 1978). Lungfish burrow into the bottoms of seasonally dry lakes. Their lariat-shaped burrows are distinctive and known in very ancient lakeside paleosols (Dubiel *et al.* 1987).

Amphibians

Spadefoot toads (*Scaphiopus hammondi*) bury themselves deeply in desert soils for extended periods (about 9 months) while conditions are dry (Ruibal *et al.* 1969). They excavate simple vertical shafts 2–5 cm in diameter with an inclined living chamber at the bottom [similar to the trace fossil genus *Macanopsis* in Häntzschel (1975)]. Extinct lepospondyl amphibians also have been found fossilized in aestivation burrows (Olson & Bolles 1975). Limbless amphibians such as the caecilians (Gymnophiona) burrow through soil like the earthworms and insects on which they feed.

Reptiles

Snakes and tortoises retreat from extremes of weather in shallow, sloping burrows. During cool weather, many individuals and species of snakes huddle together in the same burrow. This may be an explanation for the rare occurrence of interwoven, articulated skeletons of snakes in paleosols (Breithaupt & Duvall 1986). Other snakes, lizards, and snake-like lizards (Amphisbaenia) excavate long, narrow burrows (Kühnelt 1976).

Extinct reptiles also appear to have burrowed and otherwise modified soils. Articulated skeletons of the extinct therapsid *Diictodon* have been found in helical burrows within Late Permian paleosols in South Africa (Smith 1987). Dinosaur footprints are widely known in lake-margin sediments (Lockley 1986). These enormous creatures probably had a profound effect on mixing and compaction of soils. Shallow nests of dinosaur eggs and hatchlings have been found within calcareous paleosols (Horner 1982).

Birds

Many kinds of birds build nests in and on the ground, but these have only rarely been encountered as fossils (Lambrecht 1933). Especially distinctive and fossilizable nests in soil are the salt-encrusted pedestal nests of flamingos (*Phoenicopterus ruber*) on the shores of saline lakes, and the large (3.5 m diameter by 2 m high) incubation mounds of Australian mallee fowl (*Leipoa ocellata*). Other traces of birds include fossil footprints, which are especially well known as fossils in lake-margin sediments (Feduccia 1980). Local concentrations of bones of rodents and other small mammals in paleosols may represent regurgitated pellets of owls or other raptorial birds (Mellett 1974).

Mammals

Many kinds of mammals, such as moles, live their whole lives within the soil and seldom come to the surface. For others, burrows provide shelter and a safe place to raise young. Mammalian burrows range from simple inclined dens to complex systems of burrows found in prairie dog towns (Voorhies 1975). Perhaps the most complex mammalian burrows found in paleosols are the deep (up to 2.75 m) helical entrance shafts and elongate living chambers (ichnogenus *Daimonelix*) thought to be the work of extinct beaver-like rodents (Fig. 10.10). Simple shallow dens of carnivorous mammals also have been found in paleosols (Hunt *et al.* 1983).

Mammals also affect soils in other ways. Their fossil feces (coprolites) are commonly found in paleosols (Retallack 1984a). Fossil footprints have been found in weakly developed paleosols (Webb 1972, Leakey & Harris 1987). More subtle effects of large mammals have been noted in studies of buffalo wallows, i.e., areas locally trampled by North American *Bison* (Polley & Collins 1984). Compared with soil outside the wallow, trampled soil inside is more compact, clayey, alkaline, less moist, and poorer in phosphorus. As a result, wallows are overgrown with different kinds of herbs and grasses that can tolerate such different soil as well as heavy grazing. Similar effects on soil have been documented for East African elephants and hippos. Elephants destroy trees, converting bushland into

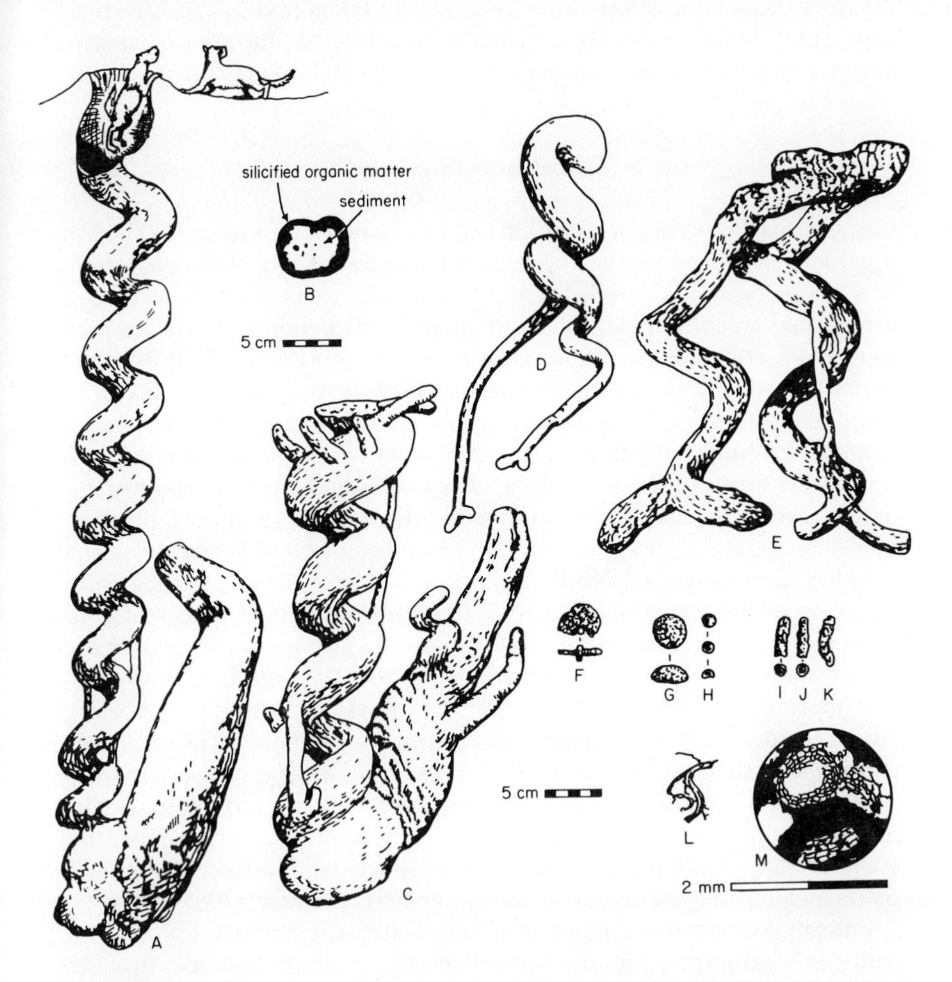

Figure 10.10 Trace fossils of rodents, beetles, ruminants and plants all drawn to the same scale, from the Miocene upper Harrison Formation, at Eagle Crags, near Harrison, Nebraska. The large corkscrew burrows *Daimonelix circumaxillaris* (A–D) are thought to have been excavated by the fossil beaver *Palaeocastor fossor* (A) occasionally fossilized within them. Smaller helical burrows (E) were probably excavated by the endoptychine gopher *Gregorymys*. Narrow burrows associated with the larger ones (C–E, I–K) may have been made by dung beetles, because flattened coprolites of ruminants (F) have associated burrows of the same size and rounded dunglike boli (G–H) are also found. The surface of all the burrows is encased in fossil root traces (L), sometimes silicified to the extent that cellular structure is preserved (M) (rescaled and redrawn from Martin & Bennett 1977, Barbour 1897, with permission of Elsevier Publishing Company and the Geological Society of America).

savanna (Buechner & Dawkins 1961, Vesey-Fitzgerald 1973). Effects of vegetation clearance and the creation of hard infertile surfaces are seen in a most extreme form in human influence on soils (Keller 1976).

Traces of ecosystems

Just as a fossil burrow or coprolite can be regarded as a trace fossil of an organism, so can a fossil soil be regarded as a trace fossil of an ecosystem. Few of the individual species of an ecosystem or their interactions leave a discernible trace in paleosols, but their more general effects may be preserved. For example, the kind of vegetation may be evident from the pattern of root traces or sequences of soil horizons (see Figs. 5.1–5.3). General features of vegetation, such as stature and spacing of plants, determine what botanists call a "plant formation," not to be confused with rock formations of geology. A plant formation is distinct from a community or association, in that it is not defined by the particular species present. Forest, woodland, and grassland are examples of plant formations. The wooded grasslands of North America, South America, Africa, and Australia all have small-leaved and often many-stemmed trees scattered in more extensive grass, although the trees and grasses in each region are different species. It is plant formations, rather than plant communities, that are reflected best in soils. It may not be possible, without supporting paleobotanical evidence, to determine from a paleosol whether the trees were oaks, acacias, or eucalypts. On the other hand, it is possible from paleosols to distinguish woodland and grassland. Plant formations can be reconstructed from fossil plant remains only under exceptional circumstances. Many fossil plants have been transported and mixed in lakes and rivers away from where they lived (Spicer 1981, Ferguson 1985).

Information on the relationship between plant formations and soil features is gathered primarily from observations of modern soils and the vegetation they support. This line of evidence is not without problems. There is a widespread misconception, for example, that the soils of tropical savanna are red, non-calcareous, base-depleted, and clayey (Oxisols). Much African grassy woodland currently grows on such soils, but these stable landscapes on Precambrian bedrock were first weathered under rainforest of Miocene or older time. Where grassy woodlands and wooded grasslands grow in young alluvium or volcanic ash in East Africa, soils are gray, yellow or brown, calcareous, nutrient-rich, and silty (Inceptisols or Mollisols). Confusion over the natural soil type of savanna has arisen because of the long time of formation of some of these soils (McFarlane 1976). Other factors also play a role. Human use of fire, for example, has created open grasslands from wooded grassland and woodlands because grasses recover from fire more readily than do trees. The widespread grassland soils (Mollisols) of the Willamette Valley of Oregon (see Fig. 6.1) are thought to be in part a product of Indian use of fire. Since European settlement, cultivated fields continue to be burnt to encourage spring growth of grasses, but fire restrictions in other parts of the landscape have allowed reforestation (Johanessen *et al.* 1971). Many of these forested soils retain the characteristic organic surface horizon of grassland soils (mollic epipedon). If these soils became paleosols, their profile form would be evidence of a long time of formation under wooded grassland rather than of their historic invasion by conifers. Conversely, the red soils of East Africa are evidence of ancient forests of millions of years ago rather than the grassy woodland of more recent origin. These situations could be viewed as confusing the true relationship between vegetation and soils, and they have taken some scientific ingenuity to unravel. In each case, however, the soil preserves a record of the vegetation that was most significant for its formation in the long run.

Early successional vegetation

In ecological succession, the first to colonize are small herbaceous forms including lichens, mosses, grasses, and ragweed (Kimmins 1987). These plants are regarded as weeds in gardens because of their ability to establish themselves on bare earth and their lack of showy flowers. Early successional soils and paleosols (Entisols, Inceptisols) retain most of the original features of their parent material, such as the bedding and ripple marks of flood alluvium, despite its disruption by small root traces and burrows.

Such very weakly developed paleosols are common in ancient fluvial sequences, especially around the margins of paleochannel sandstones (Retallack 1983a, Feakes & Retallack 1988). Weakly developed paleosols with calcareous rhizoconcretions also are known in ancient beach (Ettensohn *et al.* 1988) and desert dune sandstones (Loope 1988). These may have supported early successional dune-binding vegetation.

Forest and woodland

These are closed-canopy formations of trees that may be arbitrarily distinguished by stature: woodland is under 10 m high and forest is taller (White 1983). Large root traces are an obvious indicator of forests. Trees promote differentiation of soil horizons into a sandy, bleached near-surface (E) horizon and a clayey, sesquioxidic, or humic subsurface (Bt, Bh, or Bs) horizon, as in Alfisols. The lateral continuity of these horizons is an indication of the degree of canopy closure, because widely spaced trees promote the development of cradle knolls and basket podzols. The thickness of the soils is related to biomass because of the depth and density of root penetration and the flux of weathering chemicals produced. Forested soils generally are ca. 1 m thick, whereas those of woodlands are thinner (see Fig. 3.9). Such well differentiated paleosols with large root traces are common in floodplain facies of ancient fluvial deposits (Retallack 1983a, 1985).

Rainforest

These tall forests of humid tropical regions have two or more canopy levels, a great diversity of plant species and a profusion of vines and epiphytes. Their soils (Oxisols and Ultisols) are thick, clayey, deeply weathered of bases, and in some cases red, lateritic, or bauxitic. Many have little soil structure, apart from sand-sized spherical micropeds. In many cases the large roots of buttressed tree trunks do not penetrate the soil, but form a surface mat (Sanford 1987). Generally similar paleosols are widespread at major geological unconformities that may have supported rainforest or a number of other kinds of vegetation during their long period of formation. More convincing examples of ancient rainforest paleosols have been found in ancient fluvial deposits in association with fossil remains of tropical angiosperms (Retallack 1981a).

Humid forests of southern beech (*Nothofagus*) in Chile and New Zealand and of hemlock (*Tsuga*) in the Pacific northwestern United States are sometimes called "temperate rainforest." These form on soils (Ultisols) with well differentiated clayey subsurface (Bt) horizons that are non-calcareous and often kaolinitic. Similar paleosols in ancient red beds are thought to have formed under extinct kinds of conifers at high paleo-latitudes (Retallack 1977).

Oligotrophic forest

Quartz-rich soils low in nutrients (Quartzipsamments, Dystrochrepts, Spodosols) are now widespread in humid temperate and alpine regions where they are colonized by needleleaf forests. In tropical regions such sandy, low-nutrient soils commonly have a very deep horizon of iron and humus enrichment (Bs) and support dwarf angiospermous forest. Some sandy paleosols are identical with modern Spodosols (Pomerol 1964). Stigmarian root traces of extinct lepidodendralean tree lycopods are common in quartz-rich Carboniferous paleosols, that have not yet been found with the well developed spodic (Bs) horizon under modern oligotrophic forests (Percival 1986).

Heath

The original plant formation for which the term heath was coined is the bushy vegetation dominated by heather (*Calluna vulgaris*) in the humid granitic highlands of Britain and Scandinavia. The term also has been used for similar vegetation of thin, quartz-rich, waterlogged soils (Aquods) in humid coastal regions of Australia and South Africa (Specht 1979). Some paleosols of this second type are known from fossil leaf litters to have supported heath of extinct seed ferns (Retallack 1977).

Taiga and krummholz

Taiga is a stunted woodland found in polar regions. Krummholz is a comparable vegetation of gnarled and windswept asymmetric trees ("flag trees") found near the snowline in alpine regions (Price 1981). Both kinds of vegetation grow in soils with distinct near-surface leached (E) horizons and subsurface clayey, organic, or ferruginous (Bt, Bh, or Bs) horizons, like those of other woodlands. Most are acidic, sandy, and organic, because little moisture is lost to evaporation, and rates of decomposition of leaf litter are low in cold climates. Also limited is the activity of soil fauna and its traces in the form of burrows and fecal pellets. Most distinctive of these soils are traces of permafrost or frost heave, such as patterned ground or ice wedges (see Table 9.2). When buried in glacial sediments, these weakly developed paleosols are best identified from their stout and well preserved carbonaceous root traces (Retallack 1980).

Tundra and alpine fellfield

Open vegetation of high mountains and polar regions can be distinguished from that of dry to desert regions by the abundance of mosses, lichens, and a variety of dwarfed vascular plants, including some extremely reduced "cushion plants." Some grasslands are found also in tundra and alpine fellfields (Price 1981), but their soils are thin and weakly developed, lacking the thick clayey surface (mollic epipedon) and calcareous subsurface (Bk) horizons of grassland soils in warmer regions. Like soils of krummholz and taiga, these soils of cool climate tend to be acidic and organic and may show features betraying the presence of permafrost. As paleosols in glacial deposits they can be recognized by the size of their root traces (Retallack 1980).

Dry woodland

In many southern continents, dry, grassy woodlands (sometimes called savanna woodland) are widespread on very ancient (Cretaceous and early Tertiary) land surfaces (Cole 1986). Their soils may have red, kaolinitic, lateritic, or bauxitic horizons from ancient humid weathering, together with superimposed calcic (Bk), gypsic (By), or silicic (Bg) horizons, nodules, or crystals from their current round of soil formation in dry continental interiors. Generally similar paleosols at major unconformities showing conflicting climatic indicators are known well back into Precambrian time (Ross & Chiarenzelli 1984). A search for calcareous and siliceous rhizoconcretions is needed to establish when such paleosols supported dry woodland.

The effects of dry woodland are clearer in alluvial deposits where its soils are calcareous, with well differentiated subsurface clayey (Bt) horizons and a deeper calcareous (Bk) horizon. Calcareous and siliceous rhizoconcretions are important evidence for dry woodland from paleosols. In some cases, twigs, pollen, and permineralized stumps are preserved in these paleosols (Francis 1986), perhaps because microbial decomposition is hindered and silica is mobilized under very dry and alkaline conditions.

Fire-prone shrubland

This general term for shrublands in summer dry (Mediterranean) climates includes Californian chaparral, Chilean matorral, Australian mallee, and European maquis and garrigue (di Castri *et al.* 1981). Their soils are thin and full of charcoal over shallow bedrock. Charcoal-filled fissures of paleokarst (Harris 1957) may represent similar vegetation of the past.

Wooded shrubland

This is a plant formation with the general appearance of wooded grassland but with few, if any, grasses. Their place is taken by woody shrubs. Examples of such vegetation are the sage (*Artemisia tridentata*) and juniper (*Juniperus occidentalis*) vegetation of central Oregon (Franklin & Dyrness 1973) and the myall (*Acacia sowdenii*) and saltbush (*Atriplex vesicaria*) vegetation of inland Australia (Beadle 1981). The soils of wooded shrubland have shallow hardpans, calcareous nodules, salt crusts, and stony pavements. Root traces in these soils are more sparse and clumped than in wooded grassland, and there is no hint of the organic, granular, or crumb-structured surface layer (mollic epipedon) of grassland soils. Before the evolution of grasses, wooded shrubland may have been widespread in regions between woodland and desert (Retallack 1985).

Wooded grassland

This is a plant formation of solitary trees scattered in grass, often also called savanna or savanna parkland (Cole 1986). Wooded grassland soils have a mix of features found in soils of dry woodland, in which trees cover more than 40% of the ground, and of open grassland, with less than 10% tree cover (White 1983). Surface horizons of wooded grassland soils are organic and have a fine structure (granular or crumb). These soils are moderately calcareous, with a shallow (30–60 cm) calcic horizon. These features of subhumid to semi-arid grassland soils are mixed with others, such as weakly developed, leached, near-surface (E), or clayey subsurface (Bt) horizons more typical under woodlands. Similar moderately developed calcareous paleosols (Mollisols and Inceptisols) may preserve abundant bones of diverse mammal faunas also showing a mix of woodland and grassland affinities (Retallack 1986c).

Root traces also may be a guide to ancient wooded grassland. Most roots of grasses are less than 2 mm in diameter. Abundant fine root traces, together with a few substantial ones, are typical of wooded grassland soils. In paleosols this may be especially clear in cases where root traces have drab haloes formed by burial gley (see Fig. 3.7), which is thought to reflect the last crop of roots before burial (Retallack 1983a). Charcoal also is typical of wooded grassland, because much of it is prevented from reverting to woodland by frequent fires.

Open grassland

Treeless grassland vegetation is known by several regional synonyms, including Russian steppe, North American prairie, and Patagonian pampas. The term grassland is best used generally to include such formations, in addition to wooded grassland, parkland and groveland. The soils (Mollisols) of these grassland formations are united by their distinctive well structured organic surface horizons (mollic epipedons). This horizon is formed by the multitude of fine roots and abundant soil fauna associated with grasslands. This distinctive surface horizon may be laterally continuous and deep (1 m or more) in grasslands of subhumid climates, such as the tall-grass prairie of the western Great Plains of North America. The surface horizon is shallower and laterally variable in thickness under tussock grasses of semiarid climate, such as the short-grass prairie of the western Great Plains and intermontane rangelands of the western United States (Ruhe 1984). Abundant very fine drab haloed root traces may be important for recognizing open grassland paleosols (Retallack 1983a). Also useful is supporting evidence from fossilized opal phytoliths (Piperno 1987), grasses (Thomassen 1979), and mammals (Bakker 1983).

Shrubland

Many desert regions support a vegetation of low-growing bushes with few grasses. An example is the bluebush (*Maireana sedifolia*) shrublands of inland Australia (Beadle 1981). Individual small shrubs dot the landscape, often in intriguingly regular patterns, but there is much bare ground exposed. Root traces in these soils are sparse and strongly clumped under individual bushes. There may be evidence of wind scouring between the plants and of dunelike accumulations of wind-blown silt around them. These soils also tend to be stony, sometimes with a desert pavement of interlocking rocks. They also may be encrusted with salt and cemented with calcareous nodules. This vegetation is best recognized in paleosols by the size and distribution of calcareous rhizoconcretions (Loope 1988).

Desert scrub

Some desert vegetation has a sparse but fairly uniform distribution of woody and succulent plants (Evenari *et al.* 1986). Some of the plants of these formations are peculiar in appearance, such as the African candelabra cactus (*Euphorbia candelabra*) and the North American saguaro cactus (*Cereus giganteus*) and boojum tree (*Idria columnaris*). As with shrubland, there is a good deal of bare earth exposed. Their soils and paleosols are similar to those of shrubland but for the large root traces and stem bases found under desert scrub (Loope 1988).

Microbial earth

Salt pans and sand dunes of desert, alpine, and boreal regions may appear virtually barren of vegetation but support complex communities of microbes (Campbell 1979). Microbes may be abundant enough to discolor the ground green or purple. Apart from a few lichens, other large organisms are excluded from these communities by extreme dryness, saltiness, windiness, or cold. The soils of microbial earths show little profile differentiation. Their paleosols may be recognized by thin surface features such as rock varnish (Retallack 1986b) or mini-stromatolites (Hofmann & Jackson 1987), along with other evidence of subaerial exposure. Microbial earths were not confined to such difficult habitats before the advent of vascular plants on land (Wright 1985).

Microbial rockland

The top few millimeters of cliffs and boulders may house complex communities of microbes. In the Dry Valleys of Antarctica, sandstone and dolerite blocks have several near-surface zones of endolithic algae and lichens (Friedman & Weed 1987) that create a weathering rind like a miniature Spodosol. The small vase-shaped solution hollows of endolithic algae in limestone are even more distinctive (Folk *et al.* 1973), and fossilized examples of this have been recognized from limestones thought to have been bored both in the sea (Knoll *et al.* 1986) and on land (Folk & McBride 1976).

Bog

British high-latitude mossy vegetation of *Sphagnum* is a good example of this plant formation. The term bog is used generally for vegetation of rootless plants on acidic, freshwater waterlogged ground (Gore 1983). A characteristic of these soils is the surficial peat layer, which is an important source of domestic fuel in the northern British Isles and in Scandinavia. Moss peat forms a dull coal or carbonaceous shale, lacking vitrinite and fusinite of the kind found in coals formed under swamps. Fossil mosses are well preserved in some ancient examples of bogs (Anderson & Anderson 1985, Meyen 1982).

Marsh

The terms marsh and bog are sometimes used broadly and synonymously, but marsh is here taken as vegetation of rhizomatous or rooted, herbaceous plants of acidic, freshwater waterlogged ground (Gore 1983). If waterlogging persists for most of the year, these soils accumulate a surficial layer (O horizon) of peat in which plant material may remain recognizable. Ancient examples include cuticle coals (Krassilov 1981) and silicified peats (Kidston & Lang 1921).

Little peat accumulates in marsh soils that are seasonally drained. In these, the underlying mineral horizons remain dark and gray with organic matter and with reduced minerals, such as siderite. Root traces, rhizomes, and pollen in such paleosols are well preserved (Batten 1973) because aerobic decay is inhibited by waterlogging.

Fen

This herbaceous waterlogged vegetation of rooted or rhizomatous plants is similar to marsh but for its neutral to alkaline groundwater (Gore 1983). The natural acidity of decaying plant material is mitigated in fens by a variety of factors: groundwater draining off limestone, carbonate-fixing species such as charophytes and snails, rainfall in the subhumid range, and a pronounced dry season. Fen soils are peaty, but also may contain caliche nodules and snail shells. Ancient examples may include some calcareous nodules of coal seams containing mainly ferns (Rex & Scott 1987) and thin, gray, pellet-rich horizons, with small root traces, within sequences of peritidal oolitic limestones [Darrenfelen paleosol of Wright (1987)].

Salt marsh

This herbaceous intertidal vegetation of rooted or rhizomatous plants is found mostly at high latitudes, beyond the latitudinal distribution (currently from 30°N to 45°S) of mangal (Chapman 1977). Salt marsh is restricted to muddy bays and estuaries that are protected from wave action. Small amounts of peat may accumulate in salt marsh soils, but much organic matter is utilized by oysters, mussels, polychaete worms, and crabs. These creatures leave a variety of distinctive fossils and burrows in salt marsh soils (Frey *et al.* 1978). Other distinctive features of salt marsh soils are nodules of pyrite and marcasite formed by reduction of marine sulphate by anaerobic bacteria (Altschuler *et al.* 1983). Salt crusts are found in salt marsh soils above mean tidal level. Small carbonaceous stems and roots are characteristic of ancient salt marsh paleosols (Schopf *et al.* 1966).

Swamp

Forests and woodlands of waterlogged ground are called swamps. A well known example is the bald cypress (*Taxodium distichum*) swamp of the Florida Everglades. Considerable thicknesses of peat accumulate under permanently waterlogged swamps. These peats are transformed during burial into coals of high calorific value, full of woody fragments (vitrinite and fusinite). The mineral portion of the soil under the peat is dark with organic matter, drab with reduced minerals, and often contains well preserved large root traces. There is a copious fossil record of Histosols in the form of coal seams (Dimichele *et al.* 1987).

Peat does not accumulate under seasonally dry swamps, but their soils are generally weakly developed, gray, and contain well preserved fossil trunks and large root traces. These kinds of paleosols (Aquents, Aquepts, Aqualfs, Aquults) also are common in ancient coal measure sequences (Gastaldo 1986).

Carr

Swamps are acidic in reaction like marsh, but carr is woody wetland vegetation growing in alkaline waters. The neutralization of soil acidity is achieved in ways already outlined for fen vegetation. The woody peats of carr can be distinguished from those of swamp by the presence of calcareous shells or nodules of siderite, calcite, or dolomite. Coal balls are calcareous nodules common in some Late Paleozoic coals, and they may have formed in this way (Scott & Rex 1985). Non-coaly, gray paleosols with calcareous nodules and large carbonaceous root traces may represent seasonally dry carr vegetation (Retallack & Dilcher 1988).

Mangal

The confusion of tangled roots and gnarled stems of trees growing in the intertidal zone are the basis for the name mangal. This term is now widely used to describe this plant formation instead of the term mangrove, which refers to a particular kind of plant (Chapman 1977). Many mangroves have distinctive and potentially preservable features, such as prop roots and air roots, that aid stability and respiration in waterlogged substrates. Other plants of mangal vegetation are not so distinctive but have physiological tolerance for the anaerobic and saline condition of mangal soils. Mangal soils are similar in many ways to salt marsh soils. Both are drab colored, may contain pyrite or marcasite nodules, and contain burrows and other evidence of marine organisms, such as oyster and mussel shells. Root traces preserved in mangal soils are larger and more varied in morphology than those of salt marsh soils. There is a long fossil record both of mangal paleosols and mangrove plants (Dimichele *et al.* 1987).

Fossil preservation in paleosols

Perhaps the most obvious evidence of organisms in paleosols is fossil remains. These are abundant in some paleosols because under suitable conditions they were concentrated on ancient land surfaces. Paleosols also provide evidence for past life where no fossils are preserved. Bones, for example, are not preserved in acidic paleosols, or leaves in oxidized profiles, yet ecosystems of considerable biomass, such as tropical rainforest, grow in such non-calcareous red soils. Such thick red soils may represent breaks in sediment accumulation of many thousands to millions of years. Other kinds of soils reflect the time scales of plant succession, of the order of tens to hundreds of years. Information from paleosols on preservational biases against fossilization and on the duration of ancient ecosystems may be useful to check or enlarge inferences drawn from the fossil record of life on land. Evidence from paleosols is especially useful because it is independent of the fossils themselves.

Preservation as a function of soil chemistry

Numerous processes act to affect the fossil record of paleosols. Ecosystems provide a pool of potential fossils but, for various reasons, few of these ultimately find their way into the soil. Most are eaten or decayed, then trampled or weathered at the surface until unrecognizable. Fragments of bone or leaves also can be mixed with other remains by the action of wind and water. Once an assemblage of fossils is buried or trampled into a soil, it is subject to a new round of destructive processes. Trampling and physical compaction by swelling of wet clay result in cracking and distortion of fossils. Further microbial decay and chemical alteration provide a final filter that may determine what ultimately is preserved in paleosols.

The preservation of plant leaves and other fossil organic matter in paleosols is most clearly related to the degree of oxidation of soils. This is not so much because the fossils are directly oxidized but because the most effective microbial decomposers in soils have an aerobic metabolism. For this reason, organic matter does not accumulate in well drained soils. In soils where the availability of oxygen is limited by stagnant groundwater, thick deposits of peat form. These waterlogged soils also are the ones most likely to preserve fossil leaves and pollen. Paleobotanical research has unearthed a great deal of information about wetland vegetation but little concerning plants of dry soils. The activity of decomposer microbes is limited to a lesser extent by soil acidity. Spores and pollen may be preserved in well drained soils if they are strongly acidic (pH less than 4) or strongly alkaline (pH greater than 9). Charcoal is appreciably more resistant to decay and it has longer residence time in soils than other kinds of organic matter (Retallack 1984a).

Plants also produce a number of mineralized hard parts that can be fossilized in paleosols. Much of the coarse texture of grass and its ability to form hay when the leaves are dead are due to encrustation with silica in the form of opal. Small opal bodies accumulate in soils as the plants die or are burnt. Some soils are as much as 60% phytoliths near the surface (Dixon & Weed 1977). Plant opal is more soluble than quartz at a pH of less than 9. Both are readily dissolved at more elevated pH (Leo & Barghoorn 1976). The woody pits (endocarps) of hackberry fruits (*Celtis occidentalis*) contain 25–64% dry weight of calcium carbonate and 2–7% silica (Yanovsky *et al.* 1932). These stones of fruits may accumulate in soils, provided the soil is not acidic (pH less than 7). Rainwater may have a pH as low as 5.7 before it is considered to be polluted or acid rain (Baas-Becking *et al.* 1960) and thus dissolves calcareous phytoliths on the surface of soils.

Like calcareous phytoliths, the remains of calcareous snails and the calcite-impregnated carapaces of millipedes and woodlice are also dissolved in rain and in acidic soils. Dissolution of snail shells proceeds rapidly once the organic outer coating (periostracum) decays aerobically (Evans 1972). Millipede exuviae are even more ephemeral because few species are heavily calcified and many millipedes consume their abandoned exoskeletons (Neville 1975).

Bones and teeth are composed of a variety of calcium phosphate minerals (Dixon & Weed 1977), which are also dissolved in acidic soils. They can persist under more acidic conditions (down to about pH 6) than the calcium carbonate of snail shells. In seasonally dry subtropical Zimbabwe, human bones in 700-year-old graves were well preserved under alkaline (pH 6.2–7.9) termite mounds but had been completely destroyed under adjacent acidic (pH 4.1–5.4) soils (Watson 1967). The so-called "bog people" are corpses interred in acidic bogs of northwestern Europe for thousands of years (mostly from about 100 B.C. to A.D. 500).

They show varying degrees of bone decalcification and, in one case, complete loss of bone within well preserved skin and other soft tissues (Glob 1969). Teeth persist in acidic conditions for a longer time than bone because they are denser and less porous and so present a smaller internal surface area to dissolving solutions (Shipman 1981). Smaller bones have a higher surface to volume ratio than large bones and so are more readily dissolved. This may be the reason why bones of small and young animals are rare compared with those of adults in many paleosols (Carpenter 1982, Gordon & Buikstra 1981).

Coprolites, or fossil feces, have compositions as varied as the diets of the creatures that produced them. Coprolites of organic matter, such as plant fiber or hair, are preserved in anaerobic, waterlogged environments, but they also can be preserved by freezing, by acidity in peat bogs, or by desiccation in deserts. These extreme situations inhibit the activity of aerobic microbial decomposers (Heizer & Napton 1969). Coprolites of birds and other carnivores often contain appreciable amounts of bone and other broken-up materials of phosphatic composition (Edwards & Folk 1979). These may be preserved under the same kinds of alkaline conditions in which bones are preserved. In alkaline-oxidizing paleosols, where organic coprolites of herbivores are destroyed, almost all the preserved coprolites may be those of carnivores, despite the much greater original abundance of herbivores.

From these considerations, each kind of fossil can be considered to be chemically stable under certain general conditions of pH and Eh (Fig. 10.11). This model predicts only the most usual case. Depending on their pH and Eh, paleosols tend to be dominated by the kinds of fossils favored for preservation. This is not to say that an occasional fossil skeleton overwhelmed by a flood may not be found within a non-calcareous paleosol. Such fossils will be rare and show other signs of being exceptional, such as full articulation of the bones.

Preservation as a function of time

The accumulation and destruction of fossils in paleosols takes time. In the calcareous and saline soils of Amboseli National Park, Kenya, it has been suggested that 10 000 years would be needed for the accumulation of bones of large mammals to a density of 100 bones per 1000 m^2 (Behrensmeyer 1982a). In acidic soils of Zimbabwe, in which human skeletons in 700-year-old graves had been destroyed, the skeletons of farm animals only 20 years old showed evidence of corrosion and flaking (Watson 1967). This is not to imply that the accumulation or destruction of fossils in paleosols is constant. Soil development and ecological succession go hand in hand. The differential preservation of fossils can be related to the time over which the paleosol formed, insofar as both community composition and soil

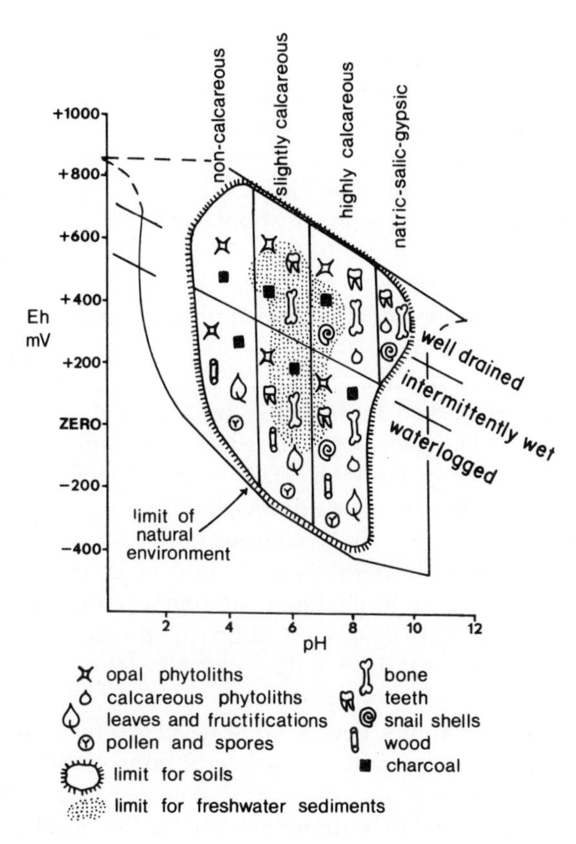

Figure 10.11 Theoretical Eh-pH stability fields of common kinds of terrestrial fossils preserved in paleosols (from Retallack 1984a, reprinted with permission of the Paleontological Society).

chemistry change during plant succession and soil development (Retallack 1984a).

The Early Eocene Willwood Formation of Wyoming is a good example of a sequence of paleosols within which plant localities are separate from vertebrate localities. Fossil leaves are found in nearstream shales and shaley paleosols (Entisols) and in coal-bearing paleosols (Histosols) and so reflect early successional and swamp vegetation (Wing 1984). Fossil snails and mammal bones are found mainly in well drained red Alfisols formed under stable floodplain forests (Bown & Kraus 1981b). Only fossil hackberry (*Celtis phenacodorum*) pits in the paleosols and rare and badly preserved Juglandaceous (walnut family) pollen grains in nearstream shale remain as direct paleobotanical evidence of these stable floodplain forests. The forests in which the mammals lived are thus poorly represented by fossil plants. It would be a mistake to assume that the diet of the fossil mammals is represented by the well preserved leaf assemblages or that this kind of vegetation was an important selective pressure in their evolution. The fossil plant assemblages represent an ephemeral and local vegetation. Vegetation that covered the region for most of the time is better represented by the red paleosols.

Completeness of the fossil record

It has long been appreciated that the fossil record provides a very biased impression of the past, but without evidence independent of fossils it is difficult to determine how biased it is. Paleosols provide evidence of the kind needed. Regardless of whether they contain fossils or not, they are themselves evidence of a former ecosystem. Since most current land communities produce leaves, charcoal, phytoliths, land snails, bones, teeth, and coprolites, then all of these categories of fossil should be preserved in every paleosol of a sequence if it had a complete fossil record. Real sequences of paleosols fall far short of this expectation for reasons such as differential preservation due to paleosol chemistry (Retallack 1984a). The most complete record of fossil plants is found in paleosols of swamps (Histosols and gleyed soils of other orders). The most complete record of fossil mammals is in well drained alkaline paleosols of dry climates (Alfisols, Mollisols, and Aridisols). These findings are equivalent to the common observation that fossil plants are best preserved in coal measures and fossil mammals most common in calcareous red beds.

Paleosols may also be useful for assessing how completely an assemblage of fossils represents a former community. Fossil leaf litters preserved where they fell on the soil provide the most unbiased record of former vegetation. These contain leaves showing great variation in their degree of decay, and also organs of differing robustness, such as flowers, fruits, and logs, that would be segregated if the remains had been transported. Fossil leaf litters are especially common in mangal, swamp, and early successional lowland paleosols (Retallack 1977, Retallack & Dilcher 1981b). Similarly, accumulations of fossil bones within paleosols on which the animals lived present a truer picture of past animal communities than those special cases where herds of animals are overwhelmed by volcanic ash or other catastrophes. Such catastrophic assemblages may be beautifully preserved and attract admiration in museum displays, but often reflect circumstances, such as seasonal migrations or aggregation around a salt lick, that are as special as those surrounding their preservation. Assemblages within paleosols, on the other hand, provide a time-averaged view of what was most abundant during formation of the soil (Shipman 1981).

This is not to say that the fossil record of communities in soils is unbiased. Some leaves are destroyed preferentially in leaf litter (Ferguson 1985). In my own observations of the leaf litter of a living scribbly gum (*Eucalyptus haemastoma*) and red gum (*Angophora costata*) woodland near Sydney, Australia, the high lignin and phenol content of the two dominant tree species favored their preferential preservation over a host of other perennial trees and shrubs with smaller and more fleshy leaves. Such differential preservation may be the reason why fossil leaf litters in nearby Triassic rocks are dominated by large, robust leaves of *Dicroidium zuberi*, whereas associated shale deposits contained a greater diversity of fossil leaves (Retallack 1976, 1977).

Studies of the abundance of modern bone on soils of Amboseli National Park, Kenya, have shown that the animals died and their bones were preserved where they lived, but there is a strong bias against preservation of small bones (Behrensmeyer *et al.* 1979). This can be attributed to the higher surface to volume ratio of smaller bones, which are thus more prone to acidic dissolution in rain and soil solutions than are larger bones. Indeed, the slope of the line relating abundance to the size of animals was 0.68, very close to the value expected from surface to volume scaling (0.67). Such size biases can be seen also in Oligocene fossil assemblages from Badlands National Park, South Dakota, USA, where bones are generally regarded as beautifully preserved. This bias against small animals can be contrasted with the situation in living populations of mammals, in which smaller animals are generally more abundant than larger ones (Damuth 1982). Using this yardstick, the fossil assemblage of the more calcareous silty wooded grassland paleosol (Conata Series of Fig. 10.12) is more representative of past communities than that of the less calcareous clayey gallery woodland paleosol (Gleska Series). The size bias in each assemblage has been removed (by Damuth 1982) using the correction factor from studies at Amboseli, with the result that Conata and Gleska assemblages are made to look more like modern assemblages. Interestingly, chevrotains (*Leptomeryx evansi*) and rabbits (*Palaeolagus haydeni*) are still well above the line of equal biomass in the Conata assemblage, as are oreodons (*Merycoidodon culbertsoni*) and three-toed horses (*Mesohippus bairdi*) in the Gleska assemblage, confirming the dominance of these taxa in the raw collections. In this case, taphonomic corrections did not overturn a prior impression of ecological dominance of these species.

These difficulties in interpreting paleocommunities from fossils in paleosols pale in comparison with the problems of reconstructing assemblages from fossils in deposits of rivers and lakes. Studies comparing leaves in modern lakes with the vegetation growing around the lake have shown that leaves of understory shrubs are under-represented in lakes compared with leaves of canopy trees (Ferguson 1985). Decay, abrasion, and mixing of leaves from different communities means that ecological reconstruction of the surrounding vegetation is difficult from the material in the lake alone. The movement of fruits and other plant parts in large deltas is even more complex (Sheihing & Pfefferkorn 1984). Although the

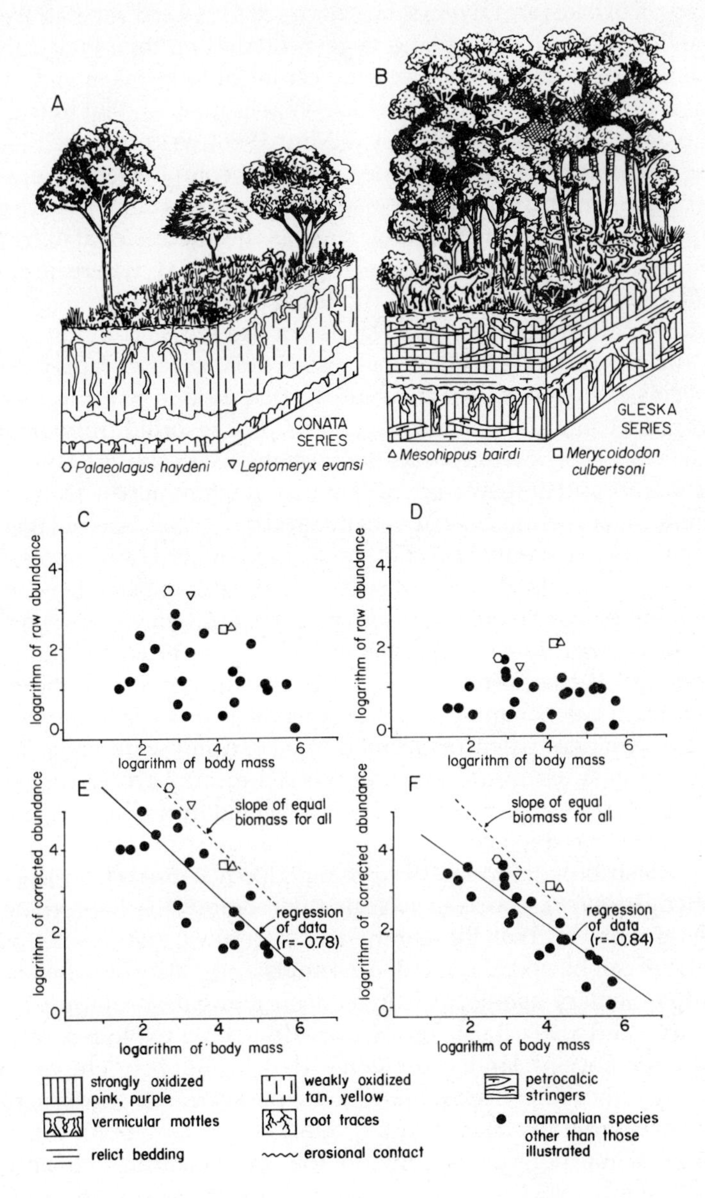

Figure 10.12 Restoration of interfluve wooded grassland on Conata Series paleosols (Andic Ustochrepts) (A) and of streamside gallery woodland on Gleska Series paleosols (Petrocalcic Paleustalfs) (B) during Oligocene deposition of the Scenic Member of the Brule Formation, in Badlands National Park, South Dakota, with plots of the abundance (number of specimens) for various species of fossil mammals in collections from each kind of paleosol (C and D, respectively), and after correction for size-related bias caused by acidic dissolution of bone (E and F, respectively) (from Retallack 1988b, reprinted with permission of the Society of Economic Paleontologists and Mineralogists).

fossil record of lakes and rivers is biased toward lowland vegetation, some far-travelled fragments of upland vegetation do find their way into lakes and rivers. The rain of pollen into the center of large lakes and into the ocean is more representative of regional vegetation than that found closer to shore or in a paleosol (Chaloner & Muir 1968). Without such records, nothing at all might be known of the taxonomic composition of vegetation in well drained inland soils. Lakes also may preserve a more precise record of temporal variation in vegetation, such as the successional eutrophication of small lake basins (Smiley & Rember 1981). There are, then, advantages to studying paleosols and fossil plants together.

Land animals are preserved in lakes but are not as common as fossil plants and fish (Grande 1980). Most fossils of land vertebrates are found in paleosols or in deposits of streams. Bones found in streams are mixed from various communities in the watershed or resorted from geological deposits exposed in stream banks (Behrensmeyer 1982a). Although there is some mixing and time-averaging of animal remains in paleosols, it is not so extreme as in assemblages of stream deposits. Fossil assemblages from paleochannels are nevertheless of value in providing an impression of regional vertebrate diversity, particularly of small mammals. In sequences in which the paleosols are unfossiliferous because they were too acidic, streams may provide the only evidence of past vertebrate life. In the Miocene Dhok Pathan Formation of northern Pakistan, bone is more easily collected from paleochannels and streamside sediments than paleosols. There is a variety of paleosols formed under swamp woodland, tropical deciduous forest, early successional woodland, and wooded grassland (Retallack 1985). Such habitat variety is reflected in the high diversity of fossil mammals in the paleochannels, but it is difficult to determine how they were distributed through the mosaic of likely habitats from the fossils alone (Badgley 1986). The paleosols and their scrappy fossil record deserve closer attention, so that the extent and nature of mammalian habitat preferences can be specified. Paleochannels and paleosols offer distinct but complementary perspectives in reconstructing past communities.

Paleosols and their trace fossils provide a little exploited source of evidence for ancient land ecosystems that is independent of fossils. Conflicting indications from paleosols compared with the fossils in them are not necessarily mistakes of interpretation, but may point the way to fundamental questions concerning the role and evolution of life on land.

11 Topographic relief as a factor

Variation in the nature of soils with topography can readily be appreciated by comparing thin, rocky soils of mountain tops with thick, fertile soils of lowland plains. However, even in relatively featureless lowlands, the nature of soil varies profoundly, depending on whether it is well or poorly drained. This aspect of soil formation is not completely independent of the others discussed here because vegetation, microclimate, and age of land surfaces vary in different parts of the landscape. In small areas, however, other factors may be limited to such an extent that variation across the landscape constitutes a well constrained set of soils (toposequence) for obtaining mathematical expressions for variations in topographically related soil features (topofunctions). The undulating till ridges abandoned from the last glacial advance (Cary Till, ca. 14 000 years old) in north central Iowa provide examples of toposequences that are all the more striking because of the subtle topography involved, i.e., a relief of only 5 m from well drained summit to boggy bottoms. Soils from the top of the ridge to the bog show marked decreases in gravel and mean grain size and increases in profile thickness and amounts of clay, organic matter, and carbonate (Fig. 11.1). As an example of a topofunction, the increase in thickness (y) of the soil with distance from the summit down the slope (x) can be expressed by a polynomial equation: $y = 1.41 - 0.91x + 0.49x^2 - 0.034x^3$ (Ruhe 1969).

Other landscapes are rather more intimidating and inspiring, as when one is gazing over row upon row of alpine ridges and peaks. Such bold landscapes can also be resolved into particular elements characterized by distinctive slope-related processes. On steep alpine slopes, vegetation is sparse, the soils are eroded by snow melt and churned by frost heave, and rocks fall from cliffs above. Such processes result in thin, shallowly rooted, little-weathered and rocky soils, adjusted to the local environment of mountain slopes. The scale and scope of such processes do not lend themselves to strict analysis as topofunctions because of the great variation in climate, vegetation, parent materials, and age of land surfaces up and down mountainsides. A common term for lateral variation in soils across landscapes is "catena." Derived from the Latin for "chain," its links are

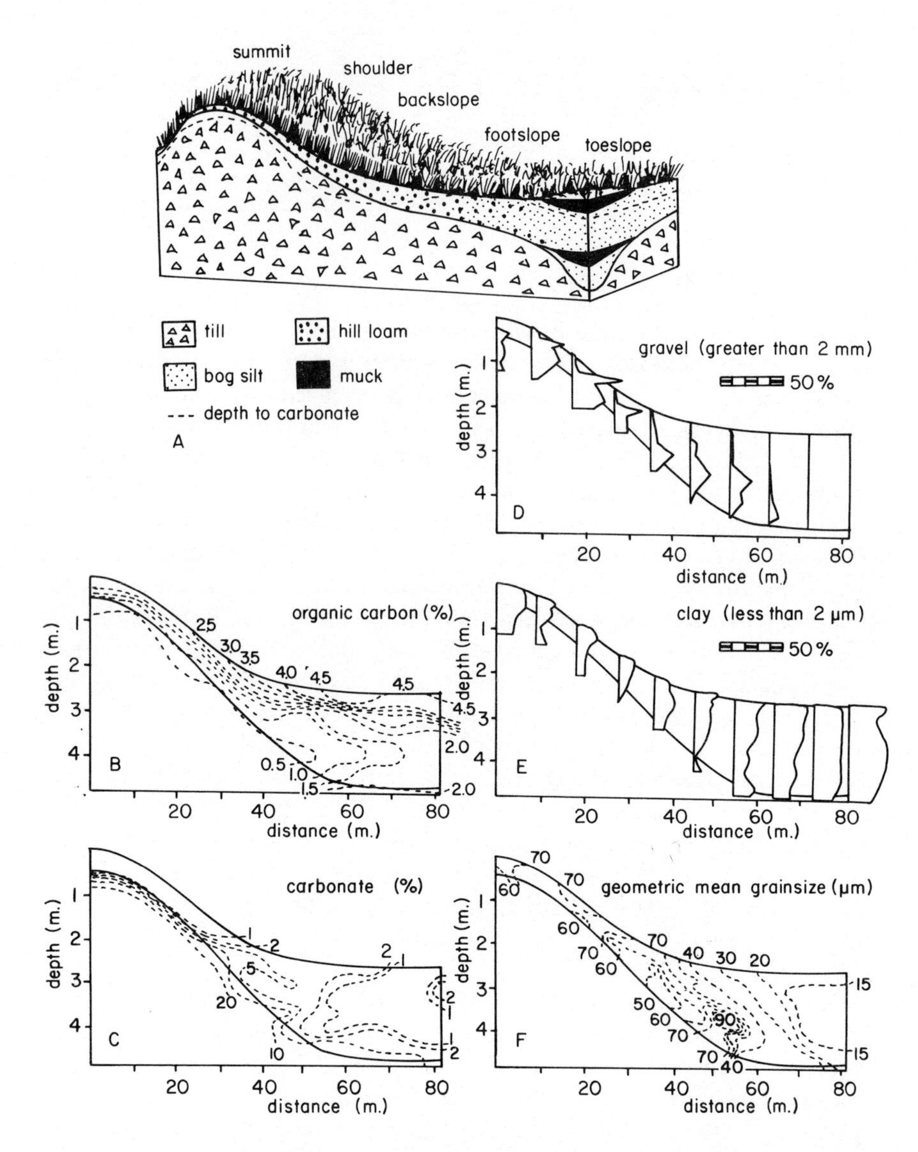

Figure 11.1 Downslope variations in mean grain size and percentage of gravel, clay, organic carbon, and carbonate in a soil catena on gently undulating Wisconsinan (ca. 14 000 years old) ridges of Cary Till around a bog near Jewell, northcentral Iowa (adapted from Walker, P. H. 1966, with permission of Iowa State University Press).

metaphors for the adjacent soil types. Catenas include any group of soils side-by-side on a landscape, regardless of whether they are, or are not, well constrained toposequences. The usual unit for a soil of a particular kind found in a catena is the soil series, as has already been discussed (see Fig. 6.1). Paleosol series and paleocatenas are merely buried soil series or catenas within alluvial sequences or along bedrock uncomformities.

Indicators of past geomorphic setting

Within modern landscapes, the following very general variation in soil features is found on descending from high to low elevations. Soil profiles are thicker, more organic, more moist, more strongly colored, less differentiated into horizons, more alkaline, more likely to contain soluble salts, less likely to have well developed hardpans of truncated horizons, and more likely to be formed on alluvial or colluvial parent materials (Buol *et al.* 1980). These are all major differences in soil character. They are, however, relative differences. They may be observable in paleocatenas but call for exceptional lateral exposure to be discernible. Differences in age of different parts of a paleocatena make their interpretation difficult. Only features that independently indicate a specific geomorphic setting are considered here.

Soil creep

Soil creep can be envisaged as a kind of landslide but with a slow, steady movement of the surface soil rather than a rapid movement of a large mass of soil and rock. Like fenceposts that gradually tilt downhill, rocks and other erect objects in the soil curve toward a prone position near the surface of the soil. This constitutes the best evidence of soil creep in paleosols (Williams 1968). Such indications of creep imply a former slope, in addition to the former direction downhill. The steepness of the former slope may be difficult to determine, because creep can be initiated on slopes as gentle as 2.5° during the summer melting of frozen ground (Washburn 1980).

Unconformities

The irregular contacts between different sedimentary sequences may preserve the lower parts of rugged landscapes of the past. A rise in sea level or the damming of a valley by a landslide can lead to burial of the landscape by sediment. Fossilized landscapes of this kind are widely known and include sea stacks, coastal cliffs (Dott 1974), box canyons (Leary

1981), radial drainage (Stewart *et al.* 1986), and glacial valleys (Andreis *et al.* 1986, Herbert 1972). The most spectacular examples of these are preserved at the contact between rocks differing in age by millions of years, such as sandstone overlying granite. Such situations may preserve original topographic relief well (Figs. 11.2 & 11.3).

Less spectacular are smaller disconformities within sedimentary sequences. Paleochannels of ancient streams are the most obvious of these because they are filled with contrasting materials, usually sandstone and conglomerates, compared with the clayey floodplain deposits. Nevertheless, stream-channels simultaneously erode from cut banks and deposit a point bar, so that the erosional disconformity at their base is not a true reflection of paleotopography at any one time. Certain features of sediments, such as the height of low-angle cross-sets of alternating shale and sandstone formed in point bars and levees, can be used as a guide to local topographic relief around stream channels (Bridge 1985), and so can the tracing of individual paleosols over levees and point bars into the floodplain (Retallack, 1981c).

Unconformities that represent gully erosion of floodplains are more subtle than paleochannels but can be recognized from associated paleosols. Each paleosol represents an unconformity, the duration of which is reflected in its degree of development. Longer periods of non-deposition may be marked by abrupt changes in the nature and spacing of paleosols. In Badlands National Park, South Dakota, the base of each rock unit has weakly developed, well spaced paleosols each overlain and separated by a thick increment of sediment. In contrast, the top of each rock unit has strongly developed, closely superimposed, sometimes overlapping and eroded paleosols. These differences represent a kind of equilibrium between subsidence and sedimentation attained near the end

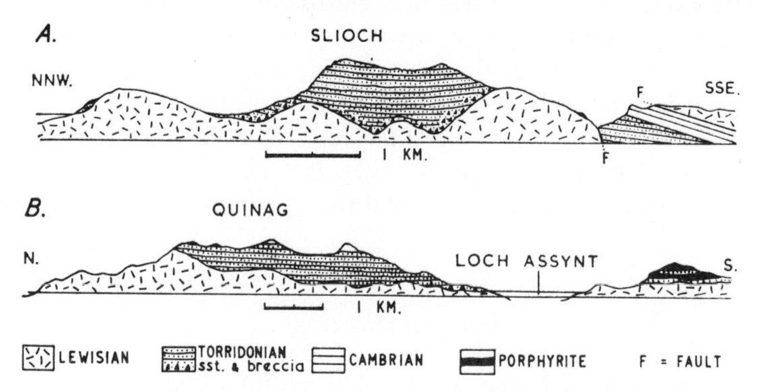

Figure 11.2 Geological sections drawn with no vertical exaggeration of paleotopographic relief on Lewisian gneiss (at least 1450 million years old) and the unconformably overlying Torridonian alluvial deposits (ca. 1000 million years old) in Slioch and Quinag Mountains, northwest Scotland (from Williams 1969, reprinted with permission of the University of Chicago Press).

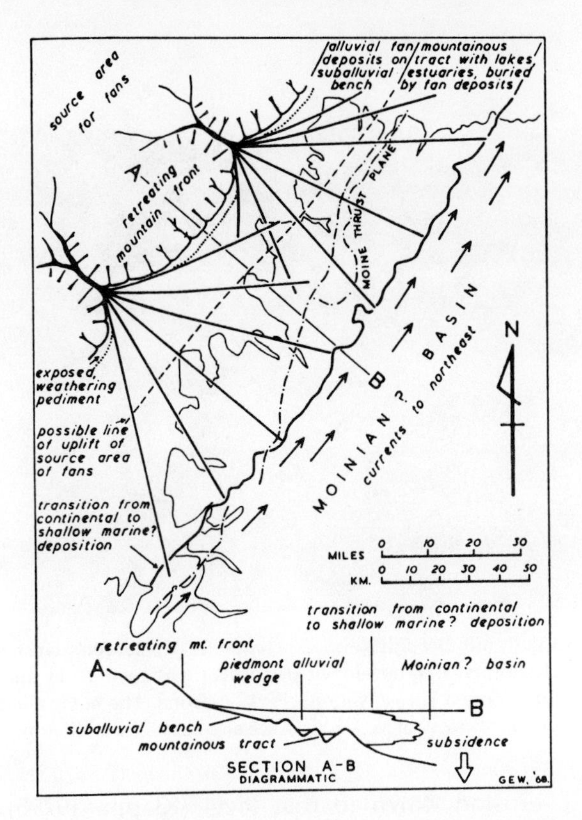

Figure 11.3 Suggested paleogeography of northwest Scotland about 1000 million years ago, during deposition of large alluvial fans of the Torridonian sequence from crystalline highlands to the west, now rifted off into Greenland (from Williams 1969, reprinted with permission of the University of Chicago Press).

of deposition of each unit until a period of gully erosion ushered in a time of more rapid sediment accumulation (Retallack 1986d). Such "aging upwards" sequences of paleosols are recognized also in other sequences (Fig. 11.4). The abrupt boundaries between such genetic units represent erosional landscapes.

Erosional planes

Some buried paleosols have erosional planes both within them and on top. These tend to form in paleosols developed in higher parts of the landscape and are therefore prone to erosional truncation followed by addition of new material. Such features are difficult to interpret even in modern surface soils (Johnson & Watson-Stegner 1987). The abrupt and laminated tops of many hardpans, such as petrocalcic horizons, have been thought

Figure 11.4 Paleogully into strongly developed sequence of paleosols (dark coloured), filled with alluvium including weakly developed paleosols (light colored) in the Late Triassic, Chinle Formation in Petrified Forest National Park, Arizona. The hill in the foreground is 11 m high (photograph courtesy of M. J. Kraus: see also Kraus & Middleton 1987).

to be due to erosion down to that level (Klappa 1979b), although dissolution by perched groundwater is also a possible explanation (Machette 1985). Erosion also may be indicated by a line of stones along an erosional plane in an otherwise stone-free soil (Ruhe 1959). It takes a powerful stream or flood to transport pebbles, compared with finer grained soil. Another way in which stone lines can form is by the burrowing activity of gophers, because stones in the soil fall through burrows to a zone beneath that is less actively bioturbated (Johnson *et al.* 1987). Such stone lines, however, are irregular in shape and often festoon-shaped under individual burrow systems. Erosional stone lines, on the other hand, form near planar surfaces.

Gilgai microrelief

Gilgai not only is a form of topographic relief on a small scale (see Fig. 4.3), but distinctive kinds of gilgai also form on level and on sloping land. There are two main kinds of gilgai (Paton 1974): those in which the depressions are equant in plan (nuram gilgai) and those in which they are elongate (linear gilgai). Nuram gilgai are common on level areas. This kind of micro-relief can itself be complex, with shelves and inner depressions, in places up to

3 m deeper than the mounds. Linear gilgai have not been found to have such great relief but they form perpendicular to contours on gentle slopes of between 15″ and 3°. Fossil examples of linear gilgai have been found (see Fig. 9.9), but extensive exposures are needed to recognize it.

Salt crusts

Soluble salts form in soils of dry climates in a variety of geomorphic positions, but the thickest surficial crusts are generally found in lower parts of the landscape. Soluble salts accumulate where ponded waters evaporate. These salts would continue downhill and out to sea were they not so close to the surface in the shallow water tables of desert playas, supratidal flats, and over-irrigated lowlands (Jenny 1941). Beds of salt minerals in sedimentary sequences commonly have been interpreted as subaqueous precipitates of saline lakes and seas, but careful analysis of associated sedimentary facies and structures has shown that some have formed in supratidal mudflats (sabkhas), like those around the Persian Gulf (Evans *et al.* 1973), or dried-out desert playas, like Lake Magadi in southern Kenya (Eugster 1969). Mudcracked dolomitic marls riddled with rossetes of trona in the Eocene Green River Formation of Wyoming, once considered lacustrine, are now thought to have formed at the surface of a dried-out lake (Eugster & Hardie 1975) and can be considered a kind of desert paleosol (Salorthid). Examples of marine-influenced evaporitic supratidal paleosols also are known (West 1975).

Cumulative horizons

Although soil formation and sediment accumulation are in some ways antagonistic processes, small amounts of windblown dust and floodborne alluvium do accumulate in soils. The rate of influx of these materials in well drained soils is commonly so slow that they are incorporated into the fabric of the soil or are eroded away, and do not form distinct surficial layers (Muhs 1983). In lowland soils, on the other hand, sedimentary layers may form at the top of the profile as vegetation grows progressively upward. These are called cumulative horizons (Birkeland 1984) or cumulic horizons in the US soil taxonomy (Soil Survey Staff 1975). There is a fine line between soils with thick cumulative horizons and ordinary sedimentary deposits. The distinction is largely a matter of root and burrow density, which reflects the relative influence of pedogenic, as opposed to sedimentary, processes. Cumulative horizons in paleosols may be a clue that they formed in low-lying parts of the landscape.

Indicators of past water table

The geomorphic setting of paleosols may not always be clear, but their relationship to the water table often is obvious and has a clear bearing on position in the landscapes of the past. The water table is that level below which the ground is permanently saturated. It separates a surface region of active leaching and more or less oxidizing chemical alteration from a stagnant region of little chemical change below. The water table may seem level when exposed as a pool of water at the bottom of a well. On a broader scale, however, it is not tabular. The water table is deeper on hillslopes than in valley bottoms, and follows elevation to a certain extent. It is perched on shaly impermeable subsurface layers and deeper within friable sandy materials. It is closer to the surface in humid forested regions than in barren deserts and varies in depth with the availability of moisture through the year, and even daily. There are three possible situations for a soil with respect to the water table (Fig. 11.5). It can be entirely above the water table, partly or wholly within the zone of water-table fluctuation, or entirely below the water table. Soils of the first kind are well drained. The second kind is flood-prone or seasonally swampy soil. The third is a permanently waterlogged soil of swamp or marsh. A variety of features allow discrimination between these general kinds of paleosols, as outlined below.

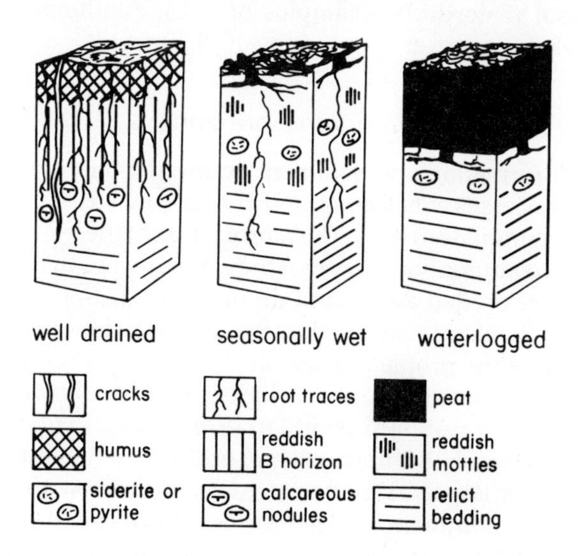

Figure 11.5 Characteristic patterns of root traces, cracking, peds, nodules, color, coal distribution, and relict bedding in formerly well drained, seasonally wet and waterlogged paleosols.

Soil horizons

The nature of horizons formed above the water table, within the zone of its fluctuation, and below it are distinct. Below the water table, conditions are chemically reducing. Many minerals are drab and gray with iron in the ferrous state or admixed organic matter, including peaty (O) horizons. Because water in this part of the soil is slow-moving and carries little material away, weathering of minerals below the water table is slight.

Within the zone of water-table fluctuation, the alternation of oxidizing and reducing conditions can produce strongly developed nodules or mottles of gleyed (Bg) horizons. Streaks of red and orange are striking in modern soils of this kind (see Fig. 4.7). Most paleosols have some of this character, because of color changes upon burial due to burial gley and reddening of ferric oxyhydrate minerals (see Fig. 7.7). Some care must be taken to disentangle features of the original soil from those imparted by conditions of burial.

Above the zone of water-table fluctuation are formed most other kinds of soil horizons, such as clayey subsurface (Bt) horizons (Cremeens & Mokma 1986). Characteristically these are yellow, brown, or red with iron oxyhydrate minerals, such as goethite or hematite. These warm colors may be altered during burial, but evidence of washed-in clay skins and the deep penetration of root traces will remain as evidence that the profile was well drained.

Calcic horizons differ in form, depending on whether they formed above or below the water table. Groundwater calcrete, precipitated below the water table, has a simple cement that fills open spaces between larger grains (Mann & Horwitz 1979). This is best observed in thin sections (Fig. 11.6B). Caliche, on the other hand, forms above the water table in aridland soils and has a more complex appearance (Wright 1982). It is usually micritic and can be seen to replace and surround clastic grains (Fig. 11.6A). Even such weather-resistant grains as quartz may show irregular dissolved margins (caries texture) and abnormally loose packing when viewed in thin section. In hand specimens, caliche is generally nodular to irregular in shape. Multiple generations of cement and local colloform or sparry fillings of cavities may indicate that the nodule could move slightly in the soil.

Root traces

The leaves of green plants require carbon dioxide to produce organic matter and oxygen, but roots are aerobic respirers. Unless roots have special aeration structures, they do not penetrate the water table. This is most obvious in waterlogged soils, in which roots spread out laterally (see Fig. 3.3). The water table indicated by root traces in a paleosol is not always in the same place indicated by soil horizons because of two common

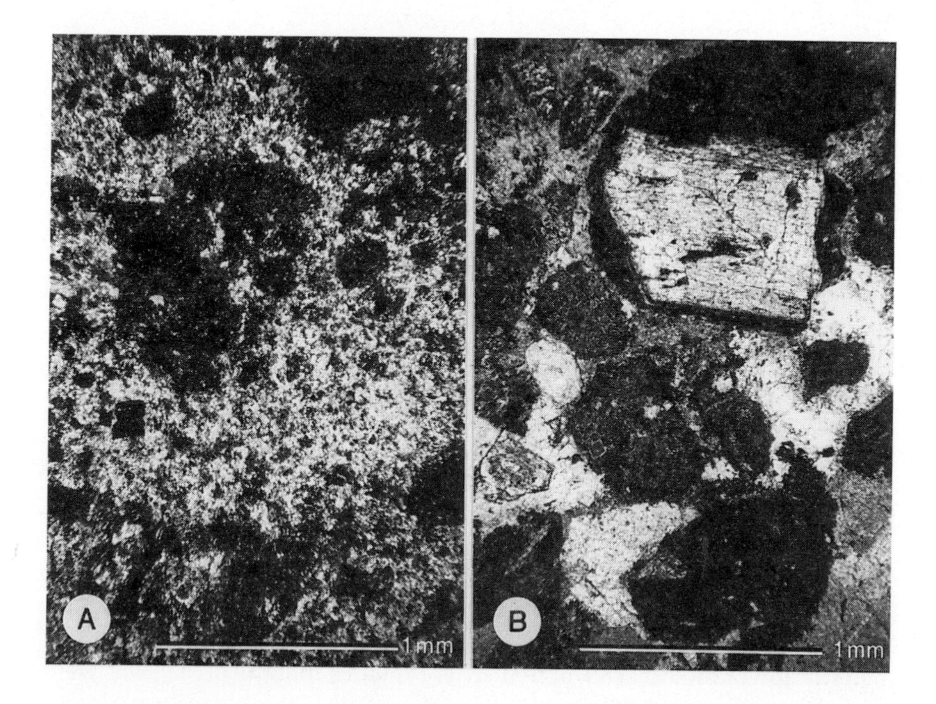

Figure 11.6 Petrographic thin sections of caliche (A) formed above the water table and groundwater calcrete (B) from middle Miocene (14 million years old), Fort Ternan Beds, near Fort Ternan, Kenya. Scale bars 1 mm (Kenyan National Museum, specimens FT-R14 and FT-R1 respectively).

complications. Where the soil is waterlogged for most of the year, except for a short dry season, root traces may extend deeply into a gleyed horizon that is dry for part of the year. In addition, it could be that fossil roots were preserved from a time of better drainage by the general rise of the water table with continued subsidence and sediment cover over the soil (Coleman 1988). These two situations can be resolved from the relative proportions of horizontal versus vertical roots and their quality of preservation.

Burrows

A few creatures, such as crayfish and crabs, live in water-filled burrows (Thorp 1949), but most soil animals are air breathers. Earthworms, insects, and mammals do not burrow below the water table. Where burrows are dense, their lower termination can be an indicator of the usual level of the water table (Retallack 1976).

Soil structure

Cutans are most characteristic of those parts of soils above the water table (Brewer 1976). Especially common in well drained parts of soils are clay skins (illuviation argillans), washed down into cracks from higher in the profile. Cutans formed above the water table tend to be yellow, brown, and red in color with sesquioxides. In contrast, waterlogged parts of soils have weakly developed cutans that tend to be dark with organic matter or manganese stain.

Cutans are the main evidence in paleosols of peds because open spaces are crushed out of paleosols during burial. Peds are largely a phenomenon of well drained upper portions of soils, where the soil can shrink and swell and where there is transport of materials. Waterlogged soils and parts of soils, in contrast, have weakly expressed soil structure.

Soil nodules

By their chemical composition, soil nodules and concretions reflect the availability of oxygen at the time they formed (see Fig. 3.13). Pyrite, marcasite, and siderite nodules are found in permanently waterlogged soils (Ho & Coleman 1969). Goethite and hematite concretions and nodules are found in better drained soils. Calcareous nodules form in dry, well drained soils. Concretionary banding, in some cases, reflects periodic change in chemical conditions. Iron delivered to concretions in a reduced state can be oxidized in concentric bands in ferric concretions. Complex nodules with reduced centers, oxidized rinds, and septarian cracking also may reflect fluctuations of oxygenation in paleosols (Sehgal & Stoops 1972). In each of these cases, care must be taken to ensure that the nodules are an original part of the soil rather than a phenomenon of deep burial or recent exposure.

Microfabric

Soil structures are just as diagnostic of the influence of water table when examined in thin sections as with the naked eye. The general microfabric of soils is different under waterlogged and well drained conditions. Bright clay fabric (sepic plasmic fabric, see Fig. 3.15) is, mostly a feature of well drained parts of soils (Brewer & Sleeman 1969). Highly birefringent segregations of oriented clay form in much the same way as clayskins, by washing down cracks or by the shearing associated with clay swelling. Waterlogged soils, on the other hand, have microfabrics similar to the parent material of a soil (asepic) or massive, nearly isotropic, fabrics (undulic and inundulic, see Fig. 3.16). The massive fabrics may be due to flocculation of unoriented clay and organic matter (Brewer 1976).

Interpreting paleocatenas

A soil catena can be characterized easily enough as the soils over the present land surface. Understanding how that particular suite of soils formed is not so easy, nor is it easy to determine which of a sequence of paleosols in sedimentary deposits constitute an actual ancient landscape. Soils are very much a part of the landscape, and their history is just as complex and intriguing.

Lateral variability of paleosols

One of the most irregular of buried landscapes is paleokarst developed on limestone or dolostone (James & Choquette 1987). Paleokarsts show considerable topographic relief, steep cliffs, and caves. The limestone surface is usually little altered beneath a sharp weathering front. The filling material may range from manganous to bauxitic in composition, depending on the degree of drainage of the karst depressions (Bardossy 1982). An added complexity of the evolution of karst topography is the collapse of limestone caves. The filling material in this case is a complex mix of brecciated fragments of roof rock, talus cones of breccia, silty deposits of streams, and shaly deposits of lakes. Complex sequences of cave fill have been reconstructed in detail for fossiliferous cave deposits of Swartkrans and other comparable early (1–2 million years old) human fossil sites in South Africa (Brain 1981).

Paleosols also vary laterally at major geological unconformities. A variety of paleosols have been documented on the ancient basement of granite and greenstone underlying the 2300-million-year-old Huronian sedimentary sequence north of Lake Superior, Ontario, Canada (Gay & Grandstaff 1980, Kimberley *et al.* 1984) and underlying the 1000-million-year-old Torridonian sedimentary sequence on the northwestern coast of Scotland (Williams 1969). In this second example, regional topographic relief has been measured to be at least 1000 m. Hills of Lewisian gneiss have been exhumed from under their Torridonian cover by more recent erosion (Fig. 11.2). On a regional scale, paleosols are much thicker (3 m) on the undulating pediment to the southwest, whereas paleosols are thin (only 10 cm) on hilltops. These differences, to some extent, reflect differences in age of the surfaces. Differences in age of different parts of this ancient landscape are difficult to estimate. The sub-Cambrian unconformity of the western United States is another ancient landsurface with considerable local topographic relief (Dott 1974) and strongly developed paleosols (Sharp 1940, Cummings & Scrivner 1980), whose burial age varies from Early to Late Cambrian (Lochman-Balk 1971), a duration of ca. 65 million years. The ages in this case are provided by fossil trilobites found in marine shales that overlie the basal Cambrian quartzites

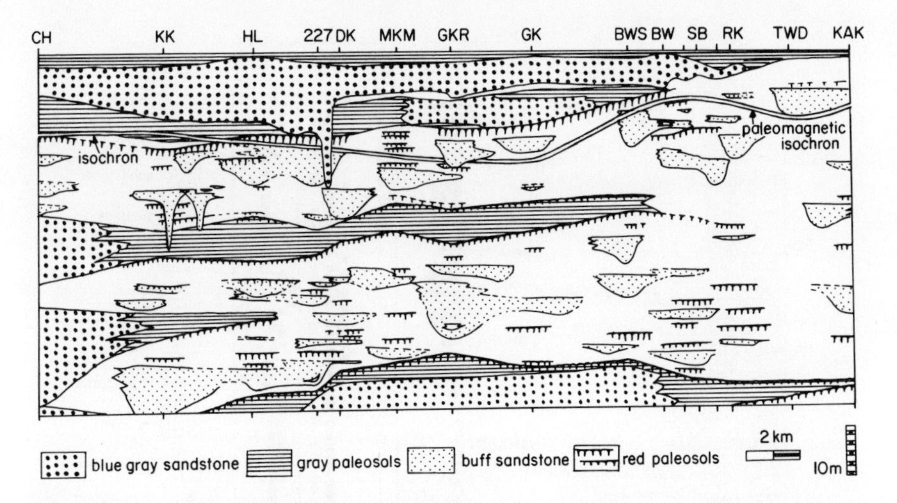

Figure 11.7 Lateral distribution of especially well developed paleosols, and channel deposits of an ancestral Ganges River (blue–gray system) and of drainage of the Himalayan foothills (buff system) and its relationship to a paleomagnetic reversal in the Late Miocene (8.3 million years old), Dhok Pathan formation, near Khaur, northern Pakistan (from Behrensmeyer & Tauxe 1982, with permission of the International Organization for Sedimentology).

of rivers and beaches overlying Precambrian metamorphic igneous and sedimentary basement rocks. Although stratigraphic marker beds and biostratigraphically useful fossils can be used to resolve age differences in such ancient landscapes, their resolution is poor (millions of years).

Lateral variations in paleosols of alluvial sequences have been widely observed but are difficult to document and interpret. Some of the strongly developed paleosols have been mapped laterally as geosols in Miocene (8.3 million years old) alluvium of the Dhok Pathan Formation of the Siwalik Group in northern Pakistan (Fig. 11.7). My own examination of these sequences has revealed many more paleosols than this (Fig. 11.8), but this does not alter the demonstration that geosols vary in age laterally. This is revealed by the truncation of a geosol by a paleomagnetic isochron, a reversal of the orientation of magnetic minerals that reflects a short-term reversal of the earth's magnetic field at the time these sediments accumulated (Tauxe & Badgley 1984). The varied paleosols along the time plane revealed by this paleomagnetic reversal are a better guide to a former catena of soils than the lateral tracing of geosols in the sequence. Ash beds deposited in a short time are also useful indicators of ancient land surfaces. Such a situation is known in Plio-Pleistocene (1–2 million years old) alluvial deposits bearing early human fossils east of Lake Turkana in northwestern Kenya (Burggraf *et al.* 1981). Without such exceptional indicators of time planes, paleocatenas cannot be precisely reconstructed.

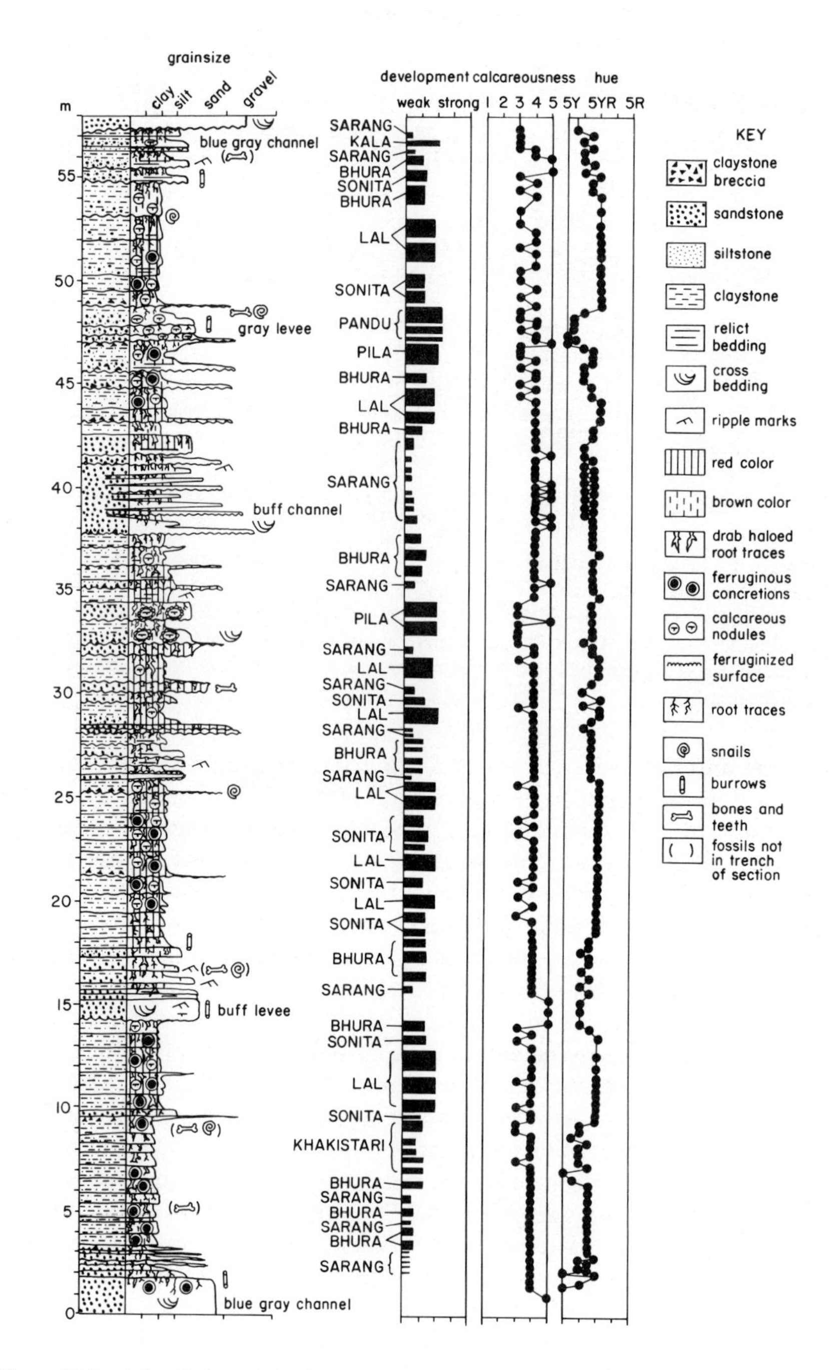

Figure 11.8 A detailed stratigraphic section showing different kinds of paleosols in the Late Miocene (8.3 million years old), middle Dhok Pathan Formation, in Kaulial Kas, near Khaur, Pakistan.

Vertical variability of paleosols

Another way of characterizing paleosols is by measuring a vertical section of the kind used in stratigraphic logging. The past geomorphic setting and relationship to the water table of each paleosol can generally be established from the various criteria already outlined. With information from associated sediments, such as sedimentary structures of paleochannel or lake deposits, these can be assembled into a general model of ancient landscapes. It may at first seem paradoxical that the lateral relationships of paleosols can be reconstructed from their vertical relationships, but this is an old technique of sedimentological interpretation, usually called Walther's Facies Law (Hallam 1981). Simply stated, the idea is that deposits formed side-by-side in nature are usually preserved on top of each other because of lateral shifts in the environments in which they form. A falling sea level, for example, results in a superposition of river, delta, and beach deposits over those of the ocean. A lateral shift of a meandering stream channel away from its point bar results in the superposition of soils in the silts and clays of its levee. Soils are thicker and better developed the further they are from the stream and the higher they are above the paleochannel deposits. The key to this common kind of geological inference is the genetic relationship between different parts of the landscape. This technique cannot be used across major unconformities that separate unrelated depositional systems. Major unconformities are obvious when they separate granitic basement from overlying sediments, but profound breaks within sequences of paleosols can also be detected by the degree of development, spacing, and erosion of paleosols (Retallack 1986d).

The Miocene (about 8.3 million years old) Dhok Pathan Formation near Khaur in northern Pakistan is a genetically related sequence of paleosols (Fig. 11.8) that can serve as a basis for reconstructing an ancient landscape (Fig. 11.9) independently of the lateral relationships already discussed. A carefully excavated and recorded stratigraphic section of these deposits has revealed eight different kinds of paleosols. Two of these (Pandu and Khakistari Series in Fig. 11.8) are drab-colored and associated with deposits of the ancestral Ganges River ("blue-gray sandstones" of Behrensmeyer & Tauxe 1982), and the other six kinds of paleosols are associated with deposits of streams draining the Himalayan foothills ("buff sandstones" of Behrensmeyer & Tauxe). The drab paleosols are interpreted as waterlogged lowland soils formed under tropical forest. The better drained foothill streams were flanked by two kinds of very weakly developed paleosols with sparse or small root traces, like that of early successional vegetation. One of these (Sarang Series) has abundant relict bedding and reflects the most disturbed near-stream areas. The other (Kala Series) has a subsurface horizon stained with manganese (placic horizon), an indication that it formed in poorly drained swales within the levee and

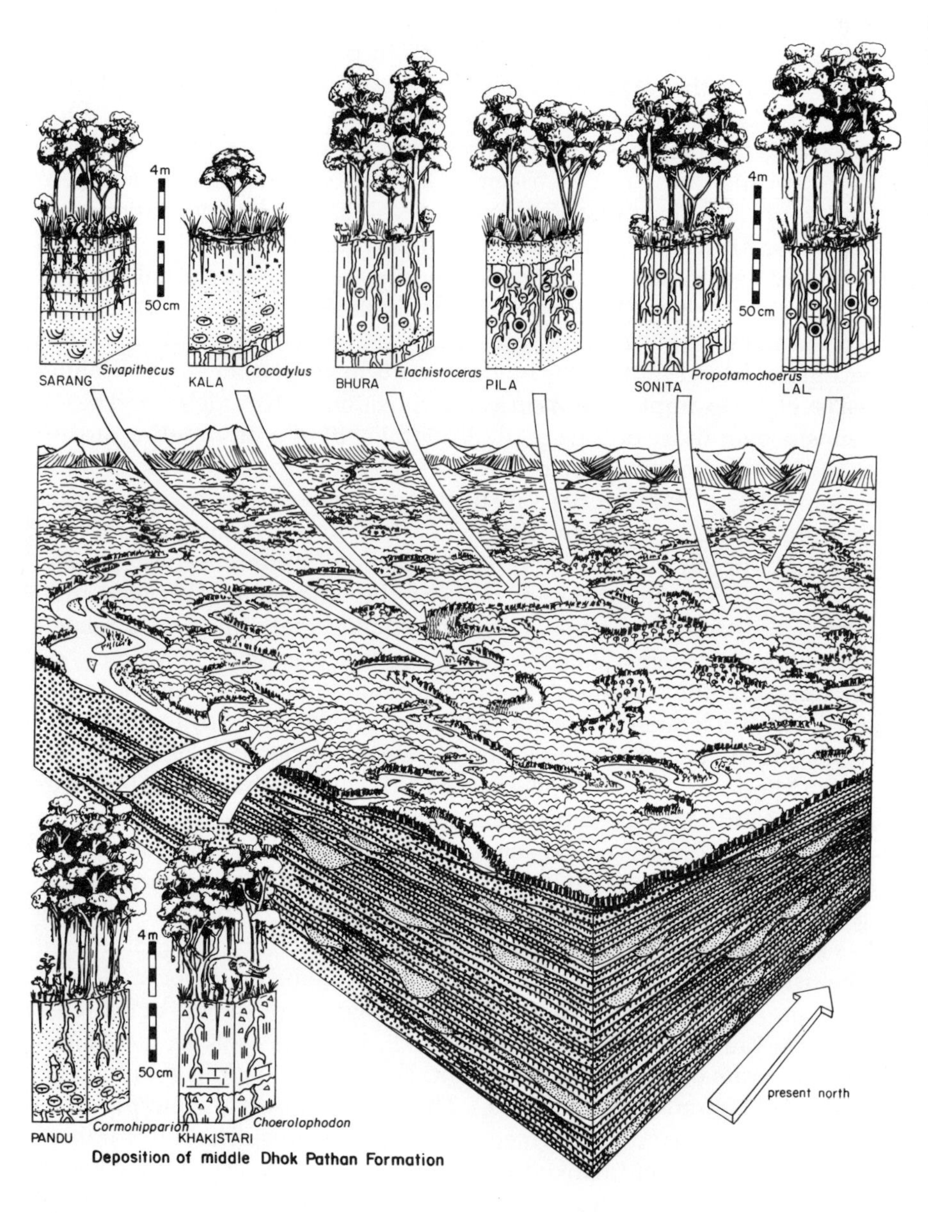

Figure 11.9 Reconstructed paleoenvironment of northern Pakistan during the Late Miocene (8.3 million years ago) deposition of the Dhok Pathan Formation of the Siwalik Group.

point bars of these streams. Weakly developed reddish brown (Bhura Series) and red (Sonita Series) paleosols have less obvious relict bedding and represent woodlands and forests at intermediate stages of plant succession. Another kind of paleosol (Lal Series) forming more elevated and better drained floodplains of the foothills drainage system is thick and red, with broad clayey B horizons and numerous large root traces. These may have supported tropical forest, similar to those of modern Uttar Pradesh. A final kind of paleosol (Pila Series) of floodplains are silty and yellow, slightly less developed, but still with recognizable clayey B horizons. Their root traces include scattered large roots and copious fine roots. This, together with their general profile form and degree of weathering, is evidence that they supported wooded grassland that was generally well drained. This is not a precise reconstruction of a particular landscape but rather a general concept of the relationship between observed sedimentary environments and paleosols.

A part of the charm of landscapes is that they are as individual as people, with a combination of characteristics inherited from parent materials and geomorphic position, or acquired from climate and vegetation. Soils are an important part of landscapes, and paleosols provide a partial narrative of landscape development.

12 Parent material as a factor

The rock or sediment in which a soil develops is its parent material and it is a starting point for the process of soil formation. In the early stages of their formation, soils are not much different from their parent materials. With time, fewer and fewer features of the parent material persist, and ultimately the soil takes on an identity of its own. The nature of a parent material is a base line that must be known in order to assess the amount of soil formation that has taken place.

Perhaps the clearest case where parent materials are known is soil developed on igneous rock, such as granite. In a completely unweathered granite, there are no clay or carbonate minerals. Even small traces of these detected around the margins of crystals in thin sections are a clue that the

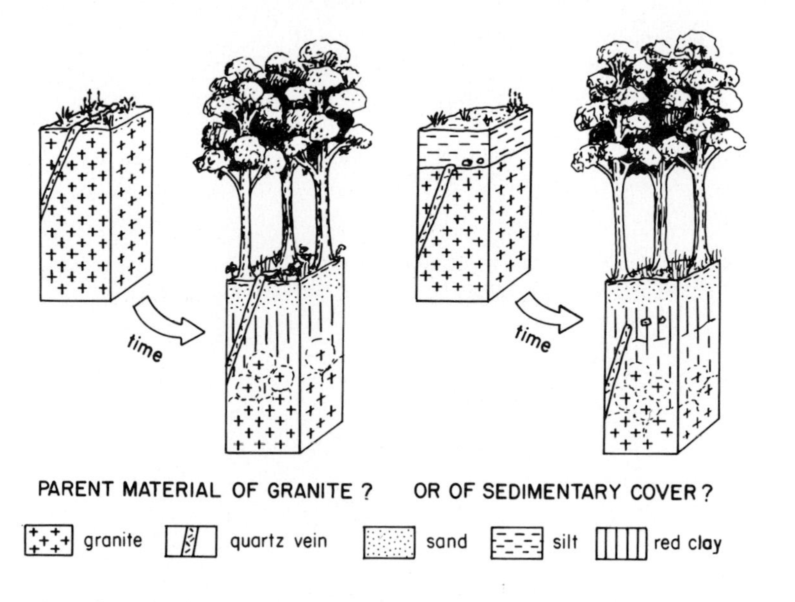

Figure 12.1 Soils developed directly on bare granite can appear similar to soils developed on a sedimentary cover on granite, although the latter may show quartz veins truncated well below the surface and a sharper base to the clayey subsurface (Bt) horizon.

rock is weathered. Small amounts of clay, carbonate, and iron oxyhydrates may be found in a transitional horizon (C) between fresh bedrock and the soil proper. This may look like parent material in its crystalline texture and other features but is considerably altered in comparison with true parent material. Material above this transitional horizon (in horizons A, E, and Bt) may show more characteristic soil textures and structures. Such soils with a deep transitional saprolite to unweathered bedrock are often assumed to have formed on fresh bedrock. This may be a fairly safe assumption if especially weather-resistant parts of the soil, such as quartz veins, persist to the top of the profile (Fig. 12.1), but what about the addition of airborne dust, landslide debris, or floodborne silt on top of bedrock? Could these additional layers of very different materials have been obscured by soil formation? Even more difficult to detect would be additions of salts or other materials from groundwater solutions.

All these doubts that could be harbored about the parent material of soils on bedrock apply with equal force to soils developed on alluvial parent materials. The differentiation of soil horizons in alluvium could be as much a reflection of the alluvial deposition of sandy over clayey layers as due to the washing down of clay into a subsurface (Bt) horizon during soil formation (Fig. 12.2). If the clayey subsurface horizon were primarily sedimentary, then the soil would be regarded as young and weakly developed. If, on the other hand, the clay is entirely pedogenic, then the soil would be interpreted as old and strongly developed.

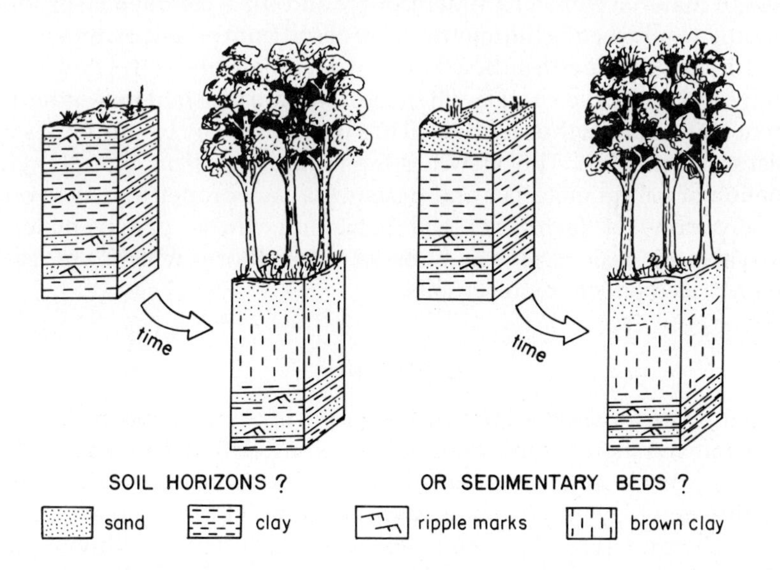

Figure 12.2 Soils developed on fairly uniform alluvium can appear similar to those developed on a less homogeneous sedimentary sequence, although the latter may have more sharply defined lithological boundaries and more obvious relict bedding.

In most cases, parent material is an independent variable in soil formation. Plutonic, metamorphic, and volcanic rocks especially are formed by processes well below the Earth's surface. Sediments are often parent material for soils and are derived from pre-existing soils, but the ways in which sediments are sorted and distributed are so variable that these, too, are independent of soils. Geological maps of many regions reveal a great variety of bedrock and sediment that is reflected in variations of soils formed on them. Certainly there are situations where one rock type is more weather-resistant than others and so forms steeper slopes, such as those of a metamorphic hornfels around a large granitic intrusion (Ollier 1969). These weather-resistant ridges of rock also may encourage different microclimatic conditions and vegetation and have younger cliffs and scarps than nearby deeply weathered rocks. Therefore, in only a few lithosequences are soil-forming factors well enough controlled to yield useful lithofunctions. The sheer chemical and physical variation of soil parent materials also is a problem for studies of their role in soil formation. The roles of some general properties of parent materials in soil formation are becoming well understood, and these will be considered before common kinds of parent materials.

General properties of parent materials

If parent material were a natural resource and soil a commercial product, economy would dictate the manufacture of soil from parent materials most like it in bulk properties. Some parent materials are converted to soil rapidly and with little energy, whereas others are resistant to weathering. Parent materials can thus be viewed in a functional way by properties that assist soil formation. This view has the virtue of glossing over the great variation of parent materials and focusing on key properties. As in other considerations of factors of soil formation, those parent materials observable in paleosols are stressed over those which are easily demonstrated only in modern soils.

Uniformity

Few parent materials are completely uniform in composition or structure. Almost always there is some irregularity, some foliation, veining, jointing, or layering that in some cases aids and in other cases hinders soil formation.

Sedimentary layering in some cases assists soil formation, as in a silty cover on bedrock (Fig. 12.1) or a sandy over a clayey layer of alluvium (Fig. 12.2). In both cases, friable material, easily penetrated by roots and weathering solutions, has been created by non-pedogenic means. Other situations of exposed bedrock or of stiff clay are not so conducive for soil formation. In some cases, non-uniform parent materials may be difficult to detect in soils and paleosols. Anomalous mineralogy may be a clue. A few grains of primary minerals that are not found in the parent material can provide an indication of later additions. Quartz, for example, does not occur in phonolite, nor does olivine in granite. Grain size irregularities also may be an indication. Formation of clay skins and oxidized grain coatings may obscure in the field what are obvious sedimentary layers of different grain size under the petrographic microscope. Perhaps the most obvious signs of grain size irregularities are stone lines, i.e., very thin zones of scattered pebbles in an otherwise fine-grained paleosol. Stone lines often can be seen to lie on erosional planes (Ruhe 1959). These and any other sharp contacts in a soil should be grounds for suspicion that the soil developed in several stages on different layers of parent material (Johnson & Watson-Stegner 1987).

Another source of non-uniformity that aids soil formation is joints, for example, deep cracks in large bodies of granitic rocks formed by contraction on cooling and expansion following unloading during exposure at the surface. These cracks allow the deep penetration of air and water so that material on either side of the joint is more deeply weathered than the rock which they enclose. The progressive weathering of blocks of rock defined by joints leads to the formation of unweathered corestones with concentric rims of weathered rock (Augustithis & Ottemann 1966). Liesegang banding can form in a similar way or by reaction fronts in slowly diffusing pore waters (Ortoleva *et al.* 1986). This onion-skin weathering is common in soils and paleosols (Williams 1968, Boucot *et al.* 1974).

The effect of foliation, folding and veining can be more complex. Quartz-rich parts of gneissic foliation, quartzite beds folded into schist, and quartz veins through granite all resist weathering more than their matrix. They commonly stand out as ribs on the ground or within the soil or paleosol (Williams 1968). These situations are not difficult to identify because the non-uniformity commonly is oriented at a high angle to soil horizons.

Induration

The formation of soil proceeds more rapidly on materials already friable and loose, such as sediment or volcanic ash, because spaces between the grains allow free movement of weathering fluids. Roots and burrows also more readily penetrate soft materials and are modeled around corestones, nodules, or other originally hard parts of the soil. In paleosols in which the whole profile has been buried and altered to solid rock, features such as root traces and burrows are evidence of formerly unindurated parts of a profile.

The ways in which indurated bedrock is converted to soil vary considerably. Foliated or jointed rocks may be loosened by preferential weathering along these planes of weakness (Ollier 1969). Crystalline rocks, on the other hand, may be loosened by destruction of those minerals, such as feldspar, most susceptible to weathering. This local unevenness of soil formation can be seen in both paleosols and soils (Grandstaff *et al.* 1986).

Effects of induration on soil formation are especially well illustrated by soils and paleosols developed on carbonate rocks of varied induration. Soils formed on limestone and marble develop by chemical dissolution through the action of acidic rain and groundwater. Soil material accumulates from insoluble residue of the limestone or sediment washed or blown in from nearby. This soft non-calcareous soil material is usually sharply divided from hard limestone by an abrupt weathering front (Jennings 1985). Numerous fossilized examples of these "Terra Rossa" or Orthent paleosols are known (Wright & Wilson 1987). In contrast, soil formed on already fragmented carbonate sediments, such as unconsolidated sand formed from oolites, foraminifera, or shell fragments, is deeply and thoroughly weathered. They have nodular and other complexly cemented zones (Wright 1982). Similar differences can be seen between Miocene paleosols formed on welded and non-welded carbonatite tuffs in southwestern Kenya (Pickford 1986a, Retallack 1986c).

Grain size

The size of crystals of igneous rocks or of grains of sedimentary rocks controls the rate of weathering in several ways. The larger the grains or crystals, the larger are the spaces between them opened by weathering. This allows greater and deeper flow of weathering fluids. The differential weathering of coarse- and fine-grained parent materials is especially well seen in soils and paleosols formed on till or conglomerate containing pebbles of different lithologies. Individual pebbles weather first on the outside. This clayey and oxidized zone gradually extends inward as a weathering rind. In tills older than 200 000 years of the eastern Sierra Nevada of California, granitic boulders were weathered to a loose sediment of angular coarse grains called grus. Coarse-grained pebbles and cobbles of porphyritic andesite also are weathered to the core, but those of fine-grained basalts have weathering rinds only 2 mm thick (Fig. 12.3). Similarly, paleosols on fine-grained basaltic dikes are usually thinner than those developed on the coarse-grained granitic rocks along major unconformities (Williams 1968). In sedimentary sequences also, grain size plays a role in the flow of water and air through the soil. Well drained gravelly soils are more deeply leached than well drained sandy soils, and the latter more than clayey soils (Birkeland 1984).

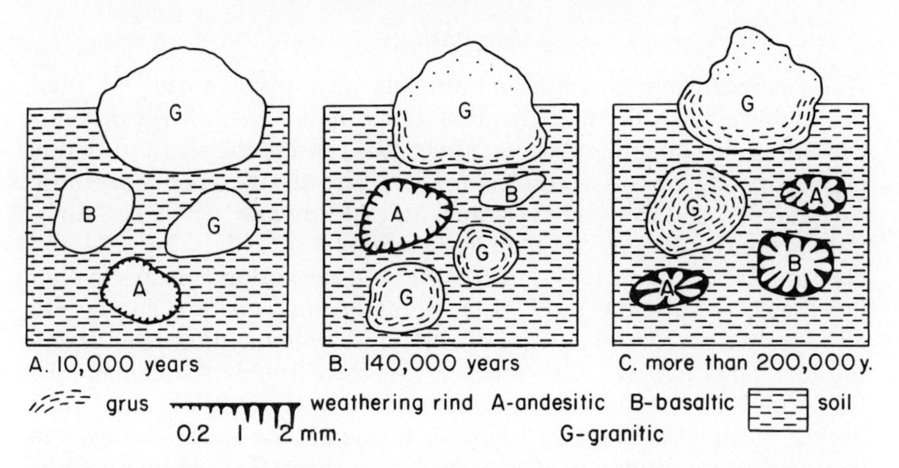

Figure 12.3 Differential development of weathering rinds over time in boulders of different internal grain size in Pleistocene tills near Lake Tahoe, California (adapted from Birkeland 1984, with permission of Oxford University Press).

Crystallinity

Resistance of minerals to weathering is to some extent conferred by the rigidity of, and proportion of defects in, their crystal structure (Eggleton 1986). One of the most stable minerals is quartz, which is a three-dimensional network (tectosilicate) of silica tetrahedra. Next in stability are the sheet silicates (phyllosilicates). Those with an open and expanded (2 : 1) structure, such as biotite, muscovite and smectite, are more easily weathered than those with a non-expanded (1 : 1) structure, such as kaolinite. Minerals in which silica tetrahedra are in chains, such as hypersthene or augite, or isolated, such as olivine, are even more readily weathered (Birkeland 1984). Volcanic glass lacks crystalline regularity. Like window glass, it can be regarded as a chilled fluid. This process of chilling, along with bubbling of frothing volatiles, is responsible for the sharp-edged fragments of the rims of bubbles called volcanic shards. Soils formed of such glassy volcanic airfall materials as ash and pumice have distinctive properties. In the humid temperate climate of the North Island of New Zealand, the half-life of andesitic glass in soils has been estimated at only 7000 years (Neall 1977). The initial alteration products also are amorphous materials, such as allophane, imogolite, opaline silica, and humus complexes with iron and alumina. In time, clay minerals form. Even when partly weathered to clay, ashy soils retain a low bulk density and a porous structure. Soils developed on glassy volcanic materials may be more clayey, humic, or deeply weathered than soils formed on more highly crystalline materials under comparable conditions of weathering.

Mineralogy

The chemical configuration of minerals also plays a role in their vulnerability to weathering. Chemical compounds vary in energy of bonds between their constituent atoms (Keller 1954) and in the size and charge of participating ions (Mason & Moore 1982). The strongest bonds are those between Si^{4+} and O^{2-} (3110 kcal/mol) and Al^{3+} and O^{2-} (1793 kcal/mol), but the bonds between common basic cations (Ca^{2+}, Mg^{2+}, Fe^{3+}, K^{+} and Na^{+}) and O^{2-} are weak (839, 912, 919, 322 and 299 kcal/mol, respectively). For these reasons, the most stable common minerals are quartz and gibbsite. Within the framework of silica and alumina in other aluminosilicates, large ions (K^{+} Ca^{2+} and Na^{+}) fit more snugly than small ions (Mg^{2+} and Fe^{2+}), and are less easily dislodged by hydronium (H^{+}) during hydrolysis. Once in solution, however, the ratio of charge to ionic radius (or ionic potential) is such that most of these ions remain in solution (ionic potential of K^{+} = 0.75, Na^{+} = 1.0, Ca^{2+} = 2.0, Fe^{2+} = 2.7, Mg^{2+} = 3.0), whereas other ions (Fe^{3+} = 4.7 and Al^{3+} = 5.9) are precipitated as hydroxides and oxyhydrates. These differences in ionic potential account for the ready hydrolysis of olivine and hypersthene, and also the accumulation of iron-rich and aluminous

Figure 12.4 Petrographic thin section of the surface horizon showing largely quartz and microiline in sericitic matrix (A), compared with the fresher and more diverse minerals of its gneissic parent material (B) in a Late Proterozoic (1000 million years old) paleosol under Torridonian alluvial deposits near Sheigra, Scotland (A is Retallack specimen R318 and B is R330). Scale bars 1 mm.

oxyhydrates in well drained soils. They also are reflected in the familiar mineral stability series (see Fig. 4.5), which is based on empirical observations of the persistence of minerals in soils and paleosols (Goldich 1938).

Granitic rocks contain potassium feldspar and quartz, with few mafic minerals, often biotite and hornblende. Clayey soils developed on granite may contain only quartz and microcline, with the mafic minerals all altered to clay. This is as much the case for modern deeply weathered soils on granitic rocks as for paleosols on such parent materials (Fig. 12.4), some as old as 3000 million years (Grandstaff *et al.* 1986). Basalt, on the other hand, consists largely of sodic and calcic plagioclases and mafic minerals, such as olivine, with few resistant minerals. Soils on basalt often are almost entirely clay, and this also is the case for very ancient examples (Gay & Grandstaff 1980).

Differences in weathering according to mineralogical composition can also be seen in sandstones. Sandstones rich in volcanic rock fragments weather deeply to clay, whereas those rich in grains of quartz or chert weather to sandy soils. These differences also are well shown in some lithified paleosols (Retallack 1977).

Chemical composition

A simplified chemical model of weathering can be constructed on the assumption that gains and losses of material in soils are largely a consequence of two common classes of soil-forming chemical reactions. First, weatherable ions are released from silicate minerals by acidic hydrolysis in carbonic acid formed from soil carbon dioxide. Second, iron and manganese are fixed in soils if oxidized by soil oxygen but are washed out of the profile if they remain in the reduced state in which they occur in most silicate minerals (Holland 1984). The carbon dioxide demand and oxygen demand of different parent materials can be approximated by the amounts (in moles per kilogram of soil) of major oxides and other compounds that could participate in hydrolysis and in oxidation reactions (Table 12.1).

Carbon dioxide demand is a crude measure of the acid "titer" of a parent material. Rocks richest in carbonates or weatherable bases will require more soil acid for weathering than those low in these constituents. The most consistently acidic environments are those of humid non-seasonal climates under biologically productive ecosystems (Brook *et al.* 1983), where even limestone and basalt can be rendered non-calcareous and base-poor, given time. In less humid and more sparsely vegetated regions acid production is lower, but soils of quartz sand may remain acidic into climates as dry as subhumid.

In an analogous fashion, the oxygen demand of parent materials also

Table 12.1 Estimating carbon dioxide and oxygen demand of parent materials from chemical data (formulae derived from common hydrolysis and oxidation reactions in soils by Holland 1984)

Carbon dioxide demand	$= D_{CO_2} = 2[m_{\text{"CaO"}} + m_{\text{"MgO"}} + m_{\text{"Na}_2\text{O"}} + m_{\text{"K}_2\text{O"}}] + m_{CaCO_3} + 2m_{CaMg(CO_3)_2} - m_{C^\circ} - 4m_{FeS_2}$
Oxygen demand	$= D_{O_2} = 0.25m_{\text{"FeO"}} + 0.5m_{\text{"MnO"}} + m_{C^\circ} + 3.75m_{FeS_2}$
Parent rock compositional reducing power (Holland's R)	$= D_{O_2}/D_{CO_2}$

where demands and m = moles of component per kilogram of rock

can be approximated by a formula based on common reactions of oxidation in soils (Table 12.1). Reduced ions of iron (Fe^{2+}) and manganese (Mn^{2+}) are soluble and generally lost from soils, whereas oxidized ions (Fe^{3+}, Mn^{4+}) are fixed in the soil as insoluble oxyhydrates. The oxidized form of these elements impart red and purple colors (respectively) to soils. Both color and iron and manganese content of soils depend on the amount of oxygen and other oxidants available at the time these ions are released by acidic hydrolysis of the minerals containing them. Swampland soils are generally drab colored and strongly leached of iron compared with reddish soils of elevated ground. This pattern can be seen from both modern soils and paleosols (Huddle & Patterson 1961, Besly & Turner 1983). Parent material plays a role in that greater amounts of oxygen are required to redden a soil developed on iron-rich parent material, such as smectitic shale or basalt, than to redden a soil on iron-poor material, such as sandstone or granite. This effect may be reflected in the common occurrence of drab iron-depleted early Precambrian paleosols in basalt parent materials on the same unconformity as slightly oxidized iron-enriched paleosols on granitic parent materials. These very ancient paleosols show no sign of gleying, and their deep profile development would not be possible in a swampy habitat (Retallack 1986c). This distinctively different response of different parent materials to the same weathering regime probably reflects very small amounts of atmospheric oxygen during Precambrian time (Holland 1984). Today there is so much oxygen in the atmosphere that well drained basaltic soils are much more ferruginized than well drained granitic soils.

Some common parent materials

Potential parent materials for soils include a wide range of rocks and sediments whose classification is the domain of geology. Many of these are very limited in distribution and so do not figure prominently in the formation of soils. Here attention is focused on widespread kinds of parent materials and the kinds of soil features that are peculiar to them.

Till

Heaps of rocky rubble left by glaciers are called till, or tillite when compacted or cemented to form a rock. Till is incompletely milled rock, ranging from rock flour produced by grinding of rocks entrained in glacial ice to huge boulders transported on top of the ice. Although they have been physically fragmented, these rocks are often little altered by chemical weathering. In composition, till is extremely variable because it consists of the various rocks found in the mountainous collecting basins of the glaciers. Pleistocene tills of the state of New York vary considerably depending on whether they were derived from the quartz-rich sandstones and granites of the Adirondack and Catskill Mountains or from the limestones and dolomites near Niagara Falls in the western part of the state. Sandy acidic soils (Spodosols) under conifer forest have developed in the non-calcareous tills of the eastern mountains, but clayey neutral-to-alkaline soils (Alfisols) under broadleaf forest in calcareous tills of the western hills (Cline 1953).

Loess

The sparsely vegetated terminus of glaciers is dusty with fine-grained sediment blown in from floodplains and glacial lakes. This windblown silt, or loess, consists of rock fragments that are finely ground but chemically little weathered. Unlike till, however, loess is very uniform in grain size because of sorting by wind.

From the point of view of plant and animal nutrition, loess is an ideal parent material (Fehrenbacher *et al.* 1986). It is rich in unweathered minerals already finely comminuted to expose their nutrient cations. The natural fertility of loess soils is also assisted by their ability to retain water (high field capacity). The soil-forming processes of decalcification and lessivage prevail, resulting in the formation of dark, clayey, base-rich soils (Mollisols and Alfisols) in regions of reasonably abundant rainfall. The high fertility and diverse biota of loess soils give them great agricultural potential, as demonstrated in the North American mid-continent and the central western Russian steppe.

Alluvium

The shales, sands, and gravels of rivers are derived in large part from soils of their drainage basins. Unlike till and loess, this parent material is not merely comminuted but also has been altered by weathering. Some easily weatherable minerals, such as olivine, are destroyed in soils and so do not generally contribute to alluvium. In many regions, however, this loss of potential fertility for alluvial soils is not severe. In arid regions and in hilly areas of temperate climates, soils of drainage basins are not so thick and strongly weathered that alluvium is seriously depleted in nutrient-rich minerals. As a result, the floodwaters of the Nile River of Egypt and of the Mississippi River and its tributaries in North America leave a legacy of nutrient-rich silt for streamside agriculture after each flood (Gerrard 1987). Near streams, there are mainly young silty and sandy soils (Entisols and Inceptisols). Further from streams where flooding is less frequent, other kinds of clayey soils (Aridisols, Mollisols, Vertisols, Alfisols, and Ultisols) form according to regional climatic conditions (Walker & Butler 1983).

The situation is different in wet, tropical regions of low relief. Here the soils may be thick and deeply weathered, and only the most weather-resistant minerals, such as quartz, hematite, and kaolinite, may be available in alluvial sands and clays (Sanchez & Buol 1974). Young soils (Entisols and Inceptisols) flank tropical rivers also, but soils away from streams include clean quartz-rich sands of abandoned stream channels (Spodosols) supporting stunted broadleaf forest and red clayey kaolinitic soils (Oxisols) supporting tropical rainforest. This extreme situation is not found everywhere in tropical regions. Little-weathered alluvium from highland regions is carried by some tropical lowland streams. Volcanoes also can renew the landscape and contribute little-weathered minerals. Nevertheless, it is useful to consider the great variety of alluvial parent material as a continuum from material almost as rich in weatherable minerals as loess and till to materials as deeply weathered and inert as some strongly developed tropical soils.

Marine sediments

Non-calcareous shales and sandstones deposited in and around the ocean can be almost as variable in composition as alluvium. This is not surprising because the thickest accumulations of marine sediments in deltas, continental shelves, and submarine fans are delivered by rivers. Once in the sea, however, this material is not appreciably weathered further. Removed as they are from surficial weathering, marine sediments tend to reflect the tectonic setting during their accumulation rather than their paleoclimate. This is clearly seen in marine sandstones, the mineralogy of which is readily determined by petrographic studies. Sandstones formed around volcanic mountain chains have principally volcanic rock fragments and ash. Those from fold mountain ranges, on the other hand, consist to a large extent of sedimentary and metamorphic rock fragments (Dickinson & Suczek 1979). Soils formed on these kinds of sandstones include abundant weatherable minerals that maintain soil fertility and promote the formation of clayey soils (Aridisols, Vertisols, and Alfisols, depending on climate and vegetation). On the other hand, sandstones formed around continental block mountains or from stable continental areas tend to be rich in quartz and kaolinite. Soils formed on these materials tend to be acidic and sandy (Spodosols) or clayey (Ultisols) in regions of reasonable rainfall.

Marine shales are more base-rich and uniform than alluvial shales for a variety of reasons. Inert clays, such as kaolinite and chlorite, accumulate close to the seashore. Base-rich clays, such as smectites and illites, on the other hand, are carried further out into deep ocean basins (Griffin *et al.* 1968). In addition, many marine shales have become more potassium rich during deep burial because of the dissolution of associated detrital potash feldspar and muscovite (see Fig. 7.8). Smectitic marine shales favor the development of heavy-textured clayey soils that are prone to waterlogging by perched water in lowland situations (Aqualfs, Aquolls, and Histosols) or to cracking in seasonal climates (Vertisols).

Many marine sediments are more uniform and laterally extensive than alluvial deposits. Deltaic and submarine fan deposits approach the complex lateral variation of alluvium. In general, marine shales and sandstone are found in discrete beds that are thick and laterally extensive. Even after uplift, deformation, and exposure, tracts of a single kind of marine sediment may cover large areas of land. The monotonous black shales of the Late Cretaceous Pierre and Bearpaw Shales of North America, for example, are parent material to large areas of Mollisols and Aridisols in North and South Dakota and Montana (Aandahl 1982). Soil types developed on such extensive parent materials may also be widespread, and individual profiles show little evidence of internal variation in parent material.

Schist

Most schists are highly metamorphosed marine shales because these accumulate to greater thickness than do non-marine shales. Easily weathered micas, particularly iron- and magnesium-rich chlorites, are the principal mineral of schists. The main weather-resistant material of schists is quartz, which is finer grained and more easily weathered than quartz in sandstones. There is a gradation in composition from schist through metasiltstone to quartzite comparable to the gradation from shale through siltstone to sandstone. The schist end of this range has the potential to form red, fertile, clayey soils rich in nutrients (Alfisols) given sufficient rainfall. The cleavage planes of schist and intercalated quartzites are almost always at high angles to the land surface rather than parallel to it like the bedding of some marine shales. This fabric allows deep leaching, unhindered by shallow impervious layers.

Limestone

Many limestones are made of skeletons of foraminifera, coccolithophores, pteropods, molluscs, echinoderms, and corals. Such fossils are obvious in thin sections of limestone (Scholle 1978) and can be used to distinguish parent material from soil carbonate. Dolostone also may contain fossils, but is less soluble than limestone, and so less easily karstified or micritized. Limestone and dolostone can by definition contain as much as 50% of other minerals. Soils formed by the dissolution of limestones consist of these residual minerals separated from bedrock by a sharp and irregular boundary (as in Rendolls and Orthents). Insoluble residue from limestone may include weather-resistant material such as kaolinite, gibbsite, and hematite, but these are buffered from extreme acidity by the nearby limestone. Some clayey impurities in limestone are smectitic and produce swelling clay soils (Vertisols), such as the Houston Black Clay of the Gulf Coast of Texas (Kunze & Templin 1956).

Granitic rocks

Here the phrase granitic rocks is used in a general sense to include coarse-grained igneous and metamorphic rocks, consisting mainly of quartz, potash feldspar, muscovite, and hornblende. The tectonic setting of granitic rocks over large areas of stable continental regions predisposes them toward the formation of thick and deeply weathered soils, and so does their grain size, which is usually coarse. The rapid weathering of hornblende and biotite in these rocks leaves a loose aggregate of angular grains called grus. It requires little additional weathering to create acidic sandy Spodosols, which are widespread on young granitic landscapes in humid regions of high elevation and latitude (Isherwood & Street 1976, Dixon & Young 1981). Soils of ancient granitic peneplains and plateaus, such as are widespread in Brazil, West Africa and central Australia, may be much more clayey and red (Ultisols and Oxisols, Lepsch & Buol 1974).

Basaltic rocks

These volcanic rocks generally have a fine grain size owing to chilling of lava flows on eruption. They are rich in iron and magnesium and in minerals such as pyroxene and olivine. They also have moderately high amounts of calcium, potassium, and sodium, mainly in feldspar. Most of their silica is in the form of silicate minerals other than quartz. Most basaltic rocks have little free quartz (tholeites and andesites), and in some there is none (phonolites). Compared with soils developed on granite, those on basaltic rocks tend to be more clayey from decay of feldspar, more fertile from release of nutrient cations, and more shallow because of lower porosity of both rock and soil. Because of their high iron content, freely drained soils developed on basalt are mostly red or brown from iron oxyhydrates (Alfisols), although basaltic soils under grasslands (Vertisols) or marshes (Histosols) may be gray or black from admixed organic matter. Many Precambrian paleosols on basaltic rocks are a distinctive lime-green color (Green Clays of Retallack 1986c), presumably because of the very weakly oxidizing atmosphere at that time (Holland 1984).

These general features of soils developed on basalt are shared by soils of other base-rich parent materials, such as marine shales and schists, but those on basalt have a different structure. The vesicular blocky- or ropy-textured tops of basaltic lava flows are more readily altered to soil than the massive underlying flow. The transition between soil and massive basaltic parent material is often only a few centimeters wide and can be irregular around corestones and vertical cooling joints (Boucot *et al.* 1974).

Ultramafic rocks

These are all characterized by unusually high concentrations of minerals rich in magnesium and iron (mafic minerals), such as pyroxene and olivine. Some of these rocks, such as peridotite and pyroxenite, are coarse-grained and form large intrusions. Others, such as serpentinite, are fine-grained and crop out as narrow bands along major faults. The exchange complex of soils on ultramafic rock is dominated by Mg^{2+}, with very low levels of the plant macronutrients K^+ and Ca^{2+}. There also are high levels of trace elements, such as Cr and Ni, which are toxic to plants (Brooks 1987). Soil formation may be limited on ultramafic rocks compared to adjacent, more densely vegetated rocks. In northern California and southwestern Oregon, USA, for example, the open shrubby vegetation of serpentinite soils contrasts strongly with the dark conifer forests on adjacent diorites and metamorphosed volcanic rocks. As a result, serpentinite soils are thinner and more rocky than those nearby.

Volcanic ash

The chemical composition of volcanic ash ranges widely to include that of both basaltic and granitic rocks. Some volcanic ash contains crystals and rock fragments of equally varied composition, but a large part of it is glassy and non-crystalline (Fisher & Schmincke 1984). Scoria and pumice are two

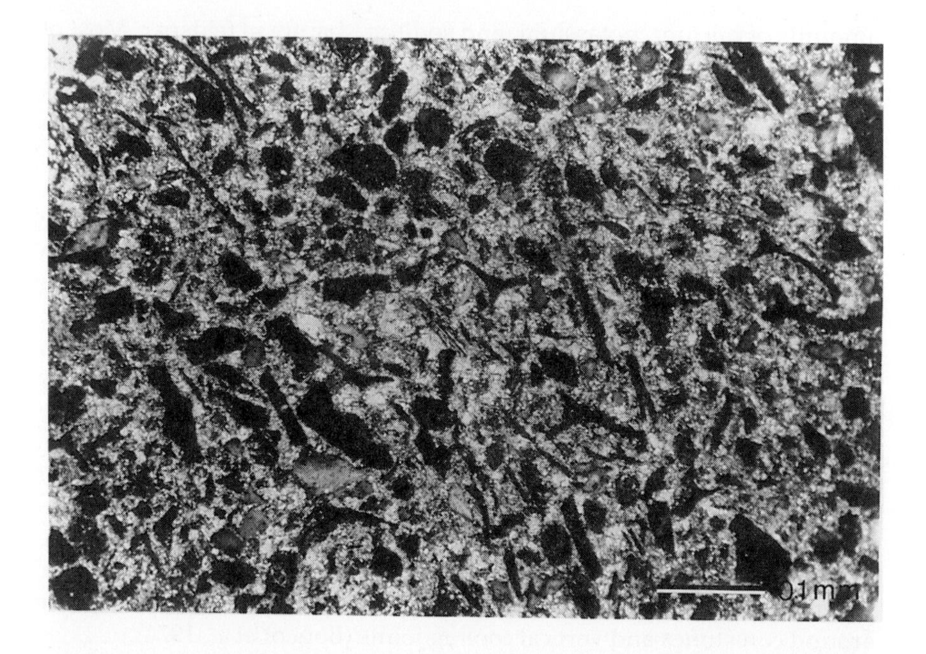

Figure 12.5 Unreplaced volcanic shards (black) observed in petrographic thin section under crossed nicols, of a caliche nodule (lower Bk horizon) of type Samna silty clay loam (Ustollic Eutrandept), of Late Oligocene age, Badlands National Park, South Dakota. Scale bar 0.1 mm (Indiana University specimen 15682).

kinds of fine-grained volcanic products riddled with bubbles formed by gases escaping from the lava during eruption, in a similar way to frothing of an uncapped beer bottle. Expanding gas bubbles also create volcanic shards. In thin sections these are isotropic to polarizing light and bounded by concave curved surfaces (Fig. 12.5). The high surface area, loose packing, and ease of weathering of volcanic ash are so distinct that ashy soils are commonly classified in distinct categories (Andepts of US taxonomy and Andosols of FAO 1971–81). These soils have a low bulk density (less than 0.9 g/cm^3), high water and organic matter retention, weakly developed soil structures, and a high content of amorphous alteration products, such as allophane, imogolite, opaline silica, and complexes of Al, Fe, and humus (Tan 1984). They may develop thick clayey soils in which the outlines of volcanic shards are preserved by replacement with other minerals. These soils are of economic importance around tropical volcanoes, where they are markedly more fertile than soils on other parent materials.

A base line for soil formation

Parent material is the starting point for a soil, and the degree of development of a soil is measured by the amount of change compared with its parent material (see Table 4.5). This way of defining parent material is logically necessary for parent material to be considered an independent factor in soil formation (Jenny 1941). There is a difficulty, though, in that this means that the exact properties of a parent material of a soil or paleosol can seldom be measured, because the actual parent material no longer exists. Instead, its nature must be estimated from nearby materials. Such estimates typically are based on four critical assumptions that should be carefully assessed when evaluating particular soils and paleosols.

Unweathered parent material

A first simplifying assumption is that the material identified as approximately the parent material has been completely unaffected by weathering. In the case of paleosols developed on igneous and metamorphic bedrock, one can be fairly confident in petrographically identifying a parent material unaffected by weathering by the absence of clays and oxyhydrates. Some of the clearest examples of the role of parent materials in the formation of paleosols are reported from major unconformities of sedimentary rocks overlying igneous and metamorphic basement.

A Precambrian (1000 million year old) paleosol developed on granodiorite near Marquette, Michigan, (Kalliokoski 1975) is typical in its chemical composition: enriched in sesquioxides ($Fe_2O_3 + Al_2O_3$) and silica (SiO_2) and depleted in alkaline earths (CaO and MgO) compared with parent material. These data normalized to TiO_2 as a stable constituent show sesquioxides and silica retained at near-constant levels. They appear enriched because of more severe losses of weatherable bases. This reassessment of the nature and degree of weathering is based on the assumptions that material deep below the profile is unweathered, that parent material of the profile was uniform and of the same composition as this lower material, that TiO_2 was stable during weathering, that the soil has the same volume as the parent material, and that compaction after burial was negligible. The compaction assumption is not so bad for granitic soils in which weather-resistant quartz grains provide a rigid framework as feldspars and micas are lost by weathering. If there were a way to estimate compaction, it might show that even the apparently stable materials (Fe_2O_3, Al_2O_3, SiO_2) had been lost by weathering, although to a much lesser extent than other materials.

Uniformity of parent material

A second common assumption is that parent material was uniform in composition within and below the paleosol. Many igneous intrusions are uniform in composition, but there is no such assurance of uniformity in alluvial sediments because of their variable composition and grain size. Deposits of meandering streams, for example, vary from sand to clay upwards within a single bed. These fining-upwards sequences reflect deposition under conditions of waning stream velocity and power (Allen 1965). Once a graded bed has been colonized by plants and animals, it may be difficult to distinguish clay originally deposited from that formed in the soil. One way around this problem for soils and paleosols with clear illuvial horizons is to compare the enrichment of illuviated clay, sesquioxides or carbonate with both overlying and underlying materials (Muir & Logan 1982). Another way is to accept the extremes of parent material composition as a kind of systematic error, as in the following example.

The Juniata Formation of central Pennsylvania is a fluvial sequence with paleosols of late Ordovician age (Feakes & Retallack 1988). Some of the paleosols are more densely penetrated by burrows than others and have more abundant caliche nodules and less obvious sedimentary relicts, indications that they formed over a longer period of time. A precise parent material composition for these paleosols cannot be found, but the compositional extremes of very weakly developed paleosols can be taken to approximate the range of parent materials of moderately developed paleosols. Concentration ratios for various oxides calculated for the moderately developed paleosol represent soil formation beyond that seen in the weakly developed one (Fig. 12.6). The range of concentration ratios obtained depended on whether clayey or sandy parts of the weakly developed paleosol were used as a parent material in the calculations. This range can be envisaged as a kind of systematic error (depicted as heavy horizontal lines in Fig. 12.6). Chemical differentiation beyond this error could conceivably represent an unusual composition for a parent material but is more likely a product of soil formation. In this case ferruginization, desilication, and calcification are indicated.

A variety of assumptions are implicit in this analysis. Fresh parent material has not been identified, only a range of compositions that may include it. Uniformity of parent material has been assumed but only within broad compositional limits of the weakly developed paleosol. The stability of titania has been assumed, because the soil was alkaline, considering its caliche. The assumption of constant volume during soil formation is reasonable because the parent material was initially unconsolidated, and the moderately developed paleosol is not extremely weathered. In contrast, volume change on burial probably was significant. Vertical burrows in sandy parts of these paleosols show little evidence of

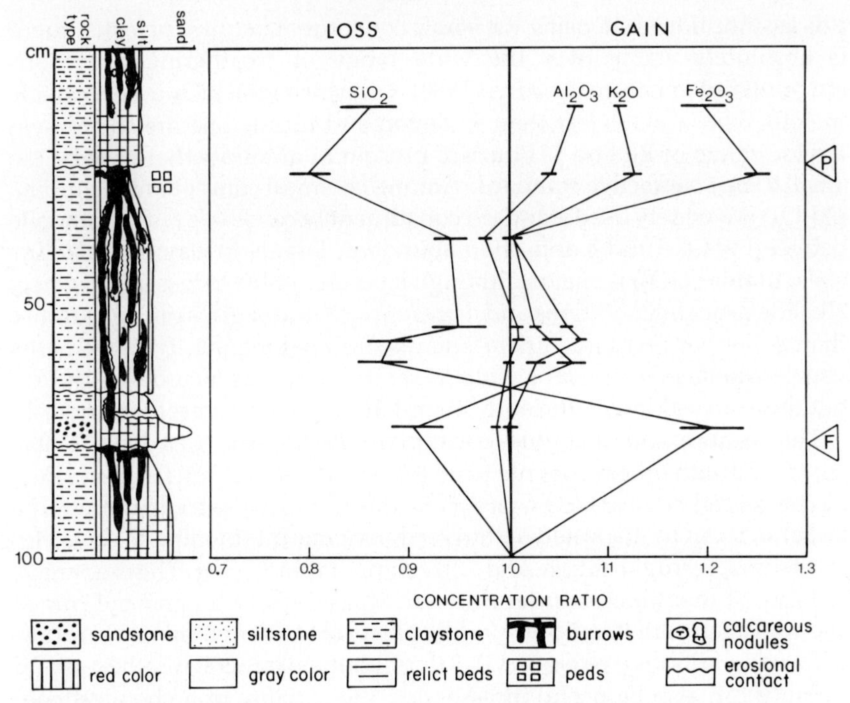

Figure 12.6 Concentration ratios showing gains and losses of major oxides relative to TiO_2 in a moderately developed paleosol (Oxic Ustropept, below P) compared with a weakly developed paleosol (Fluvent, below F) in the Late Ordovician Juniata Formation, near Potters Mills, Pennsylvania (from Feakes & Retallack 1988, reprinted with permission of the Geological Society of America).

concertina-like compaction, but the subhorizontal burrows in more clayey parts of the paleosols are crushed from a presumed original round outline, by either soil settling or burial compaction. If there were an accurate way of calculating compaction and including it in this analysis, it would make the degree of development seem less severe. This analysis reveals the direction of past soil-forming processes but falls short of an accurate assessment of their degree.

Stable constituents

A third simplifying assumption is that a particular constituent of the parent material has remained unaltered by weathering in the soil. This is a necessary assumption for calculating the volume, weight, and thickness of parent material that has weathered to create a soil or paleosol (see Table 4.5). This is also a helpful assumption for calculating gains and losses of minerals or chemical elements in situations where volume, weight, and thickness may have changed during weathering. The main difficulty with this assumption is choosing the stable constituent because no constituent is absolutely stable over the wide range of weathering conditions encountered in nature (Gardner 1980). Quartz or total SiO_2 is fairly stable in soils whose pH is less than 9. Zircon and tourmaline are stable over a wide range of Eh and pH but are present in amounts that may be too small to be an effective standard. Among chemical components, alumina (Al_2O_3) is a widely used standard constituent because it is fairly immobile between pH 4.5 and 8 and often abundant, largely in clay. For alkaline soils, titania (TiO_2) is useful, although it is susceptible to loss in extremely alkaline or acidic conditions and if present in volcanic glass or augite rather than a weather-resistant mineral such as rutile or ilmenite. Trace elements usually stable in soils over a wide range of conditions include Pb and Zr, but these are seldom sufficiently abundant to provide a useful standard.

The assumption that one constituent of the soil has been stable compared with others does not need to be made entirely on faith; a variety of checks can be made to ensure its validity. If the stable constituent is a mineral, it can be inspected in thin section or mounts of mineral separates for etching, ferruginization, and other signs of weathering. The theoretical stability of the mineral under different conditions of Eh and pH can be checked against that indicated for the paleosol by other minerals sensitive to these conditions (see Fig. 3.13). Information on mineral weathering and stability can also be useful in assessing the stability of a chemical constituent that it harbors.

Titania proved to be a suitable stable constituent in the Late Precambrian and Ordovician paleosols already considered. However, in paleosols developed on volcanic ashy alluvium of Oligocene age in Badlands National Park, South Dakota, USA, it was difficult to find any suitable stable constituent because of the abundance of easily weathered glassy volcanic shards in the parent materials (Retallack 1983b). Quartz is not a suitable stable mineral in these paleosols; it is rare, noticeably embayed within caliche nodules and much less abundant within the nodules than in non-calcareous soil nearby (Retallack 1986d). Pseudomorphs of evaporite minerals are evidence that some of these soils were so alkaline (pH ca. 9) that quartz should have dissolved. Titania also is a poor choice as it is strongly depleted toward the top of moderately to strongly developed paleosols. This could be due to unusually alkaline conditions of soil formation and because titania was dispersed in volcanic glass rather than present in the usual weather-resistant titanium minerals. The least variable chemical constituents were alumina and the trace elements Pb and Th. In view of these difficulties, and also problems of identifying unweathered parent material, of heterogeneity of these alluvial sediments, and of assessing volume changes during soil formation and burial, a different approach was used. Molecular weathering ratios were calculated from chemical analyses of each specimen. These do not assume any particular parent material but do suggest soil-forming processes between horizons, such as lessivage (silica to alumina), calcification (alkaline earths to alumina), salinization (soda to potash) and leaching (barium to strontium).

Compaction effects

A fourth assumption is that loss of volume of soils and also their compaction during burial is entirely a matter of thickness decrease. This may seem contrary to common sense because volume is a measure of three dimensions and thickness only one. However, observations on a variety of materials, including fossil plants (Walton 1936), suggest that under conditions of static vertical load, the horizontal or cross-sectional area of sediments, soils, and fossils is maintained by pressure at the side. In the absence of folding, thrusting, or other lateral compression, the pressure exerted by sediment on either side is at least equal to that exerted by overburden. Similar arguments can be made for volume loss during soil formation because weakening of a uniform soil by weathering decreases downward from the surface in such a way that one piece of soil has a similar strength to another beside it. Problems arise if soils are not uniform in strength because of deep cracks, corestones, hard nodules, burrows, or other lateral variations in composition. These difficulties are minimized if samples of the soil studied are large enough to be homogeneous in these respects. A fist-sized specimen may be representative of the microscopic crack system or very small nodules, but not of larger features. The assumption that thickness change is representative of volume change is useful because direct evidence for thickness change is easier to find in soils and paleosols, as has already been discussed under the heading of compaction after burial.

Loss of volume during soil formation reflects the washing out of chemical constituents so that those remaining become more abundant by comparison. Compaction during burial also has the effect of making differentiation within a given volume of paleosol seem more profound. It is only rarely that evidence of volume change during weathering and compaction is preserved in paleosols. Hence it is not surprising that many studies of chemical development of paleosols adopt the assumption of constant volume.

One paleosol which has preserved evidence of burial compaction is a profile 2200 million years old near the village of Waterval Onder, Transvaal, South Africa (Retallack 1986b). This paleosol has pronounced gilgai microrelief and mukkara structure (see Fig. 9.9), which includes wide (up to 2 cm) cracks filled with sandstone from an overlying paleochannel deposit, and also narrow cracks stained and filled with iron and manganese. Both the sandstone and ferromanganiferous dikes

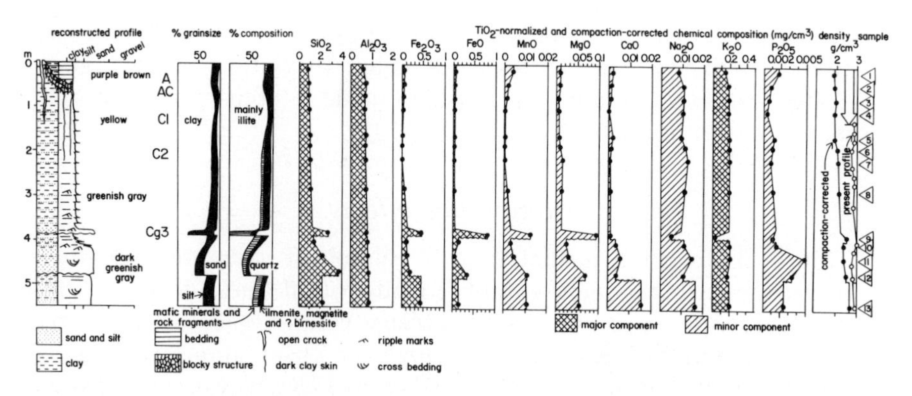

Figure 12.7 Compaction corrected columnar section, grain size, minerals, and chemical composition, as they may have been before burial of the Waterval Onder clay paleosol (Vertisol), from the early Proterozoic (2200 million years old), basal Dwaal Heuvel Formation, near Waterval Onder, South Africa (from Retallack 1986b, with permission of Elsevier Publishing Co.).

are now strongly contorted by compaction. If they are assumed to have been straight originally, their length within a selected vertical height provides an indication of compaction during burial. One estimate of compaction (0.67) was gained from a sandstone dike near the surface and another estimate (0.73) from a ferromanganiferous dike within the profile. These compaction factors (x) may be interpolated and extrapolated for other levels of the profile (y) using a linear equation between these two values ($y = 0.0074x + 0.62$). These values can be used to normalize the chemical analyses to a stable constituent. In this case, titania was chosen. Although the profile is virtually noncalcareous, it has a very slight surficial enrichment in CaO, suggesting a near-neutral pH. The parent material was assumed to have been a fluvial shale (the eighth sample down in Fig. 12.7) with local silty and sandy laminae of more mafic composition than the shale. By this analysis silica and alumina in the reconstructed soil can be seen to have remained fairly constant, but there was a slight gain in iron and manganese and losses of weatherable bases. No attempt was made to account for volume losses during weathering. Because the parent material was probably an unconsolidated sediment and the surface horizon was partly cumulative, little change in volume during formation of this paleosol can reasonably be assumed. Estimation of compaction due to soil formation in paleosols and distinguishing this from compaction during burial remain knotty problems.

In most cases of estimating the degree of weathering of a paleosol compared with its parent material, unknown quantities outnumber those which can be quantified. Despite the need for one or more simplifying assumptions, the true nature of a soil or paleosol can only be understood by comparison with its parent material.

13 Time as a factor

As in physics, chemistry, and biology, it is common also in soil science to study changes with time as the independent variable, or x axis in a graphical representation. These kinds of relationships are called chronofunctions. Some processes of soil formation, such as podzolization, are simple and rapid enough to be studied by means of laboratory experiments (Fisher & Yam 1984). Most soil-forming processes are too slow for such an approach, and chronofunctions are derived by comparing changes in soils of different age. If such studies are to reflect purely the effect of time on soil formation, then all the studied soils should also be comparable in climate, organisms, topography, and parent material. Such a study set of soils is called a chronosequence. The best known kinds of chronosequences are on flights of terraces abandoned on valley walls by downcutting streams (Harden 1982a) and on concentric moraines dumped successively by retreating glaciers (Burke & Birkeland 1979). Chronosequence studies are also useful for estimating rates of deformation and recurrence intervals of earthquakes along faults. The security of large permanent engineering works, such as dams and nuclear power plants, in tectonically active areas depends on such studies of soils (Shlemon 1985).

Soils form over time scales ranging from ecological (days) to geological (millions of years), and the establishment of how long it took to form a particular soil or soil feature calls for a variety of techniques. The weathering of tombstones in very old graveyards is perhaps the most reliable method of estimating initial rates of weathering of fresh bedrock because most tombstones show the date when they were erected. Studies of tombstone weathering demonstrate different rates of weathering for different climatic conditions (Rahn 1971). However, those tombstones so weathered that they could be regarded as soils no longer have any recognizable evidence of their age. Old buildings and foundations can sometimes be dated by associated artifacts and historical records. Such estimates of time may range as far back as several thousand years. Some ancient Roman ruins are covered in moderately developed soils whose age may be known by associated coins. Further back in time, historical records

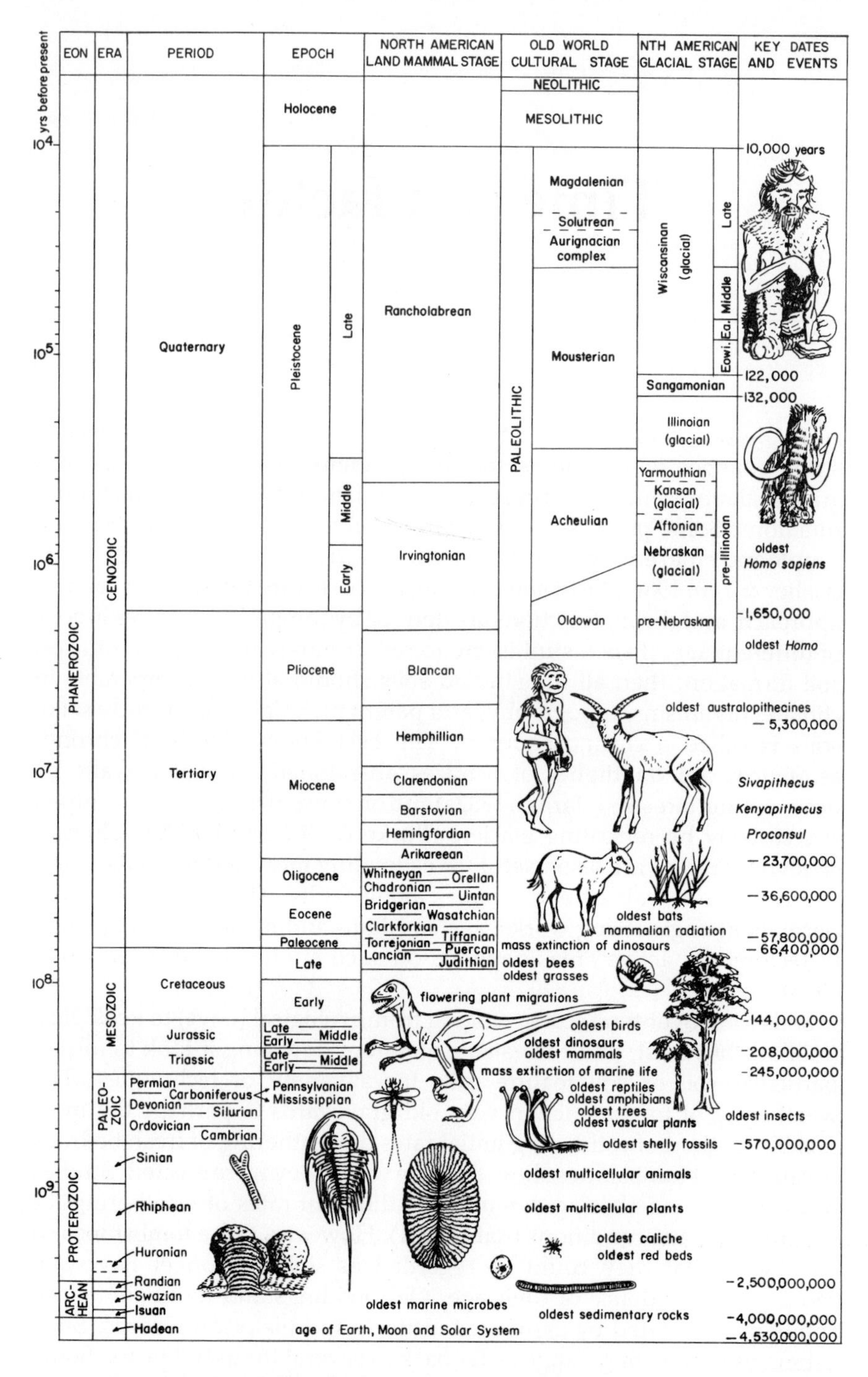

Figure 13.1 A geological time scale, plotted on a logarithmic scale to emphasize detailed understanding of mammalian evolution, human cultural stages, and glacial chronology (based on data from Nilsson 1983, Palmer 1983, Woodburne 1987, Richmond & Fullerton 1987).

become less and less accurate. It has been more convenient to give periods of time distinctive names rather than attempt to determine age in years: for example, the Middle Kingdom of ancient Egypt, the Paleolithic culture of stone tools and the Pliocene epoch of geological time. Their order is determined from superposition of artifacts or fossils in sedimentary layers, the older ones beneath the younger ones (Fig. 13.1). The resolution of these relative time scales corresponds to the duration of the named time interval. Further, the indicators of a time interval may not have changed at the same time in all areas. Paleolithic cultures still exist in remote parts of the world, despite the technological revolutions that have allowed you to read this book.

Fortunately, a variety of methods of gaining ages in years have been devised as a supplement and calibration of these relative time scales. The constant radioactive decay of the carbon isotope ^{14}C in wood, charcoal, and sea shells can be used to date soils as old as 50 000 years by conventional means of assay in a mass spectrometer and as old as 100 000 years by using a particle accelerator to assay very small amounts of the radioactive isotope still remaining (Rucklidge 1984). The accumulation of the cosmogenic isotope ^{10}Be also has promise for dating clayey soils up to 200 000 years old (Pavich *et al.* 1986). Other isotopic systems, such as uranium–lead and potassium–argon, have been used to calibrate older parts of the geological time scale. Calibration of the geological time scale gains in accuracy each year as new dates become available and can be related to the sedimentary record of Earth history. Radiocarbon dating of late Quaternary sediments is now so commonplace that the Holocene/Pleistocene boundary has been arbitrarily set at 10 000 years.

The time it takes to form various soil features varies considerably. Organic surface horizons form in only tens to hundreds of years, but millions of years are needed to form a horizon as deeply weathered as an oxic horizon (Fig. 13.2). One problem with using such data for estimating time for development of paleosols is the idea that soil properties reach a steady state: a kind of equilibrium between climatic and other forces

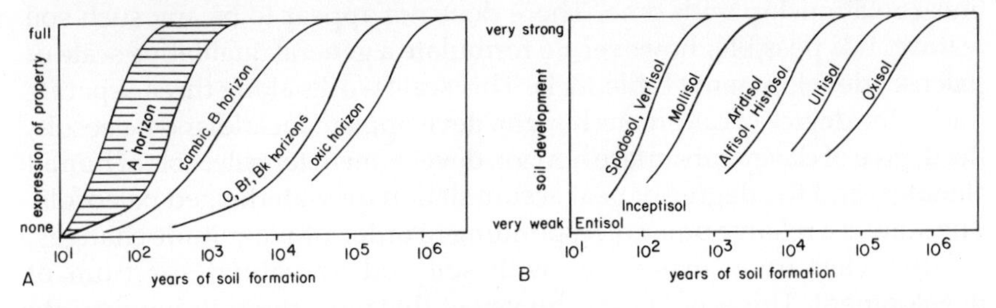

Figure 13.2 Schematic representation of the times needed to attain various properties of soils (A) and orders of soils recognized by the Soil Conservation Service of the US Department of Agriculture (B); modified from Birkeland 1984, with permission of Oxford University Press).

maintaining the soil arrayed against erosion and other forces destroying it. Steady state can also can be conceived as the ultimate potential of the location and parent material. The notion of steady state in soils has a direct parallel in the idea of climax vegetation after plant succession (see Fig. 10.1). Just as ecologists are questioning the idea of equilibrium in undisturbed vegetation, the concept of steady state of soils has also been questioned. It could be that climatic variation over the past several hundred thousand years has been such that rates of soil formation have been negligible during ice ages compared with soil-forming intervals. It could also be that the odds of disturbance by climatic change, erosion, or burial are such that very few soils overreach a usual degree of development. It could also be that apparent steady state of soils reflects difficulties in estimating the age of soils. With the greater array of dating techniques now available and a better understanding of Quaternary climatic change and geomorphic processes, these questions may be more satisfactorily answered. Recent studies of change in soil through time (Harden 1982a, Rockwell *et al.* 1985a) indicate that some soil properties reach a kind of steady state, but many do not. Some soil features do not develop appreciably until late in soil formation when some limiting condition or threshold has been exceeded (Muhs 1984). Few soil features vary linearly with time, as would be ideal for unique estimates of the time required for the formation of paleosols. If soils do reach a steady state, or if there are variations in the rate of soil formation, then the best estimates that can be attained for paleosols from Quaternary soil chronosequences are minimum times for formation.

Indicators of paleosol development

Research on soils in chronosequences provides an essential data base for the interpretation of paleosols. An ideal indicator of time for the formation of a paleosol would be present in all soils, be easy to measure and show a linear relationship with time. There does not appear to be any such soil feature. It is possible, however, to formulate a general qualitative scale of paleosol development (Table 13.1). This scale works along three separate tracks: the degree of calcareous horizon development for aridland paleosols, the degree of clayey subsurface horizon development for paleosols of humid climates, and the degree of peat accumulation for waterlogged paleosols. The scale is a relative one, corresponding to order-of-magnitude changes. Its use relies upon experience with soil and its natural spectrum of development. This is not to say, however, that more rigorous inquiry into the age of a paleosol should be neglected. Careful attention to detail and comparison with modern soil chronosequences can yield more accurate estimates of the age of paleosols, as outlined in the following sections.

Table 13.1 Stages of paleosol development (Retallack 1988a)

Stage	Features
Very weakly developed	Little evidence of soil development apart from root traces: abundant sedimentary, metamorphic or igneous textures remaining from parent material
Weakly developed	With a surface rooted zone (A horizon) as well as incipient subsurface clayey, calcareous, sesquioxidic or humic, or surface organic horizons, but not developed to the extent that they would qualify as U.S.D.A. argillic, spodic or calcic horizons or histic epipedon
Moderately developed	With surface rooted zone and obvious subsurface clayey, sesquioxidic, humic or calcareous or surface organic horizons: qualifying as U.S.D.A. argillic, spodic or calcic horizons or histic epipedon and developed to an extent at least equivalent to stage II of calcic horizons (Table 13.2)
Strongly developed	With especially thick, red, clayey or humic subsurface (B) horizons or surface organic horizons (coals or lignites) or especially well developed soil structure or calcic horizons at stages III to V (Table 13.2)
Very strongly developed	Unusually thick subsurface (B) horizons or surface organic horizons (coals or lignites) or calcic horizons of stage VI: such a degree of development is mostly found at major geological unconformities

Calcic (Bk) horizon development

In dry climates calcium carbonate is liberated by the limited weathering of minerals or accumulates in soils from windblown dust, but moisture from rainfall is insufficient to remove it completely from the soil (Fig. 13.3, Table 13.2). On silty and sandy parent materials, the carbonate in very weakly developed soils is a fine, white powder lining soil peds and slender tubular structures, such as coatings of roots and fungal hyphae. With time in weakly developed soils, discrete, hard nodules form within the calcareous horizon. In thin sections, the growth of micritic carbonate can be seen to replace matrix and skeleton grains, which are "nibbled" and irregular around the edges (caries texture) and abnormally loosely packed (Wieder & Yaalon 1982). Nodules become larger and eventually coalesce to form a solid horizon toward the top of the calcareous zone that defines moderately developed soils. In strongly developed soils, this solid layer forms a barrier to further percolation of water, which thus flows along the top of the cemented zone to form a sharply truncated and laminated top. Very strongly developed soils have an even thicker crust, in places brecciated and pisolithic, where erosion and large roots have broken the crust. Broadly similar stages of development have been documented in soils on gravelly alluvium. In the younger examples of these soils, carbonate forms encrustations on the bottoms of the clasts.

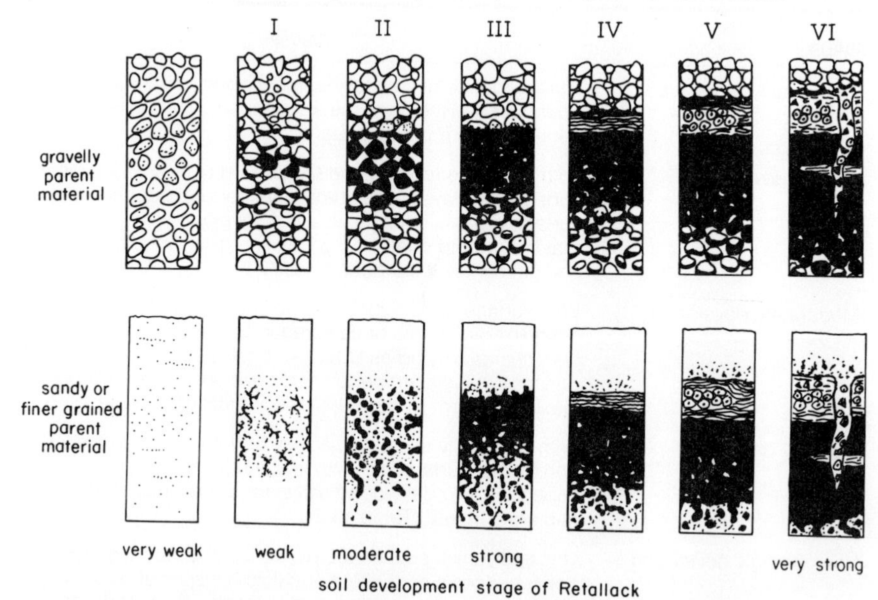

Figure 13.3 Stages of the morphology of carbonate accumulations in soils formed on gravelly and on sandy or finer grained alluvium (modified from Gile *et al.* 1966 and Machette 1985, with permission of Williams & Wilkens Co. and the Geological Society of America).

Table 13.2 Stages of carbonate accumulation in paleosols (from Machette 1985)

Stage	Paleosols developed in gravel	Paleosols developed in sand, silt or clay
I	Thin discontinuous coatings of carbonate on underside of clasts	Dispersed powdery and filamentous carbonate
II	Continuous coating all around and in some cases between clasts: additional discontinuous carbonate outside main horizon	Few to common carbonate nodules and veinlets with powdery and filamentous carbonate in places between nodules
III	Carbonate forming a continuous layer enveloping clasts: less pervasive carbonate outside main horizon	Carbonate forming a continuous layer formed by coalescing nodules: isolated nodules and powdery carbonate outside main horizon
IV	Upper part of solid carbonate layer with a weakly developed platy or lamellar structure capping less pervasively calcareous parts of the profile	
V	Platy or lamellar cap to the carbonate layer strongly expressed: in places brecciated and with pisolites of carbonate	
VI	Brecciation and recementation as well as pisoliths common in association with the lamellar upper layer	

The amount of carbonate for each morphological stage and the time it took to accumulate have been established for desert soils around Las Cruces, New Mexico (Fig. 13.4). The average rate of pedogenic carbonate accumulation in this cool desert region (mean annual precipitation 204 mm and mean annual temperature 15.5°C) is 0.26 g $cm^{-2}yr^{-1} \times 10^{-3}$. Rates of carbonate accumulation range from less than half this (about 0.09) in regions of wetter and cooler climate (near Boulder, Colorado, with mean annual precipitation 376–472 mm and mean annual temperature 9.7–11.0°C) and about twice the Las Cruces value (0.51) in a wetter and marginally warmer climate (Roswell, New Mexico, with mean annual precipitation 320–355 mm and mean annual temperature 17.2°C; Birkeland 1984). Hence, the natural variation in rates is less than the several orders of magnitude in time spans. For interpretation of paleosols, there also are problems for examples older than mid-Paleozoic of different atmospheric and biologically induced concentrations of carbon dioxide (Retallack 1986b). Nevertheless, sophisticated mathematical models are now available for estimating carbonate accumulation in soils (Machette 1985, McFadden & Tinsley 1985). These could be modified to make more accurate estimates of the time for formation of calcareous horizons in very ancient paleosols.

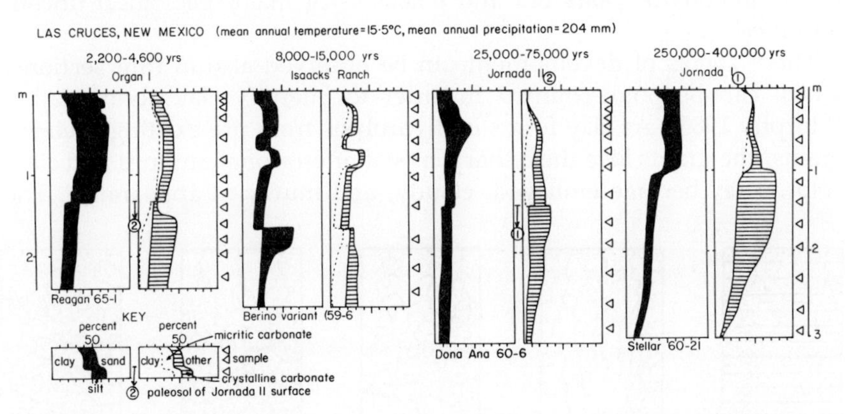

Figure 13.4 Distribution of carbonate and clay in surface soils and paleosols (indicated by arrows) of different age, formed on alluvium near Las Cruces, New Mexico (compiled from data of Gile *et al.* 1980).

Clayey subsurface (Bt) horizon development

Like the accumulation of calcareous horizons, the enrichment of clay in subsurface horizons also takes time. Clay is not so easily dissolved and moved in solution as carbonate, but its formation by weathering of parent material and of windblown dust and its physical translocation to a subsurface horizon are similar in principle. Several morphological stages of clayey subsurface (Bt or argillic) horizons can be recognized that are comparable to those of calcareous horizons. These are most readily appreciated in soils formed on mixed sandy, silty, and clayey alluvium (Fig. 13.5, Table 13.1). At the outset in very weakly developed soils, there may be little evidence of a clayey subsurface horizon, only bedded alluvium penetrated by root traces. With time in weakly developed soils, alluvium becomes cracked into recognizable peds, and some of these are defined by clay skins washed down the profile. The clay skins are at first widely spaced, and much relict bedding remains. In moderately developed soils, clay skins and peds are more pronounced, and the total clay enrichment has exceeded that required by the definition of an argillic horizon (Soil Survey Staff 1975). Some traces of relict bedding may remain within peds at this stage, but it may only be apparent from subtle grain-size variations obscured by clay enrichment in hand specimens. With further development in strongly developed soils, the total amounts of clay and of clayskins increase further so that the clayey subsurface horizon is now distinct and thick and contains no trace of relict bedding. In very strongly developed soils, the clayey horizon gains in thickness to well in excess of 1 m. It is common in soils of this stage to have a thin to non-existent surface (A) horizon, lost to erosion. These are soils of landscapes many millions of years old and paleosols of major geological unconformities.

These stages of development can be observed also in thin sections, which can be point-counted in order to quantify clay accumulation (Murphy 1983). As clay forms by hydrolysis from pre-existing skeleton grains, the grains lose their sharp crystal or erosional outlines and clear color. They become embayed, cloudy, and murky in appearance. The

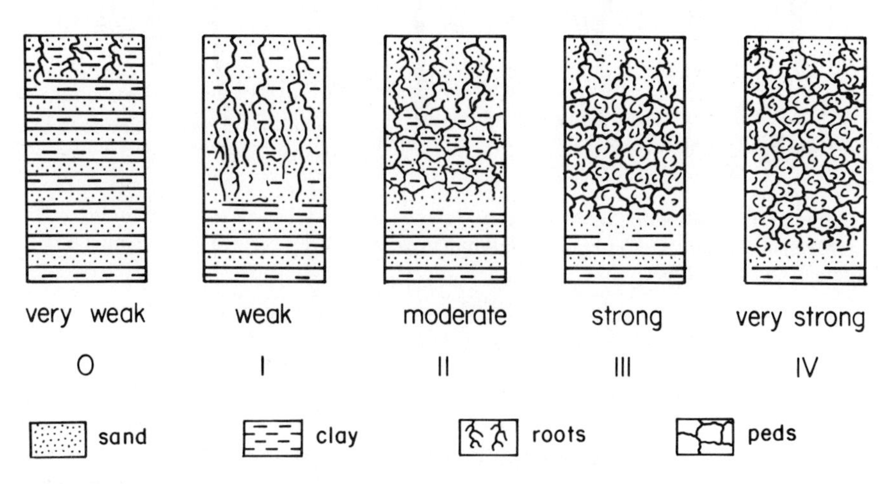

Figure 13.5 Stages in the formation of clayey subsurface horizons (Bt) in mixed clayey and sandy alluvium.

murkiness stems from fine-grained clays, which, when randomly oriented, tend to block the passage of light through the thin section. Clay that is washed down cracks, however, is deposited and squeezed into preferred orientations, which appear highly birefringent under cross-polarized light. The scattered planes of bright clay (sepic plasmic fabric) are characteristic of clayey subsurface (Bt) horizons. The streaks of bright clay are sparse and disconnected (insepic) in very weakly developed soils. In moderately developed soils, the clayey horizon has more bright clay (mosepic and masepic) until there forms a complex woven fabric of bright clay (omnisepic) in old and strongly developed soils.

The development of clayey subsurface (Bt) horizons has been calibrated along alluvial terraces of the Merced River in the San Joaquin Valley of central California (Fig. 13.6). The rate for formation of clayey subsurface horizons in this dry cool region (mean annual precipitation 410 mm and mean annual temperature 16°C) of live oak savanna is slow compared with other parts of the United States. Soils of comparable ages in the dry cool climate (mean annual precipitation 204 mm and mean annual temperature 15.5°C) of Las Cruces, New Mexico, have about 1.5 times as much clay because of the greater influx of windblown dust, which is readily weathered and washed down the profile (Gile *et al.* 1980). Argillic horizons formed at comparably fast rates are also found in wetter, cool climates (mean annual precipitation 943 mm and mean annual temperature 9.8°C) under broadleaf forests near Millport, central eastern Ohio (Lessig 1961).

In addition to climate, the texture of the parent material also plays a role in the rate of development of clayey subsurface horizons. Clayey horizons comparable to those formed on silty and clayey alluvium in 100 000 years in Ohio formed in only 40 000 years on till in areas of comparable rainfall and temperature (mean annual precipitation 1032 mm and mean annual

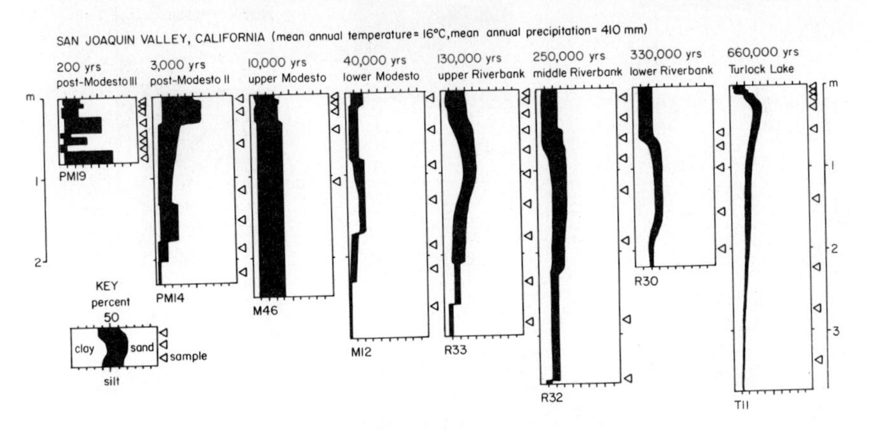

Figure 13.6 Distribution of clay in surface soils of different age on alluvium of the Merced River, San Joaquin Valley, California (compiled from data of Harden, 1982b).

temperature 10.4°C) under broadleaf forest near Williamsport, Pennsylvania (Levine & Ciolkosz 1983). One of the most rapidly formed clayey subsurface horizons documented is the accumulation of 1.7% additional clay in soil on porous loess thrown up only 100 years ago by railroad construction in regions of cool humid (mean annual precipitation 847 mm and mean annual temperature 9.6°C) prairie near Cedar Rapids, Iowa (Hallberg *et al.* 1978). Such studies on the controls and time involved in forming clayey subsurface horizons are gaining in quantity and quality and should be amenable to modelling comparable to that devised for carbonate horizons.

Peat accumulation

The formation of peaty surface horizons (histic epipedons) is more a process of accumulation than of differentiation comparable to calcareous and clayey subsurface horizons. The rate of accumulation of peat depends on a balance between the supply of organic debris from swamp vegetation

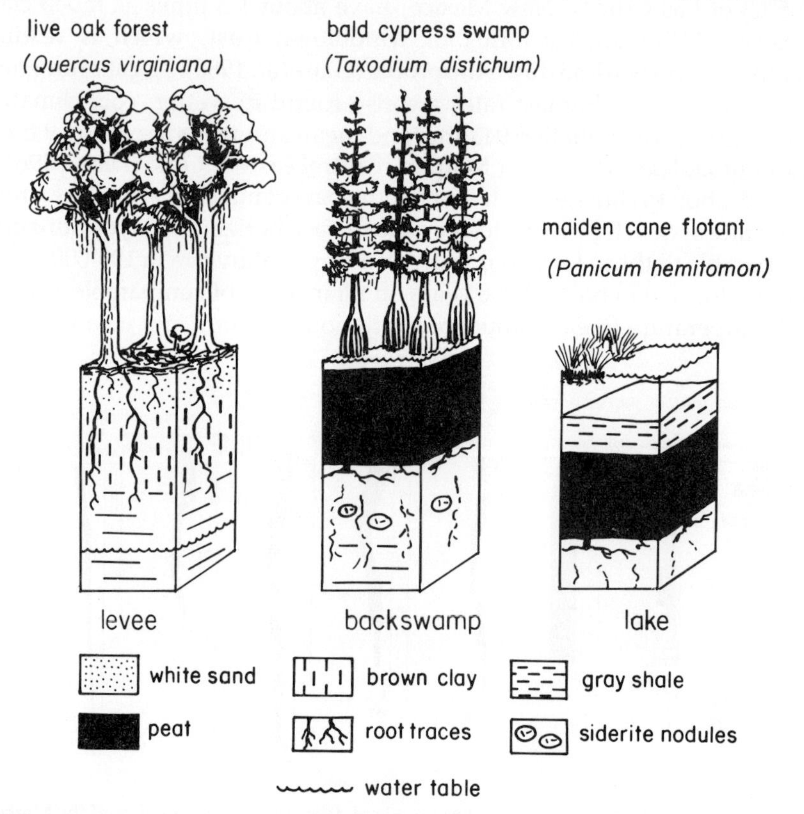

Figure 13.7 Soils and sediments in different parts of the Mississippi Delta plain, near Baton Rouge, Louisiana (based on data of Lytle 1968).

and its destruction by aerobic decay. In well drained soils, no peat accumulates because leaf litter is destroyed by aerobic decay. In subsiding swamps, however, peat accumulates around the vegetation that provides the organic debris (Fig. 13.7). The rate of accumulation of peat in swamps is not limitless. Mosses and marsh grasses can cope with fairly rapid subsidence and still generate peat, but with greater subsidence marshes become lakes or bays. For swamp trees, which must maintain some aeration of their roots, the rate of peat accumulation is scaled close to the growth rate of the trees.

Considering these constraints on rates of peat accumulation, it is useful to consider qualitative stages of peat accumulation based on experience with modern soils. Peaty surface horizons (histic epipedons) in the classification of the US Soil Conservation Service (Soil Survey Staff, 1975) must be at least 60 cm thick for mossy peats and 40 cm for other kinds of peat for the profile to qualify as a Histosol. These figures reflect practical experience with typical peat thicknesses encountered in nature, in the same way as the definitions of calcic and argillic horizons. Unlike the other horizons with characteristic morphology, stages of peat accumulation may be based on original thickness. An order-of-magnitude scale for woody peats has proven useful (Table 13.3): very weakly developed (less than 4 cm), weakly developed (4–40 cm), moderately developed (40 cm–4 m), strongly developed (4–40 m), and very strongly developed (more than 40 m). The virtue of simplicity in such a scheme can be seen by the necessity for further corrections for compaction in applying this scale to paleosols (Fig. 7.6). For example, bituminous coals are commonly compacted to about one twentieth of their former thickness (Elliott 1985). A moderately developed paleosol in this case would have a surficial coal 2–20 cm thick.

Table 13.3 Stages of development of peaty soils

Soil type	Development	Peat thickness	Time
Entisol	very weak	0-4 cm	0-80 yrs
Entisol	weak	4-40 cm	40-800 yrs
Histosol	moderate	40 cm-4 m	400-8,000 yrs
Histosol	strong	4-40 m	4,000-80,000 yrs
Histosol	very strong	more than 40 m	more than 40,000 yrs

Estimates of peat accumulation rates provide a quantitative perspective on the problem (Falini 1965). Taking 50 kg m^{-2} yr^{-1} as an unlikely maximum production of fresh vegetable matter in a hypothetical tropical forest, together with observations that living plant matter is about 80% water and that pure peats contain 50 kg m^{-3} of dry fuel, one obtains a maximum conceivable rate of peat accretion of about 20 cm yr^{-1}. This is exceedingly unlikely because it is based on unrealistically high rates of subsidence and negligible losses to decay and herbivory. It can be reduced to a more realistic rate if there is evidence for (a) the amount of mineral matter in the coal, (b) decay in the coal from the proportion of degraded materials (collinite and semifusinite), and (c) the rate of growth of plants in the form of fossil stumps whose age can be estimated from growth rings. It is common to find that at least half of a coal consists of partly decayed organic matter or mineral matter, implying a maximum rate of peat accumulation closer to 10 cm yr^{-1}. This may be reasonable for peats formed under moss and marsh grass, which can grow upwards at this rate. Peat formed under trees, on the other hand, can accumulate at rates of no more than 1 mm yr^{-1}, and where there are large trees thousands of years old, at rates of about 0.1 mm yr^{-1}, if the trees are to survive anoxia.

Rates of accumulation in nature are at the conservative end of this theoretical range (Stach *et al.* 1975). Peat of *Sphagnum* moss in a bog northwest of Lake Ringsjön in southern Sweden has been radiocarbon dated at 18 points in a thickness of 4 m above buried birch stumps. It accumulated at rates of 0.3–1.6 mm yr^{-1} (Nilsson 1964). Peats formed under bald cypress (*Taxodium distichum*) swamp in Okefenokee Swamp of southwestern Georgia have accumulated to a thickness of 5.9 m in 7000 years (Cohen 1985), giving a long-term accumulation rate of 0.84 mm yr^{-1}. Rates of peat accumulation of 0.5–1 mm yr^{-1} are common (Falini 1965). At these empirical rates, the maximum age of a very weakly developed peat would be 40–80 years, but 400–800 years for a weakly developed peat, 4000–8000 years for a moderately developed peat, 40 000–80 000 years for strongly developed peat and, even more for very strongly developed peat. These estimates are of the same order as those for formation of calcareous and clayey subsurface horizons. They may have to be tailored for the interpretation of paleosols by consideration of compaction, botanical nature, amount of degraded plant matter, and mineral content of coals. Comparison of a modern analog may be useful in interpreting the time represented by a particular coal-bearing paleosol.

Mineral weathering

Soil development proceeds as new soil minerals are created in place of old minerals. The gradual loss of mineral grains can be seen under the petrographic microscope by the progressive etching of weatherable minerals (Fig. 13.8). As weathering proceeds, the grains lose sharp edges inherited from the interlocking crystal faces of igneous parent rock or the smoothing of abrasion during sedimentary transport. Several stages of mineral etching can be recognized (Table 13.4). The status of minerals with respect to these stages or the depth of etch pits are two ways of quantifying mineral weathering in soils of different age (Hall and Michaud 1988). In alluvium of the San Joaquin Valley of California, hypersthene is fresh in

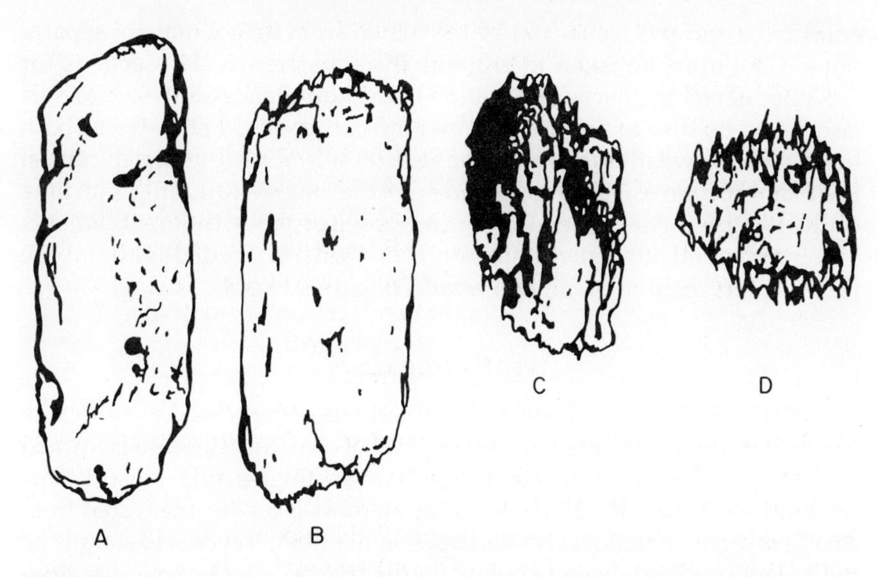

Figure 13.8 Variation in etching of grains of hypersthene from soils developed on tills 10 000–25 000 years old (A, B) and 135 000–145 000 years old (C, D) in the eastern Sierra Nevada, California (from Birkeland 1984, reprinted with permission of Oxford University Press).

soils only 10 000 years old, somewhat altered in those 130 000 years old, and completely altered in those 600 000 years old. Pyroxene in these same soils is somewhat altered in soils 10 000 years old and altered in the 130 000-year-old soils. A compilation of such data from soil chronosequences of wetter or warmer climate reveals that in both cases alteration is more rapid than in the San Joaquin Valley (Birkeland 1984).

Another way to quantify such observations is to compare the abundance of easily weathered minerals with those that are resistant. Commonly used mineral-weathering ratios include quartz/feldspar, which can be estimated by point-counting thin sections, preferably after staining for feldspar (Houghton 1980). Another ratio, zircon and tourmaline over

Table 13.4 Stages of mineral alteration in soils (from Birkeland 1984)

Stage	Description
Fresh	Crystal worn smooth or with sharp crystal faces
Pitted	Scattered etch-pits and frosting of the surface of the grain
Etched	With deep and extensive interconnected pits leaving sharp coxcomb ridges between
Nearly gone	Traces of coxcomb ridges or of lamellae or only a pseudomorphous outline of the grain remaining

amphibole and pyroxene, can be calculated from heavy mineral separations. Care must be taken to examine the separates or thin sections for evidence of etching because variation in the abundance of these minerals also could be due to original sedimentary processes. Estimates of both these mineral-weathering ratios for soils on tills of various age in central Iowa show progressively deeper and more thorough weathering with time (Ruhe 1969). Similar studies of other chronosequences in different climatic, topographic and other environmental conditions are needed as a data base for the interpretation of mineral weathering in paleosols.

Weathering rinds

Pieces of rock forming the pebbles and boulders of gravel and till represent locally indurated and less permeable parts of coarse-grained soils formed on them, and so they weather slowly from the outside inward. The thickness of these red, clayey, weathering rinds is readily measured from a large sample of boulders broken open in the field. It is best to sample all the boulders a fixed distance below the surface in order to avoid the rind-destroying effects of fire-spalling, sand-blasting or wear from animal tracks.

Differential rates of development of weathering rinds with rock type are strikingly seen in lithologically heterogenous tills, such as those around Lake Tahoe in the eastern Sierra Nevada of California (see Fig. 12.3). Weathering rinds are slower to form on fine-grained basalt than on coarse-grained andesite. Coarse-grained granitic rocks do not develop such a distinctive rind but are deeply weathered so that grains are loosened to a friable grus. Over some 200 000 years in this cool desert region (mean annual precipitation 785 mm, mean annual temperature 5.8°C at Tahoe City), granitic and andesitic boulders are deeply weathered, but cores of fresh rock remain in basaltic pebbles. These results are at the slower end of the spectrum of rind development (Truckee in Fig. 13.9), which can proceed at rates three times as fast in wetter climates.

Combinations of features

In attempts to assess overall soil development, a number of indices have been devised to reduce combinations of soil features to a single value. By using several time-dependent features, inaccuracies in individual features are minimized.

The soil development index of Harden (1982a) is a good example of the technique applied to a well understood chronosequence of soils in alluvial parent materials of the Merced River in the San Joaquin Valley of California. The index is derived from field estimates of eight different soil properties: redness (or rubification), total texture (or clayeyness), clay

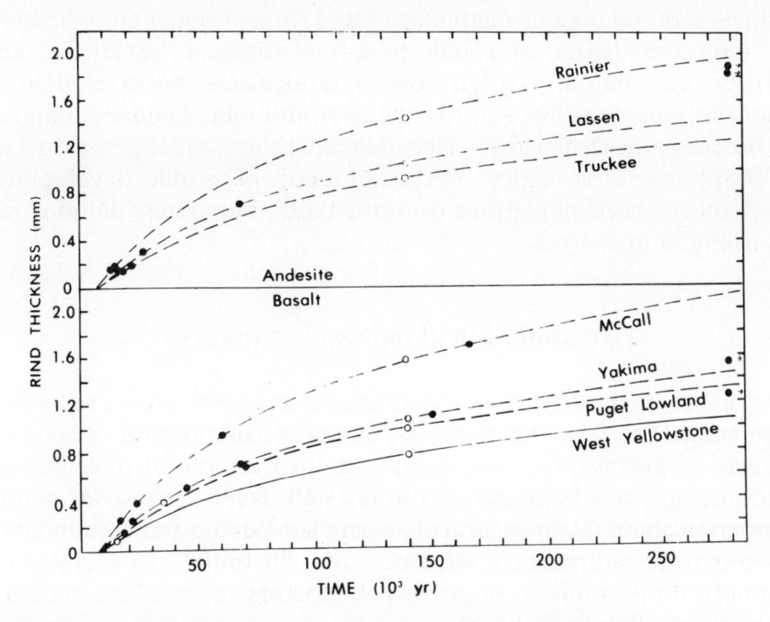

Figure 13.9 Increased thickness of weathering rinds of andesite and basalt clasts with age of their containing tills and in regions of different mean annual precipitation (from Colman 1986, reprinted with permission of Academic Press).

films, structure, dry consistence, moist consistence, darkness (or melanization), and chemical reaction (pH). The last feature is measured in both soil and parent material using a standard field meter, but others are quantified in various ways. Total texture, for example, is quantified by assigning ten points for each outline of a grainsize class within the textural triangle (see Fig. 4.2) that separates the horizon from its parent material, ten points for increases in stickiness, and ten for increases in plasticity. These values are then added together and normalized by dividing them by the maximum attainable value (90 in this case). Similar normalized values for all eight properties are averaged to calculate the development index for each horizon. These horizon indices are then averaged to calculate the overall soil development index. For some features, such as profile structure and melanization, steady state was achieved within about 10 000 years on the Merced River chronosequence. For other features, steady state does not seem to have been attained after 0.5 million years.

Despite what appear to be a number of arbitrary assumptions in the formulation of soil development indices, they include ingenious ways to quantify field observations and do seem to work. They are unlikely to be applied to paleosols in their present form, however, because many of the properties on which they are based are not preserved in paleosols. Diagenetic dehydration of ferric oxyhydrate minerals (see Fig. 7.7), destruction of soil organic matter (see Fig. 4.9), and cementation irreversibly alter the degree of rubification, melanization, stickiness, and plasticity after burial. Harden (1982a) recalculated the overall index without certain measures and found it was still robust and meaningful. If all the properties listed above were deleted, this would represent a very significant loss of accuracy. Reconsideration of profile development indices on the basis of features preserved after burial may demonstrate their utility for paleosols.

Accumulation of paleosol sequences

Soils are clues to the way in which landscapes and sedimentary sequences are put together. The structure and chemical composition of soils are products of geomorphic processes, and their degree of development reflects the age of a land surface (Ruhe 1969). Paleosols provide similar information about the formation of ancient landscapes and accumulation of non-marine sedimentary sequences, as illustrated in the following sections by the example of Oligocene paleosols of Badlands National Park, South Dakota (Retallack 1986d).

Hillslope development

There are three divergent views on the development of hillslopes over time (Fig. 13.10). The idea of slope decline was popularized by William Morris Davis at the turn of the century (Young 1972). By this view, hillslopes lose their steepness because the tops of hills are eroded more energetically than their lower slopes, which are protected by a colluvial mantle. Soils on a declining slope should be of about the same thickness and degree of development everywhere but a little thicker and better developed where the cumulative colluvial mantle is thin above the base of the slope. An alternative view of parallel retreat promulgated by Lester C. King emphasizes the constant nature of geomorphic processes on different segments of slopes. By this view, soils within the slope are of approximately the same degree of development, but exceptionally well developed and thick soils may persist on the divides. A final view of slope replacement, advocated by Walter Penck, is conciliatory. By this scheme, originally steep portions of hillsides are eroded back into gentle slopes by parallel retreat and upward extension of the lower, more gentle slope at an angle characteristic for the prevailing conditions. By this view, soils of upper replaced slopes are thinner and less developed than those of the lower replacing slopes, but soils of the ridge tops may be extremely well developed.

Few studies of soils on modern slopes have been reported that test these

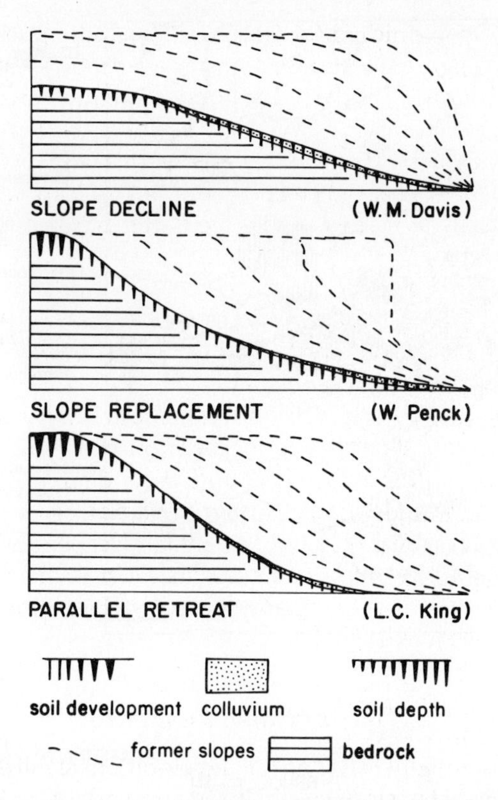

Figure 13.10 Contrasting views on the development in time of hill slopes (based on data of Young 1972).

models of slope development. One well known study around Jewell bog in north-central Iowa (see Fig. 11.1) supports a model of parallel retreat with only modest slope decline (Ruhe 1969). Other catenas contain soils better developed in downslope positions than in interfluves, compatible with slope decline (Al-Janabi & Drew 1967).

Ancient hilly landscapes can also be reconstructed from their paleosols. A part of a hilly landscape is buried at the base of the sequence of mid-Tertiary alluvial deposits in Badlands National Park, South Dakota, USA (Fig. 13.11). A thick, well developed paleosol is present in many places at the disconformity between Late Cretaceous marine rocks and the overlying alluvium of latest Eocene to Oligocene age (Retallack 1983b). By Late Eocene time, when this low rolling area was covered in humid tropical forest, it was an ancient stable landscape, as indicated by the thick, clayey, deeply weathered, kaolinitic paleosol. This situation changed during latest Eocene to Oligocene time, when a major stream draining the emergent Black Hills incised deeply (27 m in section of Fig. 13.11) into Cretaceous marine rocks, below the zone of Eocene weathering. The red clayey

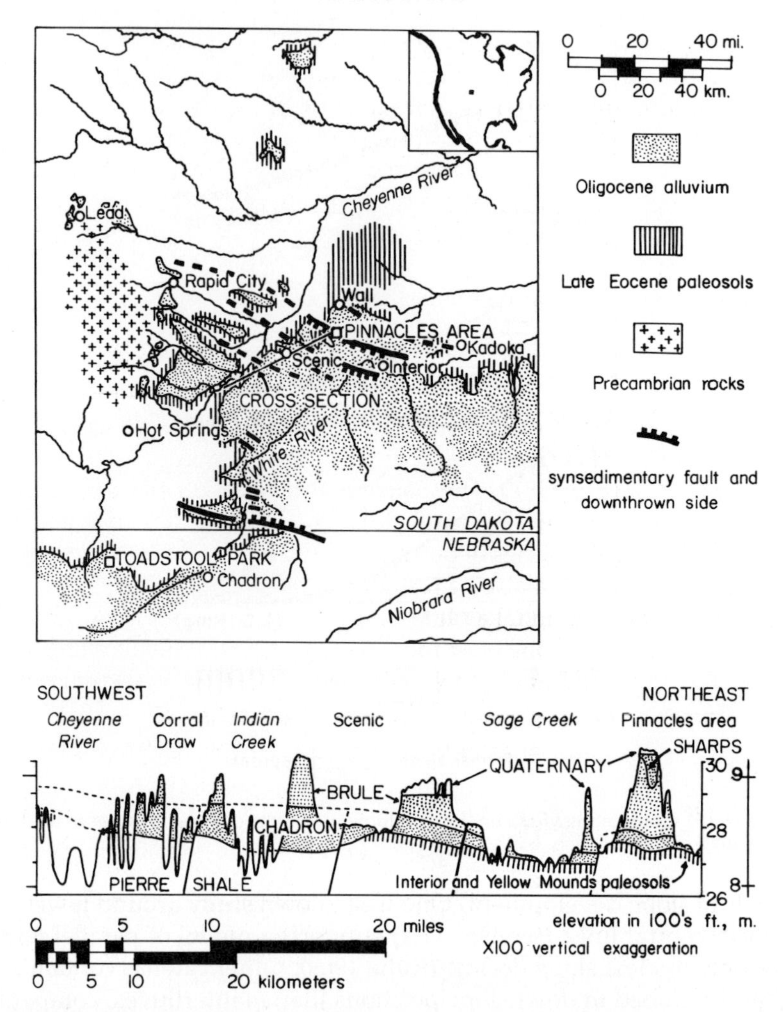

Figure 13.11 Outcrops of deeply weathered Late Eocene paleosols, of Oligocene alluvium of the White River Group, and distribution of synsedimentary faults (above), and reconstructed cross-section of a deep paleovalley (below) in southwestern South Dakota and northwestern Nebraska (from Retallack 1988b, reprinted with permission of the Society of Economic Paleontologists and Mineralogists).

paleosol and its thick, yellow saprolite were truncated by steep scarp-like slopes around this deeply incised valley. Some destabilization of the older soil surface on the high drainage divide may be indicated by bedding in deeply weathered materials above the undisturbed clayey subsurface (Bt) horizon of the basal Eocene paleosol (in measured section of Fig. 6.2). This disturbance was not frequent or profound enough to prevent substantial recolonization by forest vegetation resulting in a complex and very strongly developed paleosol (Yellow Mounds silty clay loam, a Paleudult).

With progressive filling of the youthful valley topography, what was formerly an extensive plateau between valleys became an alluvial terrace. Some time after this there was a catastrophic disturbance and recolonization by forest, this time producing another thick, well developed paleosol (Interior clay, a Paleudalf of Retallack 1983b). The persistence of very strongly developed paleosols on upland divides and terraces through two cycles of valley cutting and the weak development of paleosols within sediments of the paleovalley support the idea that this partial ancient landscape formed by parallel retreat.

Alluvial architecture

Just as buildings are composed of discrete structural elements, river deposits can be regarded as built from sediment types of characteristic appearance (Hallam 1981). The paleochannel facies is the most striking and best understood architectural element, with its cross-bedded gravel and sand. Distinctive facies of point bars, floodways, levees, oxbow lakes and floodplain soils can also be recognized.

Several different alluvial architectures can be imagined (Fig. 13.12). In the architecture of deposits of meandering streams in subsiding floodplains, the channel sandstones are narrow because the streambed is

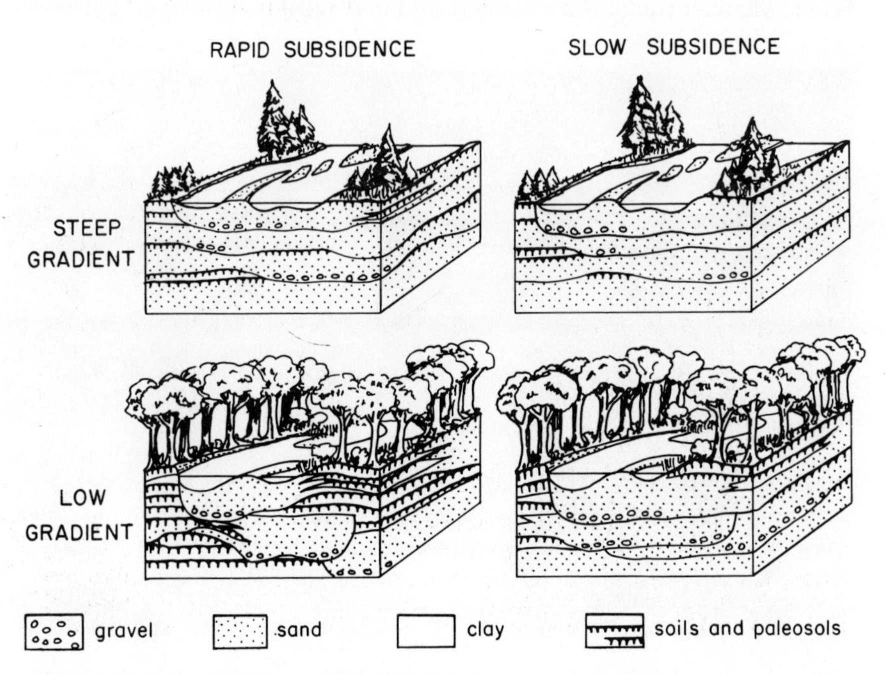

Figure 13.12 Four alternative alluvial architectures created by meandering and braided streams under different regimes of subsidence.

restricted by a lush growth of vegetation and clayey banks. A second architecture is produced by meandering streams under conditions of subsidence so low that clay and silt are lost downstream. This architecture is mainly composed of sandy and conglomeratic paleochannels. Other architectures are created by streams of steeper gradient channels that are broad, shallow, and braided by islands of coarse sand and gravel. Braided stream sequences are dominated by sand and gravel whether subsidence is fast or slow. The most distinctive architecture of these four possibilities is the first situation of asymmetric paleochannels of meandering streams separated by thick floodplain deposits. The other three situations have proven difficult to distinguish using evidence of paleochannels alone (Bridge 1985). The evidence of paleosols can help insofar as their degree of development and spacing in sedimentary sequences reflect overall accumulation rates.

Each of the rock units of Badlands National Park, South Dakota, shows a different kind of alluvial architecture, and this is in part the reason why they were distinguished as formations. Consider, for example, the Scenic Member of the Brule Formation of Oligocene age (Fig. 13.13). Are these deposits of braided streams or of meandering stream channels that coalesced because of a low rate of sediment accumulation? The degree of development of associated fossil soils provides a clue. The lower part of the Scenic Member contains moderately and very weakly developed paleosols

Figure 13.13 Sheet sandstones of Oligocene paleochannels in the Scenic Member of the Brule Formation, Pinnacles area, Badlands National Park, South Dakota.

whereas the upper part contains strongly developed paleosols, like the upper part of the underlying Chadron Formation. These general impressions can be considered in more detail by estimating duration of formation for each paleosol by comparison with modern soils whose age is known. In this case, minimal estimates were used in deference to the concept of soil-forming intervals and steady state. Once ages for each paleosol are assumed, they can be added together for a particular sequence. The thickness of the sequence divided by the added time gives a rate of accumulation (Fig. 13.14). These estimates of rates turned out to be an order of magnitude greater than those based on radiometric and paleomagnetic estimates of the duration of these rock units. A part of this discrepancy may be blamed on the conservative nature of the estimates used, but a greater part is due to the incompleteness of these sequences, which will be discussed in due course. For the present purpose, it is notable that rates of sediment accumulation estimated from paleosols show the same relative change as those calculated by radiometric and paleomagnetic means. The value of paleosol estimates is that they can be applied to sequences not amenable to direct paleomagnetic or radiometric dating. For example, the lower portion of the Scenic Member accumulated at much greater rates than the middle or upper portion. Thus, sandy paleochannels of the lower Scenic Member formed in loosely sinuous and partly braided streams, unlike the asymmetric paleochannels of meandering streams found in the upper part of the Scenic Member and the underlying Chadron Formation.

Completeness of the rock record

Sedimentary sequences are undoubtedly an incomplete record of past events and times, but just how incomplete are they? How long did it take each sedimentary layer to accumulate? And how long was it between deposition of the layers? These are important questions for estimating the resolution of sedimentary sequences for studies of pace of evolutionary or of geomorphic change. They are difficult questions to answer for most sedimentary sequences but not intractable for sequences of paleosols. Flooding events that cover soils and initiate soil formation are known to be rapid (a matter of days or weeks) compared with the time it takes to form a moderately developed soil (many thousands of years). Thus, times of deposition can be disregarded as insignificant, but times between events of deposition may be represented by development of paleosols. It is not really as simple as this. The record of time provided by a paleosol could be obscured by being incorporated in a better developed paleosol. Also, the age of a paleosol could consistently be underestimated if the features used to estimate its age were reaching a kind of steady state or represented brief soil-forming intervals. However, these are minor sources of error

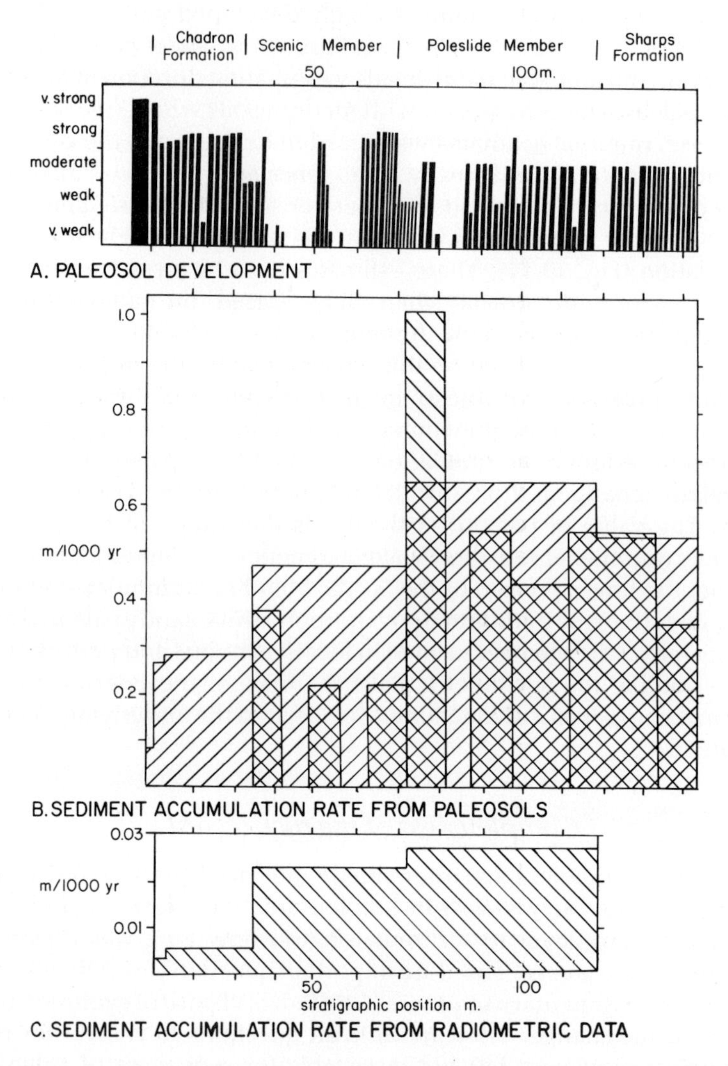

Figure13.14 Degree of development of paleosols (A) and sediment accumulation rates based on estimates from paleosols (B) and on paleomagnetic and radiometric data (C) for Late Eocene and Oligocene formations of Badlands National Park, South Dakota (from Retallack 1984a, reprinted with permission of the Paleontological Society).

compared with those introduced by large-scale erosional cutting followed by periods of filling. The erosional disconformity at the base of a depositional unit could represent a very long period and appear no different from the disconformity at the top of each paleosol. These cutting and filling cycles may be revealed by abrupt changes in the nature of paleosols (Retallack 1986d) or by lateral mapping (Kraus & Middleton 1987). With due regard for these important sources of error, paleosols may be useful records of the temporal resolution and completeness of non-marine sequences.

Resolution is a question of scale. Fine resolution for sequences of paleosols is hundreds of years and coarse resolution is millions of years. Completeness, on the other hand, is a measure of the reliability of a sequence at a given resolution. A complete sequence at a resolution of 1000 years should have on average a bed, paleosol, or other record for every interval of 1000 years. In general, a sequence of paleosols can be regarded as complete at a resolution that is equivalent to the average time of formation for the paleosols. Fractional or percentage completeness can be estimated for finer resolutions. A general impression of completeness can also be gained by casual inspection. Sequences complete at fine resolution will have numerous weakly developed paleosols with abundant relict bedding and little pedogenic clay, structure, or color differentiation. Sequences complete at coarse resolution would have few, strongly developed, paleosols, which were clayey, well structured, and strongly differentiated into red or dark colors.

A more complex way of estimating completeness of a sequence of paleosols or of sediments is by comparing rates of sediment accumulation of a sequence with rates usual for that environment and time span, as estimated from a large compilation of rates (by Sadler 1981). The rates used in this compilation include those from direct observation of sediment accumulating from floodwaters, from measurement of survey points, from radiocarbon dating, and from other forms of isotopic dating. The rates decline for estimates made over longer time spans because of correspondingly longer gaps in the record. Medial rates define an expectation of the rate for a complete section at a particular resolution.

Estimates of completeness and resolution may be compared by considering their application to the sequence of paleosols in Badlands National Park, South Dakota (Fig. 6.2). A simple way to estimate resolution from paleosols is to average the estimated number of years for formation of each paleosol in the sequence. From these values, the probability that the sequence is complete for intervals of 1000 years can be estimated by the fraction of 1000 years over the time for which it would be complete. For the Chadron Formation above the Interior paleosol, for example, there is a one-in-five chance that a given 1000-year interval will be represented by sediment capped by a paleosol (Table 13.5). Sadler's (1981) equations, based on expected rate of accumulation of fluvial sequences, can also be used (Retallack 1984a). An appearance of greater completeness is given by the extraordinarily high rates of accumulation revealed from paleosol development compared with low completeness revealed by the low rates of accumulation estimated geophysically from radiometric and paleomagnetic data (Table 13.5). According to paleosol estimates, the resolution of the sequence is only a few thousand years, but according to radiometric

Table 13.5 Temporal resolution (in years) and completeness (% of 1000-year time spans represented) in rock units in Badlands National Park, South Dakota

	From paleosol formation times		From paleosol based sediment accumulation rates		From paleomagnetic & radiometric sedimentation rates	
	Resolution	Completeness	Resolution	Completeness	Resolution	Completeness
Brule Formation	2,887	35	2,100	69	3,400,000	3
Poleslide Member	2,364	42	1,800	81	1,700,000	3
Scenic Member	3,571	28	3,000	59	1,700,000	3
Chadron Fm. above Interior paleosol	4,706	21	6,200	35	4,800,000	6

Note: Temporal resolution = minimum time span certain to be represented by some sediment in the sequence. Completeness = % of a given time span (1000 yrs used here) represented by sediment - from Sadler 1981, Retallack 1984a

and paleomagnetic estimates a few million years. The truth probably lies somewhere between these extremes because paleosol estimates of time were deliberately chosen to be minimal, and the radiometric and paleomagnetic estimates include large systematic errors and long periods of time not recorded in disconformities between the recognized units, as outlined in a later section.

By the paleosol view, the sequence is complete enough for studies of events on time scales of hundreds of thousands of years, such as evolutionary speciation or climatic change related to Milankovitch cycles. By the radiometric and paleomagnetic view, it is not. Because there are other sequences with greater completeness, studies of short-term evolutionary or climatic change would be more profitably undertaken there rather than in Badlands National Park. There is room for improvement of both kinds of estimates, but despite their disparity, each method preserved the same ranking of formations: Poleslide Member of the Brule Formation more complete than Scenic Member, and these more complete than the Chadron Formation above the Interior paleosol. The general appearance of these rock units also bears out this ranking. Paleosols of the Chadron Formation developed on the same ashy parent material as the other units, but they are clayey, with green surface and pink subsurface horizons. The Poleslide Member, on the other hand, is silty and more uniformly yellow and brown in color, reflecting its less-developed paleosols.

Episodicity of erosion and deposition

Before the advent of accurate information on the timing of events of deposition, sedimentation was widely considered cyclical. Familiar examples range from point bar cycles (Allen 1965) to glacio-eustatic cyclothems (Heckel 1986) and the tectonosedimentary "pulse of the earth" (Grabau 1940). In geomorphology, there was the erosional cycle of Davis (1899) with its youthful, mature, and senile stages of valley and river form. More recent quantitative studies of changes in landscapes and rivers (Schumm 1977) indicate that erosion and sedimentation are not as regularly cyclical as is sometimes assumed, and may be more productively considered episodic. Paleosols can be important clues to this periodicity and its causes at a variety of scales.

The finest level of episodicity preserved in most sequences of paleosols is the way in which they are found one on top of another in long sedimentary sequences. This appears to be a record of periodic destruction and sedimentation over former ecosystems and their soils. The degree of development of paleosols in such sequences provides crude estimates of the recurrence interval of such events. For flood discharges, and perhaps also other geomorphic agents, events of long recurrence intervals are events of greater magnitude. Large floods occur at long intervals, small floods at short intervals. Both recurrence and magnitude are important elements for understanding how some sedimentary sequences are put together. Regular recurrence intervals are evidence of a landscape in a kind of dynamic equilibrium between forces that preserve it, such as slow subsidence and dense vegetation, and those forces that destroy it, such as high subsidence rates and sparse vegetative cover. Erratic recurrence intervals are evidence of a system out of balance and seeking a new equilibrium. Recurrence intervals revealed by paleosols are clues to the nature of sedimentary or erosional events and whether they were normal parts of a sedimentary system, such as the avulsion of a stream channel, or something else entirely, such as a major climatic change.

Not much can be done to analyze the causes of superposition of individual paleosols in Badlands National Park because the completeness and resolution of this sequence are inadequate (Table 13.5), as discussed in the preceding section. Such studies are better undertaken in sequences of paleosols that can be dated with greater precision, such as Quaternary paleosols preserved in Czechoslovakian loess (Kukla 1977). In this case, glacially related climatic change on time scales of tens to hundreds of thousands of years has led to superposition of tundra, then grassland, then forest paleosols. It is conceivable that paleoclimatic disruption also acted in creating superposition of paleosols in Badlands National Park, but the cause of each event of paleosol superposition there remains beyond analysis.

A broader pattern of episodicity in paleosols of Badlands National Park can be related to erosional valley cutting and filling. Because each cutting phase ushered in different soil-forming conditions, each sequence of paleosols has a different appearance. Many of these features already have been used to establish rock formations in the area. When stage of

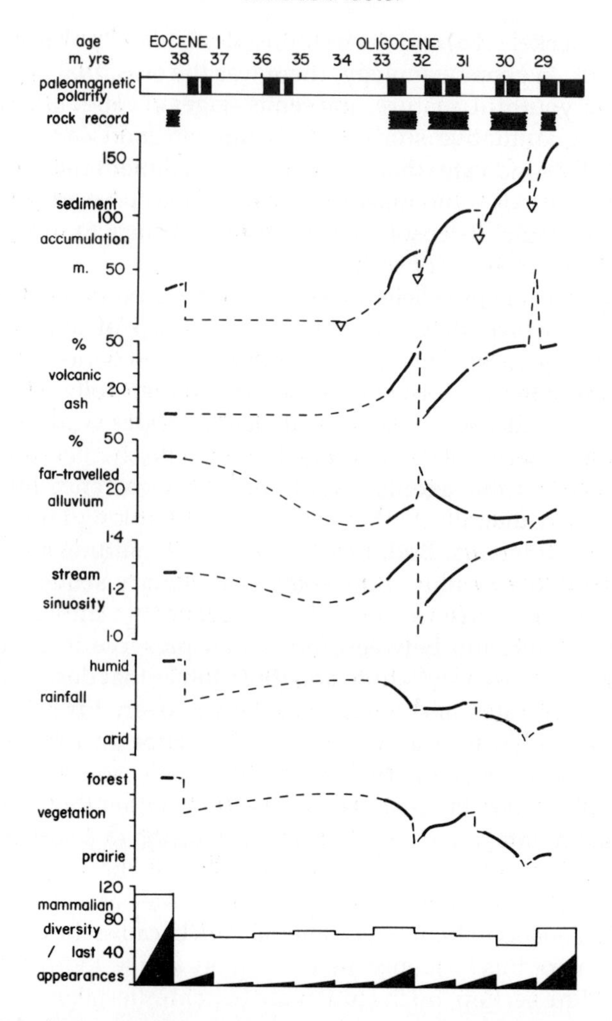

Figure 13.15 Factors controlling fluvial deposition in a measured section of the Pinnacles area (Fig. 6.2), Badlands National Park, South Dakota (from Retallack 1986d, reprinted with permission of the Society of Economic Paleontologists and Mineralogists).

development of each paleosol is plotted against stratigraphic level (Fig. 13.14), the times of erosion stand out as dividing sequences of closely superimposed, better developed paleosols below from those of distantly spaced, poorly developed paleosols above. The lower parts of each filling phase were times of rapid sediment accumulation and an erratic pattern of degree of development of paleosols. These appear to be times of disequilibrium when the system was yet to reach a balance between controlling forces. The upper parts of each sequence, however, have numerous paleosols showing about the same degrees of moderate to strong development. These are too regular to be due to chance and reflect a kind of equilibrium. During these relatively stable times, vegetation, subsidence rates and other environmental conditions are filtering possible destructive events for those of a particular magnitude. Such progressive changes in paleosols provide a clear example of geomorphic thresholds, of a system driven to a breaking point every few million years.

The resolution of the sequence is fortunately adequate to address the causes of these long-term environmental crises. The factors controlling fluvial systems include the same ones important to soil development, such as time, initial topographic relief, base level, local geology, climate, and vegetation, in addition to upland drainage net, hillside morphology, downstream deliveries, channel behavior, and pattern of deposition (Schumm 1981). Many of these factors can be quantified in ways already outlined in other parts of this book. Only a summary of the most important factors is outlined here (Fig. 13.15). By plotting the sedimentary sequence against real geological time, as estimated paleomagnetically, the cutting phases can be seen to represent as much time as the filling phases for which there is a record. The cutting phases can be recognized from breaks in

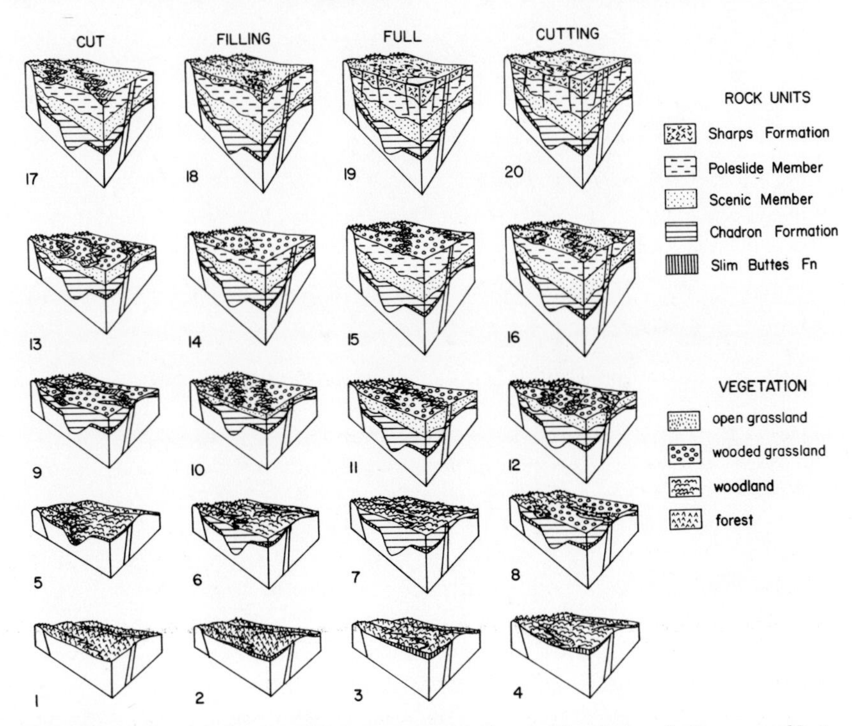

Figure 13.16 A reconstruction of changes in vegetation and landscape during several Late Eocene and Oligocene cutting and filling episodes that created the rock units now exposed in Badlands National Park, South Dakota (from Retallack 1986d, reprinted with permission of the Society of Economic Paleontologists and Mineralogists).

sediment accumulation and in the downward limit of channel incision. Some also were times of extinction of local mammals (Protherto 1985). In one case, a cutting phase coincides with an increase in far-travelled alluvium (quartz in C horizons of these dominantly volcanic-ash soils), perhaps implicating tectonic uplift of quartz-rich source terrain of the Black Hills. In another case, an especially thick volcanic ash is present, perhaps an indication of volcanic destabilization. Each cutting phase, however, corresponds to a period of drier climate and sparser vegetation (Fig. 13.16). Although contributing factors were instrumental in some erosional phases, and each was different in some way, climatic deterioration can be identified as a forcing mechanism in each case. There is independent stratigraphic, sedimentological, isotopic, and paleobotanical evidence from other parts of the world for climatic fluctuations at these times (Wolfe & Poore 1982).

In these ways, paleosols provide insight into the robustness of landscapes in the face of external environmental change. Interpretation of time for formation of paleosols opens the way to a variety of studies concerning historical aspects of landscapes and life.

PART THREE

FOSSIL RECORD OF SOILS

14 A long-term natural experiment in pedogenesis

In assessing the effects of a particular factor in soil formation, field situations are sought in which most of the variables are comparable except the factor chosen for study. In many cases this is difficult. The effects of climate, for example, may not be easy to tease apart from those of vegetation. For the study of the role of particular factors in the formation of paleosols, it is also important to select examples where other factors of the paleoenvironment and diagenetic alteration were comparable. This may seem a daunting exercise, but it is in principle no different from interpretative paleoecology or sedimentology. There are many questions of general scientific interest concerning conditions of soil formation that only can be addressed using paleosols. In the geologically distant past, for example, climatic conditions were not always linked with the same kind of vegetation as that of today. The fossil record of soils can be considered a long-term natural experiment in which many of the factors of soil formation operated in fundamentally different combinations.

The fossil record of soils is long. The oldest paleosol yet studied in detail is 3000 million years old (Grandstaff *et al.* 1986). Others have been noted as old as 3400 million years (Lowe 1983). Highly metamorphosed rocks as old as 3500 million years have some of the chemical characteristics of soils (Serdyuchenko 1968, Reimer 1986), but are too altered for confident identification as paleosols. This is not to say that such old paleosols are unlikely. The search for very ancient paleosols has just begun (Retallack 1986a) and is bound to be surprising. Until such information is forthcoming, one can only speculate about soil formation on the early Earth back to its origin some 4500 million years ago. Such speculation becomes better informed with each passing year as very ancient paleosols are better understood, as computer modeling of soil formation becomes more sophisticated and as the soils of other planets and moons come under scientific scrutiny. The processes of soil formation on the Moon and Venus are very different from those on Earth. Surface processes on these other heavenly bodies provide new models for soil formation against which likely soil formation on the early Earth can be reassessed.

Major events in the history of soils on Earth are already apparent from the rock and fossil records: the appearance of life on land, the oxygenation of a primeval reducing atmosphere, the advent of large plants and animals on land, the evolution of forests, the spread of grasslands, and human modification of the land. Their surficial effects can be anticipated only to a limited extent from what is known of analogous Quaternary soils because they remain in essence historical events. Evidence from paleosols is especially useful in establishing when they occurred. The evolution of grasslands in dry continental interiors is not reflected in the fossil record because swampy and wet conditions are needed for the preservation of plants (Retallack 1984a). Fossil teeth adapted for grazing tough grass presumably postdate the appearance of grasslands, and such ecosystems leave little trace in sediments. Grassland paleosols, however, can be distinctive (Retallack 1986c).

Paleosols also can be expected to provide insights into how major kinds of surface environments and ecosystems evolved. Organisms and their environments are bound in a complex web of interactions that has evolved through time to favor those organisms and environments best able to persist. The selective pressures acting at the initiation of what are now major ecosystem types may have been very different from those that sustain those ecosystems. A comparable phenomenon of biological evolution is pre-adaptation in the evolution of major new biological structures. Feathers, for example, may have evolved primarily for insulation and only later were pressed into service for flying (Ostrom 1974). Similarly the origin of a major ecosystem such as open grassland may have had little to do with their present ability to cope with frequent fires set by humans (Retallack 1982). Questions of the origination of different kinds of soils and ecosystems can be addressed by the study of paleosols.

As a list of major events in Earth history, those cited above are rather few. This is not to deny the importance of other events such as the Cretaceous dispersal and rise to dominance of flowering plants (Retallack & Dilcher 1981a) or the Cretaceous–Tertiary extinction of dinosaurs (Retallack *et al.* 1987). There were profound changes associated with these events that did much to fashion subsequent ecosystems, but they were not reflected in fundamental changes in the nature of soils on Earth. As soil-forming agents, angiosperms do not yet appear to have been different from seed ferns before them. Nor were dinosaurs discernibly different from other large animals, such as therapsids before them, or titanotheres and elephants after. Wider consequences for soil formation would be expected from both events, but what is known about the fossil record of soils does not bear this out.

Like organisms and communities, soils have evolved over the geological history of the Earth. For the most part, this evolution has consisted of the addition of new kinds of soils as new ecosystems and environments appeared on Earth (Fig. 14.1). There has been an overall diversification of

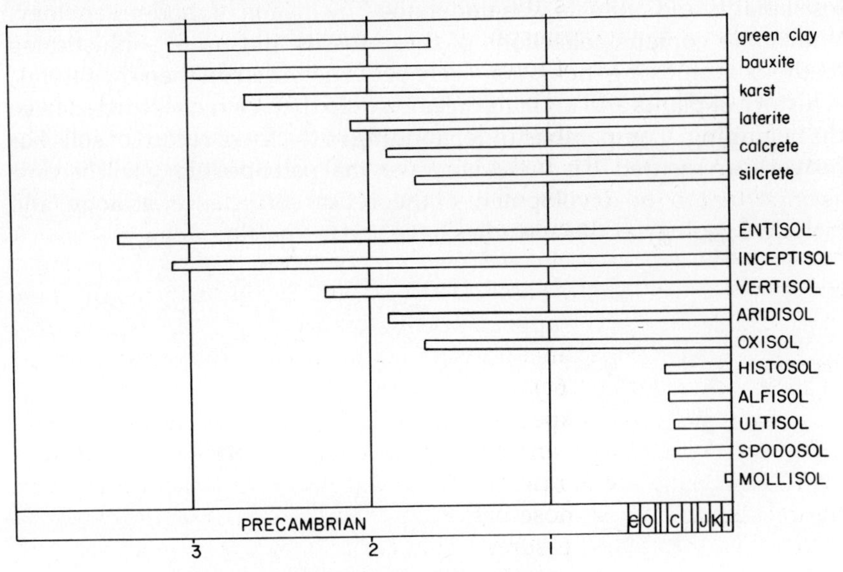

Figure 14.1 Geological range of soil features and soil orders recognized by the Soil Conservation Service of the US Department of Agriculture. Entisols and Inceptisols are assumed to be precursors of other Precambrian paleosols, even though difficult to recognize in Precambrian paleosols because they lack root traces and burrows.

soils with time because ancient kinds of soils have persisted in other regions of an increasingly heterogeneous world. Extinct kinds of soils are theoretically possible, but the only extinct soils known are Green Clay paleosols of early Precambrian time. These are green, clayey, alumina-rich and base-poor paleosols developed on iron-rich parent materials such as basalt. Their composition may reflect lower oxygenation of the atmosphere because basalts now weather to red, iron-rich soils (Holland 1984). The rarity of extinct soil types could reflect the vastly greater amount of information available on modern soils compared with those of the past. Considering difficulties in unraveling the alteration of paleosols after burial, it has been easier to identify a paleosol with a modern soil type rather than make more controversial claims that it represents a unique kind of extinct soil.

The progress of paleontology has been similar. Leafless, branching stems (*Psilophyton princeps*) in early Devonian rocks were originally compared with modern whisk ferns (*Psilotum nudum*). Large teeth (*Iguanodon mantelli*) from early Cretaceous rocks were compared with those of modern lizards (*Iguana iguana*). It took much research to establish that these fossils represented extinct kinds of organisms, i.e., trimerophytes (Stewart 1983) and dinosaurs (Carroll 1988). At first, the nature of fossil plants and animals was understood by means of modern analogy. With more copious collections of these fossils and more sophisticated methods of studying them, they were better assessed on their own terms. A history of plants and animals on land could then be reconstructed from the beginning. Comparable understanding of the fossil record of soils has not yet been secured. It is likely, however, that paleopedology will become as important to the development of theoretical soil science, ecology, and historical geology as those studies have been to paleopedology.

15 Soils of other worlds

The fertile imaginations of writers of science fiction have seldom devised life forms of other planets that were totally unlike those of Earth. Most extraterrestrial creatures of fiction are recognizably like insects or humans, complete with legs, eyes, ears, and mouths. Imagined landscapes of other worlds also owe much to human experience with deserts, jungles and cities here on Earth. Life beyond Earth, although eagerly anticipated, remains to be demonstrated. Our present understanding of landscapes of the Moon, Mars, and Venus, on the other hand, confirms the old adage that fact is stranger than fiction. Soils on the Moon, for example, are produced principally by micrometeoroid bombardment of those of Venus form by what would be regarded as metamorphic alteration on Earth. These strange landscapes formed by sand blasting and glazing have no clear analogs on Earth.

A consideration of soils and landscapes of these strange new worlds is valuable for several reasons beyond the general one of broadening intellectual horizons. These other systems of soil formation provide a basis for reassessing surficial processes on Earth. Soils and soil-forming processes on Earth vary widely, but nowhere are there processes and soils like those on the Moon or Venus.

A second reason for the study of soils of other worlds is that they may give clues to conditions during the first 700 million years or so of Earth history unrecorded in sedimentary rocks. The oldest well preserved sedimentary rocks are ca. 3500 million years old in the Barberton Mountain land of South Africa (Lowe *et al.* 1985) and the Pilbara region of Western Australia (Lowe 1983). Rocks 3800 million years old near Gőthaab (Nŭk) in southwest Greenland were once sediments (Allaart 1976), but are now strongly metamorphosed. Even older single crystals of zircon have been found, apparently dating back to 4276 million years. These were eroded from igneous rocks of that age into the younger (3350 million year old) sandstones in which they are found near Narryer, Western Australia (Compston & Pidgeon 1986). On the Moon and Mars there are land surfaces as old as 4000 million years and rocks perhaps as old as 4500 million years (Greeley 1985). Some meteorites have been interpreted as formed by soil formation, and these are 4530 million years old, close to the age of the Earth and Solar System. Hence, ideas concerning soil formation on the primordial Earth depend on information about extraterrestrial soils.

One problem for the study of altered planetary surfaces is what to call them. They are called soils in this book. This common English term is also used in many scientific accounts of Lunar, Venusian and Martian surfaces. Other reports show preference for the term regolith. Compared with soil, regolith includes a greater variety of surficial materials such as sediment and saprolite. Soil can be distinguished from saprolite by the intensity of its alteration and from sediment by its formation in place without appreciable lateral transport. Gradation is another vague term for all the sedimentary, pedogenic and other surficial processes that create regolith from bedrock. This vagueness of definition has commended both terms for geological reconnaissance of planetary surfaces based on images taken from orbiting satellites. The terms regolith and gradation deliberately gloss over the roles of soil formation, sedimentation, and biological processes. They will not be used here because the teasing apart and evaluation of these various processes is the main purpose of this account of extraterrestrial soils.

Soils of the Moon

On the Moon, soil-forming conditions are very different from those on Earth. Much of this difference can be traced to the Moon's lack of an atmosphere or free water at the surface (Table 15.1). Thus micrometeoroids

Table 15.1 Physical data on selected planetary bodies (from Greeley 1985))

	Moon	Mars	Venus	Earth
Diameter (km)	3,476	6,794	12,100	12,756
Mass (Earth = 1)	0.0123	0.1074	0.8150	1
Density ($g.cm^{-3}$)	3.34	3.95	5.27	5.52
Surface temperature (°C)	~80	~-50	~470	~20
Atmospheric composition	none	CO_2	CO_2	N_2, O_2
Atmospheric surface pressure (bars)	0	0.008	93	1
Sidereal period of revolution (yrs)	0.075	1.88	0.616	1
Equatorial surface gravity ($cm.s^{-1}$)	162	371	890	978

and cosmic radiation bombard a surface that is unaltered by chemical weathering (Fig. 15.1). Seasonal and "daily" variations in temperature are extreme. At the Apollo 17 site, the soil reached temperatures above boiling (111°C) by day and well below freezing (−171°C) by night (Taylor 1982).

Life has not yet been discovered on the Moon, despite exploration by automated vehicles and astronauts. Life as we know it on Earth would be very unlikely in all areas of the Moon now explored because water and carbon are extremely scarce.

The surface of the Moon has a topographic relief of about 16 km. Mare basins produced by the impact of large meteoroids and filled with flood basalts are on average 3–4 km below the surrounding old cratered highlands.

Compared with those on Earth, a limited array of rock types has been discovered on the Moon. Many are basaltic and ultramafic in composition and consist largely of plagioclase, pyroxene, and olivine. At the other compositional extreme are anorthosites and anorthositic breccias of the highlands. These are richer in alumina (Al_2O_3, 9–13 wt. %) and lime (CaO, 10–12%) than basalts, but have silica (SiO_2, 38–46%) within the basaltic range. Lunar basalts have some free quartz, but compared with basalts on Earth they are rich in iron (FeO, 16–20%) and titania (TiO_2, 0.3–13%) and poor in soda (Na_2O, 0.1–0.5%) and potash (K_2O, 0.02–0.4%, but in some up to 1.11%).

Radiometric dating of highly brecciated rocks from the Lunar highlands

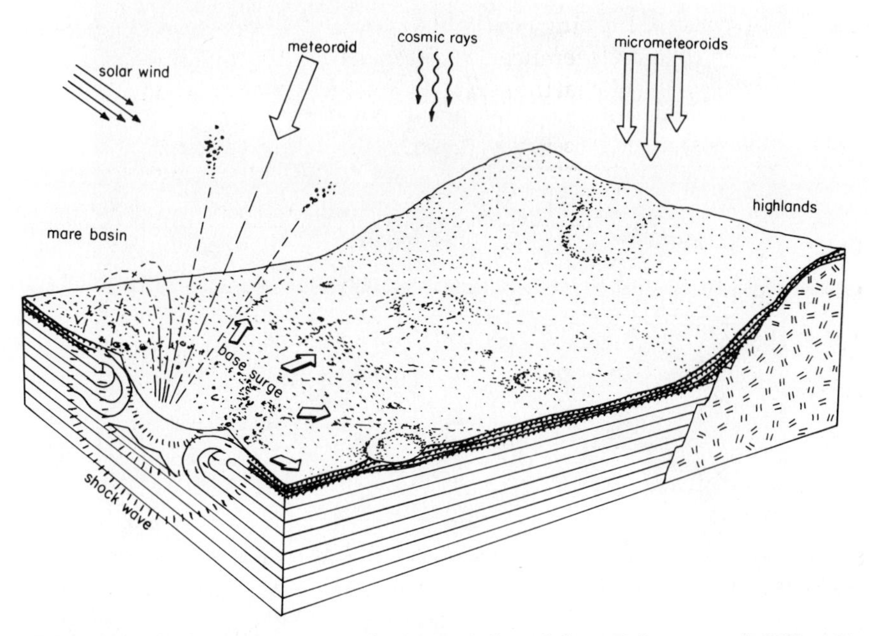

Figure 15.1 Surface processes on the Moon (adapted from Eglinton *et al.* 1972, with permission from W. H. Freeman & Co.).

indicate that most of the anorthositic crust was differentiated between 4400 and 4500 million years ago. A part of this early differentiation is thought to have been aided by massive meteoroid bombardment as the Moon orbited through an early Solar System littered with abundant rocky debris. The rate of bombardment has declined since then and the relative ages of different surfaces can be reconstructed from crater densities visible in images of the lunar surface. Mare filling lavas range in age from almost 4000 to 2000 million years. Craters as old as 2000 million years still have fresh outlines and clear rays of ejecta. The persistence of such features is an indication that rates of erosion, soil formation, and tectonic change on the Moon are all much more sluggish than on Earth.

Soil composition

The unconsolidated material at the surface of the Moon is typically fine grained and very poorly sorted. Its porosity is high (41–70%) and density low (0.9–1.1 g cm^{-3} at the surface) compared with the density of its

Figure 15.2 Lunar Soil near the Apollo 15 site on the margin of Mare Imbrium, overlooking Hadley Rille, a collapsed lava tube which contains conspicuous talus blocks. The Apollo deep core was drilled on the plain in the left distance, 3.8 km from the astronaut (NASA photograph AS15-85-011451, courtesy of F. J. Doyle and National Space Science Data Center).

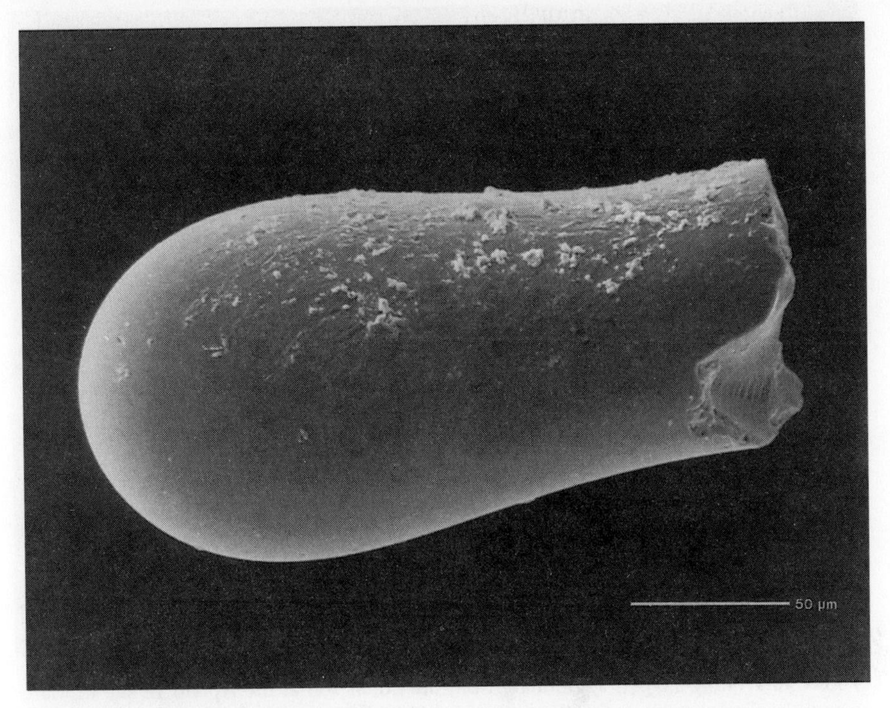

Figure 15.3 A fractured rotational homogeneous glass particle from the Lunar soil near the Apollo 15 site and Hadley Rille. Scale bar 50 µm (NASA photograph S72-52308, courtesy of F. J. Doyle and National Space Science Data Center).

constituent particles (2.9–3.2 g cm^{-3}). There are three main kinds of soil components: rock fragments, mineral grains, and glass particles.

Rock fragments are the most variable in size, ranging from silt-sized grains to large boulders (Fig. 15.2). Fragments of highland anorthosite and mare basalt have been mixed by some large impacts.

Mineral grains constitute most of the silt-sized fraction (0.06–0.03 mm), because of the fine grain sizes of crystals in lunar basalts. Plagioclase and pyroxene are present in most soils in substantial amounts (from a few to 40%). The olivine content is only ca. 1% and sporadic in occurrence. Much less common are occasional grains of ilmenite, spinel, metallic particles (kamacite and taenite), phosphide (schreibersite), and sulfide (troilite).

Glass particles are of two main kinds: the homogeneous glasses and the inhomogeneous glass-bonded aggregates called agglutinates. Both are variable in size, ranging down to fine silt size (0.02 mm) and up to 1 mm in the case of agglutinates or 2 cm in the case of some homogeneous glass particles. Homogeneous glass particles vary in shape from spheres to dumbells and teardrops (Fig. 15.3). These homogeneous glasses are thought to be volcanic ash or rock melted by the heat of meteoroid impacts.

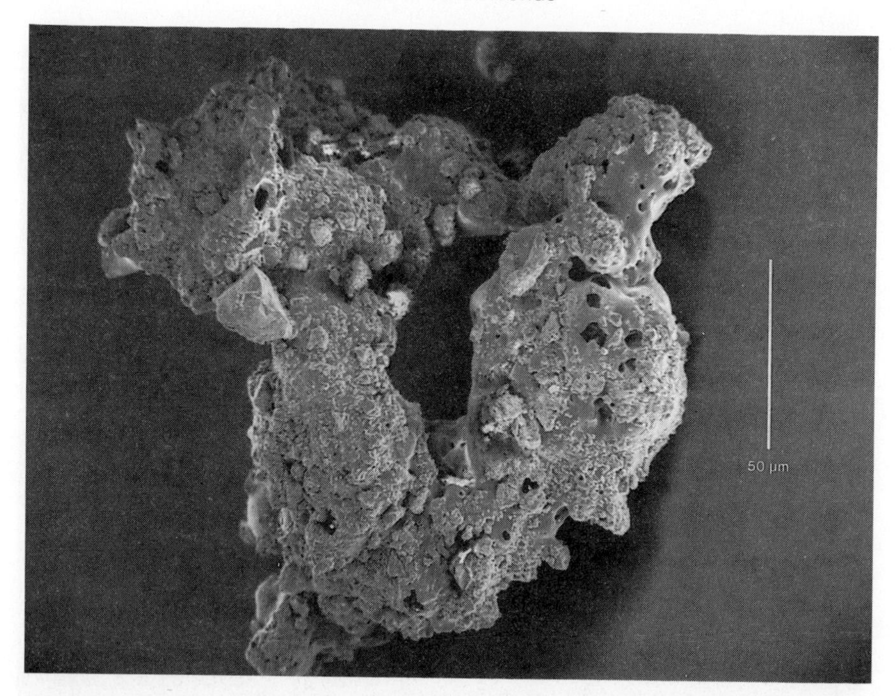

Figure 15.4 Ring-shaped agglutinate from the Lunar soil at the Apollo 12 site in Mare Cognitum. Scale bar 50 μm (NASA photograph S71-24575, courtesy of D. McKay, F. J. Doyle and National Space Science Data Center).

Agglutinates are highly irregular masses of glass and crystals that are held together by bridges of glassy cement between a variety of mineral grains and rock fragments. Some agglutinates have a ring or bowl shape, like a miniature crater. These are thought to have formed where impact melt has spread outward to cement surrounding grains of soil (Fig. 15.4). Other agglutinates that are less distinctively shaped may have been cemented by scattered drops of impact melt or may be parts of agglutinates broken by later impacts.

The chemical composition of Lunar soils is very close to that of nearby parent material. Soil formation on the Moon appears to be more of a physical than a chemical process. Nevertheless, a few anomalously calcareous and aluminous soils have been found in areas of mare basalt. These can be explained by admixture of impact ejecta from the more calcareous and aluminous highlands. If there is such extensive lateral mixing it would be difficult to identify the role of a particular parent material in soil formation.

Soil development

The unconsolidated material at the surface of the Moon has been estimated from the geometry of craters to reach 16 m in thickness and from seismic

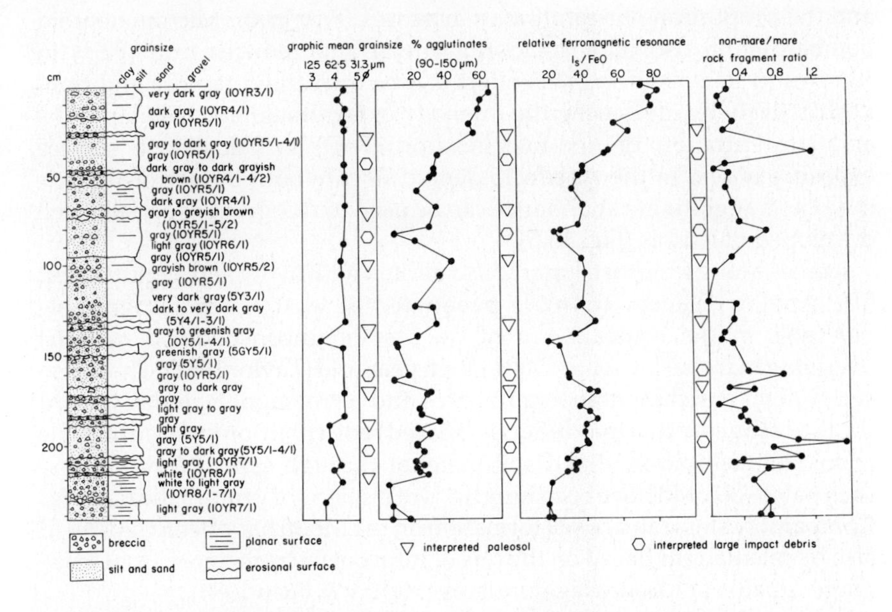

Figure 15.5 An interpretation of the most prominent buried soils in the Apollo 15 deep drill core, based on soil maturity indices (mean grain size, percentage agglutinates, and ferromagnetic resonance) and on evidence for large impacts (non-mare-to-mare rock fragment ratio; data from Heiken *et al.* 1973, 1976).

observations at the Apollo stations to reach 12.2 m. Not all this material can be considered to be soil in the sense espoused here. Processes acting to form the soil include micrometeoroid bombardment, mixing due to solar-induced charge separation, churning associated with temperature changes, and solar wind sputtering. Even the most deeply penetrating of these, micrometeoroid bombardment, affects only the uppermost few millimeters. Larger impacts broadcast a blanket of ejecta that interrupts soil formation. In the Apollo 15 deep drill core, large impacts are recorded by erosional planes and by large rock fragments, including numerous rocks from outside the mare basin in which the core was drilled (Fig. 15.5). Beds in the core can be considered as a long sequence of depositional events, some of which were punctuated by soil formation in a similar fashion to a sequence of alluvial paleosols in a fluvial sedimentary sequence on Earth. Layers on the impact-cratered Moon are not so laterally continuous as on Earth, and cannot be correlated between cores a few kilometers apart (Basu *et al.* 1987).

Developmental stages of Lunar soils are based on their degree of reworking by micrometeoroids. The starting material is a coarse-grained and poorly sorted blanket of impact ejecta. Under micrometeoroid bombardment this is broken down to a more even and smaller grain size, and the proportion of agglutinates increases (Fig. 15.6). Micrometeoroid bombardment also adds meteoritic metal and reduces iron (Fe^{2+}) in silicates to metallic iron. This effect can be quantified rapidly and non-destructively by measuring the intensity of ferromagnetic resonance (I_s) and concentration of non-metallic iron (FeO) to calculate a surface exposure index. In the Apollo 15 deep drill core both the ferromagnetic index and agglutinate abundance can be used to recognize especially well developed paleosols (Fig. 15.5).

It takes a long time for a lunar soil to darken with metal and agglutinates. The Apollo 15 deep drill core penetrated at least seven recognizable paleosols in the upper 2.4 m of ca. 5 m of unconsolidated material overlying mare basalt about 3300 million years old (Taylor 1982). There also is a general enrichment in agglutinates and ferromagnetic index toward the top of the core that may reflect increased redistribution by large impacts of micrometeoroid-modified soil material through time. Nevertheless, each paleosol could have been hundreds of millions of years in the making. Comparably slow rates of soil formation on the Moon have been confirmed also by calculations based on the rate of influx of micrometeoroids of a size range capable of forming agglutinates (McKay & Basu 1983).

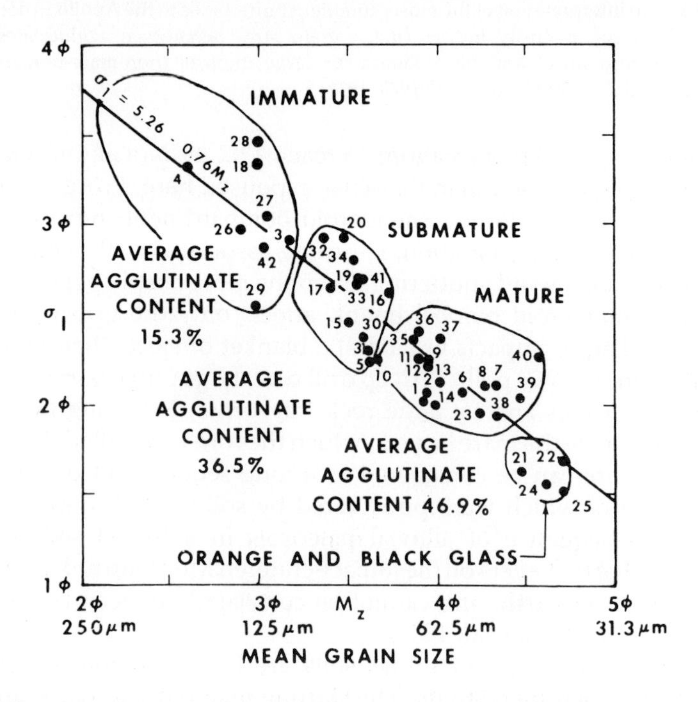

Figure 15.6 Stages in the development of Lunar soil, as shown by changes in abundance of agglutinates, mean grain size and its standard deviation (from McKay *et al.* 1974, with permission of Pergamon Press).

Soils of Venus

Venus is a permanently clouded planet. Its surface remained a mystery until landings of the Venera probes and until radar mapping from orbiters and Earth-based observatories. The climate of Venus is so hostile that few landers have been able to operate for more than a few hours (Table 15.1). The upper cloud deck, 70–56 km above the surface, is an aerosol (300 particles per m^3) of small (0.1–2.8 μm) crystals and droplets of sulfuric acid. The bottom cloud layer, 49.5–47.5 km above the surface, is similar in composition, but is denser (400 particles per m^3). Below that the atmosphere is clear enough for images of the Venusian surface to be made under natural lighting. The cloud layers trap and circulate heat so effectively in a kind of "greenhouse effect" that there is little difference in heat between the equator and the poles or between day and night. At the level of the lower clouds the temperature is close to 72°C and the pressure about 1 bar. Temperature and pressure in the highlands (elevations 10–12 km) are oppressive (374°C and 41 bar) but discernibly lower than in the lowlands (465°C and 96 bar; Warner 1983). The chemical composition of the Venusian atmosphere is mainly carbon dioxide (96%) with small amounts of nitrogen (1–3%) and water vapor (0.1–0.4%) and traces of helium (250 ppm), neon (6–250 ppm), argon (20–200 ppm), sulfur dioxide (240 ppm) and oxygen (60 ppm) (Barsukov *et al.* 1982).

At the high temperatures on the surface of Venus, neither liquid water nor organic compounds are stable, making life as it exists on Earth highly unlikely. No trace of any other kind of life or biological process has been detected. Venus has the kind of atmosphere that would be predicted on the basis of equilibrium thermodynamics, completely unlike the peculiar, biologically influenced atmosphere of Earth (Lovelock 1979).

Radar mapping of Venus has shown that it has less topographic relief than Earth, despite its comparable size. About 60% of the planet is within 500 m of the modal planetary radius. The highest point is Maxwell Montes, which rises 10.8 km above the modal radius. The lowest point is Diana Chasma, ca. 2.9 km below this datum. Images transmitted by Venera landers 9 and 10 depict bouldery slopes around a large volcanic upland, but Veneras 13 and 14 landed in broad plains (Fig. 15.7).

Two partial chemical analyses of the Venusian surface are available (Table 15.2) and these are from soil rather than bedrock. The soil at the Venera 13 site is comparable in chemical composition to alkaline melanocratic basalt and that at the Venera 14 site to tholeiitic basalt of Earth (Basilevsky *et al.* 1985). Given the chemically aggressive nature of the Venusian atmosphere, these are almost certainly altered from the parent material. Their composition is reasonable, however, considering the location of the alkaline rocks near a large rift valley and of the tholeiitic rocks on a plain near a large shield volcano. These are the kinds of tectonic

Table 15.2 Chemical analyses of extraterrestrial soils and parent materials (weight percent)

Oxides and elements	Venusian soils		Martian soils					Meteorites			
	Venera 13	Venera 14	dust	Chryse salic horizon	salic horizon	Utopia dust	Absolute error of Viking XRF	Shergotty	Chassigny	Murchison	Allende
SiO_2	45.1 ± 3.0	48.73 ± 3.6	44.7	44.5	43.9	42.8	5.3	50.4	37.0	27.22	34.23
TiO_2	1.59 ± 3.0	1.25 ± 0.41	0.9	0.9	0.9	1.0	0.3	0.81	0.7	0.099	0.15
Al_2O_3	15.8 ± 3.0	17.9 ± 2.6	5.7	-	5.5	-	1.7	6.89	0.36	2.05	3.27
Fe total as FeO	9.3 ± 2.2	8.8 ± 1.8	16.38	16.2	16.88	18.27	2.9	19.1	27.44	26.29	30.68
MnO	0.2 ± 0.1	0.16 ± 0.08	-	-	-	-	-	0.50	0.53	0.22	0.18
MgO	11.4 ± 6.2	8.1 ± 3.3	8.3	-	8.6	-	4.1	9.27	32.83	18.90	23.91
CaO	7.1 ± 0.96	10.3 ± 1.2	5.6	5.3	5.6	5.0	1.1	10.1	1.99	1.75	2.61
Na_2O	-	-					-	1.37	0.15	0.57	0.45
K_2O	4.0 ± 0.63	0.2 ± 0.07	<0.3	<0.3	<0.3	<0.3	-	0.16	0.03	0.034	0.03
SO_3	-	-	7.7	9.5	9.5	6.5	1.2	-	-	2.38	-
Cl	-	-	0.7	0.8	0.9	0.6	-	-	-	0.14	0.0224

Table 15.2 *Continued*

Oxides and elements	Venusian soils		Martian soils					Meteorites			
	Venera 13	Venera 14	Chryse dust	Chryse salic horizon	Chryse salic horizon	Utopia dust	Absolute error of Viking XRF	Shergotty (SNC)	Chassigny (SNC)	Murchison (CM)	Allende (CV)
Ni	-	-	-	-	-	-	-	-	-	1.26	1.03
S	-	-	-	-	-	-	-	-	-	2.30	2.09
C	-	-	-	-	-	-	-	-	-	1.91	0.29
CO_2	-	-	-	-	-	-	-	-	-	1.00	-
H_2O	-	-	-	-	-	-	-	-	-	12.06	<0.1
Totals	94.91	95.91	90.28	77.5	92.08	74.47	-	98.2	101.27	98.28	98.84

Note: Data for Venusian soils from Basilevsky *et al.* 1985, Martian soils from Toulmin *et al.* 1977 with their Fe_2O_3 results (left to right = 18.2, 18.0, 18.7, 20.3) recalculated to total iron as FeO to facilitate comparison, Shergotty meteorite from Stolper & McSween 1979, Chassigny meteorite from McCarthy *et al.* 1974, Murchison meteorite from Fuchs *et al.* 1973, with total Fe (20.44) calculated to FeO for comparison and NiO (1.73) recalculated to Ni, and Murchison from Bunch & Chang 1980, Allende meteorite from Clarke *et al.* 1970, with total Fe (23.85) recalculated to FeO, NiS (1.60) to Ni and FeS (4.03), NiS (1.60) and CoS (0.08) to S. Dashes signify data not yet available.

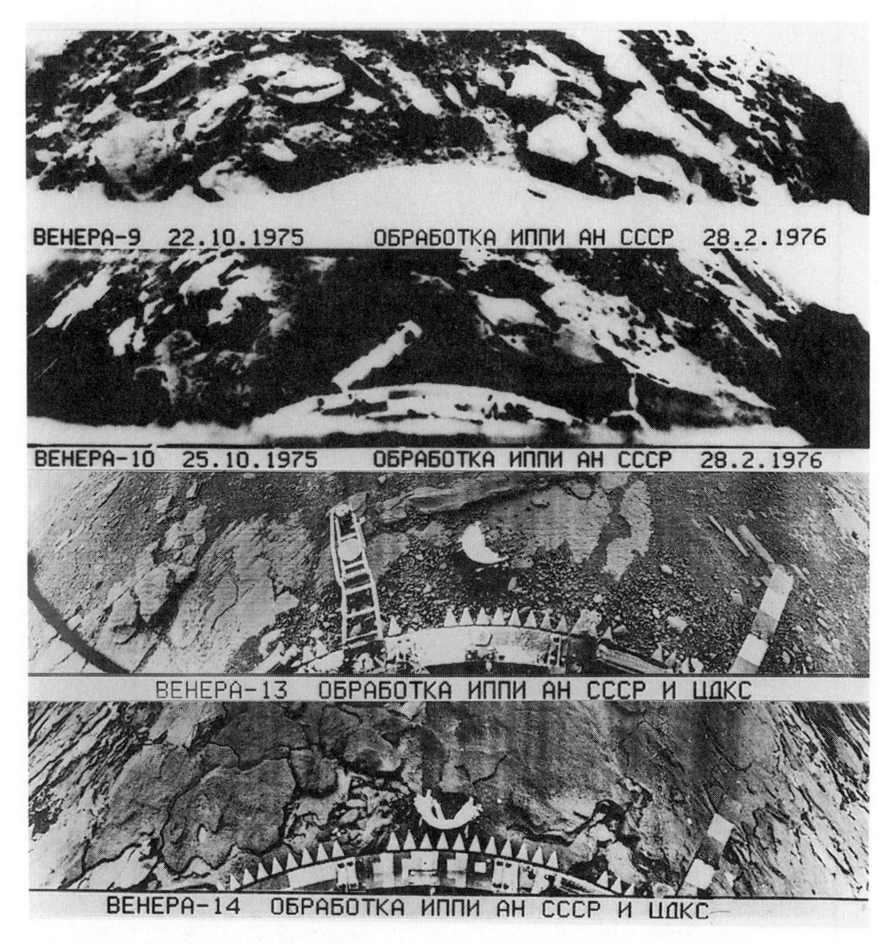

Figure 15.7 Images of the surface of Venus relayed by Soviet Venera 9, 10, 13, and 14 (photographs courtesy of R. J. Allenby and NASA National Space Science Data Center).

locations where such basaltic rocks would be found on Earth. In other areas of Venus there are elevated regions of about the size of Australia and Antarctica (Ishtar and Aphrodite Terra, respectively). These include banded regions that may contain a variety of rocks as in superficially similar orogenic belts on Earth (Crumpler *et al.* 1986), but they remain little known.

The density of cratering on Venus is an indication that its surface is fairly young. Only stony meteoroids larger than about 80 m in diameter and iron meteoroids larger than 30 m survive passage through its currently thick atmosphere to impact with the surface. The smallest surviving meteoroids would make craters about 300–400 m in diameter for stony meteoroids and 150–200 m for irons. These crater diameters are well below the optimal resolution (about 1–2 km) of present radar mapping. Despite the limitations in such data, the density of large craters observed is lower than those on Mars and the Moon. Age estimates based on the well calibrated impact frequencies on the Moon have indicated that Venusian surfaces are no older than 600 million years (Greeley 1985).

Soil composition

Information on the chemical composition of Venusian soils was relayed back to Earth by Venera 13 and 14 landers, which were equipped with miniaturized X-ray fluorescence analytical laboratories (Table 15.2). Calculations based on the observed impact and deformation of the landers calibrated with experimental dropping of comparable landers on Earth (Basilevsky *et al.* 1985) indicate that the surface material at both sites is low in density (1.2–2.5 g cm^{-3}) and highly porous (50–60%). This is surprising because the ground in the images (Fig. 15.7) looks like indurated sedimentary rocks, with internal lamination and undulations somewhat like wind ripples. Also unusual is the very low electrical resistivity of the material (only 73–89 ohm) and the darker color of the loose dust compared with the lithified-looking fragments. Dry rock or dust should have a high resistivity and dust should be lighter in color than the rocks from which it was derived.

These meager data and observations are permissive of a number of interpretations, each of which is based on such a complex series of assumptions that they should not be taken too seriously. Although the chemical composition of the soils at the Venera 13 and 14 sites is basaltic, their density, resistivity, and general appearance are more like those of siltstone. Cementation similar to that of a potter's glaze would be expected on Venus, considering the high temperature and pressure. A cement of light-colored salts around and bridging silicate grains could explain the lighter color of the slabs compared with the dust, their low electrical resistance, and their low bulk density. The fretted, glazed, and pitted appearance of these indurated-looking rocks could be due to sublimation or melting.

Theoretical models for the mineralogical composition of Venusian soils have also been proposed based on the available partial analyses and other information about the surface environment. Especially critical to such calculated reconstructions, yet poorly constrained, is the abundance of oxygen and water at the surface. In the model of Barsukov *et al.* (1982), both water (5×10^{-3}%) and oxygen (3.3×10^{-21}%) are assumed to be scarce compared with sulfur trioxide (1.85×10^{-2}%) and carbon dioxide (96%). Under such conditions, sulfate cement would be more abundant than cements of hydrated silicates or silica. The mineral assemblages favored on basaltic soils such as those at the landing sites would be primarily albite

(28%) and anorthite feldspar (24%) with smaller amounts of clinoenstatite (13%), pyrite (10%), microcline (10%), anhydrite (8%), and quartz (7%). Different mineral assemblages would be favored at higher elevations where it is thought there are slightly greater amounts of gases produced photochemically in the atmosphere, such as SO_3, CO, and O_2. Here basaltic rocks would be altered to albite (27%), anhydrite (18%), and quartz (17%) with smaller amounts of clinoenstatite (13%), microcline feldspar (10%), hercynite (10%), and pyrite (5%).

Soil development

For the sake of argument, let us adopt the theoretical model for soil composition already outlined, with all its complex assumptions. By this view, Venusian soil formation consists of solid-phase mineral transformation together with cementation of loose parts of the soil with calcium sulfate. These processes of soil formation in lowland soils on basaltic terrains are interrupted by thin layers of windblown dust from the highlands. The degree of soil development can be assessed by the extent of glazing and cementation of the originally dusty surface.

In both the highland and lowland models for basaltic soils, weathering produced quartz and feldspar at the expense of more mafic minerals so that soils become more siliceous and aluminous than their parent materials. Weathering on Venus also may produce sulfates and sulfides which may be buried and recycled in sediments (Warner 1983). The tendency toward more felsic surficial composition and the crustal recycling of the volatile sulfur are general features of geochemical cycles on Earth. A great geochemical difference between Earth and Venus is the instability of carbonates and organic matter at the surface. These cannot accumulate and be recycled into sediments of the crust, so carbon has accumulated in the atmosphere as carbon dioxide almost to the exclusion of other gases.

By both of these chemical and physical criteria, the Venera 13 site has a less well developed soil than that of the Venera 14 site. Such a relative age of these landscapes also would be expected from their geomorphic position. The Venera 14 site is farther out into a large basin (Nauka Planitia) than the Venera 13, site which is in rolling foothills of an extensive highland (Phoebe Regio).

It is unclear how much time is involved in converting a dusty surface to one that is as glazed and corroded as that visible at the Venera 14 site. Considering the high temperature and pressure on Venus, a thin layer of dust could be altered more rapidly than the hundreds of millions of years likely for the formation of comparably thin soils on the Moon. If Venusian soil formation can be compared with pottery glazing or experimental simulation of rock metamorphism, the times involved in altering such thin dust layers may take less than a few Earth days.

Even within this theoretical framework for Venusian surface conditions, other kinds of weathering can be envisaged on bedrock rather than in depositional landscapes and on felsic rather than basaltic parent materials. Soil formation on Venus is clearly unusual compared to that on Earth. Only time and more information from space exploration will demonstrate exactly how unusual.

Soils of Mars

The Martian atmosphere has regional and seasonal weather patterns and ice caps like those on Earth, but the conditions are far from Earthlike. The atmosphere of Mars is two orders of magnitude thinner than that of Earth (Table 15.1). This contrast may be related to Mars' smaller size and gravity, which allow gases to escape into space. The gases that have remained are largely carbon dioxide (95%) with minor amounts of nitrogen (3%), argon (2%), and only traces of oxygen (0.13%), carbon monoxide (0.07%), and water vapor (0.03%). Although water vapor is scarce, it is close to saturation in the thin atmosphere and large fluvial channels attest to its flow in the distant geological past. Most of the water remaining on the planet is frozen in permafrost and in polar ice caps. The low surface temperatures of Mars are due to its thin atmosphere and large distance from the Sun. The Viking landers recorded northern summer temperatures at a low northern latitude (Chryse Planitia) varying from −90°C at dawn to a maximum of −30°C at noon. Temperatures to the north in midlatitudes (Utopia Planitia) were 5–10°C cooler. In Chryse Planitia there was a slow seasonal cooling of about 22°C as the northern winter approached. This was interrupted by great dust storms which reduced the diurnal temperature fluctuation to only 10°C compared with 50°C at the start of the mission (Carr 1981).

These weather patterns and supposed "canals" of Mars have stimulated the idea that the planet was alive with a highly technological civilization. Now that there are thousands of detailed photographs of the Martian surface, this possibility seems remote. The canals were probably optical illusions of early telescopes. The channels and chasms observed appear more like fluvial and tectonic features rather than engineering works. Organic compounds were not detected by gas chromatography or mass spectrometry in the Viking landers, nor was any life seen in images of the Martian surface. The somewhat lifelike experimental results on the landers were therefore most surprising. In the pyrolytic release experiment, small amounts of radioactively labeled carbon monoxide and carbon dioxide were incorporated into organic matter that was synthesized in contact with illuminated Martian soil within the lander. In the gas exchange experiment, oxygen was rapidly released from Martian soil samples incubated in a humid environment. In the labeled release experiment a labeled nutrient solution of simple organic compounds was broken down to carbon dioxide when in contact with Martian soil. The progress of the first two experiments was little affected by high temperatures (ranging up to 145°C and 90°C, respectively) that would have curtailed a similar metabolic response of microbes. In contrast, the breakdown of organic matter was reduced (by 70%) at moderate temperature (50°C) and ceased at high temperature (160°C). This lifelike result for the labeled release experiment, and also the results of the pyrolytic release experiment, have both been duplicated in the laboratory under simulated Martian conditions using iron-rich montmorillonite as a substitute for Martian soil (Banin 1986). The examined landing sites of Mars show no signs of carbon-based life, but two of the experiments (gas exchange and pyrolytic release) can be construed as demonstrating a potential for abiotic "photosynthesis" and the third (labeled release experiment) demonstrated a strong oxidizing power.

Mars is a topographically rugged planet, ranging from elevations of 27 km above the planetary radius at the top of the large volcano Olympus Mons to 4 km below its radius in the floor of the Hellas Basin (Greeley 1985). Most of the volcanoes are low-relief mountains similar to those of the Hawaiian Islands. In contrast, many of the deep canyons have cliffed walls. Even in the relatively flat regions chosen for the Viking Landers, there is local relief around boulders, dunes, and channels (Fig. 15.8).

Little is known about the nature of bedrock on Mars. Even the most solid-looking pebbles analyzed by the Viking landers had a low density and high sulfur content, more like soil than rock. The general composition of the soil is more or less basaltic and such a composition is compatible with what is seen in images of the surface. Steep-sided composite volcanoes and cinder cones and fretted and smooth plains of pyroclastic materials are rare on Mars. Most of the volcanoes have a topographically low profile and are mantled with very long lava flows characteristic of basaltic lavas, in some cases perhaps as ultramafic as komatiites. Some kinds of basaltic and ultramafic meteorites (shergottites, nakhlites, and chassignyites) may represent rock samples from Mars that were ejected from the surface of Mars by large impacts, stored in solar orbits, and then captured by the gravity of Earth (Vickery & Melosh 1987). These are anomalously young (only 1300 million years) compared with other meteorites. They have an unusual oxygen isotope composition, negligible magnetic paleointensity, and evidence for long (0.5–10 million years) exposure to interplanetary cosmic rays that rule out most sources other than Mars (Wood & Ashwal 1981). Whether these are representative of Martian rocks will be seen when samples are returned after further space exploration.

The surface of Mars includes some heavily cratered plains that are thought to be 3500–4000 million years old. The abundance of craters on old

Figure 15.8 US Viking images of the Martian surface at (A) Chryse Planitia (NASA Viking 1 event number 11A079) and (B) Utopia Planitia (Viking 2 event number 21A024) (both courtesy of R. A. Arvidson and National Space Science Data Center).

runoff and outflow channels is an indication that these fluvial features also are old (Baker 1982). It is unlikely that water has flowed through them during the past 2000 million years. Although most land surfaces on Mars are much older than are those of Earth or Venus, there are many younger parts of the planet, especially around volcanoes, craters, scarps, eolian dune fields, and polar layered terrains.

Soil composition

Both Viking landers carried a miniaturized X-ray fluorescence spectrometric laboratory which completed chemical analyses of Martian soils (Table 15.2). Magnets of the landers also extracted loose mineral grains with the magnetic properties of maghemite. Both carbon dioxide and water were detected by the Viking spectrometers after heating soil samples to 500°C and these bound volatiles could amount to a few percent of the sample. The Viking labeled release experiment demonstrated that unheated fine-grained material contained an oxidant strong enough to decompose organic matter to carbon dioxide.

To these experimental data can be added information from images of the areas around the sites (Sharp & Malin 1984). Both sites show boulders surrounded by what appears to be windblown material of fine silt to sand size (10–100 μm in diameter). Orange pebbles that were analyzed had a low density (1.1 ± 0.15 g cm^{-3}) and a high sulfur content (Table 15.2) like the fine material. Although these looked like rocks, they probably are fragments of a cemented soil layer. The blue–gray and green–gray, vesicular and coarse-grained boulders look like blocks of basaltic or ultramafic lava flows. They may also be clasts in a mudflow breccia. Freshly exposed deep layers of the fine-grained soil are a similar blue–gray color to the rocks from which they were presumably derived (Fig. 15.9). The overall reddish color of the surface is due to a thin (1 mm) oxidized surface layer of dust. Oxidized layers also can be seen within freshly excavated

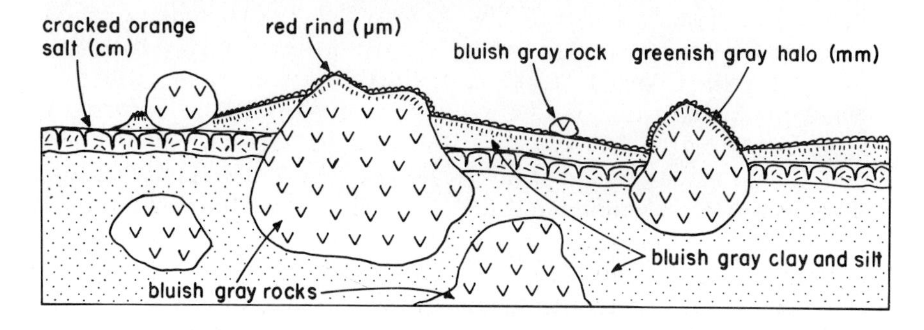

Figure 15.9 A reconstructed soil profile at the Viking 1 landing site on Mars (based on data of Sharp & Malin 1984).

dust between the boulders. The exposed surface of some of the rocks also shows a dusty oxidation rind, but others showed a gray surface. Beneath the landers where the loose dust and rock were most disturbed by touchdown, there is an orange layer. This is cracked into fragments several centimeters across in a way that suggest it is only ca. 1 cm thick. This layer has been called a duricrust, but does not appear to be nearly as thick as terrestrial duricrusts (see Fig. 6.7). It is better termed a salic horizon (Soil Survey Staff 1975), considering its high sulfur content.

These various observations of the Martian soil at the two known sites permit various speculations on its mineral composition. Reconstructions using suites of primary igneous minerals are implausible because of the high amounts of sulfur, the high oxidation state of the minerals and the significant amount of chemically bound water. This bound water is also evidence against all the minerals being simple oxides. Interpretation of the dust as iron-rich volcanic glass (palagonite) does not satisfy the reactions observed in the Viking biology experiments. The interpretation of Martian soils now widely favored is that they consist largely of clay and evaporite minerals (Banin 1986). Using these assumptions and the chemical analyses at the Viking 1 or Chryse site (Toulmin *et al.* 1977), a likely composition is one in which nontronite is the main clay mineral (47%), with less montmorillonite (17%) and saponite (15%). In this model the evaporitic minerals were kieserite (13%) and calcite (7%), and there were traces of heavy minerals such as rutile and maghemite. The salic horizon does not differ greatly in composition from the loose dust and is presumed to be more thoroughly cemented with kieserite and calcite.

Soil development

The parent material for the assumed smectites and evaporites on Mars appears to have been underlying basaltic or ultramafic igneous rocks or breccias. The degree of alteration is difficult to gauge because no analyses of Martian bedrock were returned by the landers. There is a case already outlined for regarding a distinctive group of meteorites called the SNC meteorites (after Shergotty, Nahkla, and Chassigny) as Martian rocks (Wood & Ashwal 1981). If one of the least mafic of these meteorites (Shergotty in Table 15.2) is taken as a parent material of the Martian soil, the only differences clearly beyond the experimental errors of the Viking analyses are an enrichment of sulfur and loss of lime in the soil. Although sulfite was not determined in the SNC meteorites, it cannot have been a major component judging from the totals of the analyses (Table 15.2). An assumption that known Martian soils formed from materials such as the more mafic SNC meteorite Chassigny gives an impression of more profound weathering, with increases in silica, alumina, and lime and losses of iron and magnesia. Martian soils revealed by the Viking images are shallow and the degree of weathering is slight, even compared with Antarctic desert soils on Earth (Campbell & Claridge 1987).

Leaching of salts and formation of clays by hydrolysis in water is another implication of the assumed hydrated smectitic and evaporitic composition of the Martian soil. The formation of these materials on the present Martian surface is unlikely to be rapid or widespread because of extremely low temperatures. The runoff and outflow channels of Mars provide evidence of some surface water more than 2000 million years ago. It could be that most of the salts and clays of the Martian soil formed at that time and have been blown around or held in a deep freeze since then. Other less effective and local sources of water can be imagined. Volcanic eruptions or large meteorite impacts through ground ice could generate large amounts of water for relatively short periods of weeks or months. More effective long-term alteration could be achieved by volcanically heated groundwater insulated from the Martian surface by frozen soil. It has also been found that hydrolytic weathering reactions can occur in an unfrozen monolayer of water molecules around mineral grains at temperatures well below zero in Antarctica (Ugolini 1986). Although the effects of volcanoes, impacts, and monolayer water films would be extremely slow compared with weathering under surface water, these processes cannot be discounted because of the thousands of millions of years available for the alteration of some Martian landscapes.

A paradoxical feature considering that the Martian atmosphere has very little oxygen is the thin surficial ferruginized layer of the soil responsible for the red color of the planet. Deeper and more pervasive oxidation would be expected if Mars had had a more oxygenated atmosphere at some time in the geological past. A likely mechanism is direct photo-oxidation under ultraviolet light. This process has been shown experimentally (Braterman *et al.* 1983) to be effective in acidic to neutral aqueous solutions in which the oxidation of iron is coupled with the reduction of water to hydrogen gas. On Mars this process can be imagined as occurring slowly over thousands of millions of years within water films or other local sources of water, or the oxidation could be a remnant of that distant geological time when there was free water on Mars. Photo-oxidation affects only a thin surface layer because the penetration of ultraviolet light would be curtailed by the surficial accumulation of opaque oxidized minerals.

Using these ideas concerning soil formation on Mars, the degree of development of soils at the two landing sites can be compared. At the Viking 2 (Utopia) site the boulders are angular and lying on the surface (Mutch *et al.* 1977) rather than embedded in the soil. They are blue–gray and show few red or greenish gray weathering rinds. The salt crust is smooth to lumpy and only ca. 1 cm thick. It was penetrated in places to reveal blue–gray soil below. This is a less developed soil than that at the Viking 1 (Chryse) landing site. Here the salt crust exposed was cracked into blocks, but was not penetrated. The half-buried boulders have extensive greenish gray and red alteration rinds. Most of the boulders were rounded and pitted (Binder *et al.* 1977). Such differences in degree of weathering correspond to the presumed geological age of the two sites. The Chryse site is on an extensive plain near where outflow channels end. This plain is ca. 3200 million years old with a range of 1200–3800 million years judging from crater counts (Hartmann *et al.* 1981). The Utopia site is in a region only 1800 million years old with a possible range of 2300–600 million years. The landing site is on the debris apron of the large (104 km) impact crater Mie only 170 km away. Boulders seen from the lander appear to be ejecta from this crater. Mie is a fresh-looking crater, but not the freshest on Mars (Vickery & Melosh 1987), and its debris apron bears many subsequent smaller craters (Mutch *et al.* 1977). Its age probably is comparable to that of its wider region.

If these geological ages of Utopia and Chryse Planitia apply also to soils at the landing sites, then the pace of soil formation on Mars is very slow. Soil formation also is slow in cold deserts of the Antarctic Dry Valleys on Earth where the annual temperature is a little less than −10°C, the summer maximum is 30°C and the mean annual precipitation is 45 mm. Here comparably thick (4 cm) salic hardpans and nontronitic clays are found only in soils more than 2 million years old. Oxidation rinds (primarily rock varnish) are not found on boulders exposed for less than 800 000 years (Campbell & Claridge 1987). Similar hydrolytic formation of clays, salinization, and oxidation may have been possible on Mars more than 2000 million years ago when there was surface water available to create channels. The two Viking lander sites are on land surfaces that approach that age and now endure much colder and drier climatic conditions than in Antarctica (Gibson *et al.* 1983). Hence, it is likely that these are two relict cold desert (Cryic Salorthid) paleosols formed under conditions like those in the Antarctic Dry Valleys and then deep frozen for many thousands of millions of years with only minor alteration in the form of surficial oxidation and of added eolian dust and ejecta from impacts.

Meteorites

In the long history of human fascination with meteorites, they have been regarded in many ways. Some meteorites were worked for iron and others were worshipped as divine gifts. Their fiery passage through the atmosphere as a meteor or shooting star was widely regarded as an important omen. It is now known that meteorites are fragments of rock orbiting the sun that fall within the influence of Earth's gravity. They are still esteemed as evidence for the nature and origin of our solar system.

Three main kinds of meteorites are recognized (Table 15.3). The iron

Table 15.3 Abundance and characteristics of different kinds of meteorites (from Wasson 1984)

Kinds of meteorites	Selected characteristics	Finds	%	Falls	%
	I. Stones				
Carbonaceous chondrites (CV)	Al/Si (atom %) = 12 metallic Fe/Si = 0.6-19 C = 8-50 mg/g: H_2O = 20-220 mg large (1 mm) chondrules	3	0.3	8	1.1
Carbonaceous chondrites (CO)	Al/Si (atom %) = 9 metallic Fe/Si = 2.3-15 C = 2-10 mg/g: H_2O = 3-30 mg/g small (0.2 mm) chondrules	1	0.1	6	0.8
Carbonaceous chondrites (CM)	Al/Si (atom %) = 10 metallic Fe/Si = 0.1-0.5 C = 8-26 mg/g: H_2O = 20-160 mg/g small (0.2 mm) chondrules	0	0	14	2.0
Carbonaceous chondrites (CI)	Al/Si (atom %) = 9 metallic Fe/Si < 0.1 C = 30-50 mg/g: H_2O = 180-220 mg/g lacking chondrules	0	0	5	0.7
Ordinary chondrites (LL,L,H)	Al/Si (atom %) = 7 metallic Fe/Si = 2.7-52 C < 10 mg/g: H_2O < 30 mg/g intermediate size (0.4 mm) chondrules	438	45.0	558	78.9
Enstatite chondrites (EL,EH)	Al/Si (atom %) = 5 metallic Fe/Si = 47-72 C < 10 mg/g: H_2O < 30 mg/g intermediate size (0.5 mm) chondrules	8	0.8	13	1.8
Achondrites (EUC,HOW, DIO,URE, AUB,SNC)	basaltic and ultramafic igneous rocks	15	1.5	64	9.0
	II. Stony irons				
Mesosiderites (MES)	breccias with iron and minerals mainly pyroxene	14	1.4	6	0.8
Pallasites (PAL)	breccias with iron and minerals mainly olivine	33	3.4	2	0.3
	III. Irons				
	entirely metallic	459	47.3	31	4.4
	Totals:	971	100	707	100

meteorites consist mainly of metal. The stony irons contain a mixture of iron metal and silicate minerals. Stony meteorites consist of silicate minerals with only small amounts of metal. The stony meteorites can be divided into those that have a chemical composition approaching that of the Sun (chondrites) and those that have a chemical composition more like basaltic igneous rocks (achondrites). Most of the chondrites contain distinctive spherical or ellipsoidal mineral grains called chondrules (Fig. 15.10), from which they are named. Achondrites have an igneous or brecciated texture lacking chondrules. Different kinds of chondrites are most effectively discriminated by their chemical composition and secondarily by petrographic features. Especially diagnostic chemical properties are the abundance of volatile elements most common in the Sun (mainly C, H, N, and S) over refractory elements most common in Earth's crust (mainly Al, Ca and Ti). Also used is the degree of oxidation (inversely proportional to the ratio of metallic iron to silica, among other measures). By a widely accepted system of shorthand, many kinds of meteorites are named from the initials of a locality where a particularly well known example was found. For example, CM meteorites are named from the carbonaceous Murchison meteorite found near the town of that name in Victoria, Australia. Other shorthand names of meteorites are contractions of names, such as EUC for eucrites, or acronyms, such as SNC for shergottites–nakhlites–chassignyites.

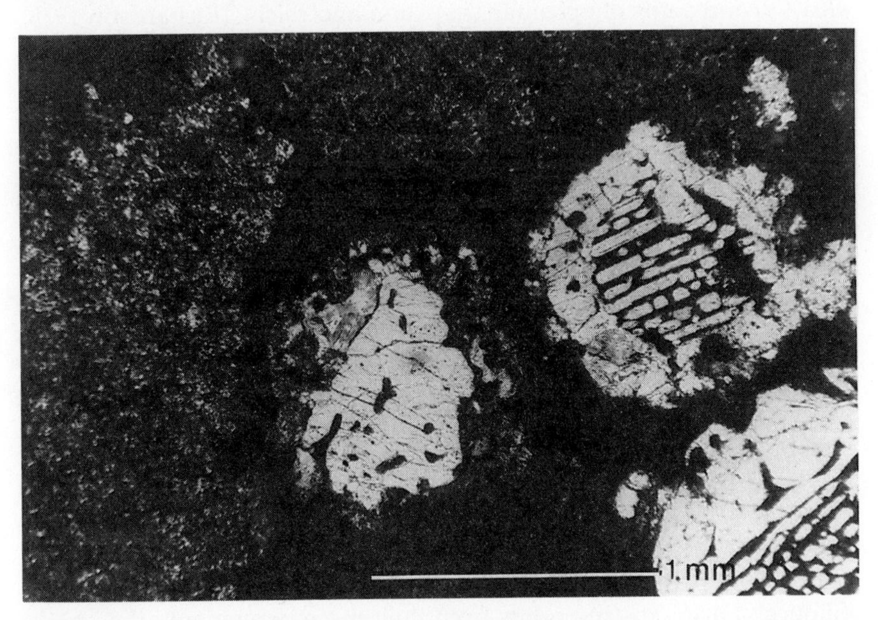

Figure 15.10 Petrographic thin section of Allende carbonaceous chondrite under crossed nicols, showing barred chondrules and clayey matrix. Scale bar is 1 mm (Arizona State University Center for Meteorite Studies specimen 818).

How these various kinds of meteorites may have formed remains debated. Some of them have been regarded as accreted from early condensates of the original cloud of gas, dust, and rocks from which the solar system was formed (Wasson 1984). Others may be parts of the interiors of small planets, moons and asteroids that had begun to differentiate magmatically. One small group of meteorites (SNC) has been interpreted as pieces of Mars conveniently delivered by a large impact there well in advance of our technological capacity to return samples (Wood & Ashwal 1981). Others have been interpreted as parts of soils of planets, moons, and asteroids. It is these that are of special interest in understanding possible processes of soil formation elsewhere in the solar system.

Carbonaceous chondrites

Unlike rocky or metallic meteorites, these rare chondrites are so soft and friable that they can be crushed between the fingers. In addition to chondrules and mineral grains that formed at high temperatures (1300–1500°C), carbonaceous chondrites also have a matrix of minerals that formed at very low temperatures (0–58°C) in the presence of water. This includes a variety of iron-rich smectitic and serpentinelike clays (Tomeoka & Busek 1988) and organic molecules (Nagy 1975). Other evidence of hydrous alteration are framboids and plaquettes of magnetite (Kerridge *et al.* 1979) and veins of epsomite, gypsum, and calcite (Richardson 1978). These low-temperature phases were not created by reaction with air and water after falling to Earth. Many of these meteorites have a thin protective fusion crust formed at high temperature during passage through the atmosphere. In some cases they have been recovered very soon after they fell. Moreover, the matrix material has been radiometrically dated, using $^{207}Pb : ^{206}Pb$, at no more than 60 million years younger than associated chondrules and inclusions which are 4565 $\pm$ 4 million years old (Chen & Tilton 1976). The $^{129}Xe : ^{127}I$ radiometric ages of magnetite in carbonaceous chondrites are the oldest known, but close (within 1 million years) to those of similarly dated minerals in other chondritic meteorites (Lewis & Anders 1975). The $^{87}Rb : ^{86}Sr$ ratios of carbonate and sulfate are evidence that they crystallized no more than 170 million years after formation of the parent body (MacDougall *et al.* 1984).

The source of carbonaceous chondrites is likely to be small (about 500 km diameter or smaller) asteroids beyond the orbit of Mars. The spectral reflectance of carbonaceous chondrites is similar to that of some (C type) asteroids, such as 130 Electra (Cruikshank & Brown 1987). These are periodically perturbed into unstable (about 10 million years duration), elliptical, Earth-crossing orbits from their more permanent orbits in the asteroid belt between Mars and Jupiter and within the Oort cloud at the outer reaches of the Solar System.

Two divergent views on the origin of carbonaceous chondrites can be caricatured as the frozen solar nebula (Wasson 1984) and the earliest soil hypotheses (Bunch & Chang 1980). By the frozen solar nebula hypothesis, the various kinds of chondrites formed by cooling and differentiation of a solar nebula of dispersed gas and dust. In this view the nebula became chemically and physically differentiated according to distance from the sun. The most highly differentiated meteorites, stones and irons, formed closest to the sun on magmatically differentiated asteroids. Among the carbonaceous chondrites, the volatile-poor refractory types (CV) formed closest to the sun (near the orbit of Jupiter) and the more volatile-rich types (CI) at the farthest reaches of the solar system. An appealing feature of this model is that carbonaceous chondrites represent the least altered products of nebular condensation and the most volatile-rich types (CI) can be used as a baseline for assessing chemical evolution of planetary bodies.

By the earliest soil hypothesis, however, some carbonaceous chondrites owe their peculiar composition to surficial alteration of planetismals during their earliest differentiation and degassing. Such soils may have been widespread in the earliest phases of formation of the solar system, but have only survived on those planetary bodies too small to have undergone significant internal magmatic differentiation and too far from the Sun for surficial modification. An appealing feature of this model is that carbonaceous chondrites represent fragments of the oldest land surfaces in the solar system at a stage even earlier than possible ancient relict soils of Mars.

The most convincing evidence for the origin of carbonaceous chondrites as soils comes from their petrographical and mineralogical features. If they represented frozen solar nebular material, then the grains and chondrules would be little altered and have sharp boundaries. In contrast, carbonaceous chondrites contain pseudomorphs of chondrules, clasts, and aggregates replaced by calcite, septechlorite clays, and iron oxides, which imply extensive chemical alteration in aqueous solution (Bunch & Chang 1980). There are also cross-cutting veins, cavity-lining concentric zones of colloidal material (colloform structure), interlocking crystal growths of septechlorite clays, clayey alteration along cleavage planes of mineral grains, and mesh-like bright clay microfabrics (lattisepic plasmic fabric). These are all more like alteration of materials by water in a soil than individual reactions between dispersed gas, liquid, and dust.

If we assume for the moment that carbonaceous chondrites are pieces of primeval soils, their conditions of formation can be reconstructed (Chang & Bunch 1986) under the familiar headings of climate, organisms, topographic relief, parent material, and time. Temperatures were probably within the range of liquid water and below the upper limit for stability of gypsum (0–58°C). Water was freely available at times, but not always. Shrinking and swelling due to wetting and drying may have produced new microfabric (lattisepic plasmic fabric) and the several generations of veins filled with salts in some carbonaceous chondrites. From comparison of carbonaceous chondrites showing different degrees of alteration, there appears to have been depletion of Ca^{2+}, Na^{+}, SO_4^{2-} and Cl^{-} from the primary minerals and chondrules to form clay by hydrolysis and salts by salinization. The hydrolytic solutions were presumably acidic, but strong evaporation and salinization most of the time would have given the soils an overall alkaline pH like that of desert soils on Earth. Weathering solutions also produced hydrous sulfide in colloidal materials, presumably by alteration of pyrite also present. Magnetite plaques may be aqueous alteration products of iron oxyhydrate. Conditions could not have been very oxidizing because a variety of organic molecules are present. From these considerations and the overall similarity of carbonaceous chondritic clays and salts to those of Martian soils, an atmosphere predominantly of carbon dioxide is likely.

Although a case has been made for biological origin of the organic matter in carbonaceous chondrites, most scientists now think that it was abiotically produced. Especially good evidence against a biological origin is the even mix of left- and right-handed versions of asymmetric molecules, which in organisms are more consistently of one kind for any particular compound (Nagy 1975). Many experimental studies (Rao *et al.* 1980) have shown that wet, clayey, salty, friable, planetary surfaces could have produced a variety of organic compounds without the aid of living creatures.

Topographic relief of the parent body is unlikely to have been very great because salts were not separated from the clays by groundwater flow to any great extent. Instead, they were deposited in cracks nearby so that the overall composition of carbonaceous chondrites remains like the non-volatile portion of the Sun.

Refractory carbonaceous chondrites (CV such as Allende of Table 15.2) can be regarded (Bunch & Chang 1980) as possible parent materials for volatile-rich chondrites (CM such as Murchison). The main differences between the two kinds of meteorites are the great gains in volatiles (H_2O, CO_2, C, and SO_3) in the supposed soil (Murchison) compared with parent material (Allende). These may have accumulated in the soil as they were degassed either directly during accretion or later during impact melting. By this line of argument, the best developed of these soils would be represented by the most volatile-rich (CI) of the carbonaceous chondrites. The planetary parent materials altered to form them could have been as refractory as ordinary chondrites. It is also possible to envisage these different kinds of meteorite materials as soil horizons, with the least altered ordinary chondritic materials at the base of a profile and the most altered materials at the surface in contact with the atmosphere and hydrosphere.

Some limits on the time for soil formation during this earliest period of soil formation can be gained from radiometric dating of carbonate veins which could postdate the chondrules of carbonaceous chondrites by as much as 170 million years (MacDougall *et al.* 1984). Such slow development is reasonable if their atmosphere were thin and rapidly lost to space. Weathering under such circumstances would have been much slower than on Earth where volatiles are abundant and recycled. Alteration could have proceeded much faster around hydrothermal vents, but this kind of high temperature alteration would not have produced the kinds of minerals observed and would have separated salts from clays. Moreover, such vents and soils associated with volcanoes would have been more contaminated with igneous materials than are carbonaceous chondrites.

Mesosiderites and howardites

These two kinds of meteorites are stony irons and achondritic stones, respectively. Both are breccias with their clasts mainly of basaltic and ultramafic rocks. Both are thought to have formed on the surface of a planet or asteroid as soil breccias formed by the impact of meteoroids, including iron meteoroids in the case of mesosiderites (Bunch 1975, Jain & Lipschutz 1973).

The parent planet or asteroid of these soil fragments is not known with certainty, but some aspects of it can be reconstructed. The basaltic and ultramafic rock fragments of mesosiderites and howardites are the same as those found in other kinds of achondritic stony meteorites such as eucrites and diogenites. The mineralogical and chemical composition of these igneous rocks is compatible with an origin by partial melting of a source region of chondritic composition followed by varying degrees of fractional crystallization (Stolper 1977). In contrast to basalts of the Moon and Earth, eucrites and associated rocks appear to have formed within an asteroid or planet previously undifferentiated magmatically. Radiometric dating of all these different kinds of igneous meteorites has yielded ages of about 4530 million years, close to the origin of the Solar System's igneous activity. This early origin together with the low pressures needed for the formation of eucritic magmas and the similar spectral reflectance of these meteorites to that of observed asteroids (especially 4 Vesta) has led to the idea that eucritic parent bodies were of asteroidal dimensions (about 500 km in diameter).

The fragmental texture of howardites and mesosiderites is similar to that of lunar breccias (Bunch 1975). They also resemble lunar soils in containing solar rare gases, crystals riddled by tracks of solar flare particles, impact-generated glasses, and micrometeoroid craters. These features provide evidence of soil formation largely by meteoroid impact in a very thin atmosphere. Impact not only pulverized the surface but added material to the soil. The trace element composition of howardites cannot be matched by different mixtures of basaltic (eucritic) and ultramafic (diogenitic) fragments like those obvious in thin sections, but can be explained by addition of about 2–3 wt. % of material similar to CM carbonaceous chondrites (Chou *et al.* 1976). In mesosiderites there is abundant metal (17–80 wt. %; Mason & Jarosewich 1973), presumably added by the impact of iron meteorites. The volume of these iron bodies and mixing of basaltic and ultramafic rock fragments are indications of more violent impact and a more rugged, cratered topography than outlined for the otherwise similar impact-generated soils of the Moon. Soil formation on eucritic asteroids may have been as violent, as rapid, and as ancient as that envisaged for the period (before 4000 million years ago) of intense bombardment evident in the Lunar highlands (Hartmann *et al.* 1981).

Relevance to early Earth

A new scientific discipline such as planetary geology offers many excitements: the joy of unanticipated discovery, the tantalizing search for patterns among data still in many ways inadequate, and the evaluation of a variety of plausible hypotheses. The likely naivety of present views of planetary evolution and extraterrestrial soil formation (Fig. 15.11) can be appreciated well when expressed in the simple and unqualified form of a creation myth as in the following paragraphs.

The visible universe began its present expansion from a primeval fireball some 20 000–30 000 million years ago. Most of the matter released by that initial "Big Bang" condensed to hydrogen upon initial cooling. Eddies of hydrogen contracted gravitationally to form the first stars, lit by nuclear reactions producing helium. In the dense interiors of very large stars complex nucleosynthetic reactions were possible, producing heavy elements such as carbon, nitrogen, sulfur, silicon, aluminum, calcium, and potassium, and small amounts of really heavy elements, iron and nickel. These large stars were unstable and prone to explode as supernovae, suffusing these heavy elements through space. One such cloud of rock, dust, and gas began to contract gravitationally ca. 4600 million years ago to form our Sun and Solar System. Most of the matter fell toward the center of the nebula, but some larger pieces and an associated sheet of dust and debris continued in orbit around the growing but still cold Sun. Both the future Sun and numerous orbiting planetismals swept up nearby dust and debris. The grinding and impact associated with planetary accretion created the distinctive spherical chondrules still found in many meteorites.

The originally chondritic Sun, planets, and planetismals continued to evolve, depending on their mass and the influence of nearby bodies. The smallest masses (less than 50 km in diameter) changed little because of their small reserves of radiogenic heat and volatiles. Those that escaped

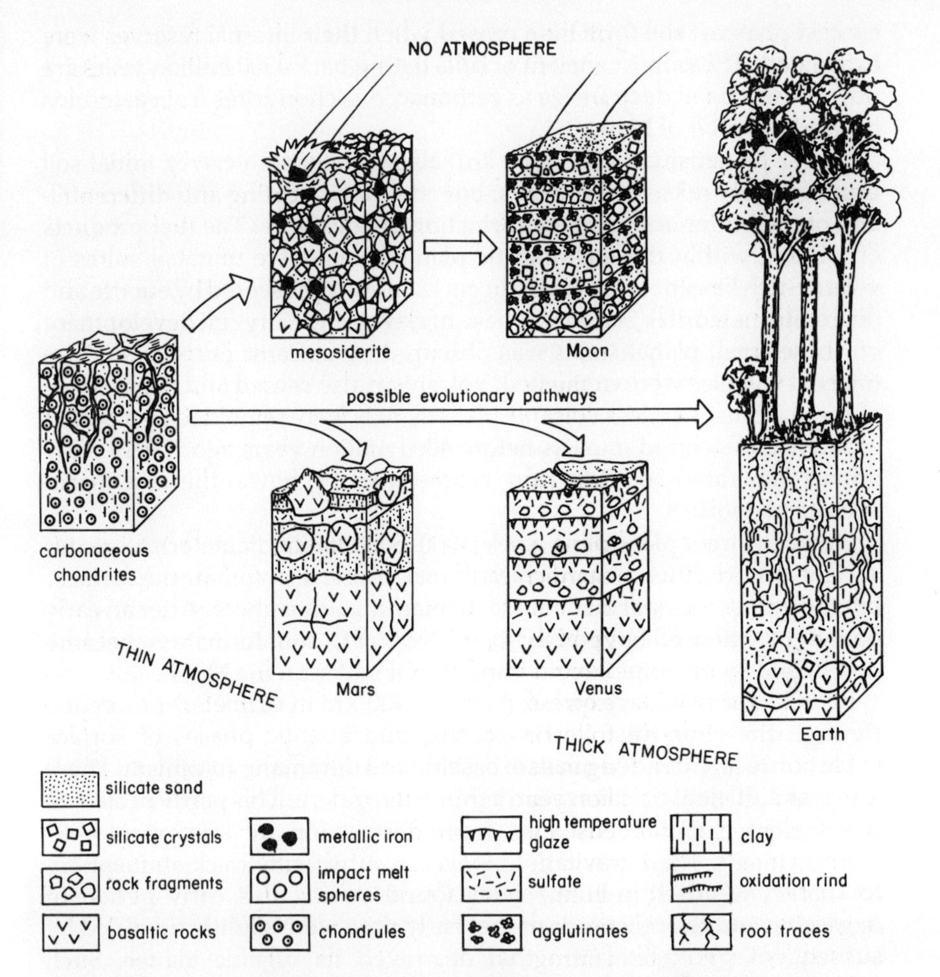

Figure 15.11 A schematic comparison of soil formation on Earth, Moon, Mars, Venus, and some hypothetical meteorite parent bodies.

the subsequent differentiation and ignition of the Sun are occasionally perturbed or fragmented from their remote orbits to land on Earth as chondritic meteorites.

On larger planetismals (50–500 km in diameter), gravitational compaction of planetary interiors, radiogenic heating, and impacting meteroroids released volatiles (H_2O, CO_2 and other gases containing S and N) to the surface. Here chondrule-rich pre-existing materials were hydrolytically altered to clay and organic matter. Also formed on evaporation of these volatiles into the thin atmosphere of these small planetismals were salts such as gypsum, epsomite, and calcite. In many instances the mass of these planetismals was insufficient to check the loss of volatiles to space and this earliest phase of soil formation ceased when their internal reserves were exhausted. These most ancient of soils dating back 4500 million years are now preserved in deep freeze as carbonaceous chondrites from asteroids beyond the orbit of Mars.

Larger planetismals (500–1000 km) also developed a clayey initial soil from the early release of volatiles, but continued heating and differentiation of their interiors led to the formation of volcanoes. The first products of melting within these chondritic planetismals were unusual suites of volatile-rich basaltic and ultramafic rocks, now represented by eucrite and diogenite meteorites. An early phase of clayey and salty soil development on these small planetismals was obscured by volcanic eruptions. Once internal volatiles were exhausted, volcanism also ceased and the surfaces of these atmosphereless volcanic planetismals were remolded by a period of intense meteoroid impacts before 4000 million years ago. These early impact-generated soils are now represented by howardite and mesosiderite meteorites.

On even larger planetary bodies (1000–5000 km in diameter), a greater variety of more differentiated volcanic rocks came to dominate the surface, but their mass was still inadequate to retain an atmosphere. After an early period of intense, large-body bombardment, soil formation became dominated by micrometeoroid impact, as it still is on the Moon.

Even larger planetary bodies (5000–10 000 km in diameter) proceeded through the chondritic, carbonaceous, and eucritic phases of surface evolution to an extended phase of basaltic and ultramafic volcanism. There also was sufficient mass to retain a thin atmosphere. This partly protected the ancient carbonaceous soils from destruction by micrometeoroid bombardment. There was not, however, a sufficiently thick atmosphere to sustain weathering at the rates found during this early period of degassing and the relict soils of that early time were slightly modified by subsequent photo-oxidation that destroyed its organic matter. Such oxidized, clayey, and salty soils are now known on the ancient land surfaces of Mars.

On yet larger planetary bodies such as Earth and Venus (10 000–15 000 km diameter) thick atmospheres of volatiles accumulated. A combination of volcanism and surface weathering that still continues created such discrepancies in rock composition and density that gravitationally driven crustal recycling was initiated. On Earth this takes the form of plate tectonics. Crustal recycling is suspected on Venus, but remains poorly understood. The ancient record of the earliest carbonaceous and clayey soils on both planets was destroyed by subsequent events. On Earth, biological and weathering systems resulted in burial of carbon dioxide as organic matter and limestone, and the burial of sulfate in salt deposits. In this way both these common and chemically aggressive volatiles were kept out of the atmosphere, which came to consist mostly of the inert gas nitrogen. Sulfate salts may be precipitated and recycled on Venus, but there is no evidence on Venus of a comparable mechanism for recycling carbon dioxide. Over millions of years this gas has built to high pressure. Like a greenhouse, it has retained incoming solar radiation to produce high surface temperatures that make it increasingly difficult for limestone to precipitate or for development of organisms to bring it under control.

The preceding selection of ideas on the evolution of soil formation in our galaxy is only one of a number of plausible combinations of currently argued hypotheses. The purpose of this account is less to establish a particular history than to reveal the potential of such studies. The rapidly accumulating data on meteorites and planetary surfaces offer fertile new ground for a paleopedological point of view.

16 Earth's earliest landscapes

It has long been suspected that during the earliest phases of Earth's geological history its surface environment was very different to that which we enjoy now. Surprisingly, Precambrian paleosols now known are similar in many ways to deeply weathered soils at the land surface today. Nothing in the Precambrian rock record has yet been found that is comparable to the soils rich in glass and agglutinates on the Moon or the glazed and metamorphosed soils thought to be forming on Venus. Known Precambrian paleosols are neither as oxidized and sulfate-rich as the soils of Mars, nor as carbonaceous and sulfate-rich as carbonaceous chondrites. By this standard, Precambrian paleosols are not as different from modern soils as they could be.

A certain degree of continuity in soil-forming processes over the past several thousand million years is also evident from information about Earth's surface environments gleaned from other lines of geological research (Fig. 16.1). Evidence of liquid water is provided by the oldest sedimentary rocks on Earth, ca. 3800 million years old in southwestern Greenland near Gο̊thaab (Nŭk) (Allaart 1976). These include metamorphosed remnants of conglomerates, quartz sandstones, limestones or dolostones, claystones, and the distinctive cherty and iron-rich rocks called banded iron formation. From these early sediments it can be assumed that there were rains, running water, and large lakes or oceans.

Even at this early time there was extensive weathering by hydrolysis. This kind of weathering can be inferred by comparing the proportion of bases ($CaO + MgO + K_2O + Na_2O$) normalized to a common igneous value of alumina (15.6%) for sediments of various ages compared with the igneous and other rocks of their presumed source terrains (Fig. 16.1A). Surprisingly, very ancient Precambrian sedimentary rocks show a near-modern degree of weathering, even though (a) vascular land plants did not exist to contribute acidic weathering solutions and (b) these sediments have not been eroded and redeposited (or "recycled") to the extent of recent sediments (Holland 1984). The main weathering acid, then as now, was probably carbonic acid. Other evidence for carbon dioxide in the Precambrian atmosphere is provided by the isotopic ratio of carbon-13 to

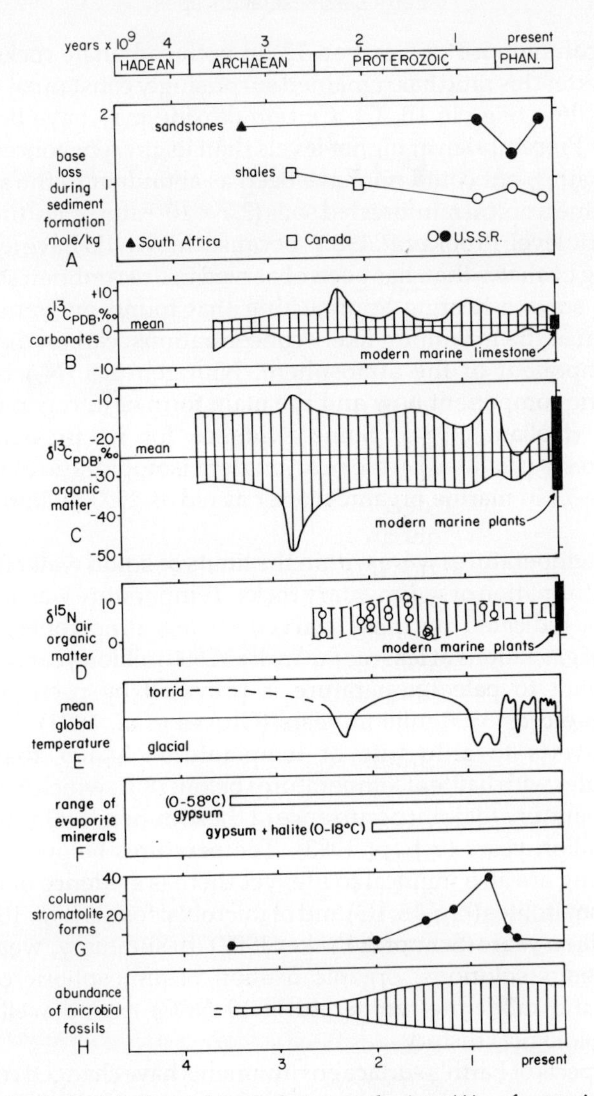

Figure 16.1 Geological indices of hydrolytic weathering (A), of organic fixation of atmospheric carbon and nitrogen (B, C, D, G, H), and of moderate temperatures (E, F, G, H) well back into recorded geological history. A, Loss of acid-titratable bases (CaO + MgO + K_2O + Na_2O) in moles per kilogram of average shales and sandstones compared with their likely source regions (data from Holland 1984); B, C, ratios of the carbon-13 to the carbon-12 isotope relative to the Peedee belemnite standard in limestones and dolostones (B) and in fossil kerogen (C; after Schopf 1983); D, ratios of the nitrogen-15 to the nitrogen-14 isotope relative to modern air in fossil kerogen [each spot represents one measurement from Schopf (1983)]; E, qualitative assessment of torrid and glacial phases of Earth's paleoclimate (after Frakes 1979); F, stratigraphic range of gypsum and of gypsum associated with halite, and the likely surface temperatures that these imply (after Schopf 1983); G, number of described forms of columnar stromatolites (after Awramik 1971), E, abundance of microbial fossil occurrences (after Schopf 1983).

the more common isotope carbon-12. In both carbonate rocks and fossil organic matter this ratio has remained surprisingly constant over recorded geological time (Fig. 16.1B, C). Carbon dioxide may have been present during the Precambrian at higher levels than its present concentration (of 3.4×10^{-4} atm), but could not have been as abundant in the atmosphere as it is in some tropical rainforested soils (3.7×10^{-2} atm or 110 times present atmospheric level; Brook *et al.* 1983), because this would have led to greater weathering of shales than has been observed in Precambrian sedimentary rocks and surface heating approaching that found on Venus (Kasting 1987). Even at the maximum likely concentrations, carbon dioxide was a minor component of the atmosphere. Nitrogen gas (N_2) is the main atmospheric component now and the main form of nitrogen released by volcanoes (Holland 1984). Some evidence for its presence in past atmospheres is provided by the near-modern isotopic ratio of nitrogen-15 to nitrogen-14 in marine organic matter as old as 2500 million years (Fig. 16.1D).

Surface temperatures were within the limits of liquid water (0–100°C) to permit the formation of sedimentary rocks. Temperature was at times close to zero, as indicated by tillite, varved shales, dropstones, and other evidence of glaciations at least as far back as 2500 million years (Fig. 16.1E). Another limit to paleotemperature is provided by pseudomorphs of gypsum as old as 3500 million years (Groves *et al.* 1981), because this mineral inverts to anhydrite at temperatures higher than 58°C. It coprecipitates with halite at temperatures below 18°C, which is a maximum likely temperature for joint occurrence of these minerals (Fig. 16.1F) as old as 2000 million years (Schopf 1983). Temperatures below freezing and above boiling are also inimical to life, yet there is evidence of organically bound stromatolites (Fig. 16.1G) and of microbial fossils (Fig. 16.1H) as old as 3500 million years (Schopf & Packer 1987). In summary, weathering by acidic aqueous solutions, organic fixation of atmospheric carbon and nitrogen, and moderate temperatures (0–58°C) extend well back into recorded geological history.

Other aspects of Earth's surface environments have changed considerably over the sweep of recorded geological history as revealed by peculiar kinds of rocks abundant in the distant geological past and now seldom formed (Fig. 16.2). Some of these changes can be understood in terms of a primeval reducing atmosphere becoming oxidized. Other features are compatible with the development of large granitic continents from an ultramafic primitive crust of the Earth. The two processes may be interrelated. Increased production of oxygen by photosynthetic organisms would have been sustained by nutrients liberated by the weathering of increasingly large continental masses (Knoll 1984). The consumption of oxygen by reduced iron produced at mid-ocean ridges would have become less important as continental weathering became more extensive (Veizer *et al.* 1982).

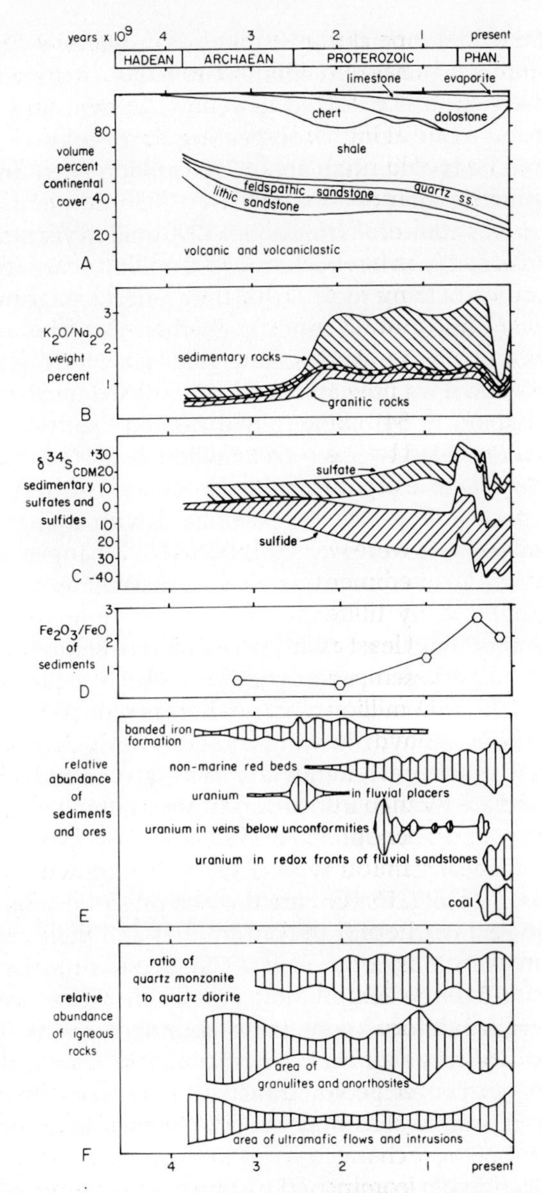

Figure 16.2 Indices of change in crustal and atmospheric composition through geological time. A, Changing proportions of sedimentary and other rocks covering continental regions (after Ronov 1964); B, ratio of weight percent potash (K_2O) to soda (Na_2O) in sedimentary and in igneous rocks (after Engel *et al.* 1974); C, isotopic ratio of sulfur-34 to sulfur-32 relative to troilite (FeS_2) from the Canyon Diablo meteorite, in sulfide and sulfate minerals in sedimentary rocks (after Schopf 1983); D, the ratio of weight percent oxidized iron (Fe_2O_3) to reduced iron (FeO) in sedimentary rocks and ores (after Ronov & Migdisov 1971); E, abundance of various sedimentary rocks and ores (after Habicht 1979, Meyer 1985); F, abundance of different kinds of igneous rocks (after Engel *et al.* 1974).

Evidence for oxidation of the atmosphere is provided by the distribution of iron, uranium, and sulfur in sedimentary rocks. In the present oxygen-rich atmosphere, iron is oxidized to yellow, brown, and red minerals. Under reducing conditions, iron-bearing drab-colored minerals are formed. No red beds older than about 2000 million years are known (Fig. 16.2E). A distinctive kind of laminated iron-rich sediment called banded iron formation was abundant from about 1700 million years ago back to the earliest record of sedimentary rocks ca. 3800 million years ago (Fig. 16.2E). One interpretation of these rocks is that they reflect a stage in Earth history under reducing atmospheric conditions when iron was removed from igneous rocks on land or in mid-oceanic ridges in a ferrous state. Once in the iron-rich ocean it was precipitated by locally oxidized surface waters, by photo-oxidation, or by biologically mediated fixation in banded iron formations (Holland 1984). Since 1700 million years ago, however, much iron has been oxidized and has remained in soils and other continental red beds. In deep oceans, it has coprecipitated with manganese to form nodules. Similar processes may be reflected in changes in the ratio of oxidized iron (Fe_2O_3) over reduced iron (FeO) in sediments over geological time (Fig. 16.2D).

Uranium is another element with very different solubilities in different oxidation states. Oxidized uranium (U^{6+}) is found in yellow minerals such as carnotite, which are readily soluble, whereas reduced uranium (U^{4+}) is found in insoluble metallic-looking minerals such as uraninite (UO_2). These reduced uranium minerals are readily oxidized in the present atmosphere and are washed into groundwater. When they reach a locally reducing region in underground sandy beds, they may precipitate as a local uranium ore of the roll type (Fig. 16.2E) known as old as latest Carboniferous. Of similar origin are the rich ores of uranium in veins or breccias associated with major unconformities and their paleosols as old as 2000 million years (Robertson *et al.* 1978). Some of the most productive uranium ores are 1900–3100 million years old. These are fluvial conglomerates and sandstones containing abundant grains of unoxidized uraninite that are rounded and concentrated in layers like the placer deposits of modern streams. The transition from placer to unconformity and roll-type uranium ores may reflect changes to a more oxidizing atmosphere.

A final indication of changes in the atmosphere is sulfur. The oxidized form of sulfur (S^{6+}) in sulfates such as gypsum is more soluble than the reduced form (S^{2+}) in sulfides such as pyrite. Rounded grains of pyrite also are found within uranium-bearing fluvial placers of 2300–2800 million years ago and are rare subsequently. However, casts of gypsum crystals have been found in sedimentary rocks as old as 3500 million years ago. Isotopes of sulfur provide additional clues. Microbes producing sulfide in stagnant parts of the ocean and swamps select isotopically lighter sulfur (with negative δ values; richer in sulfur-32). However, the evaporation of sea water to produce sulfates shows a very small enrichment of the heavy isotope (positive δ values; isotopically rich in sulfur-34). The divergence in isotopic composition of sulfide and sulfate can be traced back at least 2300 million years (Fig. 16.2C). This has been interpreted as evidence for oxygenation of the atmosphere to levels where terrestrial weathering provided unprecedented amounts of sulfate to the ocean. Alternatively, it could reflect decreased oceanic temperatures and rates of biological sulfate reduction (Ohmoto & Felder 1987).

Another fundamental change recorded in Precambrian rocks was the development of large granitic continents from an originally ultramafic and basaltic crust. The engine of crustal development operating now is plate tectonics. Recycling of rocks begins below the deep trenches of the world's oceans where cold basaltic oceanic crust plunges under its own weight beneath continental margins and island arcs. The melting and mixing of this basaltic material with more granitic material of the lower crust produces igneous rocks that are generally richer in silica, alumina, lime, and potash and poorer in iron, magnesia, and soda than oceanic basalt (Nisbet 1987). When exposed to weathering, erosion, and redistribution by streams and wind, this material is again chemically altered, generally to enrich silica and alumina at the expense of alkali and alkaline earth metals. This great rock cycle is not complete because the low-density, silica-rich continental crust continues to ride well above the high-density, iron- and magnesium-rich oceanic crust. The rock record on land is biased toward these lighter crustal rocks, so it is a surprise to find that continental rocks older than about 3000 million years are mostly mafic volcanics, intrusions, and sediments derived from them (Fig. 16.2A, F). These are the Archaean greenstone belts, narrow, highly deformed sequences of sedimentary and volcanic rocks wrapped around domes of less mafic kinds of crystalline rocks. The world in those days can be imagined as an Indonesian archipelago of global proportions. Small amounts of granitic rock high in potash and silica are found in very old greenstone and granulite terrains, but these become much more common roughly 2000 million years ago (Fig. 16.2F). This corresponds in time with a marked jump in the potash to soda ratios of igneous and sedimentary rocks (Fig 16.2B) and in other geochemical markers (Veizer & Compston 1976). Sedimentary sequences deposited on thicker and more stable continents of these granitic rocks have remained undeformed in many cases. Limestone, dolostone, and quartz sandstone, of these stable continents are much more common than earlier in the rock record (Fig. 16.2A). The appearance of these platform sequences in any particular region may be abrupt because major unconformities and structures commonly separate Proterozoic sedimentary basins from Archaean granulites and greenstone belts. With more widespread radiometric dating of these different kinds of rocks, it is now known that the advent of platform sequences on the present continents occurred at different times in different regions over a time span from 3000 to 1000 million years ago. Island arcs today also can be arranged into a comparable ranking of evolutionary stages from small andesitic arcs such as the Marianas to larger and more complex ones such as Indonesia. Volcanic arcs such as Japan approach small continents in size and complexity. These ultimately will be swept up into the more massive continental mass of Asia, as have other arcs before them. Although the formation of continents is an unfinished process, the evolution of the earliest ones heralded new kinds of environments. With these new kinds of landscapes came new kinds of soils whose fossil record may provide additional evidence for the nature and pace of early crustal evolution.

Oxygenation of the Earth's atmosphere

The idea that the atmosphere of the early Earth was reducing and became oxygenated through photosynthetic and other activities of microbes is widely accepted (Schopf 1983), but is it correct? Could red beds older than 2000 million years have been reduced by metamorphism? Could there have been a very arid climate or exceptionally high rates of sedimentation to preserve detrital grains of uraninite in river deposits? These and other objections have been raised by those who believe that there was appreciable atmospheric oxygen back to the beginning of the Precambrian rock record (Clemmey & Badham 1982).

As a record of the interaction between the atmosphere and surface sediments and rocks, paleosols are coming to play a more prominent role in the examination of this long-standing geological problem (Holland 1984). This is not to say that the interpretation of such ancient paleosols is without problems. Several difficulties arise in considering the likely oxidation state of Precambrian paleosols. For example, paleosols could have been chemically reduced by metamorphism or clay diagenesis. Their oxidation state may not represent atmospheric conditions because they were too clayey or impermeable for air circulation, too rich in organic matter or other reducing materials, or too waterlogged. Fortunately, there are Precambrian paleosols for which these concerns can be set aside.

Pre-Huronian paleosols of Ontario, Canada

Among the most thoroughly studied paleosols are profiles developed on Archaean granite and greenstone at the unconformity below 2300-million-year-old fluvial sediments of the Huronian Supergroup, exposed mainly in mine workings north of Lake Huron, Ontario, Canada (Gay & Grandstaff 1980, Kimberley *et al.* 1984, Farrow & Mossman 1988). One of these profiles developed on greenstone has been called the Denison paleosol after a nearby mine (Fig. 16.3). It consists of a green sericitic rock, lighter in color and finer grained than its parent greenstone. Its clay content (now sericite) was greatest near the surface (now compacted to a depth of 0.6 m), moderately high to a considerable depth (now to 10.5 m) and negligible within the parent greenstone. The clayey surface horizon has been interpreted as an A horizon. The less clayey subsurface horizon is a Crt horizon showing relict crystalline fabric, like saprolites of modern deeply weathered soils. This is a thick saprolite, but it is not as well developed as the mottled and pallid zones found under some Tertiary laterites (see Fig. 6.7).

Formation of clay (now sericite) from actinolite and feldspar, and to a lesser extent from chlorite, appears to have been the major soil-forming process. There is also an increase in the amount of quartz at the expense of the same minerals. Because some of the quartz grains are larger than those in the parent greenstone, some quartz may have been newly introduced or formed rather than merely a residual concentrate. Alumina and silica vary in abundance with clay and quartz content as they would in a modern soil. The leaching of magnesia from the profile is much as would be expected in a humid climate. The same could be said of lime, although there was very little of this in the parent material of the soil. There was even less soda in the parent material and it varies little within the profile. Surprisingly, potash appears to have accumulated. Total iron is strongly depleted from the profile. There is a little more trivalent (oxidized) iron than bivalent iron near the surface of the profile (above 0.6 m), but below that iron is present mainly in the reduced or bivalent state.

Another paleosol along the same pre-Huronian unconformity, but developed on pink alkali granite, has been called the Pronto paleosol (Fig. 16.3). This has an especially clayey, light-green surface (A horizon, now above 0.5 m) and a deep (now down to 5.5 m) clayey, white subsurface (Crt) horizon. The main soil-forming process appears to have been the formation of clay (now sericite) from chlorite and feldspar. Chemical variation in alumina, silica and magnesia is similar to that observed in the Denison paleosol. The amount of potash is almost constant with slight subsurface (horizon Crt) accumulation and slight surficial (A horizon) depletion. There are low values of lime throughout the Pronto paleosol and soda shows some surficial depletion similar to that seen in soils of humid climates. A notable feature of the Pronto paleosol is the surficial enrichment of total iron, largely in the oxidized state in the upper part of the profile (above 0.6 m).

These two paleosols on the same ancient land surface are both drab colored. They are surprisingly different in their degree of oxidation and amount of total iron. Just as modern soils are known to be products of many interacting factors, the interpretation of Precambrian paleosols must also

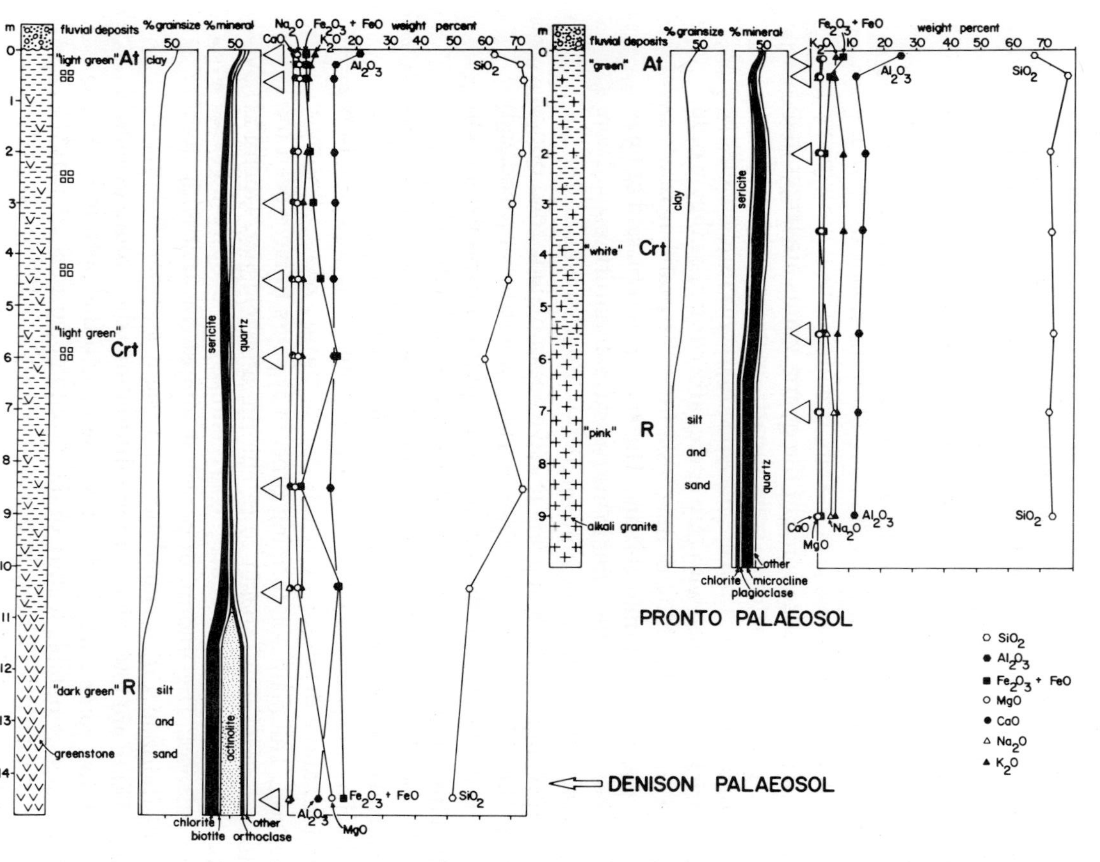

Figure 16.3 Columnar section, clay content, and mineralogical and chemical composition of the Denison (left) and Pronto paleosols (right), both Green Clays, approximately 2300 million years old, on the unconformity between Archaean basement and Huronian fluvial sediments in the area north of Lake Huron, Ontario, Canada. Lithological symbols as for Figure 16.5 (from Retallack 1986c, reprinted with permission of Blackwell Scientific Publications).

take into account complications such as metamorphism, clay diagenesis, original permeability, organic matter content, paleodrainage, and parent material.

Metamorphic environments are mostly reducing, so it is possible that the lack of red paleosols older than about 2000 million years is because most rocks of this great age have been strongly metamorphosed. However, brick red paleosols have been preserved in the Late Silurian Bloomsburg Formation of Pennsylvania despite its metamorphism to lower greenschist facies (Retallack 1985). This is a higher grade of metamorphism than suffered by the pre-Huronian paleosols (Roscoe 1968). Further, the existence of both weakly oxidized and unoxidized paleosols on the same ancient land surface is evidence against severe metamorphic alteration of their oxidation state.

Changes in the composition of clays during burial may further obscure the original nature of paleosols. Particularly problematic is possible diagenetic enrichment of potash, thought to be derived from the dissolution of muscovite and potash feldspar and exchanged in deep groundwater before compactional dewatering (see Fig. 7.8). Other explanations for potash enrichment include less intense leaching of potash before the advent of vascular land plants (see Fig. 7.9) and evaporative enrichment in a dry climate. This last explanation is unlikely for pre-Huronian paleosols considering their abundant clay, depth of weathering, lack of evaporite minerals or crystal casts, and depletion of soda in both profiles. The inconsistent behavior of potash, enriched in the Denison paleosol and depleted in the Pronto paleosol, undermines the likelihood of pervasive potash metasomatism and its possible alteration of the oxidation state of both paleosols. Further, there is no clear relationship between potash and the much more abundant iron within the profile (Denison) which would have been most affected. Potassium metasomatism has been found to have affected Late Ordovician paleosols, but it did not discernibly alter their red color or oxidized iron (Feakes & Retallack 1988). The whole question deserves closer scrutiny, but in this case the mild degree of potash metasomatism indicated is unlikely to have affected greatly the oxidation state of these paleosols.

Drab-colored soils with iron-bearing minerals containing reduced or ferrous iron are formed today in anaerobic waterlogged parts of the landscape (Ho & Coleman 1969). This explanation for drab Precambrian paleosols has general appeal because waterlogged soils are more abundant than oxidized soils in many sedimentary environments and so would more likely be preserved in the rock record. Poor drainage seems unlikely for the Pre-Huronian paleosols, however. Their depth and degree of development are well in excess of those found in modern waterlogged soils. In the Denison paleosol, clay skins deep in the profile (6 m down in the compacted paleosol) are evidence of a very low water table. It is possible that the clay and soil structure of both Denison and Pronto paleosols formed under freely drained conditions, but were then reduced under waterlogged conditions shortly before burial by overlying fluvial deposits. If this were the case, though, it would have affected both paleosols more equally.

Another possibility is textural control of the oxidation state of these paleosols. Modern clayey soils are less permeable to air and water than are sandy soils (Hillel 1980). In the case of the pre-Huronian paleosols, however, the fine-grained parent material (greenstone) is weathered to a greater depth than the coarse-grained parent material (granite), the reverse of the usual relationship in modern soils. In previous studies of pre-Huronian paleosols (Roscoe 1968) it was noted that greenstone profiles were generally thinner than granitic profiles, and it could be that the Denison paleosol includes a cumulic surface horizon. If we accept the Denison paleosol as an exceptionally thick profile, then it is an indication that grain size was not of overriding importance, nor was the texture of the resulting soil because the Denison and Pronto paleosols do not differ appreciably in the amount of clay despite their different parent materials.

Even in the present oxidizing atmosphere, complexes of clay and stable organic matter in some well drained Vertisols can impart a drab color. Neither of these paleosols is as dark or organic as such modern soils. Small amounts of organic carbon have been detected in the Denison paleosol (Gay & Grandstaff 1980). There may have been some kind of microbial life in the soil, but it seems unlikely that their effect could override the effects of atmospheric oxidants in such well drained soils over such a long period of time envisaged for their formation (Pinto & Holland 1988).

A final consideration is the amount of iron present in the parent material. Under present oxygenic conditions, most iron released by weathering is oxidized. Thus, the reddest soils form on iron-rich parent materials such as basalt and greenstone. Curiously, the situation is reversed with the pre-Huronian and other Precambrian paleosols. Those developed on iron-rich rocks such as the Denison paleosol are less oxidized and have lost iron compared with those developed on iron-poor rocks such as the Pronto paleosol. Such a relationship is easiest to understand under conditions of a very weakly oxidizing atmosphere. Generally, soil formation can be considered as a competition between acidic hydrolysis, in large part driven by carbonic acid from soil carbon dioxide, and oxidation by soil oxygen. Under atmospheres very low in oxygen, hydrolysis would release iron from iron-rich parent materials at a rate greater than it could be oxidized so that it would be lost from the profile. On the other hand, even small amounts of oxygen may be sufficient to oxidize small amounts of iron slowly released by hydrolysis of iron-poor parent materials, with the result that iron is fixed within the profile (Holland 1984).

Hence, none of the known circumstances for producing reduced modern soils can be applied easily to these Precambrian paleosols. Differential oxidation of different parent materials under low concentrations of oxygen seems most likely, as outlined in the preceding paragraph. There is a need for closer examination of these and other paleosols, particularly the possible role in them of ancient microbial communities. The behavior of other oxygen-sensitive elements in these paleosols, such as sulfur and uranium, also could prove important. For the moment, though, these drab paleosols are sufficiently distinctive that I have proposed (Retallack 1986c) calling them Green Clays. They appear to be a record of conditions much less oxygenated than the present atmosphere.

Pre-Torridonian paleosols of northwest Scotland

Another group of well known paleosols are developed above ancient biotite gneiss, amphibolite, and microcline pegmatite at the unconformable contact with Proterozoic (1000 million years old) alluvial fan deposits of the Torridonian Supergroup (Fig. 16.4) in northwestern Scotland. These former soils mantled low hills and a piedmont flanking a fault mountain range of considerable topographic relief (see Figs. 11.2 and 11.3) now rifted and drifted from Scotland to Greenland (Williams 1969). A profile

Figure 16.4 The Sheigra paleosol (bleached and reddened zone about 1 m thick to right) under Torridonian alluvial fan deposits approximately 1000 million years old, and another paleosol of comparable age on amphibolite (left-hand side) near the hamlet of Sheigra, northwest Scotland, UK. The white tape extending down from the unconformity, left of center, is 2 m long (from Retallack 1986c, reprinted with permission of Blackwell Scientific Publications).

developed on biotite gneiss has been called the Sheigra paleosol (Fig. 16.5). The surface (A) horizon (30 cm thick in the compacted paleosol) is stained purple–red. Underlying this is a zone of mottled and light-colored rock including large corestones of little-altered parent material. Alteration extends more deeply (up to 6 m from the surface) along joint planes. The profile is more clayey toward the surface where there are crude, platy peds, but there is a relict crystalline texture throughout. Quartz and microcline persist in the weathered part of the profile, but biotite and plagioclase of the parent gneiss have been extensively altered to clay (now sericite; see Fig. 12.4). Clay formation is reflected in the decreased silica and increased alumina toward the surface. Lime and soda are depleted toward the surface, but potash appears to have accumulated as is usual in Precambrian paleosols. Total iron is enriched in the surface horizon and slightly depleted in the mottled and light-colored zone compared to parent

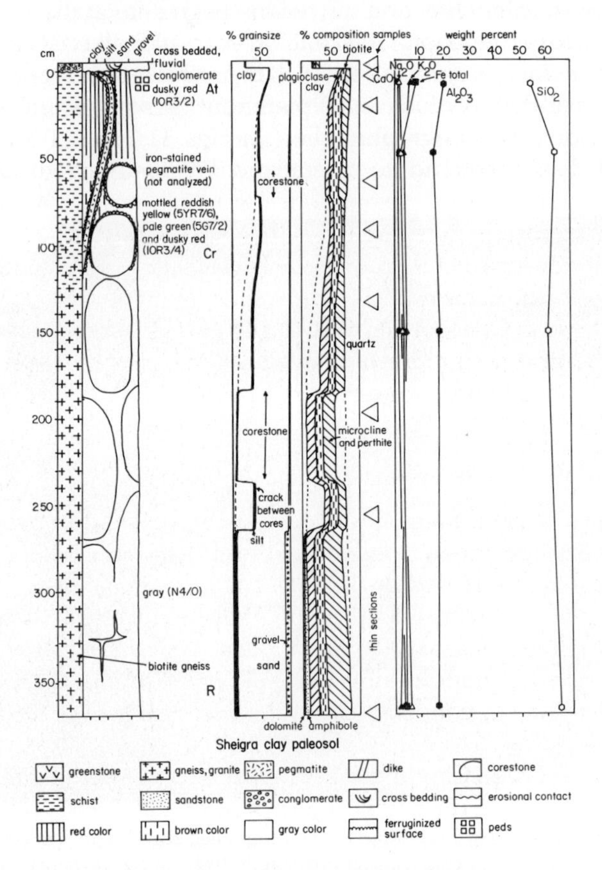

Figure 16.5 Columnar section (measured in field), petrographic composition (point counting by G. S. Smith) and chemical composition [from analyses of Williams (1968)] of the Sheigra paleosol of Figure 16.4.

material. Ferric iron (Fe_2O_3) is well in excess of ferrous iron (FeO) throughout the paleosol, in contrast to the reduced state of iron in the parent material.

Clay formation and ferruginization appear to have been the principal processes involved in the formation of this paleosol. The very weakly calcareous nature of this profile is compatible with a wet paleoclimate, but dolomite nodules deep within this and nearby profiles are evidence that rainfall was within the subhumid range. Although these paleosols may represent a long time of formation at a major unconformity, they are not profoundly weathered or differentiated into horizons and may be regarded as Inceptisols.

Other features of this paleosol provide indications of an oxygenated atmosphere. It is no more metamorphosed or potassium-enriched than pre-Huronian (2300 million year old) paleosols (Gay & Grandstaff 1980) or Late Ordovician paleosols in Pennsylvania (Feakes & Retallack 1988). Nor was it a waterlogged paleosol. The bending of weather-resistant pegmatite veins at the surface of the paleosol has been interpreted as evidence of Precambrian soil creep on a moderate slope (Williams, 1968). Its pervasive oxidation and great depth of weathering (now 6 m in places) are evidence that it was well drained. The parent material is coarse grained and a coarse crystalline texture persists throughout the whole profile. There is little direct evidence for life in the soil, but this is not likely to have been any more complex or massive than microbial ecosystems. Finally, the parent rock had a high iron content and thus a high demand for oxygen, closer to that of the Denison paleosol than the Pronto paleosol. Yet it was oxidized, perhaps reflecting a significant advance in the oxygen content of the atmosphere 1000 million years ago compared with 2300 million years ago.

Calculating atmospheric oxidation from paleosols

There are now several Precambrian paleosols pertinent to Precambrian atmospheric composition (Holland 1984), but few have been studied in such detail as the pre-Huronian and pre-Torridonian paleosols discussed here. Profiles similar to the Green Clay of the pre-Huronian unconformity are known back to at least 3000 million years ago (Grandstaff *et al.* 1986). Alumina-enriched metamorphic rocks as old as 3500 million years also may have originated as Green Clay paleosols (Serdyuchenko 1968, Reimer 1986).

The youngest Green Clay and oldest appreciably oxidized paleosols are of interest in signalling turning points in the oxidation of the atmosphere. As far as I am aware, the youngest Green Clay profile is developed on schist beneath the Athabaska Sandstone in a drill hole west of Wollaston Lake, Saskatchewan, Canada, and is a little more than 1513 million years old (Holland 1984). The oldest paleosol yet reported to be visibly ferruginized is a profile 2200 million years old on granite underlying the Lower Sosan Group in the Simpson Islands, Great Slave lake, Canada (Stanworth & Badham 1984). Ferruginized paleosols are more common in younger Precambrian rocks (Retallack 1986c, Kalliokoski 1986).

Estimates of the amount of oxygenation can be based on chemical data from paleosols, provided that one is prepared to make a number of simplifying assumptions (Holland 1984). One needs to assume that there was limited diagenetic or metamorphic alteration, negligible water-logging, circulation of air to the soil, and little effect of microbes. Also assumed are a limited range of weathering reactions common in modern soils, so that the amount of oxygen needed to oxidize fully all oxidizable materials and the amount of carbon dioxide needed to hydrolyze susceptible minerals can be calculated from molar proportions of elemental constituents of the parent material (Table 12.1). In this way the ratio of oxygen demand to carbon dioxide demand can be estimated. Because an

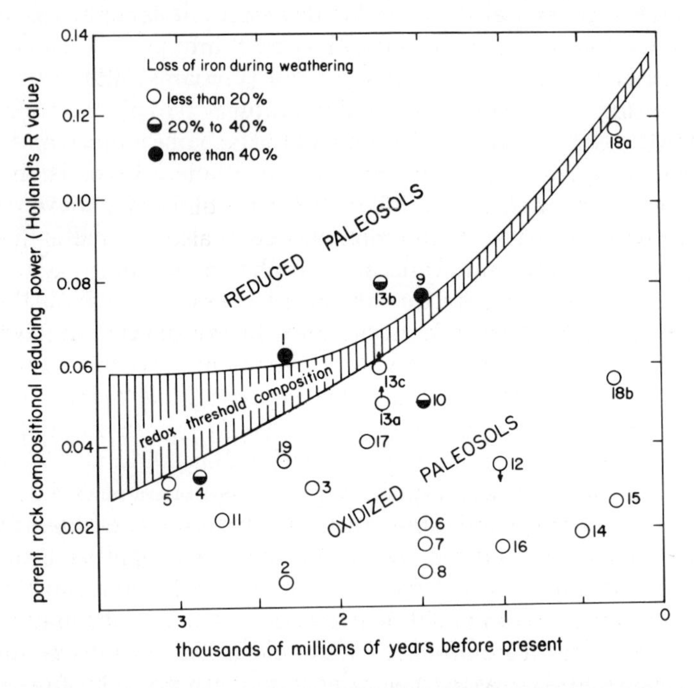

Figure 16.6 Behavior of iron in paleosols of different geological age with different parent rock compositional reducing power (Holland's *R*), revealing increased oxygenation of the atmosphere. Paleosols discussed in this book include the Denison (1), Pronto (2), Waterval Onder (3,) Jerico Dam (5), Sheigra (12), and Arisaig (18) paleosols. [Other numbered profiles and data are from Holland (1984) and Pinto & Holland (1988), Retallack (1986b) for reinterpretation of point 3, Morris (1985) for revised age of point 17, C. R. Feakes for data on points 18a and b, and unresorted NAN profile of Pronto Series of Farrow & Mossman (1988) for point 19]. An alternative culling and interpretation of these data are given by Pinto and Holland (1988).

atmosphere with a ratio of oxygen to carbon dioxide in excess of the demand of the parent rock would result in iron fixation and an atmosphere with a lower ratio in iron loss, the ratios for former atmospheres can be bracketed by comparing the distribution of iron in paleosols formed in chemically different parent materials. For the iron-depleted Denison paleosol, for example, the ratio of oxygen demand to carbon dioxide demand of its parent material is about 6.2×10^{-2} mol/kg, and for the iron-enriched Pronto paleosol it is about 0.6×10^{-2} mol/kg. Similar calculations based on a number of paleosols (Fig. 16.6) have shown that the value of this ratio at the transition between iron loss and iron gain in paleosols older than 3000 million years was about 3×10^{-2} mol/kg. Using physical constants at an assumed moderate temperature (15°C), this critical value for the ratio can be converted to a ratio of partial pressures of oxygen to carbon dioxide of about 1.3 ± 0.5 in the atmosphere before 3000 million years ago, as opposed to about 600 in the present atmosphere.

The partial pressure of oxygen can be determined if one is prepared to make additional assumptions about the partial pressure of carbon dioxide. Considering that acidic weathering of Precambrian sediments is comparable with that seen in modern sediments (Fig. 16.1A), but without the benefit of vascular plants, carbon dioxide must have been more abundant than it is today (3.4×10^{-4} atm), but not more than the highest value found in some tropical forested soils (3.7×10^{-2} or 110 times present atmospheric level; Brook *et al.* 1983). There are other reasons also for assuming a higher concentration of atmospheric carbon dioxide in the distant past, such as the greenhouse effect needed to counteract the low luminosity of the young sun (Kasting 1987). Taking these figures as a range for carbon dioxide concentration gives outer limits for oxygen content more than 3000 million years ago of 2.7×10^{-4}–6.6×10^{-2} atm. This is 0.4–0.002 times the present atmospheric level of oxygen of 0.21 atm. Even closer limits can be placed by introducing additional assumptions and evidence (Grandstaff *et al* 1986).

Such calculations have helped to reconcile those who thought that Precambrian atmospheres were anoxic and those who considered them as well oxygenated as today's. Oxygen appears to have been present in amounts much less than at present, but was far from negligible. It was present in amounts comparable to that of carbon dioxide in the atmosphere today. Both gases were probably minor components of the atmosphere. Presumably nitrogen was the main atmospheric component then as it is now.

Differentiation of continental crust

The view that the development of continental crust begins with subduction and partial melting of oceanic crust under oceanic island arcs has only emerged in the past two decades, and its application to crustal differentiation during Archaean times is even more recent (Nisbet 1987). By this model, highly deformed greenstone belts were oceanic basins and the granitic and gneissic domes were cores of volcanic arcs. The development of continents is thus seen as a process of collision and amalgamation of these island arc systems. An alternative view of crustal evolution emphasizes the unique nature of many Archaean rocks and interprets greenstone belts as rift valleys and narrow oceans that developed on an extensive, thin, primitive crust. By this view, a long period of vertical magmatic differentiation produced a thin, early crust. This crust was thickened by copious volcanic activity along greenstone belts and it was around these later continents that subduction of oceanic crust and other phenomena of plate tectonics began (Goodwin 1981). Arguments concerning the earliest phases of tectonic activity have hinged largely on interpretations of how the various chemically distinct kinds of igneous rocks may have formed and to a lesser extent on evidence from ancient sedimentary environments. Paleosols also may provide pertinent clues although they have been little exploited to this end.

The transition from a world like a global Indonesian archipelago or Icelandic rifted plateau to one with a few places like Death Valley in California or the Andean altiplano would have had profound consequences for soil formation. According to the view that a thin, primitive crust was downwarped to form greenstone belts and intruded in actively uplifted regions by large plutons, there should be few, if any, strongly developed paleosols in early Archaean terranes. By the plate tectonic view, the earliest island arcs, once formed, would remain as buoyant crustal masses after volcanic activity ceased. Hence some well developed soils might be expected even on early Archaean mafic rocks. By both models, the amalgamation of continents during late Archaean and early Proterozoic time should have resulted in some areas falling within the rainshadow of mountain ranges or rift valley scarps, and calcareous desert soils (Aridisols) would have formed on granitic plutons and alluvium of stable continental platforms. Some of these soils may have appeared like modern duricrusts such as bauxites, laterites, silcretes and other deeply weathered materials formed on stable continental land surfaces. Extensive mountain ranges and continental regions would have attracted glaciers and promoted the development of periglacial features in soils. Therefore, calcareous, very deeply weathered and periglacial paleosols should appear in the rock record at about the same time as the earliest continents. With these general ideas in mind, consider the geological record of these kinds of paleosols and products of deep weathering.

Bauxites

These alumina-rich rocks are thought to form by long-term weathering of silicate minerals at moderate pH in humid, non-seasonal climates and on stable land surfaces. They also form as a weathering residuum or deposit on depressions in limestone. Numerous examples of these karst bauxites are known ranging back in age to Precambrian (Bardossy 1982). The most ancient examples are highly aluminous rocks associated with karst landscapes 2300 million years old on the Malmani Dolomite of South Africa (Button & Tyler 1981).

Bauxites may be even older than this, judging from records of highly aluminous schists containing kyanite, corundum, and pyrophyllite. Bauxites have 45 wt. % or more alumina, well above the average for igneous rocks of 15.9 wt. % or for shales of 14.6 wt. % (Garrels & MacKenzie 1971). Alumina enrichment can be caused both by hydrothermal alteration and by weathering, so that care must be taken in interpreting aluminous rocks in highly deformed and very ancient terranes. Corundum ores in highly metamorphosed rocks of the Aldan Shield of the USSR as old as 3500 million years have been interpreted as former bauxites (Serdyuchenko 1968). Similar explanations also have been advanced for corundum ores about 2700 million years old in Zimbabwe and Western Australia, although both have now been interpreted as hydrothermally altered rocks around volcanic hot springs (Schreyer *et al.* 1981, Martyn & Johnson 1986). Other promising candidates for fossil bauxites are pyrophyllite and kyanite-rich rocks in the volcanic and sedimentary sequence of the 3000-million-year-old Nsuze Group of the Pongola Supergroup of South Africa (Reimer 1986).

Numerous other Precambrian rocks may represent aluminous paleosols, but are not as aluminous (only 20–45 wt. % Al_2O_3) as bauxites (Rozen 1967, Serdyuchenko 1968, Button & Tyler 1981, Reimer 1986). These Green Clay paleosols may have formed under a similar weathering regime, but over shorter lengths of time. A little metamorphosed paleosol below the Pongola Supergroup of South Africa provides an indication of the degree of soil development possible about 3000 million years ago. This profile was strongly enriched in alumina (22.4 wt. %) and clay (42 vol. %) at the surface and was comparably clayey down to depths of at least 6 m (after compaction). This has since been metamorphosed to sericite and (near the surface only) to andalusite (Grandstaff *et al.* 1986). This profile represents a kind of weathering that could have produced bauxite if it continued.

Laterites and Oxisols

These strongly ferruginized soil materials also are found on very stable land surfaces. Oxisols are highly weathered, red, clayey soils whereas pedogenic laterites (or plinthites of Soil Survey Staff 1975) are horizons of soils that harden irreversibly on contact with air. Evidence of hard, ripped-up clasts in associated sediments must be sought to distinguish very ancient Oxisols from laterites. It is difficult to make such distinctions in highly metamorphosed amphibolite to granulite facies, Oxisol-like rocks, such as the 2600–2900-million-year-old khondalites of peninsular India (Dash *et al.* 1987). The oldest clear examples of Oxisols are known from Canada and are at least 1660 million years old below the Thelon Sandstone in the Thelon Basin, Northwest Territories (Chiarenzelli *et al.* 1983) and ca. 1500 million years old beneath the Athabaska Group in the Athabaska Basin of northern Saskatchewan (Holland 1984). Kaolinite is an indicator of thorough weathering and has persisted in these profiles despite subsequent diagenetic alteration. These paleosols have a surficial hematite-rich zone (up to 24 m thick) overlying a transitional horizon (also 24 m thick) and then a chlorite-rich zone (up to 30 m thick) over fresh bedrock. These weathering zones are similar in thickness to the lateritic, mottled, and pallid zones found beneath modern deep weathering profiles (see Fig. 6.7). No laterite has been found, although it could have been eroded. These ancient profiles may be the oldest known Oxisols, but cannot yet be accepted as the oldest laterites.

Rounded pebbles of lateritically enriched banded iron formation have been found in sedimentary rocks as old as 1800 million years near Mount Tom Price in Western Australia, and 2070 million years at Sishen in South Africa (Morris 1985). This very ancient lateritization was younger and unrelated to the original accumulation of the banded iron formation. It was only the first of a series of surficial lateritic enrichments of the ore which were interrupted by periods of burial and metamorphism. Banded iron formations are a parent material that readily weathered to lateritic crusts. The fossil record of lateritization on other parent materials of Paleozoic and Precambrian age remains to be detailed.

Calcretes

A variety of calcareous materials fall within the general term calcrete and two of these are not of great importance to the development of continental crust. Nodular and micritic alteration of exposed marine limestone and dolostone have a long fossil record (Bertrand-Sarfati & Moussine-Pouchkine 1983, Grotzinger 1986), perhaps as old as 3400 million years in the Strelley Pool Chert of the Warrawoona Supergroup in Western Australia (Lowe 1983). Groundwater calcretes also are of little relevance, but so far have only been traced as far back as the Late Ordovician Juniata Formation of Pennsylvania (Feakes & Retallack 1988).

More pertinent is nodular calcrete formed in weakly calcareous parent materials found in dry regions of land masses as small as the Big Island of Hawaii (Cline 1955). Nodular calcrete has a long and copious fossil record (Allen 1986a, Kalliokoski 1986) extending back 1900 million years, as represented by extensive pisolitic limestone of the Mara Formation of the Goulburn Group in Bathurst Inlet, Northwest Territories, Canada (Campbell & Cecile 1981).

Silcretes

There is no shortage of very ancient cherts associated with paleosols. The critical question in the present context is how much of it formed as continental silcretes, by cementation with solutions from the deep weathering of silicate minerals in soils (Goudie 1973). Cherty rocks also form where silica is mobilized under highly alkaline conditions of evaporitic and desert environments (Summerfield 1983) and where hot water from volcanic hot springs cools as it flows away from the vent (Rinehart 1980). These latter explanations both seem more likely than deep weathering for cherty paleosols in the 3400-million-year-old Strelley Pool Chert of the Warrawoona Supergroup near Marble Bar, Western Australia (Lowe 1983) and in the 3300–3500-million-year-old Upper Onverwacht and Figtree Groups near Barberton in South Africa (de Wit *et al.* 1982, Duchač & Hanor 1987).

The oldest identifiable continental silcrete is much younger, at the basal unconformity beneath the 1760-million-year-old Pitz Formation in the Thelon Basin of the Northwest Territories of Canada (Ross & Chiarenzelli 1984). This silcrete truncates a deeply weathered paleosol formed during a long period of humid weathering before deposition of the Pitz Formation. Many of the silcretes at and above the unconformity include quartz overgrowths and vugh fillings of length-slow chalcedony similar to those found in aridland silcrete. They are, in addition, associated with evaporitic pseudomorphs and eolian dune sandstones. This most ancient sandy desert can be traced over ca. 600 km of the Canadian Shield.

Karst

This irregular dissolutional and corrosional topography characteristic of limestones and dolostones is a very ancient landform (James & Choquette 1987), now recognized as far back as 2700 million years in the Steeprock Group west of Thunder Bay, Ontario, Canada (Schau & Henderson 1983). The soil material filling this paleokarst includes local concentrations of hematite, manganese, and chert, similar to those known from another very ancient karst, about 2300 million years old, on the Malmani Dolomite of Transvaal, South Africa (Button & Tyler 1981).

Periglacial soils

A variety of distinctive soil features are found in soils of glaciated continents. Fossil ice wedges filled with massive or horizontally bedded sand are found in periglacial regions of severe cold but moderate humidity (see Table 9.2). Such structures have been found among glacial sediments as old as 2300 million years in the Ramsay Lake Formation near Espanola, Ontario, Canada (Young & Long 1976). Sand wedges have vertical layering formed by periodic opening and filling, and indicate colder and more arid conditions. Sand wedges have been found in association with glacial deposits of Late Proterozoic age (1000–600 million years) in a number of localities in South Australia (see Fig. 4.4), Scotland, and Norway (Williams 1986). This evidence of periglacial paleosols confirms that of associated tillites and varved shales. As early as 2300 million years ago some continents were glaciated. By 1000 million years ago, there were even colder and more arid glacial climates, perhaps because continents were larger so that some places were more isolated from the ocean.

Calibrating continental emergence from paleosols

Many of these paleosol indices of continental development remain poorly known in very ancient rocks, but from the meager data available they do extend surprisingly far back into the rock record. Deep weathering under a humid climate on stable land surfaces extends back at least 3000 million years, and perhaps as far as 3500 million years if some of the more highly metamorphosed examples are paleosols and not hydrothermal alteration. Silcretes formed on margins of hypersaline lagoons and volcanic hot springs and calcretes formed by subaerial modification of marine carbonates also may be as old as 3500 million years. These locally dry, coastal environments may have been downwind of large volcanic edifices. Karst can be traced as far back as 2700 million years. Karst could be older because it forms under the same conditions inferred for very ancient Green Clay paleosols, but on limestones and dolostones.

The amalgamation of continents of some size and elevation is suggested by periglacial paleosols as old as 2300 million years. Continents were many hundreds of kilometers across by about 1900 million years when paleosols on weakly calcareous parent materials are found with extensive calcrete and silcrete of the kind formed in deserts. Arid continental glacial climates are indicated by ice wedges in periglacial paleosols 1000–600 million years old. The cooling and drying of continental interiors may have been related to development of rain shadows behind marginal mountain ranges. It also could have been related to changing atmospheric composition from warmer early Precambrian greenhouse conditions to a more oxygenated late Precambrian atmosphere.

In view of these varied lines of evidence from paleosols, a likely model of crustal evolution is one in which buoyant nuclei of protocontinental crust were small areas of stability back to near the beginning of the rock record ca. 3500 million years ago. This is more compatible with the view that continental crust was generated by subduction and other processes associated with plate tectonics than the view that it was generated by rifting or vertical tectonic activity associated with a thin, primitive crust. Terrestrial weathering itself played a role in this early chemical differentiation of the Earth's surface. The growth of continents was a slow process judging from the spread in age of the earliest records of the various pedogenic indicators of continentality. Small continents existed by Late Archaean times, ca. 2900–2500 million years ago. Large continents with deserts at least 600 km across appeared by Early Proterozoic time, ca. 2500–1600 million years ago.

Precambrian scenery

It is natural to cling to familiar images in trying to reconstruct Precambrian landscapes. Commonly in popular books on Earth history they are depicted as sulfurous and smoky volcanic landscapes. In active volcanic regions of the North Island of New Zealand, of the Big Island of Hawaii, or of the interior plateau of Iceland, it is easy to imagine oneself transported back in time to the earliest Earth. But how typical were these environments at that time? Highly aluminous, clayey paleosols as old as 3000 million years are an indication that in some places nothing nearly so dramatic as a volcanic eruption happened in many hundreds of thousands, perhaps millions, of years.

Another source of inspiration for Precambrian landscapes of the imagination are desert regions. The sandur plains of Iceland, the alluvial fans of Arizona, and the channel country of southwestern Queensland are seen as possible models of unvegetated Precambrian fluvial landscapes. Desert dunes and playa lakes of Death Valley in California and of the Sahara Desert in Africa may be used to complete a mental image of Precambrian wastelands. Undoubtedly there were some braided streams and alluvial fans. There is evidence of evaporitic coastal lagoons back at least 3500 million years. Yet there were equally ancient clayey landscapes formed under a regime of deep, humid weathering. Extensive desert dunes, calcareous soils, and silcretes did not appear until ca. 1800 million years ago.

In recent years, images of the surface of the Moon, Mars, and Venus have provided further sources of inspiration for reconstructing Precambrian landscapes. Yet none of these bear especially close comparison to Precambrian paleosols.

From what little is now known about Precambrian paleosols, they are now a more promising line of evidence for reconstructing soils and landscapes of the past than these other analogues (Fig. 16.7). About 3000 million years ago areas of active volcanoes, high mountains, braided

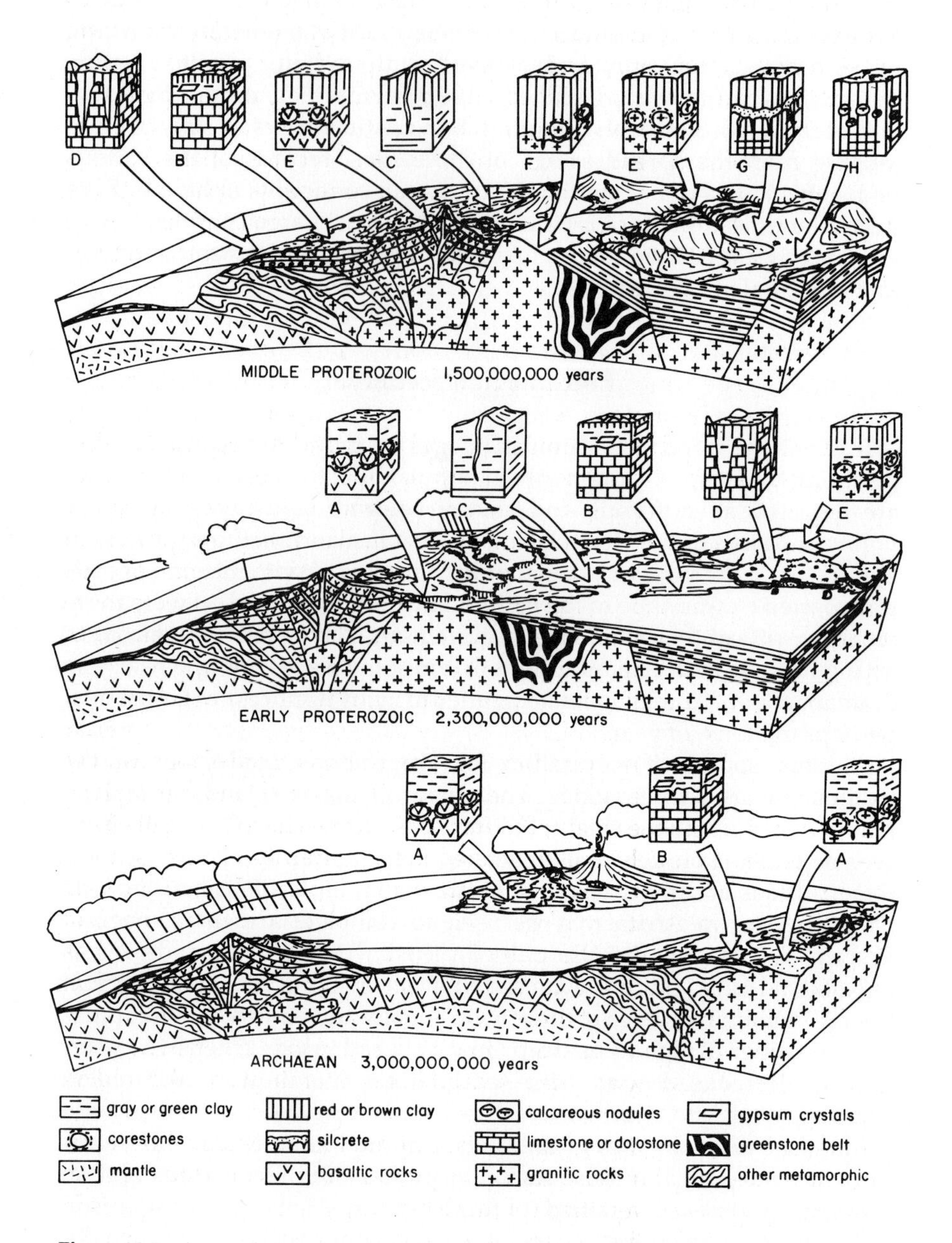

Figure 16.7 A speculative scenario for the evolution of Precambrian soils, atmosphere, and continents. The soil types illustrated are Green Clays (A), salty clay soils or Salorthids (B), swelling clay soils or Vertisols (C), karst and drab cave earth or Orthents (D), oxidized incipient soils or Ochrepts (E), red and deeply weathered soils or Oxisols (F), desert soils with silcretes or Durargids (G), and desert soils with calcareous horizon or Calciorthids (H).

stream channels, tidal flats, and playas would have seemed similar to equivalent modern landscapes. Also similar to today were salty scum soils on limestone, dolostone, and evaporites around dry coastal lagoons downwind of large volcanic edifices. In contrast, rolling terrains indicated by Green Clay paleosols would have seemed unusual by modern standards. One can imagine rounded hillocks of deeply weathered gneiss or granite traversed by low crumbling walls of more weather-resistant quartz veins. Massive spines and walls of volcanic intrusions may have towered over undulating clayey terrain of deeply weathered ash and lava. Without benefit of multicellular vegetation, some of these landscapes would have appeared barren and desolate and yet also lacked the harsh rocky outlines and dunes of modern desert landscapes. These curious extinct landscapes and soils may have occupied large areas of tectonically stable regions. Their deep weathering can be attributed to a humid, maritime climate and warm temperature assured by the greenhouse effect of elevated levels of atmospheric carbon dioxide. Leaching of silica, alkalies, alkaline earths, and iron from these soils would have encouraged the development of dolomites, banded iron formations and cherts in marine and near-marine sedimentary environments. Photosynthetic microbes had appeared by this time, but oxygenation of the atmosphere remained at low levels because the available land area and its nutrient supply was small compared with the oceans.

By roughly 2300 million years ago small continents had developed by amalgamation of early island arcs. By this time oxygen was still a minor component in the atmosphere, but it was present in amounts sufficient to oxidize iron slowly released by acidic hydrolysis from iron-poor granitic parent materials. These weakly oxidized soils were similar in profile form to Inceptisols. Iron-rich basaltic rocks, on the other hand, released too much iron for all of it to be oxidized. These iron-depleted Green Clay soils were the last representatives of a kind of well drained soil that became extinct. The carbon dioxide partial pressure in the atmosphere was lowered as more and more carbon was sequestered in organisms and buried organic matter and carbonate. Easing of the greenhouse effect allowed accumulation of ice and snow on large mountains and in polar regions. Associated with these earliest glaciations were the oldest periglacial soils and structures. Vertisols also appeared in seasonally dry continental interiors. Karst landscapes formed on extensive platform carbonates exposed on these early continents. These new kinds of soils were additions that diversified the landscape.

By about 1500 million years ago large continents had formed by amalgamation of smaller ones. Carbon dioxide had become noticeably less abundant and oxygen was present in appreciably greater amounts. Most well drained soils by this time had the warm brown and yellow hues of iron oxyhydrates. Oxisols and laterites date back to about this time. The interiors of continents isolated from the ocean by marginal mountain ranges developed into extensive deserts, including widespread playas and dune fields. A suite of desert soils also appeared, including soils with calcareous and siliceous hardpans. The world was becoming a rather more familiar place.

This cartoon-like account of early soil development of Earth is undoubtedly incomplete and may also prove incorrect. However, its inadequacies can be addressed by studies of additional paleosols which now can be seen as direct results of those elusive Precambrian landscapes.

17 Early life on land

Soils and life on Earth are so intimately interconnected that it is well to ask, "what is life?." Three general properties especially distinguish living things from inanimate open systems such as flames and eddies: living things are complexly organized, have self-reinforcing sequences of chemical reactions, and can reproduce.

At a molecular level, even such simple organisms as bacteria can be considered perpetual motion machines of incredible complexity (Fig. 17.1). As in the crudely constructed machines that extend our own livelihood, several classes of components are needed: instructions, duplicators, assembly controllers, and work stations. The assembly controllers, for example, are enzymes. These are long strands of protein folded in such a way that they interact with other molecules to promote a particular chemical reaction within the general molecular chaos of heat agitation. Not only must individual large molecules fit into the general system, but parts of these molecules must also fit within the larger ones. Sugars and amino acids are asymmetric molecules that can be produced abiotically in left- and right-handed versions that polarize light in a clockwise (D) or counter clockwise (L) direction. Remarkably, only D-enantiomers of sugars and L-enantiomers of amino acids are found in organisms. The origin of such complex and finely tooled machines by chance is such an unlikely event as to be near impossible, but it is the very complexity of organisms that is vital for keeping them going (Monod 1971).

Organisms also are constantly in motion. Even when standing still, they are processing material and energy in ways that keep these processes going. Such metabolic activities include the breaking down of organic matter in order to gain energy. This is well demonstrated by the fermentation of sugars by yeast to produce alcohol or the respiration of sugars by animals (Table 17.1). Both of these processes are called heterotrophic because they rely on organic matter for energy. Another kind of metabolic activity is termed autotrophic because organic matter is formed from simpler inorganic materials using the energy of sunlight and special catalysts. Autotrophic metabolic processes include the anaerobic photosynthesis of sulfur bacteria and the aerobic photosynthesis of green

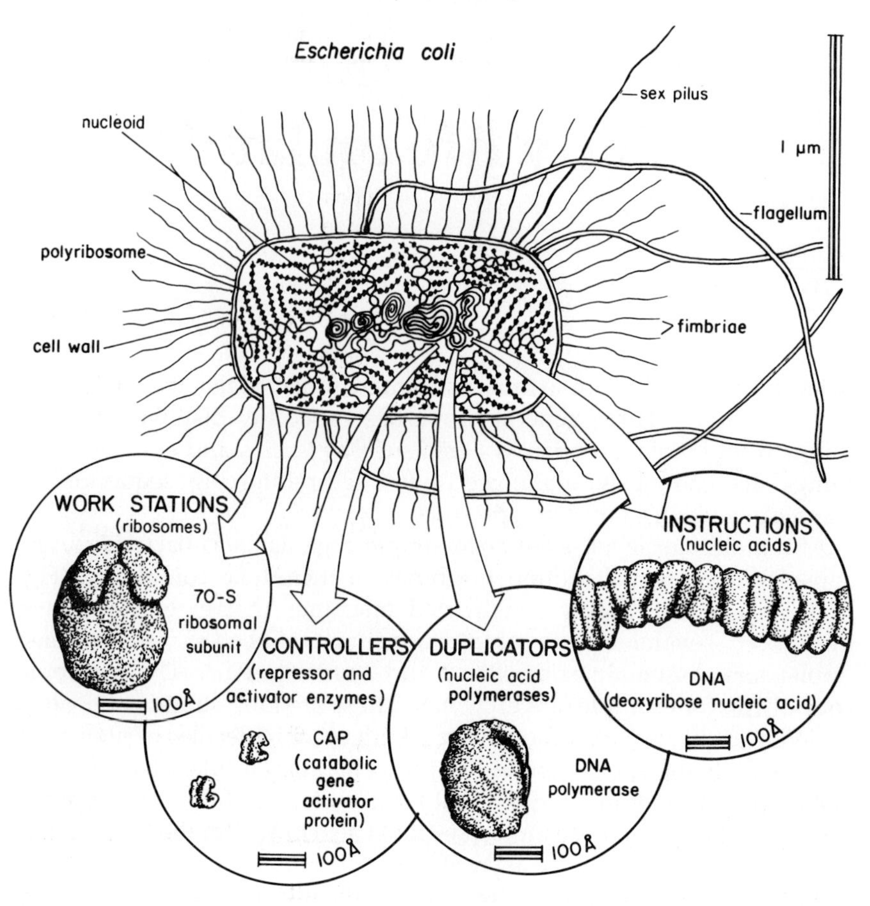

Figure 17.1 A simple cell of the prokaryotic enterobacterium *Escherichia coli*, considered as a molecular perpetual motion machine with examples of several classes of spare parts (data from Kornberg 1980, Steitz *et al.* 1982, Nanninga 1985).

plants (Table 17.1). The exact nature of the metabolic pathways is less important in defining life than the way in which they provide both energy and machinery for the enterprise to continue.

Organisms also create self-starting, small machines with coded instructions of the more complex machinery and processes needed for their survival. The instruction carriers in life as we know it on Earth are nucleic acids such as deoxyribonucleic acid (DNA) and ribonucleic acid (RNA). The small reproductive machines may be simply a divided cell or specially designed cells, such as spores of plants and sperm and eggs of animals.

Research into each of these aspects of the origin and early history of life has made spectacular progress over the past few decades. Experimental investigations have shown that a variety of carbon-based compounds form

Table 17.1 Simplified metabolic processes of living organisms

I. Fermentation

$$C_6H_{12}O_6 \rightarrow 2C_2H_5OH + 2CO_2 + \text{energy}$$

sugar	alcohol	carbon dioxide

II. Respiration

$$C_6H_{12}O_6 + 6O_2 \rightarrow 6CO_2 + 6H_2O + \text{energy}$$

sugar	oxygen	carbon dioxide	water

III. Anaerobic photosynthesis

$$6CO_2 + 12H_2S \xrightarrow[\text{bacteriophyll}]{\text{light}} C_6H_{12}O_6 + 6H_2O + 12S$$

carbon dioxide	hydrogen sulfide	bacteriophyll	sugar	water	sulfur

IV. Aerobic photosynthesis

$$6CO_2 + 12H_2O \xrightarrow[\text{chlorophyll}]{\text{light}} C_6H_{12}O_6 + 6H_2O + 6O_2$$

carbon dioxide	water	chlorophyll	sugar	water	oxygen

spontaneously upon electric discharge or ultraviolet radiation into reducing atmospheres containing such gases as H_2, NH_3, CH_4, H_2O, N_2, CO and CO_2 (Miller & Orgel 1974). Unicellular, organic-walled microfossils in stromatolitic cherts of lake and ocean margins have now been found as old as 3500 million years in the Warrawoona Group of northwestern Western Australia (Schopf & Packer 1987).

Advances also have been made in understanding the nature of metabolic processes and their fossil record. Photosynthetic production of organic matter is a reducing reaction that can proceed in either anaerobic or aerobic modes (Table 17.1). Both kinds of photosynthesis show a preference for the lighter of the available isotopes of carbon (^{12}C rather than ^{13}C) in carbon dioxide. This fractionation of isotopes is virtually unchanged in suitably preserved fossil marine organic matter back at least 3500 million years (see Fig. 16.1C). A metabolic activity known as dissimilatory sulfate reduction is a kind of anaerobic respiration linking oxidation of organic matter to the reduction of sulfate to sulfide (see Table 9.1) with a marked preference for the lighter isotope (^{32}S rather than ^{34}S). This began at least 2300 million years ago, judging from the isotopic record of sulfates and sulfides (see Fig. 16.2C).

Attempts to produce the molecules of organic reproduction, DNA and RNA, by undirected inorganic synthesis have not yet been successful although constituents of these molecules (sugar phosphates and nitrogenous bases) have been found in the residues of experiments energizing mixtures of reducing atmospheric gases (Miller & Orgel 1974). The synthesis and maintenance of DNA have been found to be such a tedious and difficult procedure that the smaller molecule RNA has been viewed as a more likely original molecule of life (Eigen *et al.* 1981). As a molecule that transcribes the code of DNA and also synthesizes proteins using this information, RNA is a more self-sufficient molecule than DNA. Parts of RNA molecules (introns) cut out of the parent molecule can act as an enzyme: they rapidly destroy the molecule by promoting hydrolysis (Zaug & Cech 1986). Some fossil unicells appear to be dividing, but such records are hardly necessary to make the point that life reproduced from 3500 million years ago to the present (Schopf & Packer 1987).

Did life originate in soil?

Despite these experimental and geological advances in understanding the origin and early evolution of life on Earth, there remain large gaps. The steps between tarry abiotically synthesized organic matter and a system that would sustain the synthesis of RNA are difficult to reconstruct by any simple scheme. In the fossil record of life there is a similar gap. Nothing is known about life between 3500-million-year-old microfossils and carbonaceous chondrites ca. 4500 million years old with their simple organic compounds that can be regarded as a geological analog to laboratory experiments on prebiotic synthesis. Both the experimental and geological studies share significant biases in addition to gaps. The fossil record of life has largely come from studies of stromatolites. These are laminated, domal structures formed by sediment-binding microbes in shallow water of seashores and lake margins. This bias toward aquatic life extends also to many experimental studies that have produced organic compounds in solution (Miller & Orgel 1974). This tradition of research is compatible with scenarios (Oparin 1924, Haldane 1929) in which life arose in a primeval soup of ancient oceans.

Alternatively, one can take a general view of the problem and try to recognize perpetual motion machines other than the carbon-based, organic-compound-recycling, and DNA-replicating creatures with which we are most familiar. For example, clay or iron oxyhydrate minerals could stabilize soils and so keep them in near-surface environments where they would be supplied with further raw materials by weathering. Soils are a crude example of a perpetual motion machine that could plausibly have nurtured life itself (Anderson & Banin 1974, Bohn *et al.* 1985). Another perpetual motion machine of a sort is the formation of sulfides and reduced organic matter in hydrothermal vents supplied by volcanic fluids (Corliss *et al.* 1981, Nisbet 1987).

An additional hypothesis popularized by the Swedish chemist Svante Arrhenius (1909) is that life evolved elsewhere in the universe and that Earth was colonized by propagules that could withstand long-distance transport in space. A related idea is that Earth was deliberately colonized with simple organisms by extraterrestrial civilizations (Crick 1981). Such views remove the problem of life's origins to another part of the universe, but do not eliminate it. The environments where that life arose are likely to have been Earth-like in many respects because life now is very well suited to Earth. It therefore remains a useful exercise to investigate how life could have originated from natural causes here on Earth.

Soup, spa, or soil?

In building the bodies of organisms as we know them now, carbon compounds are needed in considerable concentration. This is hard to imagine in the open ocean or other aquatic environments where enormous amounts of carbon are required to create and feed the first organisms. If there were local concentrations of organic matter in dried tidal pools, on sea foam or as scums on icebergs, these would need to be dehydrated for some time to states robust enough to withstand hydrolysis and dissolution on rewetting (Woese 1980). There are similar problems with an overabundance of water around volcanic vents, especially submarine vents. To this may be added the denaturing of complex organic molecules by acidic hot water (Miller & Bada 1988). Soils, on the other hand, are wet by chemically mild rain and groundwater and dry out after rain and flooding. They are stable surfaces that could accumulate organic matter produced by even the slowest and least effective of prebiotic organic syntheses.

Organisms need a variety of nutrient elements. Many of these elements, particularly H, C, N, S, O, and Cl, are widely dispersed in water and atmospheric gases. Probably from the beginning these would have been equally freely available in all of the environments considered. Micronutrient elements may not be so widely dispersed, but are needed in such small quantities that they seldom are limiting. The remaining macronutrients, on the other hand, show distinct partitioning between environments. Of these phosphorus is, and probably was, the limiting nutrient for life because it is needed in much greater quantities than is available in most rocks, soils, sediments and waters (Fig. 17.2). Phosphorus is much more abundant in basaltic (1400 mg/kg) than in granitic rocks, soils

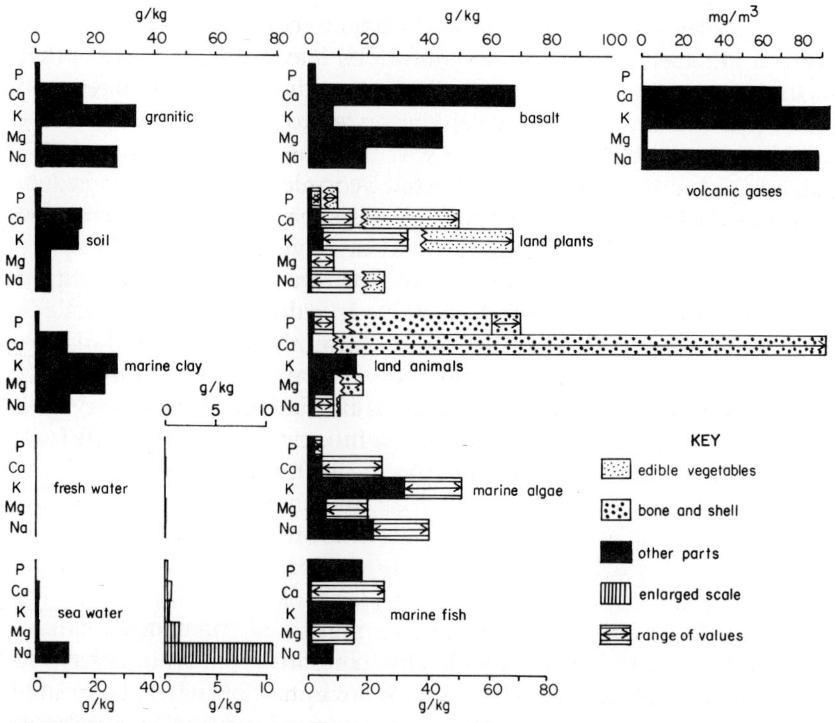

Figure 17.2 Average amounts of mineral macronutrients (P, Ca, K, Mg, Na) in organisms, soils, rocks, and natural gases (data from Bowen, 1979).

or shales (700–800 mg/kg: Bowen 1979). There is very little in fresh (0.02 mg/kg) or sea water (0.06 mg/kg). Thus, life is abundant on land and the margins of large oceans and lakes (Lieth & Whittaker 1975). Open waters away from continental sources of phosphorus tend to be biological deserts. Potassium is strongly retained in clays of soils and shales compared with other non-volatile macronutrient cations. Sodium is readily dissolved during weathering and has tended to accumulate in the ocean. The behavior of magnesium and calcium is intermediate. They are lost from many soils, but some reserves remain and in dry climates they accumulate in soils. The macronutrient element composition of organisms is variable. Edible vegetables are especially rich in them compared with other land plants, and so are mineralized tissues such as bone compared with muscle. In general, however, the pattern of relative abundance of non-volatile macronutrient elements is like that of soil, with some similarity to marine clay. These patterns of relative abundance found in organisms are completely unlike those for granitic rocks, basalt, and fresh or sea water. Further, the composition of life is compatible with the view that it evolved from commonly available materials here on Earth, or a planet very similar to it, rather than in some different part of the universe.

In addition to raw materials for life, stringent conditions are needed to sustain metabolic activity. These conditions may have been even more exacting for the earliest of life forms, before the advent of robust and varied cellular life in which metabolic processes are protected from the outside world by membranes. Temperatures within the range of liquid water, moderate pH and Eh, and protection from cosmic rays would be needed, in addition to a complex structure for the separation of reactants and products, and a source of energy. Oceanic environments mostly have appropriate temperatures. The Eh of oceans can vary considerably from anoxic bottoms to oxidized surface waters. In reduced atmospheres early in Earth history, it may have been more uniformly reducing and capable of preserving organic matter. The pH of oceans is too stable for the formation of all the compounds needed for life. Fats and some amino acids require acidic conditions whereas other amino acids necessary for life form under alkaline conditions (Miller & Orgel 1974). It could be argued that these various compounds were formed and then combined in locally isolated parts of the ocean, such as small pools on icebergs or in tidal flats. However, low-temperature or evaporitic concentration of salts would make useful combinations of these complex molecules difficult. Energy sources useful for synthesizing organic matter and driving metabolic processes include lightning, shock waves, meteorite impact, and sunlight. Most of these sources of energy, with the exception of sunlight of visible and infrared wavelengths, would have been more harmful than helpful, especially in shallow water. Micrometeorite bombardment may have been excluded by an early dense atmosphere, but large meteorites may have been very destructive towards shallow marine environments (Maher & Stevenson 1988). Ultraviolet light may have been harmfully intense before oxygenation of the atmosphere and development of the ozone screen. The protection of diaphanous materials including carbonates may have screened some radiation.

Constraints on metabolic processes in volcanic hot springs are different. The temperature of water gushing from deep oceanic vents may reach 380°C, but there is a rapid decline in temperature over only a few centimeters away from the vent (Spiess *et al.* 1980). Volcanic gases and fluids from the vents are strongly acidic and reducing (von Damm *et al.* 1985). The reducing power of hydrothermal vents, the strong Eh gradient to less reducing surroundings, and a strongly acidic pH could have helped make some compounds useful for life, but not others (Miller & Orgel 1974). Most of these organic compounds would be washed out into the open ocean, but volcanic vents on both land and sea do have networks of compartments that could have separated and slowed reactions. Examples include the porous vuggy accumulations of metal sulfides around deep-sea vents, siliceous sinter deposits found on land and associated pre-existing vesicular lavas or porous volcanic ash. Volcanic vents also receive energy in the form of subterranean heat in deep-sea vents and this is isolated from the harmful effects of hard radiation and meteorite impact at the surface.

Soil temperature, pH, and Eh, are generally more moderate than near volcanic vents. The pH of soils now varies considerably, but under a carbon dioxide-rich early atmosphere is likely to have been acidic (pH 5–6). The Eh of soils is usually equal to or lower than that of the atmosphere, which on the early Earth is likely to have been reducing (Holland 1984). Soils also contain an extensive array of compartments in the form of soil pores. Beneath the soil solum there may be additional compartments in volcanic ash, vesicular lava, deeply weathered granite or uncemented sand. There are wider ranges of pH, Eh, and porosity in soils within very small spaces than in either volcanic hot springs or the sea. The less extreme and locally heterogeneous conditions in soils are more suited to persistence of complex organic molecules. Reactions in dilute solutions of soils are controlled by the intricate geometry of water films and surface properties of mineral grains rather than by bulk chemical and physical properties of the medium. In hot springs, chemical reactions are driven to a particular outcome by flow through a narrow zone of rapidly changing temperature and other conditions. In the sea, reactions tend to a chemical equilibrium. In soils, however, a variety of complex and reversible reactions are feasible. Wetting and drying of soils promote dissolution and precipitation of salts, hydrolysis and neoformation of silicate minerals, and swelling and shrinking of clay. Soils are more like oceanic environments than hot springs in deriving most of their energy from sunlight, and with this comes a potential cost in the form of harmful ultraviolet radiation. Just as in the shallow ocean, however, there are levels where harmful radiation is screened to manageable levels by refraction through diaphanous materials or by reflection within soil pores, yet sufficient heat and light energy remains to drive photosynthesis and weathering reactions (Sagan & Pollack 1974). Life in soils also would be prone to destruction by impacts from large meteorites until such time as it extended to deep soil cracks or other protected sites from where it could recolonize devastated landscapes.

Methods of reproduction can be considered in terms of the modern system based on nucleic acids. The formation and persistence of a simple nucleic acid such as hydrolysis-resistant, guanine-rich RNA is difficult to envisage in the ocean for reasons that have already been outlined. Moreover, if such a fragile self-replicating molecule did arise against all odds, then the problem would be its regulation. In an extended aqueous medium it would either devastate other evolving molecular systems that included suitable nucleotide bases or diffuse away until dissolved. This problem is less extreme in a volcanic vent because life could have arisen in the narrow zone between abiotically produced organic molecules within the vent and the homogeneous, nutrient-poor ocean or atmosphere beyond. However, this zone is very narrow, with vigorous flow of hot, acidic water that decomposes amino acids in a matter of hours (Miller & Bada 1988). Replication is least troublesome in soils, which have numerous small spaces for the assembly of different reaction products. These dry out periodically so that reactions in solution are disrupted in a dehydrated disequilibrium. If the supply of raw materials by rain water kept pace with the time available for episodes of synthesis in evaporating water films, then reproduction of molecules could continue indefinitely. Many local replicating molecular systems could develop in a single soil adjacent to different minerals from the parent material.

In addition to the replication of individuals, the natural selection of different kinds of molecular systems would have been necessary for them to have evolved in any sense like life. The earliest "living" molecules would have had some property that promoted their survival (phenotype) in addition to the information to create that property (genotype). A particular consistency or sliminess of organic matter, for example, could create complex surface foams from wave action in the sea, or bind replicating molecular systems to an especially favored location in the stream of a volcanic vent, or stabilize a soil against destruction by erosion. The greater stability of soils over longer time spans would have been more effective in amplifying the natural selection of initially slow and inefficient synthesis of organic matter and early biological systems.

Did life originate in an oceanic soup, volcanic spa, or soil sludge? The historically acclaimed concept of a primeval soup has stimulated a great deal of excellent experimental work in organic chemistry and other pertinent areas of the physical sciences. There has been little experimental work on the origin of life in volcanic hot springs or in soils, despite their theoretical claims as flow reactors that could have nurtured the origin of life on Earth. In the ensuing sections are outlined some modeling and experimental studies of the kind needed to explore options of how life might have evolved.

Role of clays

Because of their complex physical and chemical properties and fine grain size that make detailed understanding difficult, clay minerals have figured prominently in some theories about the origin of life (Cairns-Smith & Hartman 1986). Specific roles assigned to clays include concentrators of reactants, catalysts for reactions, templates for assembly of complex molecules, compartment boundaries for reactions, and devices for the storage and transfer of energy and information.

The ability of clays to absorb organic matter is well known. The expansion of smectite clays by absorption of ethylene glycol between the aluminosilicate sheets is a standard determinative tool for identifying smectites. I use smectitic clays to pick up spills of cutting oils in the laboratory, and have often found my hands chapped and dry after digging in smectitic paleosols.

Clays have also been shown to aid in the synthesis of parts of organic molecules (monomers) and in their assembly into complexly folded chain-like organic molecules (polymers). For example, cytosine, uracil, and cyanuric acid are a few of the monomers that have been formed from carbon dioxide and ammonia solution in the presence of kaolinite clay. Long polypeptides have been formed from aminoacyl adenylate and oligonucleotides from phospharamidates in the presence of smectite (Rao *et al.* 1980). The exact role of clays in promoting these reactions is not known with certainty. It could be that they participate in reversible intermediate reactions as a kind of catalyst, that their pattern of charge distribution serves to adsorb and orient the reacting molecules as a template or that they have a more general regulatory effect as a chemostat, for example, by buffering acidity.

Clays form a variety of microscopic structures that could be regarded as compartments for chemical reactions. A "house of cards" structure is widespread in clays because of the generally negative charge on the surface of the platelets, but positive charges along the edges that develop under neutral to acidic conditions. Long narrow tubes are formed by imogolite, concentrically arranged vesicles by halloysite and flexible sheets by rectorite (Sudo *et al.* 1981). These kinds of structures are not as well designed as the glassware of a modern molecular biology laboratory or the ribosomes of living organisms. Nevertheless, differences in pH, Eh, and chemical contents of compartments could feasibly allow more complex sequences of reactions than in open water.

Some kinds of energy can be stored and transferred by clays. Considerable physical forces are unleashed when expanding clays are wet. These forces can shear and compress ped faces or throw the clayey surface of a soil into folds (see Fig. 9.9). Clay expansion also can expel soil solutions to sandier parts of the soil and temporarily seal the subsurface of the soil from the atmosphere. Light energy can be stored in clays when electrons are excited by sunlight into defect positions in the atomic structure of clays. The release of this energy can be triggered by drying out of the clay or by wetting with chemicals such as hydrazine (Coyne *et al.* 1984). Whether this thermoluminescent energy is powerful or directed enough to do useful biosynthetic work remains to be determined.

The ability of clays to store and transfer information has been claimed (Weiss 1981), but so far not proven to the satisfaction even of those scientists most receptive to the idea (Cairns-Smith & Hartman 1986). The idea remains a theoretical possibility that clays could be a kind of genetic crystal in which a specific order of crystal features is replicated by crystal growth (Cairns-Smith 1982). Several kinds of crystal genes can be imagined. A very simple kind consists of mixed layer clays in which layers of different mineral composition are arranged in a specific sequence. This one-dimensional crystal gene would grow by neoformation of clay along the edges and reproduce by breakage of complete stacks of the sequence that could nucleate further copies. Another general kind of genetic crystal could carry information as a particular pattern on the sheet-like crystal faces, for example, by substitution of iron for aluminum within the atomic lattice. Many other kinds of pattern-forming crystal defects could also carry information. Growth of this kind of crystal gene could proceed by addition of new layers that replicated the pattern. Reproduction of these crystals would be achieved by breaking along the cleavage planes to form nuclei for other stacks of the original pattern. The value of such sequences or patterns can be imagined insofar as they might affect the bulk properties of clay, such as its swelling capacity in water. A genetic crystal of clay is one that could replicate physical properties useful for its own persistence. Such vital mud has so far eluded discovery, but the concept points the way to fundamental research on the ultrastructure of clays and on its effect on their bulk properties.

Role of iron minerals

A variety of iron-bearing clays (nontronites), carbonates (siderite), and oxyhydrates (goethite) are formed by the weathering of iron and iron-bearing silicate minerals. These minerals are also of interest for studies of the origin of life because they form compartmentalized structures and facilitate the oxidation–reduction reactions that can result in the production of organic compounds.

The crystal forms of fine-grained iron minerals rival those of clays in complexity and variety. They include small hexagonal planes of hematite, sheaf-like bundles of siderite needles, hollow balls of ferrihydrite (Dixon & Weed 1977), and extensively channeled crystals of akaganeite (Holm 1985). Like comparable structures of clays, these could have separated chemical reactions important to the early evolution of life.

Organic compounds also have been formed by photochemical oxidation of iron carbonate (siderite). In these experiments, ultraviolet light photo-oxidized the iron of siderite to ferric hydroxide and reduced its carbonate to formate, formaldehyde, and other organic compounds (Joe *et al.* 1986). The photo-oxidation of iron-bearing clays is also a potential electron source for the reduction of carbon dioxide to organic compounds in solution (Braterman *et al.* 1983). These kinds of redox reactions can be considered an abiotic form of "photosynthesis." Before the Earth's atmosphere was appreciably oxidizing and before continental crust became granitic, there would have been more siderite and similar iron-bearing minerals near the surface of the Earth than at present. The photochemical production of organic matter by such surface processes could have created soil materials similar to carbonaceous chondrites. The preservation of organic matter in the face of continuing destructive ultraviolet radiation would have been facilitated by the coprecipitation of opaque iron hydroxides and oxyhydrates. These mineral factories of abiotic organic matter deserve the kind of dedicated testing that has been lavished on the study of organic synthesis in aqueous solutions.

A scenario of a selfish soil

Much of the fine experimental and geological work done on the origin of life in the sea was inspired by the hypothetical scenario proposed by Oparin (1924) and Haldane (1929). Studies of the origin of life are so full of theoretical possibilities that such internally consistent scientific fables can be a valuable conceptual focus. Similar scenarios for the origin of life in volcanic vents are now being developed (Corliss *et al.* 1981, Nisbet 1987). A detailed scheme for origin of life in the soil can be envisaged very much like that proposed theoretically by Cairns-Smith (1982). This alternative scenario is detailed here to promote its investigation and as a good story in its own right.

Stage one in Cairns-Smith's scenario is a time of complex clays, which could have been most abundant in stable, intermittently wet, surface environments of moderate temperature, such as soils. Here the growth of crystals would be slow, confused by adjacent crystals and periodically interrupted or reversed. This fickle crystallization is completely different from the externally driven crystallization of plagioclase and olivine in a rapidly cooling basaltic lava or the precipitation of quartz crystals from hot vein-filling fluids in a cooling granitic magma. Clays may have been formed initially on Earth by hydrothermal alteration and by weathering during the earliest degassing and differentiation of the planet. More complex clays would have become widespread as surface environments assumed more of the character of a chemostat. Atmospheric carbon dioxide could have warmed the Earth by a greenhouse effect despite the weak radiation from a young and not yet fully ignited Sun (Kasting 1987). Warming of surface temperature would allow liquid water. This would have been acidic with dissolved carbon dioxide and so weathered the landscape to produce residual clays and soluble carbonates. Carbonates accumulating in oceanic sediments would have buried carbon dioxide that might otherwise accumulate in the atmosphere to develop a runaway greenhouse like that on Venus. Such a fragile early chemostat could only develop on planets of a certain size and distance from the Sun, as the failed chemostats of Venus, Mars, and the Moon demonstrate. Nevertheless, such geochemical balances were vital for allowing complex reversible soil reactions.

Stage two was a time when certain kinds of clays became more common because they had bulk properties useful for their own persistence. Consider, for example, Cairns-Smith's (1971) story of the misadventures of four kinds of clay, nicknamed Sloppy, Lumpy, Sticky, and Tough because of their wetted consistency (Fig. 17.3). Imagine that these clays are formed by weathering of mineral grains by rain water percolating through a soil. Depending on the parent material and chemical composition of local solutions they form different proportions of smectite interlayers. The Sloppy clay expands readily when wet and is washed deeply into the profile during heavy rains. Tough clay, on the other hand, is virtually inert. It forms immobile rinds that surround the minerals from which it weathered and thus prevent further weathering. Because of continuing losses of Tough and Sloppy, the clays that come to dominate the surface soil will be those with intermediate characteristics, such as Lumpy and Sticky. These clays expand during rain so that they plug the soil pores and maintain their position in the soil. When conditions are drier they crack away from the parent mineral grains so that additional clay can

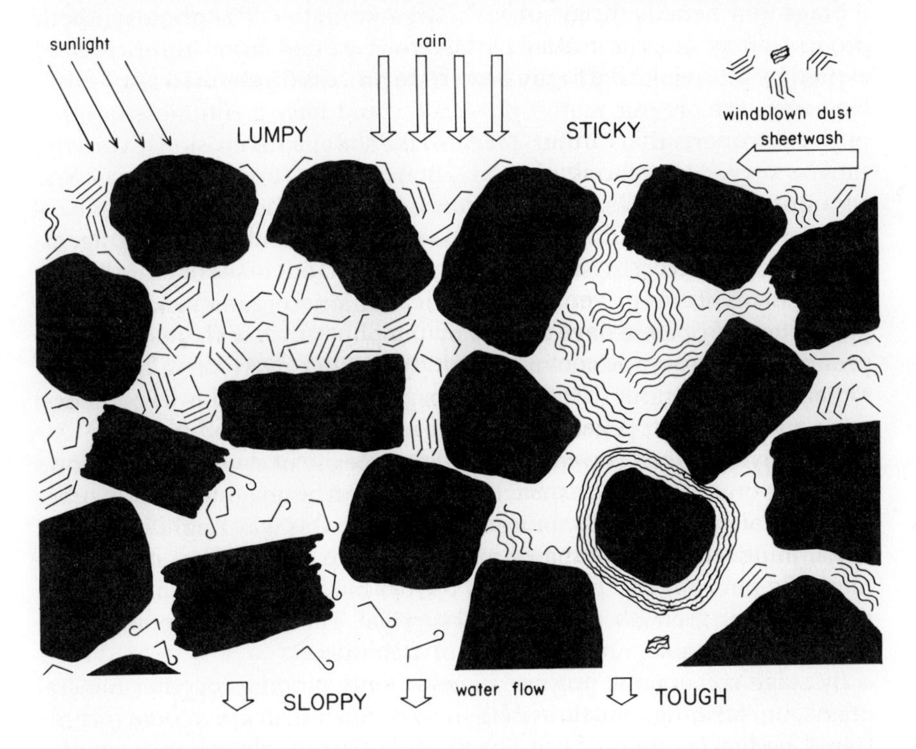

Figure 17.3 Differential survival and evolution of four kinds of clay, nicknamed Sloppy, Lumpy, Sticky, and Tough, each of different consistencies when wet, among the mineral grains of a soil (modified from Cairns-Smith 1971, with permission of the University of Toronto Press).

be made. Lumpy clay protects against erosion following mild rain, but a less easily dislodged clay such as Sticky may be needed to defend the soil against thunderstorms and sheetwash. In this way there is natural selection of particular kinds of clays for particular environmental conditions. Those clays in the zone of materials and energy transfer at the surface of the soil could potentially persist by preventing soil erosion. In contrast, clays washed out to sea or deep into the soil were not subject to such stringent natural selection.

During a third stage of more deeply weathered stable soils, particular clay minerals have been produced in abundance and are dispersed as small crystals. In our mineral melodrama of clay formation in soil (Fig. 17.3), this would promote the formation of Lumpy over Sticky clay because of Lumpy's superior ability to form propagules that could withstand transport. Differences in physical appropriateness between Sticky and Lumpy clay might be slight, but the slower dispersal of Sticky would lead to soils regionally dominated in the long run by Lumpy clay. Nevertheless, a variety of clays would be found in different places and levels of soil.

Stage four heralds the synthesis of organic matter. The photosynthetic production of organic matter by electrons gained from iron-rich clays excited by ultraviolet light may be a crude process (Braterman *et al.* 1983). However, the organic matter produced could have a sliminess or other physical property that further stabilized the soil against erosion. Even with slow and crude methods of production in soils, organic matter could accumulate to considerable abundance in time. This positive feedback allowed natural selection for the most effective soil-stabilizing agent, because less effective organic matter was washed out to sea or buried away from sources of energy and materials for additional organic synthesis. The production of opaque oxyhydrates by the photo-oxidative phase of such primitive "photosynthetic" reactions could filter destructive ultraviolet radiation from organic surface horizons. Soil material at this hypothetical stage would have appeared similar to carbonaceous chondrites.

Stage five is one in which structural complexity of clay minerals allows multistep organic synthesis. Such processes can be imagined within tubes or chains of vesicles in clay or iron minerals. In this way might be formed large, multicomponent organic molecules such as polymers of enantiomerically pure sugars. These might further bind the soil against erosion, thus ensuring their own preferential survival. The soil at this point can be imagined as a crude, abiotic version of a ribosome.

By stage six, organic polymers are not only binding together mineral grains, but forming structures of their own. Such structures could include round bodies [proteinoids of Fox & Dose (1972)], sheets and complex networks. Some of the circular organic structures in carbonaceous chondrites (Nagy 1975) could represent such locally flocculated masses of abiotically produced organic matter. Some of these organic masses may have catalyzed organic synthesis, as the earliest ribosome-like particles (Barbieri 1985).

Stage seven sees the appearance of genetic material formed from organic matter. Long molecules such as DNA or short folded molecules such as RNA may have had value as soil-binding agents themselves. Even if less effective as binding agents, their accurate replication independent of adjacent mineral grains would ensure more consistent reproduction of particular structures and physical characteristics.

By stage eight, organic genes have proven more versatile, more miniaturized and more rapidly assembled than the archaic clay machinery surrounding them. These organic genes already may have begun to parasitize some steps in the synthesis of large organic molecules such as proteins. With this newly refined manufacturing ability these complex organic molecules could begin a takeover of this earlier function of soil clays and iron minerals. The outmoded crystal machinery of the soil becomes increasingly outpaced and outclassed by the speed and efficiency with which organic molecules can commandeer the primary source of energy from the Sun and the primary materials of carbon dioxide, water, and nutrient elements.

Stage nine is one of increasing complexity of the organic machinery of life as genetic molecules begin coding and producing enzymes and lipid membranes that further facilitate their own survival. Some of these independently functioning organic systems continued in the ancient mode of photosynthetic production of organic matter. Others could have begun consuming organic matter in the soil as the first heterotrophs. The bodies of filamentous and amoeboid forms would have further strengthened the ability of the soil to withstand erosion. At this point also early organisms enveloped in protective membranes may have begun to exploit other environments less favorable to complex biosynthesis such as the ocean and volcanic vents. This stage was achieved at least 3500 million years ago, judging from the fossil record of microfossils (Schopf & Packer 1987) and of thick, clayey, organic-poor paleosols (Grandstaff *et al.* 1986, Reimer 1986).

This detailed story of ever more complex systems would seem highly unlikely were it not that at each step the result is improved soil stability that serves to perpetuate the system in the zone of energy and materials transfer. The subsequent evolution of nucleated cells by the genetic takeover of earlier chloroplast and mitochondrion precursors allowed the evolution of larger and more complex organisms (Margulis 1981), more effective at binding the soil. This natural selection was continued with the evolution of multicellular land plants and then of increasingly complex forest communities. By such a view, three-dimensional machines as complex as living things did not evolve by chance, but out of necessity imposed by the arrow of time.

Evidence for early life in paleosols

Theoretical studies of the origin and early evolution of life are useful indications of what might have happened, but ultimately there are too many theoretical possibilities, not all of which could actually have happened. Historical records are needed. Paleosols have much to offer as records of past life on land. Unfortunately, the record is severely compromised by difficulties of preservation. Organic microfossils, for example, are not preserved in oxidized paleosols. However, they may leave microbial trace fossils in the form of lamellar or ministromatolitic mineral deposits. They also could leave evidence of polysaccharides or other organic binding in the form of recognizable soil structures. Finally, there may be evidence of life in the distribution of organic carbon and of elements such as phosphorus that are important for life in the soil.

Microfossils

No well preserved microfossils have yet been demonstrated unequivocally to have lived in a Precambrian paleosol. This could be because soils were not colonized by organisms or because soils were too oxidized or biologically active for microbes to be preserved. Neither conclusion is yet justified because so few paleosols have been examined for microfossils.

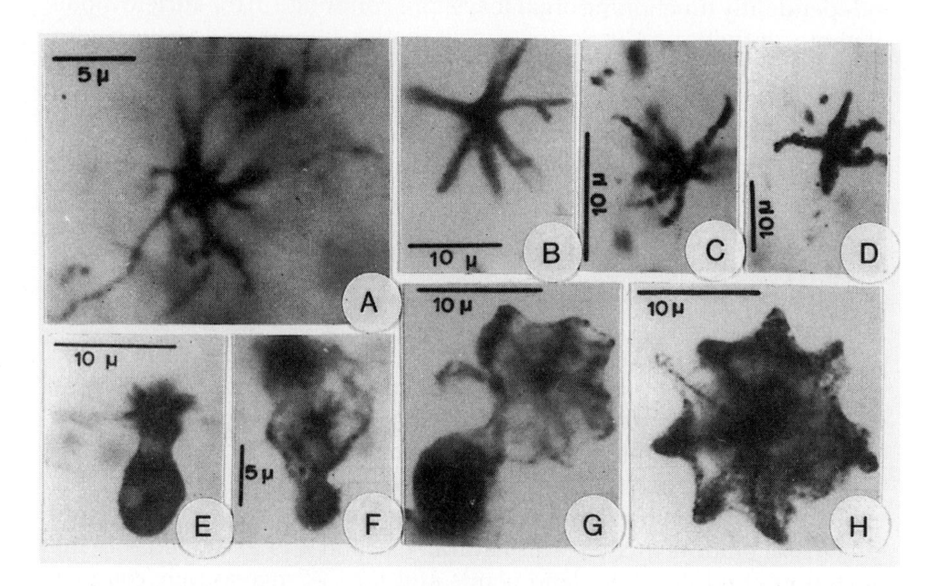

Figure 17.4 Microfossils from the 2000-million-year-old Gunflint Chert at Schreiber Beach, western Ontario, Canada: (A–D) trace fossils of an iron-fixing bacterial colony, *Eoastrion simplex*; (E–H) body fossils of budding bacteria *Kakabekia umbellata* (photograph by E. S. Barghoorn, courtesy of A. H. Knoll). Scale bars 5 and 10 µm.

One line of evidence for early microbial life on land is the similarity of Precambrian marine microfossils to modern soil microbes (Campbell 1979). *Eoastrion* is a stellate microfossil with an opaque central body (Fig. 17.4A–D) abundant in parts of the 2000-million-year-old, shallow, marine Gunflint Chert of southwestern Ontario, Canada (Barghoorn & Tyler 1965). It is identical with a modern form called *Metallogenium* which has been regarded as a manganese-fixing bacterium. There are serious doubts whether *Metallogenium* is an organism, colony of organisms, degraded remnant of organisms, or trace of their activity (Margulis *et al.* 1983). Similar structures are common in modern soils and in rock varnish (Dorn & Oberlander 1982). Other Precambrian microfossils such as *Eosynechococcus* also are similar to modern microbes that colonize bare rocks on land (Golubic & Campbell 1979). *Kakabekia* (Fig. 17.4E-H) is another microfossil from the Gunflint Chert similar to modern soil microbes (Siegel 1977).

Some microfossils have been found in situations which may have been terrestrial, but convincing examples older than Cambrian are not known. Microfossil nostocalean cyanobacteria have been found in cherts filling cracks into basement rocks underlying the 2100–1800-million-year-old Pokegama Quartzite in Minnesota, USA (Cloud 1976). Although this could have been a crevice fauna within a soil, it is equally possible that these microbes lived in the intertidal or subtidal zone of the sea in which the overlying Pokegama Quartzite was deposited. Supposed fossil fungi and lichen-like plants have been found in 2300–2700-million-year-old fluvial rocks of the Carbon Leader in the Witwatersrand Group in South Africa (Hallbauer *et al.* 1977). Although the forms of these fossils may be artifacts of laboratory preparation procedures, the carbon probably is biogenic (Barghoorn 1981). Finally, well preserved phosphatized filaments in Middle Cambrian phosphorites of western Queensland, may represent cyanobacterial colonization of subaerially exposed limestone (Southgate 1986).

Trace fossils

Possible microbial trace fossils have been found in the 2200-million-year-old Waterval Onder clay paleosol of South Africa (Retallack 1986b). These are features like modern rock varnish. Small (1–0.1 mm in width) grains in the surface (A) horizon of the paleosol have botryoidal encrustations of opaque iron and manganese (manganoferrans) on their tops but not on their bottoms (Fig. 17.5). Careful examination under the scanning electron microscope has shown that some of these opaque encrustations have internal lamination, but in most cases this is difficult to see because of their metamorphic recrystallization. There are numerous theories on how modern rock varnish forms, but a biological origin now is widely accepted. Debate continues on what kinds of organisms are responsible. Bacteria,

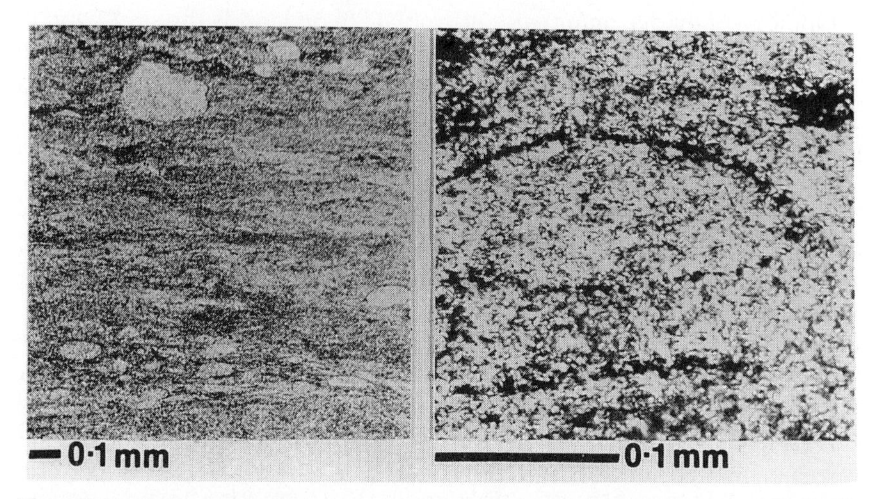

Figure 17.5 Lamination, scattered crumb beds, and possible rock varnish structure at two different magnifications in the surface (A) horizon of the 2200-million-year-old Waterval Onder paleosol near Waterval Onder, Transvaal, South Africa. Scale bars are 0.1 mm (from Retallack 1986b, reprinted with permission of Elsevier Science Publishers).

algae and fungi are found associated with modern varnish (Krumbein & Jens 1981, Dorn & Oberlander 1982, Staley *et al.* 1982)

Comparable microstructure that may be a microbial trace fossil has been found in the surface (A) horizon of the Jerico Dam paleosol beneath the 3000-million-year-old Pongola Supergroup in South Africa. These are microscopic caps of fine-grained andalusite (probably once kaolinite) with internal laminae of amorphous iron and manganese oxides on top of large quartz grains. These small colloidal caps (former manganoferriargillans) are distinct from the sericitic materials of the rest of the paleosol and could not be weathering rinds of the underlying quartz grains (Grandstaff *et al.* 1986). They are similar in some ways to ministromatolitic forms of rock varnish (see Fig. 10.2A) and could represent microbial colonies under which dust was trapped, weathered to clay, and interstratified with opaque minerals produced by oxidative photosynthesis or aerobic chemautotrophy.

Soil structure

A striking feature of the Waterval Onder clay paleosol is the very different soil structure of its surface (A and AC) horizons compared with its subsurface (C) horizons. The surface (A) horizon has a platy structure with scattered crumb peds (Fig. 17.5). The near-surface (AC) horizon has coarse, angular blocky pedal structure defined by clay and opaque oxides (illuviation manganoferriargillans) washed down the cracks (see Fig. 3.12).

These structures are similar to those in soils, although certainly not as complex as the granular and crumb structure found in modern grassland soils (Mollisols). On the other hand, these fossil soil structures show wispy edges, healed cracks, indications of slight rotation in a loose medium, and alteration haloes different from those seen in simple, abiotically produced breccias, boxworks, or systems of mudcracks. The distinctive structures of modern soils are created by coatings of polysaccharides and other materials (Griffiths 1965, Foster 1981) which smooth over the edges of peds, bind together groups of peds and localize the action of weathering solutions.

Organisms and their slimy organic products serve to stabilize the entire soil in addition to individual peds (Booth 1941). Rock varnish (Dorn & Oberlander 1982) and carbonate crusts (Krumbein & Giele 1979) armor modern desert soils against erosion. Could this be the reason why many Precambrian paleosols are so thick, clayey, and deeply weathered? The Jerico Dam paleosol below the 3000-million-year-old Pongola group in South Africa, for example, now has about 50 vol. % clay to a depth of at least 6 m (Grandstaff *et al.* 1986). The 2300-million-year-old Denison and Pronto paleosols of Ontario, Canada, are also impressively well developed (see Fig. 16.3). These represent landscapes that were stable for hundreds of thousands, if not millions, of years. This is not what one would expect on an abiotic landscape. Under abiological conditions each mineral grain, as it was loosened from bedrock by chemical and physical weathering, should quickly have been removed by surface erosion if it were appreciably above the water table (Schumm 1977). As a consequence, Precambrian paleosols on bedrock should be sandy and thin rather than clayey and thick as some of them were. Under abiotic well drained conditions also, soils formed on clay deposits should have been thin and shallowly cracked like those forming in desert badlands of the western United States (Schumm 1956). Yet Precambrian paleosols on shales such as the Waterval Onder paleosol (Figs. 9.9, 17.6) are not of this kind. Could organic matter have been a glue that held Precambrian landscapes together?

Organic carbon

Evidence for life in Precambrian paleosols could also be gained by analysis for organic carbon. Small amounts have been found in some Precambrian paleosols. The 2300-million-year-old Denison paleosol (see Fig. 16.3) contains 140–2500 ppm by weight of reduced organic carbon (Gay & Grandstaff 1980). This was scarce near the surface (140 ppm) and most abundant (2500 ppm) at a depth of 2 m. Such a low concentration and distribution are compatible with that found in Quaternary oxidized paleosols (Stevenson 1969). Very few other Precambrian paleosols have been systematically analyzed for organic carbon. It is not so much the presence of organic carbon that is critical because a wide variety of Precambrian rocks contain it (Schopf 1983). Rather, it is the down profile attenuation in abundance of organic carbon that is likely to provide information on the antiquity of life on land.

An alternative to organic carbon is to analyze isotopes of carbon in carbonate within paleosols in order to determine whether there was fractionation of the lighter carbon isotope (^{12}C rather than ^{13}C) as is characteristic of soils formed under photosynthetic organisms. Lightening upwards trends in both carbon and oxygen isotopic composition have been found in a paleosol developed on the mid-Proterozoic (about 1200 million years old) Mescal Limestone near Young, Arizona (Beeunas & Knauth 1985). These isotopic changes are similar to those seen below modern soils (Chromusterts) on limestone on the south coast of Barbados (Ahmad & Jones 1969) and in several similar Mesozoic and Paleozoic paleosols ranging back in age to a Mississippian paleosol on the Newman Limestone near Olive Hill, Kentucky (Allan & Matthews 1982). Isotopic fractionations observed in the Mescal Limestone are greater than can be accounted for by simple exchange with atmospheric carbon dioxide or by the observed degree of metamorphic alteration in the paleosol. This paleosol has been taken as evidence for photosynthetic microbes on land as least as far back as 1200 million years (Beeunas & Knauth 1985). Additional better preserved examples are needed to verify these results (Vahrenkamp & Rossinsky 1987).

Trace elements

Although information on organic matter in Precambrian paleosols is not widely available, there is information on trace elements commonly complexed by organic matter, such as Ba, Cr, Cu, Ni, and Zn (Farrow & Mossman 1988). In the 2200-million-year-old Waterval Onder clay paleosol, for example, these elements are all enriched toward the surface (Fig. 17.6). Most of these elements also are enriched in clayey parts of modern soils (Aubert & Pinta 1977), but this is not an explanation for the upper part of the Waterval Onder paleosol because it has a nearly uniform clay content throughout. These enrichments are especially striking in comparison with the immobility or depletion of other trace elements such as Zr and Rb which also are stable in modern soils.

Another element showing surficial enrichment in the Waterval Onder paleosol is phosphorus. This scarce macronutrient is depleted at the surface of modern forested soils and paleosols because of the requirements of large plants (Smeck 1973). The pattern of surficial enrichment of phosphorus in the Waterval Onder clay paleosol is more like that found in modern grassland soils or microbial earths in which much of the biomass is within the ground.

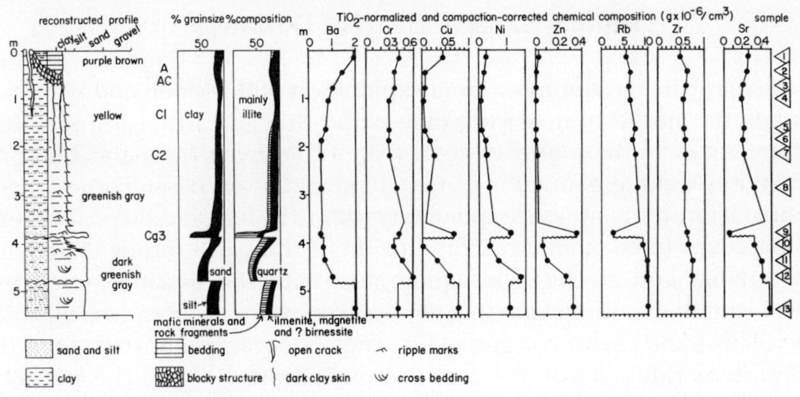

Figure 17.6 Reconstructed profile and trace element composition of the 2200-million-year-old Waterval Onder clay paleosol, near the village of that name, in Transvaal, South Africa (from Retallack 1986b, reprinted with permission of Elsevier Science Publishers).

Phosphorus in the 2300-million-year-old Denison clay paleosol shows a surficial decline parallel to that of analyzed organic matter (Gay & Grandstaff 1980). The Pronto paleosol of comparable age also shows a surficial decline in abundance of phosphorus, as do several other paleosols at major unconformities (Grandstaff et al 1986). In contrast, the Jerico Dam paleosol developed on 3000-million-year-old basement to the Pongola Supergroup shows a surface enrichment in phosphorus that does not quite reach the amount found in the parent material below a deep zone of phosphorus depletion. Such inconsistencies in the phosphorus distribution of paleosols at major unconformities need to be reassessed with studies to determine how completely the uppermost horizons of the paleosols are preserved.

Antiquity of life in soil

If evidence for life in Precambrian paleosols remains inconclusive, it is largely because such evidence has only begun to be assembled. Five avenues of research have been outlined here. For the 2200-million-year-old Waterval Onder paleosol and the 3000-million-year-old Jerico Dam paleosol, possible rock varnish structure, identifiable soil structures, and the distribution of phosphorus are compatible with the presence of life in them. For a 1200-million-year-old paleosol in the Mescal Limestone of Arizona, a lightening upwards trend in carbon isotopes is most like that produced under photosynthetic organisms. What little information is available on organic carbon analyses and microfossils in Precambrian paleosols does not provide positive evidence for life in them. The existence of life on land as far back as 3000 million years thus remains not only a reasonable speculation but also an idea amenable to further testing from the fossil record of soils.

Mother Earth or Heart of Darkness?

Of the many images returned from exploration of the Moon and planets, perhaps the most evocative was a blue, cloud-draped Earth rising over the barren, rocky horizon of the Moon (Fig. 17.7). Even from afar there is something unique about the Earth compared with other planets and moons. It is more common for planetary soils to be full of salts and for their atmospheres to be mainly carbon dioxide. If the Earth can be imagined without life and at chemical equilibrium with the gases released by volcanism, then it should have an atmosphere primarily of carbon dioxide (99 vol. %) and minor oxygen (1%), and an ocean pasty with salts of sodium chloride (35 vol. %) and sodium nitrate (1.7%). The present atmospheric composition of mainly nitrogen (78%) and oxygen (21%) with minor argon (1%) and carbon dioxide (0.03%), and its clean oceans of water (96%), a little salt (3.5%), and only traces of sodium nitrate, are products of a series of biogeochemical cycles in which life plays an important regulatory role. Not only are the Earth's surface environments maintained at this cosmically peculiar composition, but the persistence and evolution of life over the past 3500 million years can be taken as evidence for similar regulation in the past. The idea that life is in control of the Earth and that

Figure 17.7 Earthrise from the moon (NASA photograph, courtesy of R. J. Allenby and National Space Science Data Center).

Earth's surface systems are a kind of cybernetic extension of life is the central concept of the Gaia hypothesis (Lovelock 1979). Gaia was mother Earth in Greek mythology.

An opposing view can be named for Erebus, the primeval darkness of Greek myth. By the Ereban view, life arose by the most remarkable of accidents only on those planets that had a narrow range of physico-chemical conditions to allow it. Life has persisted in a generally hostile environment only where specific activities proved advantageous for scraping a living from available resources. In the Ereban view, the pervasive influence of life is illusory because life still depends ultimately on gases, liquids, and rocks erupted as a byproduct of the internal differentiation of the Earth, and on the quality and quantity of radiation from the Sun. This thin rind of the biosphere with its tenuous grip on Earth resources could still be destroyed utterly by impact of the Earth with a large extraterrestrial bolide or by full-scale exchange of nuclear weapons.

From this perspective, currently popular views of the origin and early evolution of life (Schopf 1983, Knoll 1985b) are distinctly Ereban in character. In this view, life evolved by the longest of chances in the primordial soup of the world ocean and was continuously buffeted by larger environmental forces. The productivity of the earliest autotrophic forms of life was limited by availability of abiotically produced organic nutrients such as sugars for fermenters. Once heterotrophic photo-synthetic organisms had evolved, their productivity was limited by ultraviolet radiation in shallow water, by lack of mineral nutrients such as phosphorus in deep marine environments, and by predation from earlier evolved autotrophs. Oxygen released by these struggling early photo-synthesizers was initially scavenged by reduced iron-bearing minerals and decaying organic carbon compounds. It was not until these sinks for oxygen were buried and photosynthetic productivity increased in more extensive shallow seas of continental margins that oxygenation of the atmosphere become noticeable. This early oxygenation mobilized sulfur from soils as sulfate, but it did not accumulate in sea water because it was buried as sulfide reduced by microbes in the deep ocean and as sulfates in evaporites of shallow restricted arms of the sea. Carbon also was buried in organic carbon and carbonates. Nitrogen accumulated in the atmosphere as a weakly reactive pressure-building gas. Its reduced forms (such as ammonia, NH_3) were oxidized to nitrate (NO_3^-) and this was converted to nitrogen gas (N_2) by denitrifying microbes that use nitrate to oxidize organic matter. It was only after some time that ozone accumulated in the stratosphere (10–50 km up) and filtered out harmful shortwave radiation so that microbes could colonize formerly barren land surfaces. Modern oxidizing photosynthetic ecosystems thus came to conquer formerly hostile habitats while poisoning with oxygen an earlier reducing fermentative ecosystem that lingers on in swamps, stagnant ocean bottoms, and the guts of animals.

An alternative perspective has been offered here [modified from Cairns-Smith (1982)] that is rather more Gaian. In this view, life arose in a stable, prebiotic surficial system that promoted its own persistence. Acidic weathering in an early carbon dioxide-rich atmosphere promoted the formation of clayey soils. These became more effective at holding the landscape together as organic matter was produced by the abiotic photo-oxidation of iron-bearing clays coupled with reduction of carbon dioxide. In the complex, reactive, alternately wet and dry cracks and cavities of stable clayey soils, organic matter accumulated to produce materials somewhat like carbonaceous chondrites. Increasingly complex organic compounds would be preserved as long as they promoted the persistence of the soil against erosion. Organic matter, microbial scums and tropical rainforest can be seen as a continuum of increasingly effective methods of stabilizing the landscape. Oxidation of the atmosphere from small stable areas of productive soils was minimal until the development of continents. By 3500 million years ago life had begun to colonize sea shores. Soils oxidized by organic photosynthesis had long been cleansed of organic carbon that accumulated in marine limestones and black shales, of nitrogen that accumulated in the atmosphere and of sulfur that accumulated as marine evaporites. Biological reduction of sulfate to sulfides in the ocean began 2300 million years ago, and probably earlier. Methanogenesis in stagnant swamps and oceans occurred ca. 2800 million years ago. Photosynthetic ecosystems have always been the source of primary productivity. Methanogenic, fermentative, and respiratory ecosystems are, and always were, dependent on organic matter produced by photosynthesis and have extended the influence of life to habitats where photosynthesis is not possible.

Present geological and experimental evidence does not allow a clear choice between these or other scenarios for the early evolution of life. A better understanding of Precambrian paleosols is a promising and barely tested additional approach to the problem. It will not, however, be the whole answer. The Gaia hypothesis has marshalled evidence that life has some homeostatic influence on modern surface processes. Whether this also was true for the distant geological past will only be established by considering all possible facets of Precambrian environments.

18 Large plants and animals on land

Although the greening of the land with photosynthetic microbes may have occurred well back in geological time, the appearance of multicellular creatures was a much more recent event, by current estimates Late Ordovician. The age and nature of the earliest multicellular soil creatures remain uncertain because their fossil record is so sketchy. The fossil record of early land organisms is largely known from remains preserved in nearshore aquatic environments where amphibious and aquatic creatures could have been preserved mixed with those transported from nearby land. In contrast, fossil soils offer evidence of conditions and life activities where the creatures actually lived.

The fossil record of marine multicellular organisms extends much further back than that for continental creatures. Metazoan burrows and fecal pellets in marine rocks have been reported as old as 2500 million years. Few of these records can be trusted as reliable because most have proven on closer examination to be clasts, gas-escape structures, or other abiotically produced features (Cloud & Lajoie 1980). Also not to be trusted are carbonaceous fossils that could be sheets of unicellular algae. Carbonaceous compression fossils as old as 1300 million years have the regularity of form and branching that one would expect of authentic multicellular algae (Hofmann 1985). A diverse assemblage of latest Precambrian (about 600 million years old) large, soft-bodied creatures has been found in many parts of the world, but is best known from the Ediacara Hills of South Australia. These fossils were marine multicellular organisms whose morphology and style of preservation are so unique that their taxonomic affinities remain unclear (Seilacher 1984). A variety of marine trace fossils appear along with these curious Ediacaran forms. Small shelly fossils first appear in rocks ca. 580 million years old, before the generally agreed lowest boundary of the Cambrian (Sepkoski & Knoll 1983). In earliest Cambrian rocks, there is fossil evidence for most skeletonized phyla of marine invertebrates and for many kinds of calcareous algae (Riding & Voronova 1985).

In contrast with this ancient record of multicellular marine algae, the well documented record of megafossil land plants does not begin until the early Silurian (Llandoverian). In rocks of this age in Maine, USA, are slender (1–2 mm in diameter), branching, carbonaceous tubes in position of growth, called *Eohostimella* (Schopf *et al.* 1966). In mid-Silurian (Wenlockian) rocks of Ireland are found other slender branches with globular terminal spore sacs. These and other Silurian fragments referred to *Cooksonia* (Fig. 18.1A) have been widely regarded as the most ancient vascular land plants, although this still remains to be proven conclusively (Taylor 1988). It has been surprising to find complex kinds of vascular plants such as the leafy *Baragwanathia* (Fig. 18.1C) in rocks of comparable age in central Victoria, Australia (Late Silurian: Ludlovian) and at Djebel Fezzan, Libya (poorly dated but between Llandoverian and Emsian). These large leafy plants are extinct Drepanophycales, a kind of lycopod. Small (4 mm long), helically twisted lenticular sporangia of latest Silurian age (Pridolian) in Wales referred to *Torticaulis* may be the oldest bryophytes. Also of this age from Podolia, USSR are calcified egg cells referred to *Trochiliscus* which appear to have been an aquatic alga similar to the living stoneworts (*Chara* and *Nitella*). Latest Silurian to Early Devonian (Pridolian to Gedinnian) rocks in the British Isles, Europe, and North America have yielded remains of sizeable (up to 7 cm long), thalli bearing numerous ovoid sporangia (Fig. 18.1B). These remains, called *Parka*, are superficially similar to the modern aquatic charophyte alga *Colochaete*, considered the closest relative to modern bryophytes and vascular land plants among living algae (Mishler & Churchill 1985). Latest Silurian nematophytes, *Prototaxites* and *Nematothallus*, represent bizarre

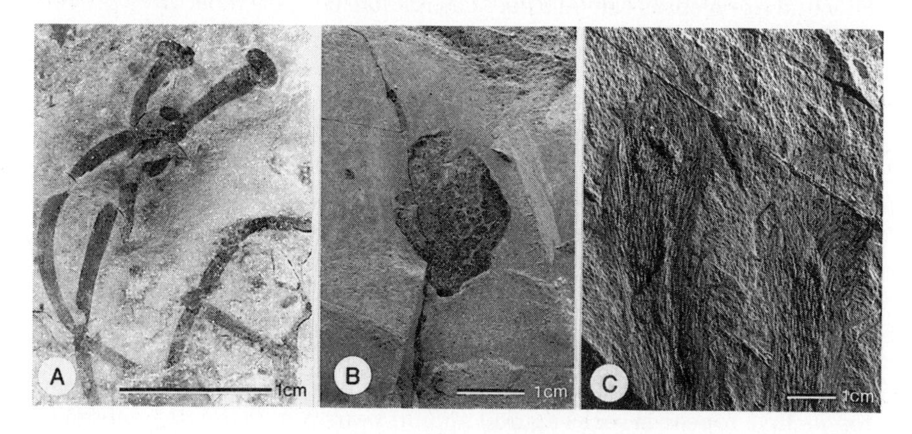

Figure 18.1 Some early land plant fossils: A, the *Rhynia*-like "cooksonoid" *Cooksonia* sp. cf *C. pertoni* of Late Silurian age (Pridolian) in the Bertie Waterstone from near Cedarville, New York, USA [photograph courtesy of H. P. Banks; plant described by Banks (1969)]; B, enigmatic thalloid plant, *Parka decipiens* of Early Devonian age (Gedinnian) in the Dundee Formation at Turin Hill, Angus, Scotland (Retallack specimen P8384A); C, a drepanophycan *Baragwanathia longifolia* of Early Devonian age (Emsian) in the Wilson Creek Shale, near Alexandra, Victoria, Australia (Retallack specimen P4731A). Scale bars 1 cm.

extinct kinds of plants. Their thick cuticles and robust internal mesh of conducting tubes would have given them a consistency like that of a mushroom. Trunks of *Prototaxites* reached the size of large trees (up to 1 m in diameter and 2 m long). In Devonian rocks, there is a greater diversity of fossil charophyte algae, nematophytes, thalloid bryophytes and vascular land plants (Gensel & Andrews 1984). By this time large multicellular plants were well established on land.

Another source of information on early land plants is microscopic organic fragments such as conducting tubes, cuticles, and spores that can be liberated from silicate rocks by dissolving them in hydrofluoric acid. This line of evidence indicates a greater antiquity for land plants than the record of plant megafossils (Gray 1985a). These differences are to be expected because microscopic propagules are more widespread and abundant than fossil plants large enough to be collected by hand. Thick sheets of cells like cuticles of land plants and thick-walled, desiccation-resistant spores permanently united in a tetrahedral configuration referred to *Tetrahedralites* (Fig. 18.2A) are now known as old as early Late Ordovician (Caradocian) in both Libya and Arabia. The spores are thought to represent early land plants because they share some features with spores of living liverworts. In Silurian (mid- to Late Llandoverian) rocks appear the first trilete spores of the genus *Ambitisporites* and other similar genera (Fig. 18.2B), which are thought to represent the oldest vascular land plants. The diversity of trilete spores increased dramatically later during the Silurian period (late Wenlockian to Pridolian). Other microfossils similar to conidia, ascospores, and hyphal fragments of terrestrial fungi have been

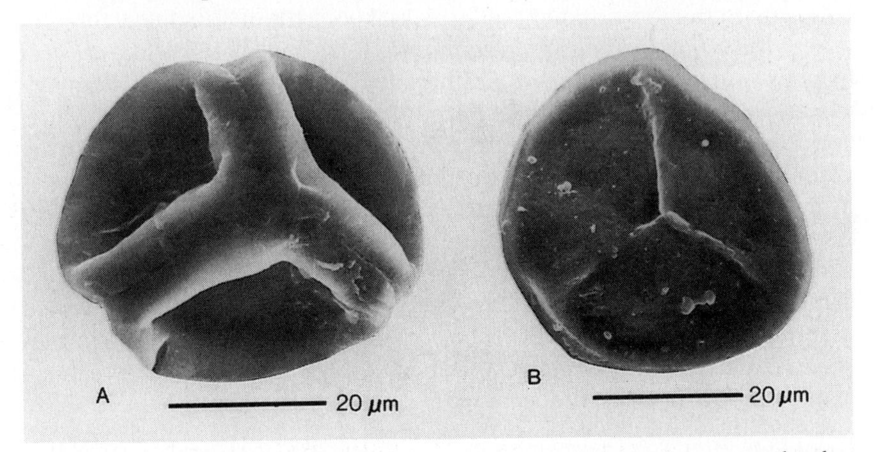

Figure 18.2 Spores of early land plants: A, permanent tetrad of an early non-vascular plant *Tetrahedralites* of Late Ordovician (Ashgillian) Elkhorn Formation, in Ohio Brush Creek, Adams County, Ohio (courtesy of J. Gray, specimen G1285); and unnamed new genus of trilete spore of an early vascular land plant from the Late Silurian (Ludlovian) Burgsvik sandstone, near Burgsvik, Island of Gotland, Sweden (courtesy of J. Gray, specimen G923). Scale bars 20 μm.

found in these Late Silurian assemblages and less well preserved in assemblages of Early Silurian age in the eastern United States (Sherwood-Pike & Gray 1985). Thus both fungal decomposers and vascular land plants were diverse by the end of the Silurian period.

The fossil record of possible early land animals is more sketchy than that of megafossil plants (Beerbower 1985). Curiously, some of the oldest fossil representatives of animal groups that now are completely non-marine have been found in marine rocks: Middle Cambrian Onychophora (Whittington 1985), and myriapod-like fossils (Robison 1986), Middle Ordovician earthworms (Morris *et al.* 1982), and Early Silurian myriapods (Mikulic *et al.* 1985). Also difficult to interpret are assemblages of fossil arthropods found in shales and dolostones that lack a normal marine fauna. The best known of these are eurypterid assemblages of Latest Silurian age (Pridolian) in northern and western Britain and western New York State, USA (Kjellesvig-Waering 1961). Other examples include diverse arthropod fragments in the Early Silurian Tuscarora Formation of Pennsylvania, USA (Gray 1985b) and tracks of arthropods from the Late Ordovician Harding Sandstone of Colorado, USA (Fischer 1978). These may all have been estuarine or lagoonal, but other early arthropods are so isolated in their geological occurrence and morphology that they have been interpreted as lacustrine, for example, the Middle Cambrian eurypterid *Kodymirus vagans* from Czechoslovakia and the Early to Middle Ordovician horseshoe crab ally *Chasmataspis laurencii* from Kentucky, USA (Bergstrom 1979). These constitute a pool of potential invaders of the land, but there is little evidence that they did in fact leave the water until fairly late in the development of terrestrial vegetation. The latest Silurian eurypterid *Baltoeurypterus tetragonopthalmus* had gill tracts like those of isopods and may have been amphibious (Selden 1985). Late Pennsylvanian horseshoe crabs such as *Euproops danae* may have climbed high into trees whose leaves are mimicked by their spines (Fisher 1979). On the other hand, several other groups of creatures now confined to land appear to have remained aquatic during early Paleozoic times. Scorpions are known as fossils as old as Early Silurian, but there is no evidence that they had book lungs enabling excursions on to land until Middle Pennsylvanian time (Kjellesvig-Waering 1986). Similarly, the earliest well accepted examples of fossil millipedes (*Kampecaris forfarensis*) of latest Silurian age have pecularities of limb design and tail shape that are more like those of aquatic arthropods (Almond 1985).

In Early Devonian (Siegenian) rocks there is at last undisputed evidence of fully terrestrial animals. This includes flat-back millipedes (*"Kampecaris" tuberculata*) from red beds between lava flows of the lower Old Red Sandstone (Almond 1985) and the varied fauna of collembolans, spiders, and mites found in the Rhynie Chert (Fig. 10.6), both in Scotland. Bristletails now are known as old as Early Devonian (Emsian; Labandiera *et al.* 1988) and centipedes as old as Middle Devonian (Givetian; Shear *et al.* 1984). Fossil winged insects are not known until mid-Carboniferous (Namurian) when they appear in the fossil record in some diversity. Perhaps the most disconcerting aspect of this sketchy fossil record of continental creatures is that major discoveries are still being made. Understanding of early land animals is based on such exceptional finds and preservation that it cannot yet be regarded as a representative sample.

Evidence of multicellular organisms in paleosols

Compared with the imperfect fossil record of continental plants and animals, paleosols are preserved in a wider range of environmental conditions and are much more abundant sources of evidence, for the antiquity of multicellular organisms on land. Fossils in paleosols would be the best line of evidence, although these require stringent conditions for preservation. Trace fossils such as burrows in paleosols also are useful evidence. Many of the earliest land plants and animals were small and would have left subtle traces, but some potentially continental creatures, such as eurypterids and *Prototaxites* were large and would have left obvious traces in paleosols if they ever had lived on land. Although few early Paleozoic paleosols have been examined from this point of view, there already is evidence for large plants and animals on land earlier than would have been expected from their megafossil record and more in accord with microfossil evidence.

Non-vascular land plants

Possible evidence for non-vascular land plants of Late Ordovician age (perhaps as old as Caradocian) has been found in red paleosols of the Dunn Point Formation in the Arisaig area, Nova Scotia, Canada (Dewey, in Boucot *et al.* 1974). The paleosols consist of about 1.3 m of red, calcareous claystone on top of weathered corestones and columnar-jointed flows of andesite (Fig. 18.3). Near the surface of one of the paleosols are irregular swales, about 1 m wide and 20 cm deep, filled with red shale redeposited from erosion of the paleosol. Small, white reduction spots within the mounds between these erosional swales are similar to drab mottles forming during diagenetic alteration of organic matter buried in oxidized paleosols (see Figs. 3.6, 7.7). Dewey suggested that the mounds were stabilized against erosion by clumps of non-vascular land plants which lacked rooting structures substantial enough to leave obvious traces. These red, weakly calcareous paleosols and associated mudflows formed on a hilly, tropical, volcanic landscape in a seasonally dry, subhumid climate.

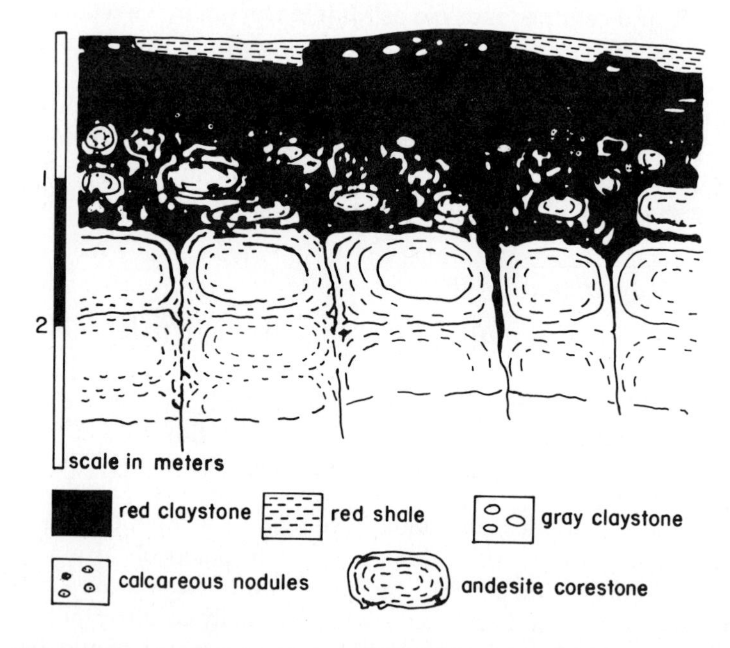

Figure 18.3 Thick red paleosol with superficial erosion scours and reduction spots, developed on basaltic andesite flow in the Late Ordovician (Caradocian), Dunn Point Formation near Arisaig, Nova Scotia (from J. F. Dewey, in Boucot *et al.* 1974, reprinted with permission of the Geological Society of America).

Such vegetation of rootless land plants in dry soils has few modern counterparts. It can be visualized as similar to the moss and lichen cover of desert or alpine boulders. This kind of vegetation does not have a distinctive name and I propose to call it polsterland, after a moss colony or polster.

A second example of early land plants in a paleosol is *Eohostimella heathana* from the Early Silurian (Llandoverian) Frenchville Formation near Stockholm, Maine, USA (Schopf *et al.* 1966). These fossil plants are small (1–2 mm in diameter), erect, dichotomously branching, carbonaceous tubes in a dark, finely laminated shale that elsewhere contains nearshore marine fossils. A case has been made that these carbonaceous tubes were organically lined worm burrows (Strother & Lenk 1983). However, their coalified outer layer appears cellular and spinose in places, and its organic chemical composition is more like other Silurian and Devonian vascular land plants than fossil marine algae (Niklas 1982). The interior of these fossils is not sufficiently well preserved to reveal for certain whether they were vascular plants, nematophytes or some other group. The laminated, carbonaceous paleosol was little developed and waterlogged (an Aquent), as is common for soils of modern salt marsh vegetation.

Invertebrates

The oldest animal traces now recognized in paleosols have been identified in the Late Ordovician (Ashgillian) Juniata Formation near Potters Mills, Pennsylvania, USA (Retallack & Feakes 1987). These burrows are so abundant and well preserved that they had hitherto been regarded as evidence for estuarine or marine incursions into these non-marine rocks. However, this part of Pennsylvania at this time was at least 260 km east of the midcontinental seaway and the sedimentary sequence containing them has all the characteristics of a purely fluvial sequence. No marine fossils of any kind have been found in association with the burrows. More convincing evidence comes from the close association of the burrows with soil features (Fig. 18.4). Their increased density toward the upper part of the profile parallels the increased development of soil structures (platy peds), microfabric (skelsepic to skelmosepic), the abundance of clay and the degradation of mica and feldspar (Fig. 18.5). The density of burrowing also corresponds to the degree of weathering, including ferruginization, desilication, and clay formation, as reconstructed by comparing normalized major oxides in a moderately developed paleosol to an underlying poorly developed paleosol (Fig. 12.6). Small dolomitic nodules ensheath about

Figure 18.4 Clayey red paleosols (dark recessive-weathering zones) and sandstone paleochannels in the Late Ordovician (Ashgillian), Juniata Formation in road cut near Potters Mills, Pennsylvania. The top of the type Potters Mills clay paleosol is at the base of the upper black band on the scale bar. Scale in foreground is calibrated in feet (from Retallack 1985, with permission of the Royal Society of London).

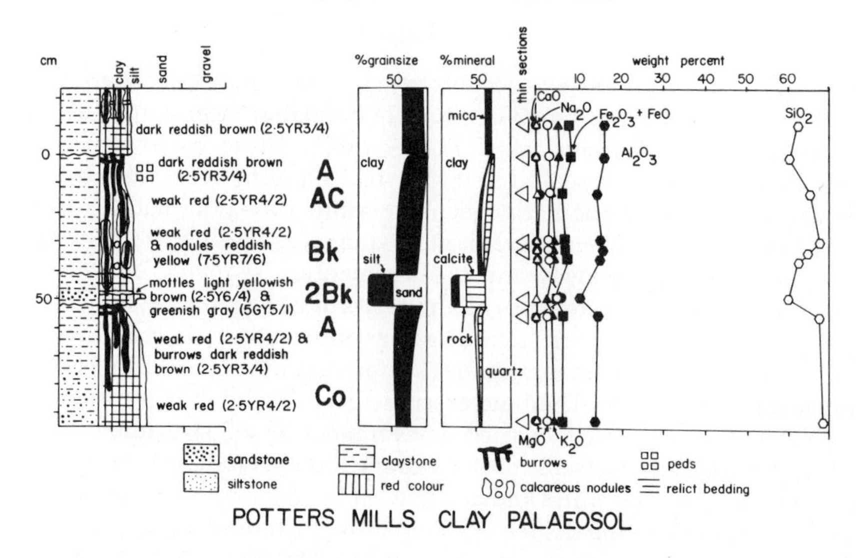

Figure 18.5 Columnar section (measured in field) and petrographic and chemical compositions (from point counting and ICP chemical analysis) of the type Potters Mills clay paleosol (Oxic Ustropept, above) and the type Faust Flat silty clay paleosol (Fluvent, below) from the Late Ordovician, Juniata Formation, near Potters Mills, Pennsylvania (from Retallack 1985, reprinted with permission of the Royal Society of London).

half the burrows in the moderately developed paleosol, but are not present around less abundant burrows in the weakly developed paleosol, or in associated burrowless lacustrine shales. These caliche nodules are thus indications that the burrows were a permanent feature of the soil rather than inherited from a pre-existing lacustrine parent material or added after soil formation had been terminated by flooding.

These late Ordovician fossil burrows are roughly tubular and about 1 cm in diameter. Their outer margins are marked by a thin zone of ferruginized clay which is strongly grooved and smeared (slickensided) by compaction. Most of the encrusting caliche is just outside, but in some cases cuts across, the ferruginized layer. Ovoid masses within some of the burrows resemble fecal pellets (Fig. 18.6B). Other burrows have bilaterally symmetrical backfill structures of alternating silty and clayey bands that are W-shaped in both longitudinal (Fig. 18.6C) and transverse sections (Fig. 18.6D) of the burrows. In order to gain an appreciation of the three-dimensional distribution of the burrows, large blocks of matrix were cut into parallel slabs at approximately 1 cm intervals and the distribution of the burrows plotted in three dimensions. The burrows form a complex network of horizontal and vertical galleries that vary considerably in width. Horizontal galleries are more common near the surface, but deeper within the profile most of the burrows are blindly ending vertical shafts. None of

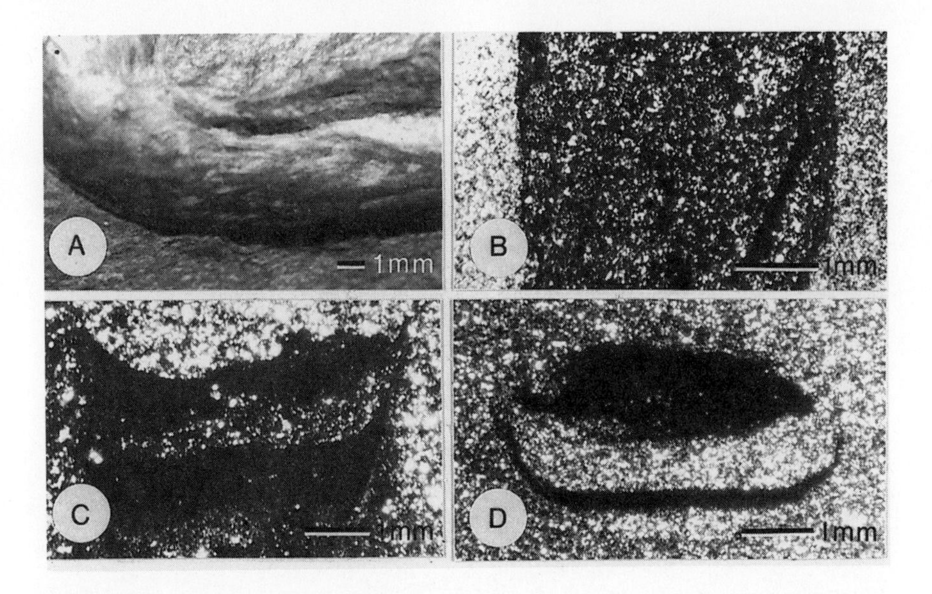

Figure 18.6 Fossil burrows from Late Ordovician paleosols near Potters Mills, Pennsylvania. A, In hand specimen, cracked open in bas-relief, showing surface striations and a hint of bilateral symmetry; B, petrographic thin section of a burrow under crossed nicols showing ellipsoidal masses (defined by black clay skins), intepreted as fecal pellets; C, petrographic thin section under crossed nicols along axis of burrow showing W-shaped backfill structures; D, petrographic thin section under crossed nicols transverse to long axis of burrow showing bilateral symmetry of backfill structure. Scale bars are 1 mm (from Retallack & Feakes 1987), reprinted with permission of the American Association for the Advancement of Science).

the burrows were observed to branch. Further information on the nature and diversity of the burrowing organisms was gleaned from variation in size of the burrows as measured on specimens exposed in bas-relief (Fig. 18.6A). The burrows range in diameter from 2–21 mm with numerous narrow modes that are parasitic on broader peaks. Such parasitic modes are common in size distributions of fossil animals that grow in marked size increments, such as arthropods. The greater range and less regular overall shape of the distribution for the weakly developed paleosol compared with the moderately developed one are an indication of greater diversity of burrows or of environmental fluctuation in near-stream habitats than away from streams.

Considering the associated caliche, parasitic modes of the size distribution and W-shaped backfill structures, these burrows were excavated by bilaterally symmetrical organisms that grew in well defined growth increments and were able to withstand dry soil conditions. Hence, they are unlikely to have been made by earthworms or velvet worms. Of other

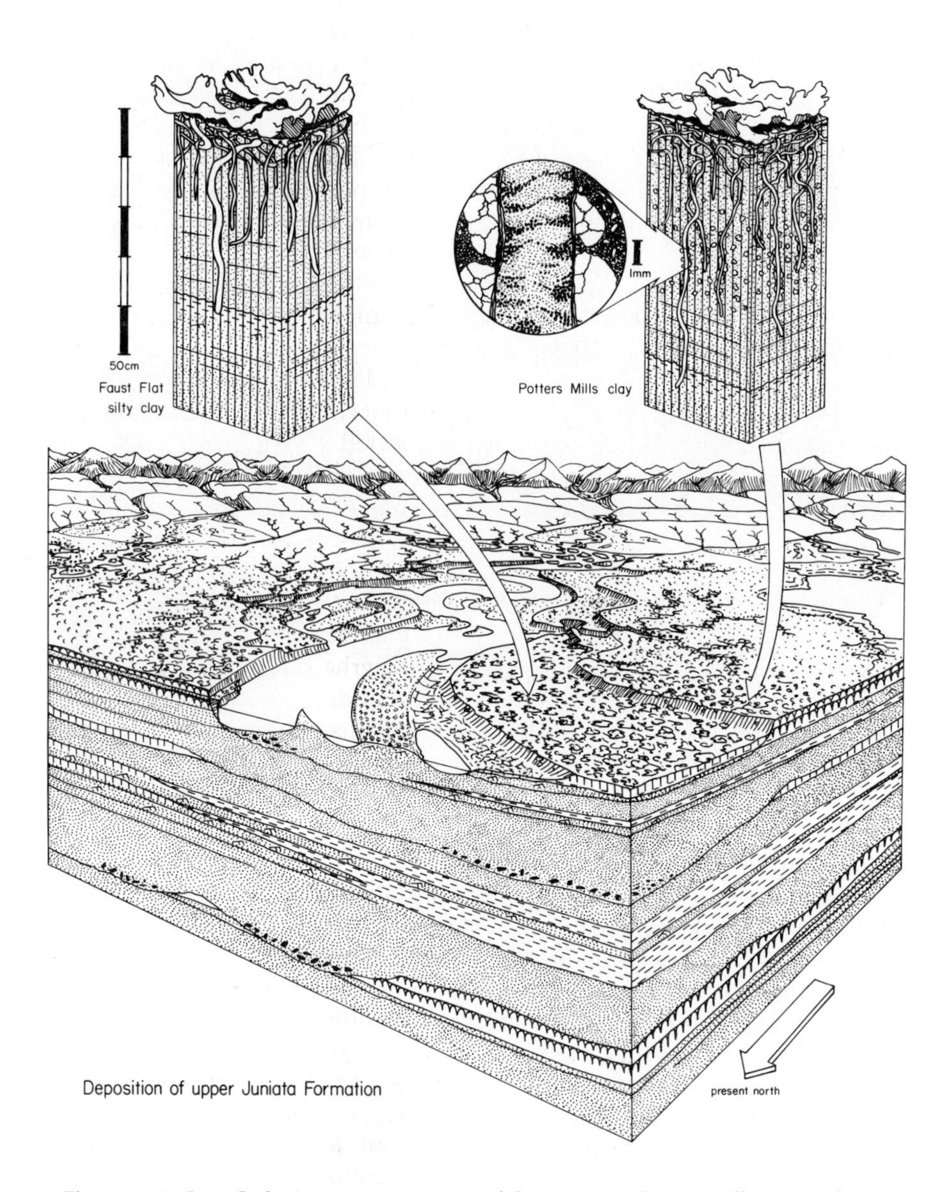

Figure 18.7 Late Ordovician reconstruction of the area near Potters Mills, Pennsylvania. The vegetation shown is conjectural, and represents the maximum likely biomass compatible with chemical and structural features of the paleosols (from Feakes & Retallack 1988, reprinted with permission of the Geological Society of America).

possible early soil invertebrates, the desiccation resistance and burrowing abilities of eurypterids, aglaspids, horseshoe crabs, and scorpions were probably inadequate for the job. Spiders and centipedes are not found nearly this far back in the geological record and no living representatives of these creatures make burrows such as these. The fossil burrows are most like those of roundback millipedes, although this also would extend considerably the geological range of this group. There remains a possibility that the burrows were made by organisms completely extinct or unknown or by organisms which by virtue of their functional morphology and modern behavior seem to be inadequate burrowers.

These early land animals may have fed entirely on unicellular soil algae, and the abundance of burrows in these paleosols may reflect a long period of soil formation without active destruction of abandoned burrows. The most elaborate ecosystem compatible with evidence from paleosols would have included liverwort-like plants lacking roots that would leave obvious traces. Hence, these soils could have supported a polsterland similar to that envisaged for the Late Ordovician paleosol of Nova Scotia (Fig. 18.3).

These paleosols formed on alluvial outwash of the Taconic Mountains to the east (Fig. 18.7) at a time of very low global sea level because of glaciation in the present region of the Sahara Desert of Africa. The climate in Pennsylvania at this time appears to have been tropical, considering the diversity of North American marine faunas and their paleolatitude. Rainfall was in the semi-arid to arid range and perhaps seasonal if the depth of caliche in the paleosols and its ferruginous bands can be taken as a guide. This was a drier and more hostile environment than usually envisaged for early land creatures.

Vascular land plants

The difference between early vascular land plants such as rhyniophytes and nonvascular land plants such as nematophytes is not always easy to determine because they can be externally similar and lack the large branching root traces of modern vascular plants (Edwards & Edwards 1986). Making this distinction is even more difficult from the indistinct traces left by plants in well drained paleosols. For example, paleosols in the Late Silurian (Ludlovian) Bloomsburg Formation near Palmerton, Pennsylvania (Retallack 1985) are generally similar to those already described from Late Ordovician rocks, but are thicker and better developed. They have indistinct wispy bioturbation of the surface horizon (Fig. 18.8, upper profile). The wisps are small (1–5 mm across), irregular, subhorizontal to vertical, branching tubular features filled with clay (ortho-isotubules) or sand (metagranotubules) different from the surrounding matrix (Fig. 18.9). They are much more irregular than the tubular burrows already described that also are found in some paleosols in the Bloomsburg

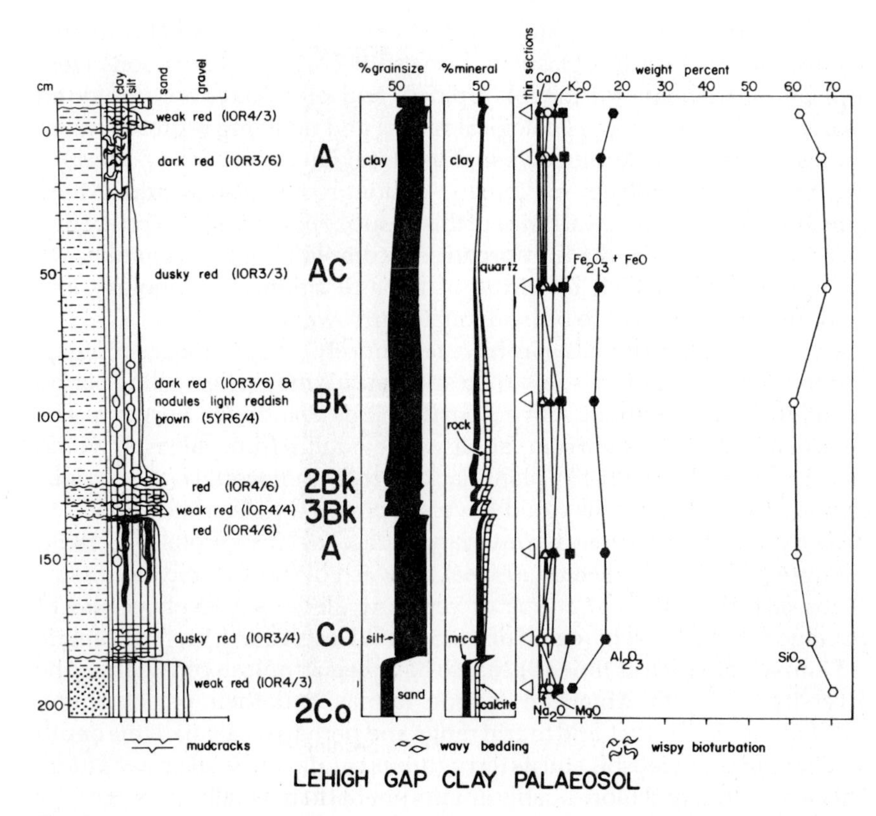

Figure 18.8 Columnar section (measured in field) and petrographic and chemical composition (from point counting and ICP chemical analysis), of the Lehigh Gap clay paleosol (an Oxic Ustropept, above) and Faust Flat clay (Fluvent, below) in the Late Silurian (Ludlovian) Bloomsburg Formation, near Palmerton, Pennsylvania. Lithological symbols as for Figure 18.5 (from Retallack 1985, reprinted with permission of the Royal Society of London).

Formation (Fig. 18.8, lower profile). The wispy bioturbation looks most like the traces of plant rhizomes. There is no feature of them yet observed that is especially distinctive of a particular kind of vegetation.

Red calcareous paleosols with irregular root-like bioturbation have been found in Late Silurian (Pridolian) red beds of Nova Scotia (Boucot *et al.* 1974) and Britain (Allen 1986a). By this time vascular land plants were abundant and widespread. Of the kinds of megafossil plants found in associated deposits, xeromorphic rhyniophytes more likely vegetated these dry soils than thallose liverworts or nematophytes. This dry land, herbaceous, rhizomatous vegetation was probably similar to modern desert vegetation of resurrection plant (*Selaginella lepidophylla*) or alpine and open woodland carpets of clubmoss (*Lycopodium deuterodensum*). This kind of plant formation does not have a name. I propose to call it brakeland, following the common English expression "fern brake."

Figure 18.9 Wispy bioturbation on a plane parallel to bedding (above) and in a fracture plane (below) of the surface (A) horizon of the Late Silurian, Lehigh Gap paleosol, near Palmerton, Pennsylvania. Scale graduated in mm (Retallack specimen R155C).

Vascular plants certainly formed soils by Early Devonian time (Siegenian) as represented by the Rhynie Chert in Scotland (Kidston & Lang 1921). This silicified peat contains exquisitely preserved remains of various vascular and non-vascular plants (see Fig. 10.6) in addition to fossil arthropods. Alternating layers dominated by *Horneophyton lignieri* and by *Aglaophyton major* may represent successive episodes of flooded marsh followed by better drained marsh.

A timetable

Despite the paucity of pertinent studies, a broad timetable is apparent from paleosols for the advent of various kinds of large plants and animals on land. Microbes probably were abundant on land for much of Precambrian time. Non-vascular land plants and arthropods may have vegetated dry, well drained soils by late Ordovician time (Ashgillian). Salt marsh vegetation of non-vascular land plants is known as old as Early Silurian (Llandoverian). Vascular land plants may have lived in well drained soils by Late Silurian (Ludlovian). Root-like structures are more widespread in paleosols of latest Silurian (Pridolian) and Devonian age. Primitive vascular and non-vascular plants of freshwater marshes are known from silicified peat of Early Devonian age (Siegenian).

The fossil record of soils thus supports evidence from microfossil spores for Late Ordovician appearance of land plants long before they appear as Silurian megafossils. The fossil record of burrows in Late Ordovician paleosols also predates Silurian records of land animals. On the other hand, records of fossil vascular plants are known as both spores and megafossils earlier than they can be clearly documented from paleosols. It remains to be seen whether this is a real failing of the paleosol record or reflects the preliminary nature of paleosol studies.

How did multicellular soil organisms arise?

Establishing exactly when various groups of multicellular organisms came to live on land has proven difficult. Investigating how it happened is even more forbidding. The diversity and adaptive features of early continental fossil assemblages provide some clues (Beerbower 1985) which can now be assessed against the fossil record of soils. Several separate problems can be identified: what were the evolutionary origins of the earliest land plants and animals, and to what extent were their life styles constrained by environmental factors such as nutrient availability, atmospheric oxygenation, flooding, volcanism, or changes in sea level?

Origin of land plants

The closest living relative of land plants are charophytes (Mishler & Churchill 1985). These green algae live in lakes and rivers. From this and the geologically ancient fossil record of other kinds of algae in marine rocks, it is widely assumed that land plants were derived from aquatic algae (Chapman 1985). They are seen as invading the land as if it were some kind of military campaign for territory. Steps in the invasion can be envisaged as a technological escalation of military hardware: the acquisition of (a) aerially dispersed spores, coated in rigid, desiccation-resistant, sporopollenin walls, (b) a thick desiccation-resistant cuticle, (c) tracheids for support and water transport, and (d) intercellular air spaces and stomates for gas exchange. The theoretical difficulties for large aquatic algae evolving such structures have prompted an alternative view that land plants evolved from unicellular or very small multicellular soil algae (Stebbins & Hill 1980). In this view, the paraphernalia that are needed for plants to survive on land could have evolved entirely within soils rather than within environments intermediate between water and land such as tidal flats and streamsides.

The idea that land plants evolved in soil is strengthened by the evidence of complex microbial communities in soils well back into Precambrian time (Retallack 1986b,c). Additional evidence is the currently earlier record of plant life in well drained soils (Late Ordovician) than in salt marsh (Early Silurian). The limited records available can hardly be regarded as representative, but they do point the way to a source of new information on this question.

Indeed these new discoveries also support some aspects of the traditional view of invasion of the land from the sea. The difficulties of conquest would have been mitigated by burrows in soils and by pre-existing microbial ecosystems. Microbes would have stabilized the landscape and initiated cycles of nutrient utilization in soils. Burrows could have provided moist and sheltered local habitats. In modern deserts, the burrows of rodents may be small, semi-autonomous communities including algae, fungi, and herbivorous and dung beetles, all protected within the cool moist burrow from a harsh external environment (Halffter & Matthews 1966). The early evolution of vascular land plants in intertidal habitats is also supported to some extent by their occurrence in marine rocks geologically older (Wenlockian) than evidence for them in paleosols (Ludlovian). Where exactly early land plants evolved will remain uncertain until further studies of paleosols are undertaken.

Origin of land animals

The early evolution of land animals is a separate problem. There has been little doubt that arthropods invaded the land from marine habitats because of their long evolutionary history in marine rocks as old as basal Cambrian. In many ways marine arthropods were pre-adapted for life on land. Woodlice, crabs, and crayfish are just a few representatives of aquatic arthropod groups that have independently exploited land habitats (Powers & Bliss 1983). Their exoskeleton is effective for support and protection both on land and sea. On land it also reduces desiccation. Their jointed limbs permit locomotion, burrowing, and predatory feeding both on land and in the sea. Such mobility also allows complex mating rituals. The main evolutionary innovations of land arthropods were methods of respiring in air without drying out, such as book lungs and tracheary systems, and methods of liquifying solid food in a preoral chamber (Størmer 1977). It is possible that some of the enigmatic Cambrian arthropods found in unusual marginal marine facies and perhaps even Middle Cambrian fossil velvet worms in deep marine rocks were washed in from lakes or land. One could also imagine that the earliest land animals were very small creatures with a jointed exoskeleton other than arthropods like living tardigrades.

Current evidence from paleosols is not adequate to address this issue, but does point to a significant new line of enquiry. In the Late Ordovician Juniata Formation near Potters Mills and the Late Silurian Bloomsburg Formation near Palmerton, both in Pennsylvania, USA, burrowed paleosols are found in inland fluvial facies near the upper part of these formations, and also in near-marine and intertidal paleosols in the lower part of these formations. Drab, shaly deposits of inland lakes have been identified in both sequences, but these so far have proven devoid of trace fossils or other indications of life (Retallack 1985). There is therefore no evidence yet of invasion of land from lakes, but a potential to evaluate trace fossil diversity in land-to-sea paleoenvironmental gradients. Whether this will demonstrate invasion of the land from the sea remains to be seen.

Ecology and environment of early land creatures

The earliest land plants and animals are usually imagined as conservative in their use of water and nutrients, slow to grow and reproduce, and in other ways making do in a generally hostile environment. This view of early fossil land plants is based in part on their apparently low diversity, slow rates of evolution, simple structure, and comparison with modern plants with broad environmental tolerances, such as liverworts. Slow-moving herbivorous animals, such as millipedes, are expected to be the earliest kind of land animals according to this stress-tolerant view. However, alternative life styles are known in modern plants (Grime 1979). In frequently disturbed parts of the landscape, for example, it may be more advantageous to produce copious small and widely dispersed propagules rather than persist in one place or grow to a large size. Such a life style found in modern plants regarded as weeds also could be argued for early land plants given their small, smooth-walled, permanent tetrads dispersed by wind and water, and capable of forming several small reproductive plantlets (gametophytes) that could fertilize each other if no others were nearby. The small size of many early continental animal fossils also supports the idea that early land animals were small and rapidly breeding, like many modern creatures regarded as pests. A third life style has been called competitive because it emphasizes the growth of a large body that commandeers resources at the expense of other organisms. This view also could be supported for early land plants if large nematophytes (such as *Prototaxites*) or large eurypterids and scorpions played a role in early continental ecosystems. These three general kinds of life style are responses to nutrient-poor, to frequently disturbed, and to nutrient-rich environments, respectively. Evidence for these selection pressures can be expected from paleosols.

One difficulty that could have been faced by early creatures on land was an atmosphere rich in carbon dioxide with little oxygen and lacking an ozone shield that filtered out strong ultraviolet radiation. Short-wave length (220–290 nm) ultraviolet radiation could disrupt nucleic acids and proteins if there were no ozone screen, as would be likely at an oxygen level in the atmosphere of less than 0.01 times the present level (Kasting 1987). About 0.02 times the present level of oxygen is needed for the production of cutin in cuticles that protect against dessication, of phenylpropanoids that act as a screen against ultraviolet radiation, and of lignin as a basic material for structures of water transport and support in plants (Chapman 1985). The track of atmospheric oxygenation since Precambrian time is not constrained by studies of as many paleosols as one would like (Fig. 16.6), but oxygen probably exceeded 0.02 times the present level by Late Precambrian time, if not earlier (Grandstaff *et al.* 1986, Pinto & Holland 1988). Paleosols developed on peridotite and on basalt as old as 1000 million years within and below the Keweenawan Supergroup of Michigan, USA (Kalliokoski 1986, Kalliokoski & Welch 1985), have retained oxidized iron and are as strongly reddened in appearance as Late Ordovician paleosols developed on andesitic basalt in the Dunn Point Formation of Nova Scotia, Canada (Fig. 18.3). These preliminary impressions do not favor the idea that strong ultraviolet radiation or low oxygen levels limited the growth of land animals or the evolution of chemicals critical to key adaptations in plant life on land. Nor is it favored by studies of the oxygen requirements of late Precambrian and early Paleozoic marine organisms (Rhoads & Morse 1970, Runnegar 1982).

Obtaining water and nutrients are closely related problems for plants. some nutrients (especially N, P, and S) can only be used when supplied by microbes as ions in solution (Stevenson 1986). Animals obtain them by eating plants or other animals. Fossil fungal hyphae and spores as old as Silurian (Sherwood-Pike & Gray 1985) show that microbial nutrient procurement systems are of considerable antiquity.

Other nutrients (Mg, K, Ca, Na, and Cl) are obtained by mineral weathering. The water and nurient resources of Late Ordovician paleosols in the Juniata Formation of Pennsylvania can be reassessed from this perspective. The depth of leaching of carbonate in these paleosols indicates subhumid conditions, probably with a severely dry season. This kind of climate is now encountered in regions of subtropical wooded grassland. The carbonate also is evidence of abundant calcium that could be used in the exoskeletons of arthropods. Detailed chemical studies of one of the profiles has shown a strong surficial enrichment of potassium. This and petrographic observations on early Paleozoic sandstones (Basu 1981) could be taken as an indication that this element was under-utilized, as is typical under modern fungi and liverworts (Shacklette 1965). However, the stoichiometrically inordinate abundance of potassium is a clue that its enrichment was in part diagenetic (Feakes & Retallack 1988). Some other elements (Ba, Cr, Sc, V, Zn, and Ti) increase in abundance toward the surface, presumably because of their well-known affinity for clay which was more abundant there. However, others (Li, Nb, Ni, Sr, Y, and P) that commonly follow clay and organic matter within soils are surprisingly depleted at the surface. This may reflect loss within vegetation that was not intimately mixed with the soil and left little humus. The clay content, degree of weathering, and evidence for nutrient recycling in this paleosol are greater than in many modern desert soils, but less than in soils of wooded grassland. Conditions were certainly not as grim as they could have been.

Additional possible limitations for early plants and animals were environmental perturbations. Floods, dry seasons, volcanic eruptions, and marine transgressions all can be imagined as destabilizing the establishment of early continental ecosystems. Such environmental disturbance leaves a record in the degree of development of paleosols in long sequences. In the Late Ordovician Juniata Formation paleosols are sparse in the sequence and for the most part are weakly developed (Faust Flat Series), representing only a few hundreds of years of soil formation at most. Caliche-bearing paleosols (Potters Mills Series) are less common and represent parts of the landscape that were stable for several thousands of years. The presence of ferruginized caliche concretions indicates a dry season. The dominance of sandstone in thick tabular beds with only local evidence of asymmetrical channel-like forms is an indication that the Juniata Formation was deposited in braided to loosely sinuous bedload streams which may have had a somewhat flashy flood regime (Cotter 1978). Although such a disturbance is more frequent than in ecosystems of humid floodplains, it is not as severe as that found in desert regions.

Volcanism is also unlikely to have been a critical limitation considering the evidence for plants and thick red calcareous paleosols on andesitic flows of the Late Ordovician Dunn Point Formation of Nova Scotia (Fig. 18.3). Dramatic sea-level fluctuations also are unlikely to have restrained continental ecosystems. Latest Ordovician glaciation was accompanied by changes of sea level, yet paleosols and microfossils are evidence for similar continental ecosystems both before and after glaciation.

In summary, conditions may not have allowed the establishment of large, competitive plants on land during Ordovician or Silurian time. The large fossil logs of *Prototaxites* more likely grew in shallows of the sea or lakes like modern kelp. Resources in known Late Ordovician and Silurian paleosols were adequate for both stress-tolerant and weedy early land plants. Life styles of the earliest land animals can be reconsidered in a similar way. The animals that formed burrows in Late Ordovician paleosols were herbivorous arthropods able to tolerate desiccation. Large carnivores of the kind that are found fossilized in lagoonal, estuarine and marsh sediments were probably not a part of soil ecosystems at that time. Evidence from paleosols supports the notion that terrestrial food chains were built from the ground up, from producers to consumers and from microbes to monsters.

Biological innovation or environmental regulation?

There are two distinct views of the early evolution of continental ecosystems. In the traditional view, the oceans were teeming with multicellular life, but conquest of the land was held at bay by hostile conditions there. When ultraviolet radiation, flood frequency, and moisture variation of soils were mitigated, then pre-adapted organisms invaded the land from lacustrine and marginal marine habitats. Once unleashed from environmental constraints, evolutionary processes filled this new adaptive zone. This view is similar in some ways to the idea that life itself evolved by chance and diversified only under the constraints of atmospheric and continental growth, a view that I have characterized as Ereban.

A contrasting view based on the Gaia hypothesis (Lovelock 1979) is that life arose as part of a self-sustaining system that continued to incorporate evolutionary innovations. In this Gaian view, evolutionary progress was not so much determined by environmental changes as by biological innovations that ultimately created new opportunities. These biological innovations may have been slow to appear because they were technically difficult. Even if they did appear, the environment that would select for them might not have been at hand. Land habitats suitably prepared by microbial communities for colonization by multicellular organisms may have been present in some places and suitably adapted multicellular plants in other places, long before the two happened together to allow exploitation of the land. In the Gaian view, plants and animals arose according to a schedule determined by the pace of coevolution rather than by the tempo of environmental change.

Although a Gaian or Ereban view of life is most critical in interpreting its earliest evolution, the two views are more effectively assessed from Phanerozoic evidence because that is where the rock and fossil record is least speculative. Perhaps the most conclusive test is the extent to which life controls various surficial cycles of volatile elements today, such as the carbon cycle (Mooney *et al.* 1987). In an extreme Gaian view, biotic control is nearly total and has been for some time. In an extreme Ereban view, it is minimal and whatever slight degree of control that can be demonstrated is of geologically recent vintage. There are a range of intermediate positions that seem more reasonable (Fig. 18.10). A case can be made that life exerted considerable control on atmospheric composition, temperature, and other aspects of the environment well back into the geological past (Nisbet 1987). It is doubtful if this was a total hegemony, however, because there is an observable Phanerozoic consolidation of control by land plants and animals as outlined in the following section.

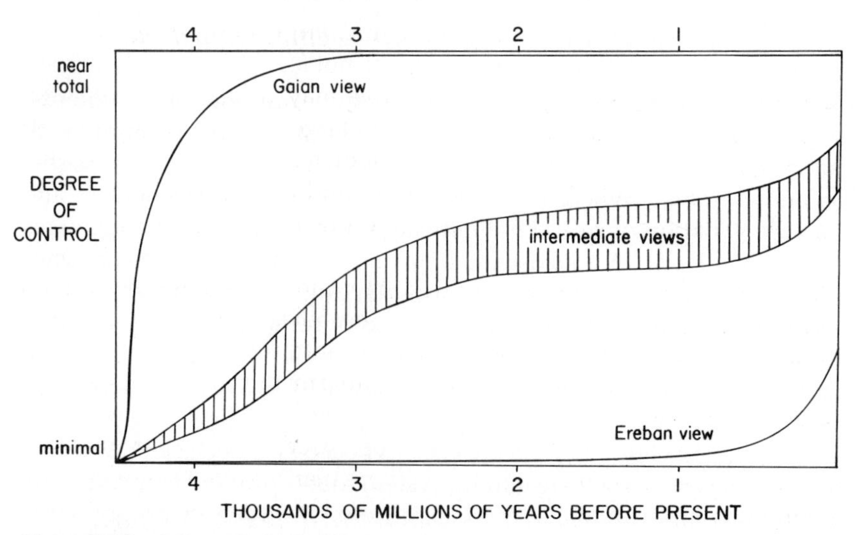

Figure 18.10 A theoretical consideration of extreme Gaian and Ereban views of the history of life on Earth over the past 4500 million years.

Putting down roots

Although the fossil record of soils remains obscure on exactly how the earliest continental ecosystems evolved, some effects of their evolution are apparent. By mid-Paleozoic time organisms had gained a considerable measure of control over continental environments. This can most clearly be seen from the thickness of coals and from alluvial architecture in Paleozoic non-marine rocks, as discussed below.

Blanketing bogs

The increased biomass of continental ecosystems is most obvious from waterlogged paleosols where a portion of that productivity was preserved as a new kind of soil, the peaty soils or Histosols. The oldest known Histosol is the Early Devonian (Siegenian) Rhynie Chert of Scotland (Kidston & Lang 1921). Only a handful of such permineralized peats of different geological ages are known (Knoll 1985a). More typically, ancient Histosols are not silicified. Coaly layers of large non-vascular plants are found in Silurian (Llandoverian to Ludlovian) rocks of the eastern USA (Willard 1938, Pratt *et al.* 1978, Strother & Traverse 1979). These do not appear to meet the organic matter content or thickness required of Histosols, but should be re-examined from this point of view. The most ancient and spectacular non-silicified Histosol is the 1.4-m thick Barzass Coal of Middle Devonian age (Eifelian) near Barzass village, Siberia (Stach *et al.*

1975). This is a cuticle coal of the enigmatic, large land plant *Orestovia devonica* (Krassilov 1981). Woody coals are not found until much later in the Devonian period. The earliest example may be a thin (10 cm after compaction) coal overlying claystone with large root traces in the Late Devonian (Fammenian) Hampshire Formation near Elkins, West Virginia, USA (Gillespie *et al* 1981). More abundant and thicker woody coals are found in rocks of mid-Carboniferous and younger geological age (Gardner *et al.* 1988).

A tangle of plant bodies in waterlogged terrain would have done much to stabilize them against floods, storms, and other perturbations. It also restricted the diffusion of atmospheric gases into the stagnant soil water below, thus promoting preservation of organic matter there. The burial of reduced carbon in organic matter results in the release of oxygen to the atmosphere that otherwise might have been used to fuel its decay. More oxygen is released in coastal swamps where marine sulfate is reduced to pyrite and other sulfides by microbes. This also prevents the accumulation of toxic sulfur gases in the atmosphere. Denitrification of nitrate (NO_3^-) to molecular nitrogen (N_2) is also a microbially mediated phenomenon of reducing swampy environments (Stevenson 1986). The engine of atmospheric oxygenation and nitrogenation that had been set in motion by Precambrian microbial communities thus received a considerable boost from the advent of stable and productive swamp ecosystems.

Taming streams

The general appearance of fluvial deposits has changed considerably over geological time, but more by the addition of new kinds of alluvial architecture than the abandonment of older ones. Land plants and the kinds of alluvial processes associated with them remained confined to some areas, whereas desert stream patterns persisted in others. Several different elements of fluvial deposits conspire to give the general impression that life was taking control of some parts of the landscape. There is an increased abundance of biogenic structures compared with purely sedimentary ones, an increase in the amount of clay compared with sand and an increased complexity of bedding both laterally and vertically.

The platy peds in Late Ordovician paleosols are a coarse kind of structure compared with those in geologically younger paleosols. Late Silurian and Devonian paleosols, in contrast, are more massive and hackly in appearance with abundant and diverse burrows, root and rhizome traces, calcareous and ferruginous nodules, and a variety of ped structures. The *Psammosteus* Limestone of the Early Devonian (Gedinnian) Old Red Sandstone in southwestern Britain (Allen 1986a) is a sequence of paleosols each so strongly bioturbated and thick that there is little sign of sedimentary structures.

A greater biomass of plants and animals also promoted clay formation in soils. In soils stabilized against erosion, primary minerals are unlikely to escape hydrolytic weathering. The amount of clay in fluvial deposits is related to many factors: the amount of shale or topographic relief in source regions, the rate of sediment accumulation, the amount of rainfall, and the stabilizing effect of vegetation. If it were possible to control these variables, then differences in clayeyness of fluvial sequences of different geological age might be apparent. This has been attempted in a crude way by comparing Late Ordovician, Late Silurian and Late Miocene sequences of paleosols, all of which formed in alluvial outwash of major mountain ranges of quartzofeldspathic composition in subhumid, seasonally dry paleoclimates (Fig. 18.11). The increased clayeyness through time evident from this comparison is compatible with the increased amount of total clay in the sedimentary rock record through geological time (see Fig. 16.2A).

Equally likely is a change in the nature of clays through geological time. Highly weathered clays such as kaolinite should be more abundant after the advent of land plants because they depleted nutrient cations more effectively than before (Knoll & James 1987). Increased abundance of smectite at the expense of illite might also be expected after the advent of vascular land plants, because they require more potassium than non-vascular plants. Both kaolinite and smectite are more common after Devonian time and rare before that (see Fig. 7.9). Potassium enrichment by illitization and sericitization of clays during burial (Curtis 1985) has done much to obscure this trend, although it still can be seen from the lack of weathering of potassium feldspars in Paleozoic sandstones (Basu 1981).

Increased complexity of bedding in alluvial sequences also accompanied the advent of land plants and animals. Early Paleozoic channels were laterally extensive, broadly symmetrical in cross section and scoured shallowly into underlying deposits (Cotter 1978), as are stream channel deposits still in desert regions. These braided streams are choked with more sand and gravel than they can effectively transport and this is deposited as islands in the channel. By mid-Paleozoic time many channels became laterally restricted and asymmetric, with a cut bank on one side scouring deeply into underlying deposits and with a sequence of interbedded sandstone and shale on the other side. As in channels of meandering streams in well vegetated, humid regions, the lateral spread of these streams is checked by stable, clayey, vegetated banks so that they carry mainly a suspended load of fine-grained materials. A number of loosely sinuous, intermediate stream patterns are known (Schumm 1981). These stream patterns are also dependent on a variety of factors other than vegetation. Meandering stream deposits are known even in Precambrian rocks ca. 2300 million years ago where they appear to have incised into prior deposited flat-lying shales (Button & Tyler 1981). In the examples of alluvial sequences chosen to keep various other environmental variables

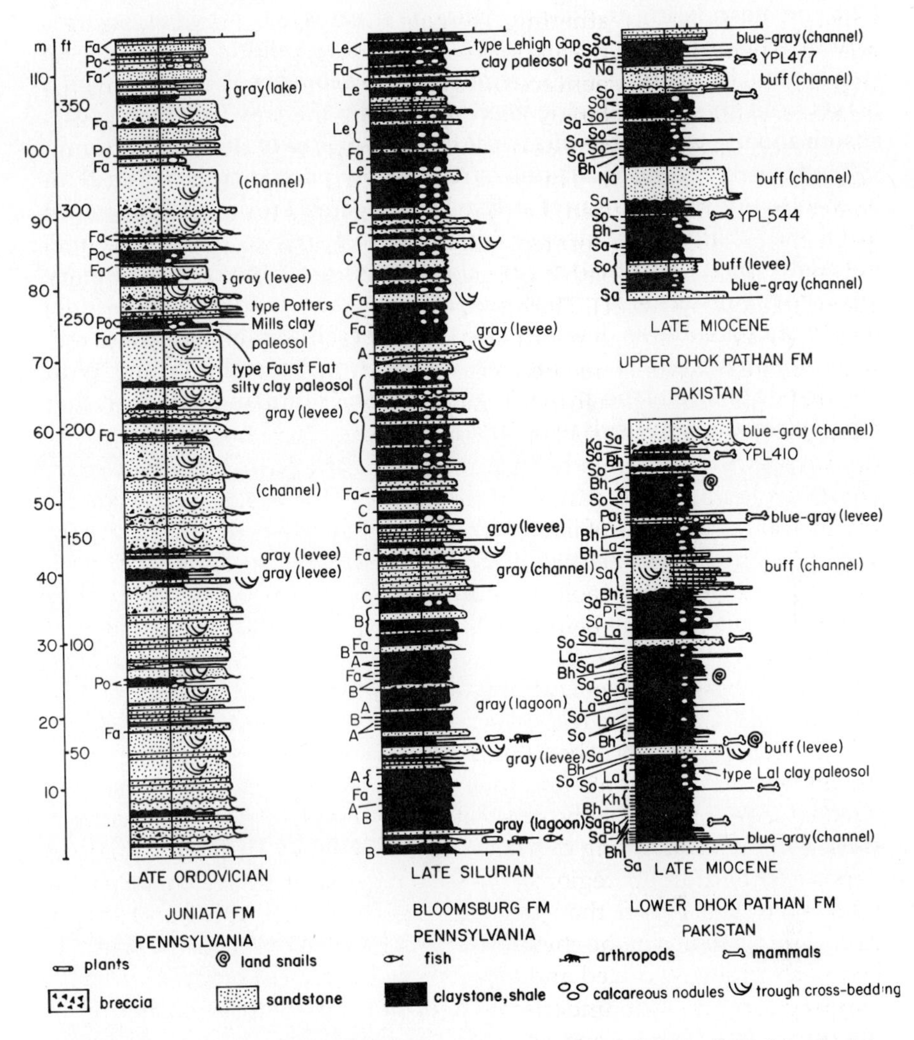

Figure 18.11 Columnar sections, drawn to the same scale, of the Late Ordovician Juniata Formation and Late Silurian Bloomsburg Formation in Pennsylvania and of the late Miocene upper and lower Dhok Pathan Formation in northern Pakistan. All three sequences formed in alluvial outwash of major mountain ranges in seasonally dry, subtropical climates. Individual paleosol series in order of stratigraphic appearance include Faust Flat (Fa), Potters Mills (Po), Lehigh Gap (Le), unnamed Silurian (A, B, C), Sarang (Sa), Bhura (Bh), Khakistari (Kh), Sonita (So), Lal (La), Pila (Pi), Pandu (Pa), Kala (Ka), and Naranji (Na).

constant (Fig. 18.11) there is a transition from Late Ordovician braided to loosely sinuous streams to Late Silurian loosely sinuous streams and to Miocene meandering and loosely sinuous streams.

Each of these aspects of alluvial architecture reflects increased soil stabilization by plants. Floods are not only contained by the way in which plants hold stream banks together but also by the way in which plants absorb and use rain water within the catchment area of streams (Schumm 1977). In well vegetated regions, streams flow permanently rather than intermittently. Stabilization of stream banks against erosion and increased water availability allow deeper weathering of soils to obtain essential nutrients such as phosphorus and potassium. Increased primary productivity results also in increased carbon fixation and oxygen production by photosynthesis. It is likely that much of this stabilizing system was set in motion within soils by microbes early during Precambrian time. With the advent of large land plants it gained momentum and direction that were increasingly difficult to reverse.

19 Afforestation of the land

The evolution of forests was in some ways a continuation of evolutionary processes set in motion with the advent of vascular land plants. Their appearance heralded marked advances in weathering and stability of soils and landscapes. With trees appeared the various kinds of soils with leached, near-surface (E), and clayey and ferruginized subsurface (Bt and Bs) horizons: Alfisols, Ultisols and Spodosols. This increased diversity of soils paralleled a diversification of life into the varied habitats found within forests.

The fossil record of trees extends back at least to Middle Devonian time (Givetian). Sandstone casts of stumps of that geological age in waterlogged paleosols near Gilboa, New York, USA, have been referred to *Eospermatopteris erianum*, a plant whose botanical affinities remain unclear (Gensel & Andrews 1984). Well preserved secondary wood in fossil trunks referred to *Callixyon* has been found almost this old (Late Givetian to Frasnian). With characteristic paleobotanical caution against the possibility of wrongfully assigning parts of different plants to the same species, the fossil leaves (*Archaeopteris*) and spores (*Geminospora*) now thought to belong to these trees were initially given different botanical names. These particular fossil plant parts have been found attached to one another so it is well established that they belonged to the same kind of ancient trees (Fig. 19.1). This combination was surprising botanically. *Archaeopteris* belongs to an extinct group of plants, the progymnosperms, that had woody anatomy similar to that found in modern conifers, but reproduced by means of spores as do modern ferns. Their wood is composed of tracheids, which form the central conducting strand of all vascular land plants. In trees, however, the living tracheids and the zone of cells that give rise to them (cambium) form a ring near the outside of the trunk (Fig. 19.1). Within this ring are the hollow xylem that form the wood of the tree. These were successive zones of tracheids that lived as the tree grew in girth. In trees that grow in seasonal climates, these concentric zones are growth rings of xylem, varying in size according to whether they grew in summer or fall. In a sense, then, trees grow by maintaining a thin layer of living tissue around a skeleton. Dead xylem tissues still play a role as conducting tubes

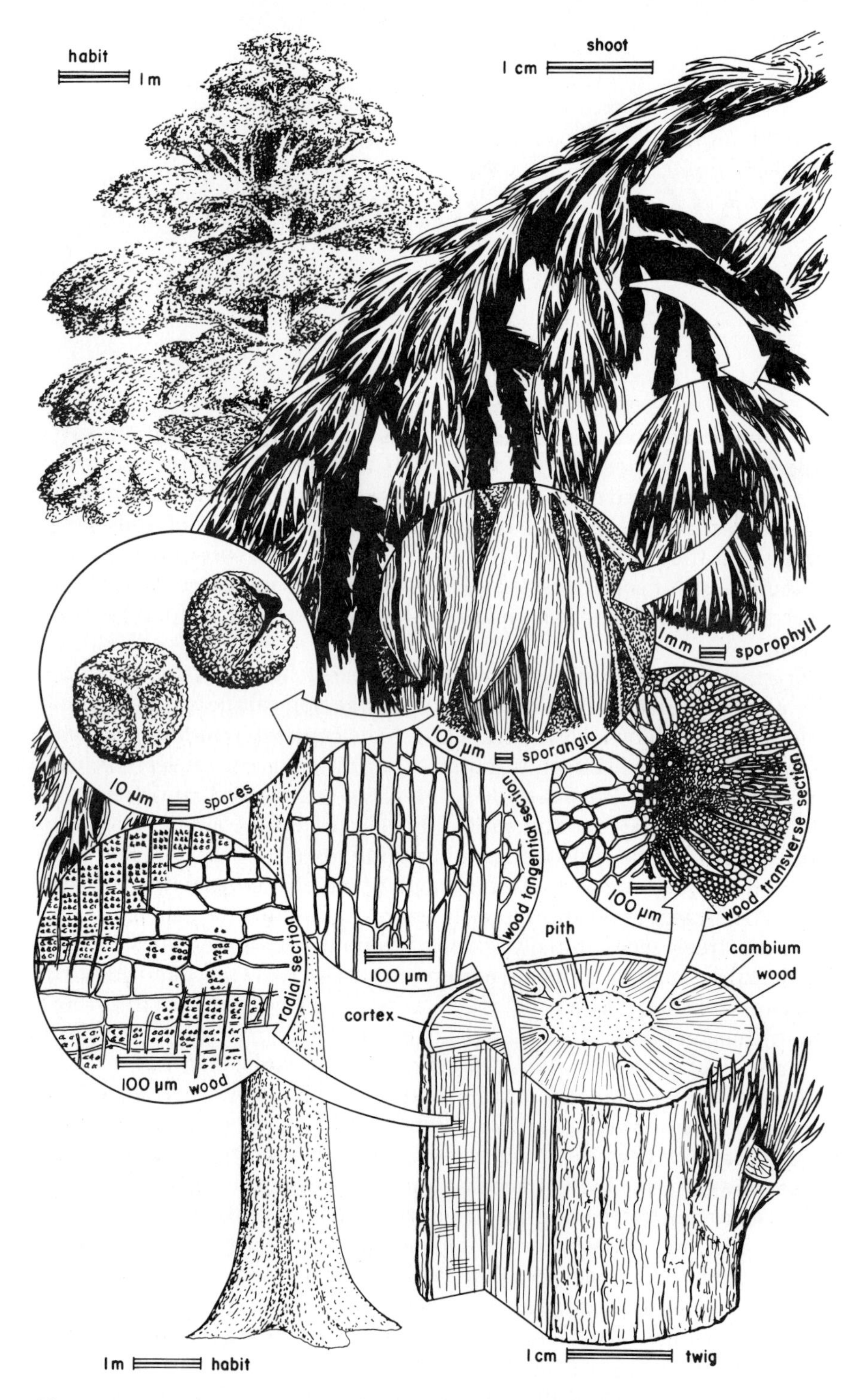

Figure 19.1 A reconstruction of *Archaeopteris macilenta* (leaves and cones), with its wood (*Callixylon zalesskyi*) and spores (*Geminospora*) from the Late Devonian (Frasnian), lower Walton Formation, near Sidney, New York, USA (data from Beck 1960, 1981, Allen 1980).

for moisture drawn up the trunk by the water potential created by transpiration. The wood can sustain significant damage without killing the tree, whereas cutting the bark and underlying living cambium all around the tree is fatal.

The spores of progymnosperms, like those of ferns and most primitive land plants, probably produced small, delicate reproductive plantlets (gametophytes) on which the reproductive gametes met in thin films of water. Gametophytes require a moist place or season for effective reproduction. For this reason ferns are most abundant in habitats at least periodically moist.

Trees with other kinds of trunk construction and reproductive capabilities also appeared in the fossil record during the Devonian period (Gensel & Andrews 1984). Extinct forms of horsetails (*Pseudobornea ursina*) and of lycopods (*Protolepidendropsis pulchra*) formed large trees as early as Late Devonian (Frasnian). Their internal anatomy has not been studied in great detail, but well known tree horsetails and lycopods of Carboniferous age had a cylinder of secondary wood with soft-walled tissue (parenchyma) scattered among the wood cells (xylem). Both their roots and shoots grew in a pre-programmed way to a determinate structure (Eggert 1961, 1962). Other weak-stemmed early trees such as the Middle Devonian (Givetian) cladoxyl *Pseudosporochnus nodosus* had isolated strips of radially arranged woody tissue in the trunk rather than a solid cylinder of wood (Leclerq & Banks 1962). Trees such as *Austroclepsis australis* as ancient as mid-Carboniferous had stems that were weak and small. Most of the trunk was made of leaf bases and narrow adventitious roots that anchored the plant in the soil (Sahni 1932; Morgan 1959). These various kinds of early trees were spore-bearing plants that reproduced by means of water-dependent gametophytes. The earliest seed plants, in which the gametophyte was protected from desiccation by enclosure in seed coats, date back to latest Devonian time (Fammenian; Gillespie *et al.* 1981). It is likely that the earliest fossil seeds were borne on shrubs, but seed ferns such as *Pitus primaeva* were large trees by earliest Carboniferous time (Tournaisian; Retallack & Dilcher 1988).

Measurements of the diameters of fossil logs ranging in age from Silurian through Devonian show a steady increase in girth from the matchstick-sized Late Silurian *Cooksonia* to trees of *Callixylon* 1.6 m in diameter (Chaloner & Sheerin 1979). Such a gradual attainment of large stature has encouraged a number of explanations for the evolution of trees (Beerbower 1985). They may have grown tall in order to shade out neighboring plants because light is a plant's chief source of energy. It also could have been to discourage competing plants by showering them with mild poisons. Flavonoids are used in this way by many modern trees and the ability to synthesize these substances is widespread among vascular plants, bryophytes, and even some aquatic charophyte algae (Chapman 1985).

Trees also could have evolved to deter herbivory and this would have been especially effective before the evolution of winged insects. The ability to digest woody tissue has evolved in very few groups of animals, such as termites. Large trunks and branches scatter the nutritious young buds and leaves in space, so that animals may expend more energy finding them than they gain by eating them. Another possible factor in the evolution of trees was the greater scatter of spores attainable from greater heights. By this argument, the taller trees would have prevailed by outreproducing shorter ones. These are all Gaian views for the evolution of trees. Alternative more Ereban views relate the evolution of trees to environmental difficulties. It could be that shortages of nutrients in the soils of unvegetated watersheds stunted the growth of the earliest land plants, or it could be that the frequency of disturbance by floods, windstorms, or landslides was such that few plants could grow for long enough to attain the stature of trees. These are other questions, in addition to the simpler one of when forest ecosystems appeared, that can be tested against the fossil record of soils.

Early forest soils

Fossil stumps in growth position are an obvious, but all too rare, indication of ancient forests because they require rapid burial in a waterlogged habitat. More abundant evidence of early forests comes from surficial disruptions of paleosols due to former windthrows, from the presence of charcoal in paleosols, from deeply penetrating large root traces, and from the differentiation of leached, near-surface (E) and enriched, clayey, or ferruginized subsurface (Bt and Bs) horizons. The evolution of forests appears to have been a geologically gradual affair (Chaloner & Sheerin 1979) and because of the sheer abundance of Devonian paleosols known (Allen 1986a) there is promise that the early evolution of forests and their soils will be revealed in some detail from the study of paleosols.

Inceptisols

As would be expected from the known Late Devonian fossil trees of New York and Pennsylvania, USA, there are abundant large fossil root traces in alluvial deposits of that age there (Barrell 1913). The degree of development of many of these paleosols is not strong, however. Many appear to have been Inceptisols. One of these weakly developed paleosols with large fossil root traces near the town of Hancock, New York (Fig. 19.2) has been studied in some detail (Retallack 1985). This paleosol has relict bedding throughout. This is especially obvious near the surface of the profile which may have accumulated from floodwaters restrained by

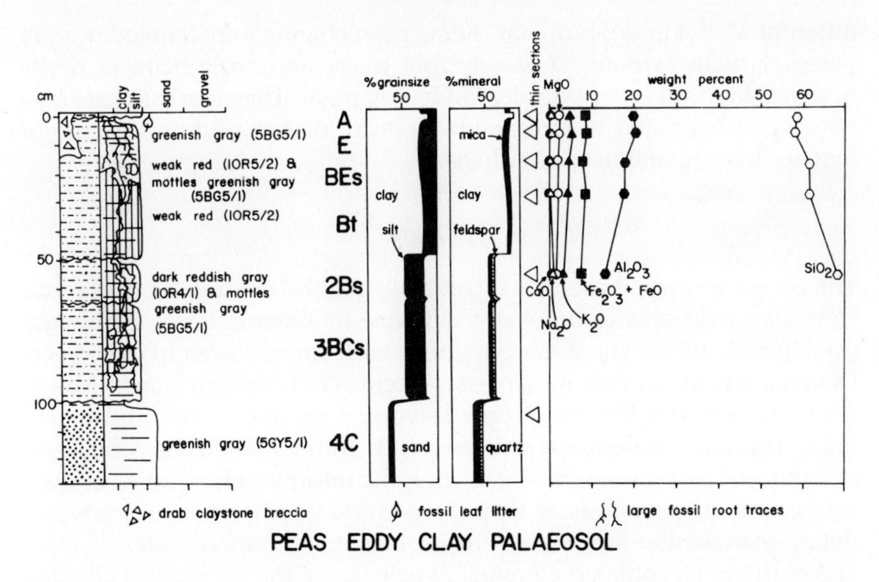

Figure 19.2 Columnar section (measured in field) and petrographic and chemical composition (from point counting and ICP chemical analysis) of the Peas Eddy clay paleosol, a Tropaquept in the Late Devonian (Frasnian) Walton Formation, near Hancock, New York, USA. Lithological symbols as for Figure 18.5 (from Retallack 1985, reprinted with permission of the Royal Society of London).

vegetation as the soil continued to form. The large root traces and a fossil leaf litter of partly decomposed remains of *Archaeopteris halliana* within the cumulative surface horizon provide evidence that the paleosol was forested. Other evidence for forest cover is the differentiation of a laterally continuous, iron-poor, gray–green surface (A) horizon over a purple subsurface (Bs) horizon slightly richer in iron. Translocation of iron to subsurface horizons can be mechanical because of channels offered by decaying roots and open burrows. Some of the fossil root traces and burrows in the paleosol are filled with iron-stained clay washed down in this way, but the larger ones are filled with drab-colored sand rather than clay. Physical mixing played a minor role in the formation of this subsurface (Bs) horizon because relict bedding has persisted throughout the profile. A more likely mechanism is the translocation of iron by the chemical action of phenolic and other substances leached from leaves by rain or from the decaying leaf litter. Although the differentiation of this paleosol is slight, it shows initial effects of the soil-forming process of podzolization.

This and other Late Devonian paleosols in New York with large root traces are non-calcareous and closely associated with sandstones of former stream channels. In contrast, associated red, nodular, calcareous Ochrept paleosols away from paleochannels lack both large root traces and well differentiated A and B horizons. Some paleochannels are flanked by gray paleosols with carbonized woody root traces and an horizon of pyrite nodules, like Sulfaquepts under modern mangal. These early forests may have been restricted to streamside galleries dissecting low shrubby or herbaceous vegetation of interfluves.

Alfisols

The oldest of these base-rich forest soils may be among the calcareous Devonian paleosols of the Aztec Siltstone in Victoria Land, Antarctica (McPherson 1979). The Aztec Siltstone was formerly thought to be latest Devonian (Fammenian), but a reassessment of its fossil fish faunas (Young 1982) suggested a Middle to Late Devonian age (late Givetian to early Frasnian). These paleosols are impressively differentiated chemically and structurally, but quantitative petrographic information, studies of their root traces, and analysis of their deep-cracking patterns are needed to determine whether they were Alfisols, Inceptisols, or Vertisols.

A younger Devonian (Fammenian) paleosol in the uppermost Old Red Sandstone on the rock platform north of Pease Bay near Cockburnspath, Scotland, can more confidently be identified as an Alfisol, probably a Calcic Haploxeralf (Figs. 19.3, 19.4). Abundant large roots and burrows

Figure 19.3 The Pease Bay loamy sand paleosol, a Calcic Haploxeralf in the Late Devonian (Fammenian) upper Old Red Sandstone near Cocksburnpath, southwest Scotland. The sledgehammer is 28 cm long.

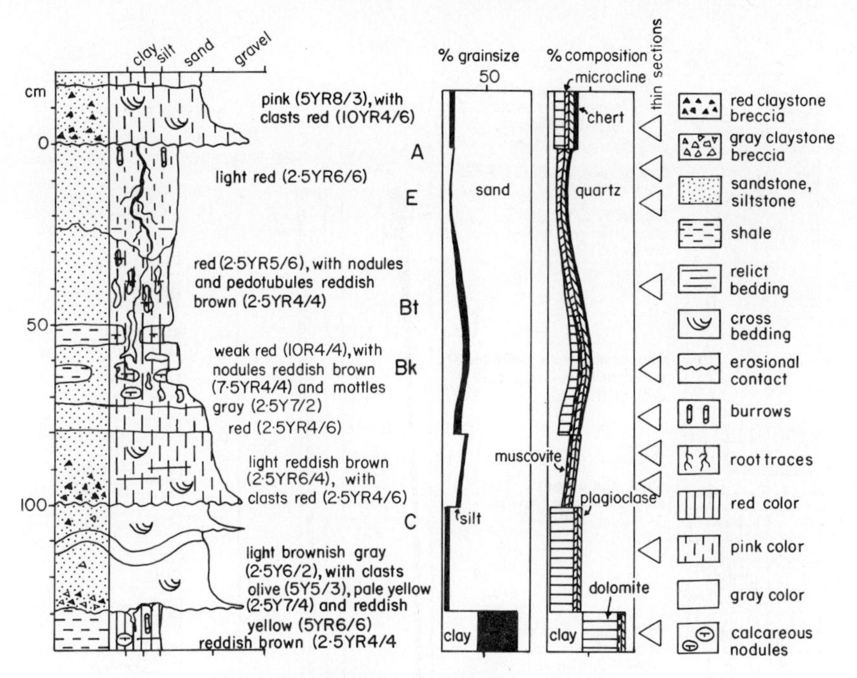

Figure 19.4 Columnar section (measured in field) and petrographic composition (from point counting by G. S. Smith) of the Pease Bay loamy sand paleosol in the upper Old Red Sandstone near Cockburnspath, Scotland.

encrusted locally with hematite penetrate deeply from the top of the profile. Its near surface horizon is sandy and light-colored (E) compared with the redder and more clayey subsurface (Bt) horizon. This clayey horizon is not entirely pedogenic, but appears to be a disrupted shale bed within the parent material. Nevertheless, it shows a wispy bright clay microstructure (skelmosepic plasmic fabric), and the overlying sandy layer is sufficiently enriched in what appears to be pedogenic clay to be regarded as an argillic horizon. A high base status of the soil is indicated by abundant dolomitized caliche nodules.

Fossil Alfisols are now known from all the succeeding geological periods up to the present (Retallack 1986c), but few of them have yet been adequately characterized.

Ultisols

It is not known for certain how far back in the geological record there are Ultisols, the base-poor clayey forested soils. There may be some fossil examples as old as Carboniferous (Retallack 1986c), but these have not yet

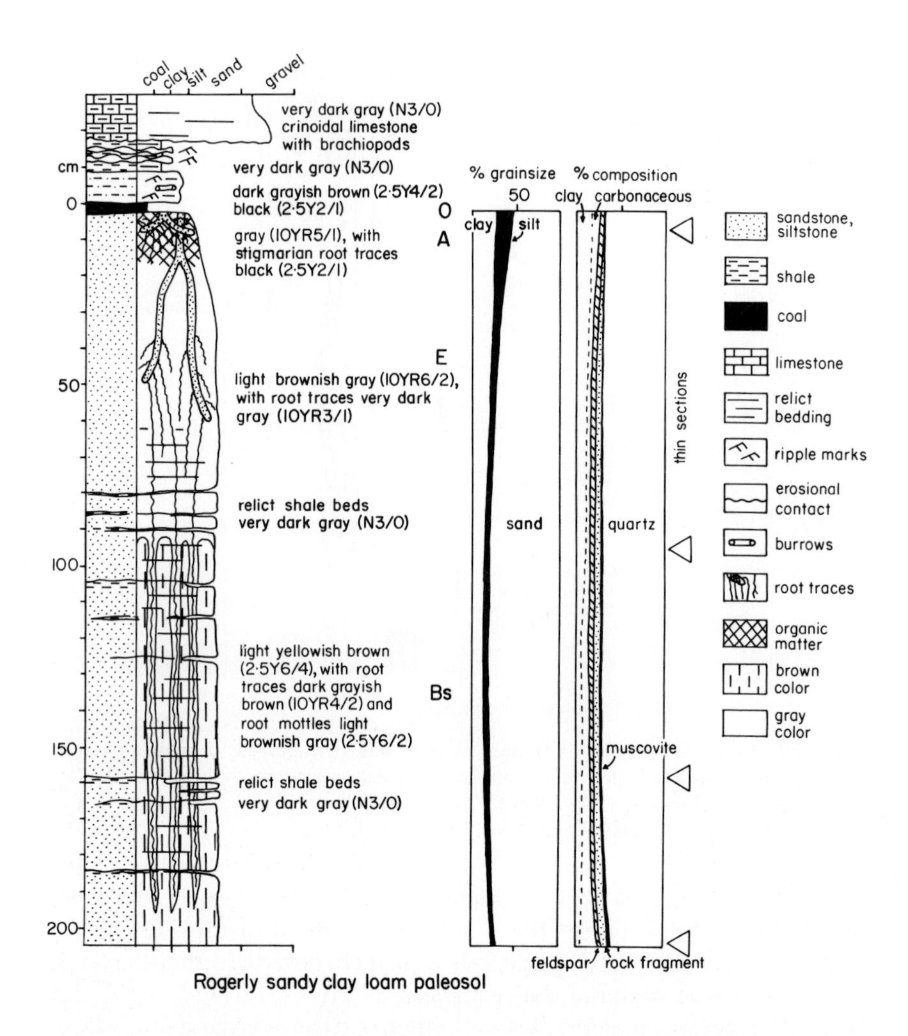

Figure 19.5 Columnar section (measured in field) and petrographic composition (from point counting by C. J. Percival) of the Rogerly sandy clay loam paleosol, a Dystropept on the mid-Carboniferous (Namurian), uppermost Firestone Sill, near Stanhope, England.

been reported in detail. Early Triassic "paleoplanosols" of southern France (Lucas 1976) also may have been Ultisols. The oldest Ultisol actually identified as such is the Yellow Mounds silty clay loam paleosol developed between the late Eocene Slim Buttes Formation and the latest Eocene to earliest Oligocene Chadron Formation in Badlands National Park, South Dakota, USA (Retallack 1983a, b). Many Ultisols now at the land surface also are near major unconformities and could have begun forming as long ago as late Tertiary (Cady & Daniels 1968).

Spodosols

These acidic soils form either on sandy parent materials or by destruction of clays in loamy parent material through the action of acidic phenolic or other compounds produced by conifers or heath plants in humid climates. The beginnings of this process can be seen in some Late Devonian Inceptisols (Fig. 19.2).

Another early Inceptisol that almost meets the requirements of a Spodosol is the Rogerly sandy loam paleosol (Fig. 19.5) near Stanhope, Britain (Percival 1986). This profile is developed on sandy levee deposits of the Firestone Sill of mid-Carboniferous age (Namurian). It is a thick profile (more than 2 m) with a surface coaly layer (O horizon) and thick underlying sandy layer riddled with stigmarian lycopod rootlets (A horizon). A near-surface (E) bleached horizon overlies a brownish sandstone (Bs) with relict bedding and with traces of opaque sesquioxides around the original sand grains, now surrounded with quartz overgrowths (Fig. 19.6). The degree of encrustation and staining in this horizon is insufficient to meet the criteria of a spodic horizon (Soil Survey Staff 1975). Although better regarded as a Dystropept, it is similar in many ways to deep podzolic soils found in tropical regions under dwarf forest (Richards 1952). Had it formed over a longer period of time it may have

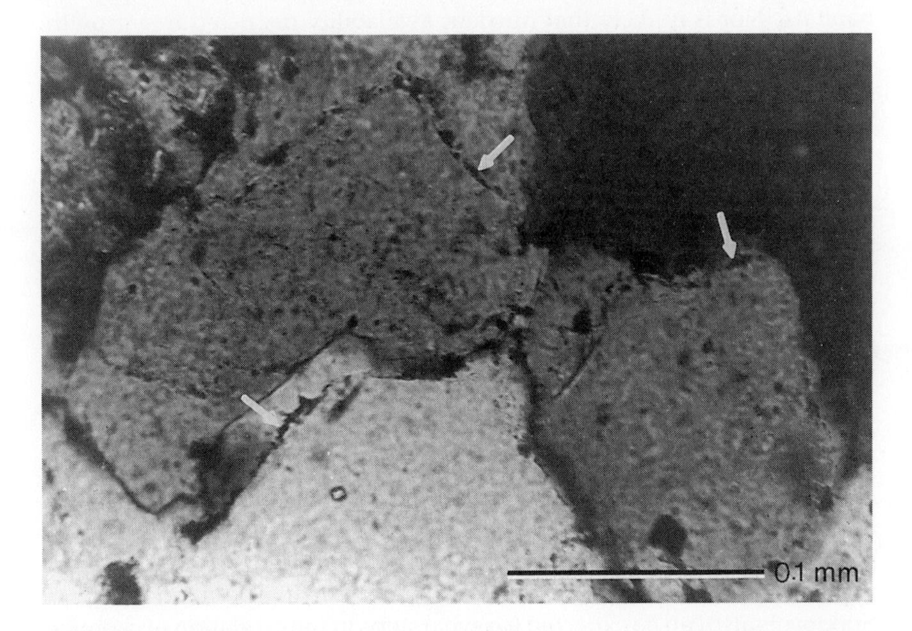

Figure 19.6 Petrographic thin section under crossed nicols of incipient spodic horizon (iron staining of grains indicated by white arrows within later diagenetic quartz overgrowth), from the subsurface (Bs) horizon of the Rogerly sandy clay loam paleosol on the mid-Carboniferous Firestone Sill, near Stanhope, England (C. J. Percival thin section Rh35). Scale bar 0.1 mm.

become a Spodosol and others may yet be discovered in rocks of comparable age.

So far, however, no Spodosols have been found among other reported Carboniferous paleosols (Percival 1986). There are also problems with regarding Early Triassic podzolized heath paleosols as Spodosols (Retallack 1986c). The oldest well documented Spodosol is between the Late Eocene (Bartonian) Sables de Beauchamps and Formation d'Ezanville near Paris, France (Pomerol 1964). The profile near Ermenonville was an Orthod.

Origin of forest ecosystems

Considering the currently inadequate data base of described paleosols, it is difficult to test hypotheses about the early evolution of forests effectively. At face value, the present paltry record of early forested paleosols suggests that they were at first weakly podzolized Inceptisols of streamsides. Forests in fertile bottomland Alfisols came later and were followed by forests of low-fertility Ultisols and Spodosols.

The idea that forests evolved only where little disturbed by flooding is not supported by available evidence from paleosols, because streamsides of early forested landscapes were very flood-prone compared with other habitats. Nor is it likely that nutrient availability restricted tree growth, because none of the early paleosols under forest or other vascular plants were especially deeply weathered. Another hypothesis not supported by existing evidence is that trees grew tall in order to scatter their spores more widely than other plants. The earliest forested paleosols currently known do not appear to have been widespread. Along streamsides the fluvial transport of spores and gametophytes would have rendered tall stature unnecessary. The idea that trees evolved to avoid herbivory is supported by indications of podzolization, because phenolic substances produced by the trees would also have suppressed herbivores. Plant competition could also have played a role because early forested soils had surface horizons of even thickness, an indication of a closed canopy.

If it is true that forests formed calcareous Alfisols later than non-calcareous Inceptisols, it may have been due to biological innovations such as the evolution of seeds (Chaloner & Sheerin 1981) permitting the colonization of drier soils by large plants. If Ultisols and Spodosols appeared geologically later, then it could be because increased complexity of forest ecosystems enabled recycling of nutrients from decaying vegetation in soils in which nutrients were scarce. The fossil record of Spodosols also can be expected to reveal steps in the escalation of chemical warfare waged by plants against herbivores culminating in copious production of phenolic compounds in modern plant formations such as heath and conifer forest.

None of these questions can be answered satisfactorily at present. They are introduced as a perspective for the further examination of mid-Paleozoic paleosols.

A diversifying landscape

With the addition of the main orders of forested soils during mid-Paleozoic time, most types of soils now known on Earth were already present. Of the ten orders of the US Soil Taxonomy (Soil Survey Staff 1975), only the grassland soils (Mollisols) had not yet appeared. All the older orders of soils persisted along with early forests and their soils as outlined in the following sections.

Entisols

These very weakly developed soils are so broadly defined that they include early stages of development of all kinds of soils. They probably formed even during Precambrian times, but are difficult to detect in Precambrian rocks because the degree of weathering is slight and there were no burrowing animals or rooted plants to leave obvious traces. This was no longer the case on subtropical alluvial flats by Late Ordovician time (Ashgillian) when the Faust Flat silty clay paleosol (a Fluvent) formed (see Figs. 18.5, 18.7). Comparable Entisols in alluvial sequences with fossil burrows, root traces, and conspicuous relict bedding are common in sedimentary rocks of every younger geological period (Retallack 1986c).

Fossil Entisols also record the invasion of woody plants into other habitats such as seashores, deserts, periglacial regions, and rocky uplands. The oldest marine-influenced paleosol now known is in the Early Silurian (Llandoverian) Frenchville Formation near Stockholm, Maine (Schopf *et al.* 1966). This contains remains of small plants of uncertain botanical affinity that formed a vegetation somewhat resembling a salt marsh. Larger plants comparable to modern mangroves had invaded near-marine habitats by Devonian time. There is evidence for this in weakly developed paleosols containing marine fossils and large tree stumps of Middle Devonian (Givetian) age in New York (Johnson 1972). Other mangal Entisols have been described from the Early Carboniferous (Tournaisian) Calciferous Sandstone near Foulden, Scotland (Retallack & Dilcher 1988) and from the mid-Cretaceous (Cenomanian) Dakota Formation near Russell, Kansas and Fairbury, Nebraska (Retallack & Dilcher 1981a,b). Saline coastal soils were another habitat colonized and stabilized by large woody plants.

Sandy Entisols with root traces have been found in eolian deposits of the Pennsylvanian (Desmoinesian) Hermosa Formation in Utah, and of the

Permian (Wolfcampian) Cedar Mesa Sandstone in Utah and Fountain Formation in Wyoming (Loope 1988). The size distribution of calcareous root traces in these paleosols are indications of desert shrublands and scrub.

The oldest periglacial paleosols (Cryorthents and Cryopsamments) dating back to Precambrian time, ca. 2300 million years ago, are implied by reports of ice wedge casts in the Ramsay Lake Formation north of Espanola, Ontario, Canada (Young & Long 1976). Other examples of patterned ground have been found associated with glacial deposits of Precambrian (Williams 1986) and Late Ordovician age (Daily & Cooper 1976, Biju-Duval *et al.* 1981).

In Late Carboniferous to Early Permian glacial deposits of the Seaham Formation, near Lochinvar, New South Wales, Australia, there are carbonaceous root traces at numerous horizons. These can be regarded as periglacial Entisols. They are associated with the fossil progymnosperm or seed fern *Botrychiopsis*. These low-diversity fossil plant assemblages may have formed a vegetation similar to that of modern polar tundra (Retallack 1980).

In the Early Permian basal Shoalhaven Group near Lue and at Kanangra Wall, both in New South Wales, there are fossil Entisols with large woody root traces and an associated flora characterized by the glossopterid seed fern *Gangamopteris* (Retallack 1980). This periglacial vegetation may have been analogous to modern taiga of Arctic Canada and Siberia. These fossil soils and floras record the colonization of periglacial habitats by large vascular plants as early as Permian.

Weakly developed, rocky upland Entisols (Orthents) had certainly begun to be colonized by latest Triassic time. Fissures of karst topography of that age on Early Carboniferous limestones near Bridgend, south Wales, are filled with cave earth and charcoalified remains of a fireprone shrubland that was in a sense a forerunner of modern chaparral (Harris 1957).

Histosols

The advent of trees had important consequences for swamp and marsh soils. Fossil Histosols can be traced at least back to Early Devonian (Siegenian) time, but the oldest woody coals are of latest Devonian age (Fammenian; Gillespie *et al.* 1981). With the advent of trees in wetlands, much thicker peats accumulated. Most of the world's economically mineable coal is of Carboniferous and Permian age (Stach *et al.* 1975). Paleosols associated with coals have been given a variety of names: underclay, seat earth, fireclay, tonstein, and ganister. Most of these represent the mineral portions of Histosols, but some are more properly identified as gleyed representatives of other soil orders (Gardner *et al.* 1988).

The present fossil record of Histosols indicates a later (Fammenian) afforestation of peaty substrates than of waterlogged mineral soils (Givetian). If this holds up to further scrutiny it would add weight to the inference from the fossil record of Spodosols that trees were slow to adapt to acidic, nutrient-poor habitats. A number of special features allow trees to persist in swamps. Hollow roots, for example, allow gas exchange in stagnant groundwater and appeared in lycopods such as *Lepidosigillaria whitei* during middle Devonian time (latest Givetian; Scheckler 1986). Such adaptations to waterlogged conditions tend to distinguish swamp vegetation from that of surrounding, better drained soils. Fossil plants of Carboniferous coal measures are long-ranging, morphologically conservative species compared with the rapidly evolving fossil plants of shales and sandstones that represent vegetation of beach ridges, levees, and other better-drained land (Dimichele *et al.* 1987). Thus, swamps appear to have been colonized late by trees and to have remained evolutionary as well as literal backwaters.

Not all early Histosols were as nutrient poor as most modern swamps. A distinctive kind of fossil Histosol found principally in Late Carboniferous coal measures of the mid-continental United States and in mid-Carboniferous coal measures of Europe and Ukraine contains numerous calcareous or dolomitic nodules. These so-called "coal balls" are original features of the soils, useful for estimating compaction of coal, as already discussed. The woody vegetation of these alkaline wetlands was thus a carr, rather than swamp.

Calcareous nodules are extremely rare in modern peats, which normally are too acidic for calcium carbonate to be a stable mineral. A modern calcareous peat from Eight Mile Swamp in southeastern South Australia is an exception (Stephens 1943). In this soil, lime is produced by charophyte algae and aquatic snails in a coastal swamp buffered by springs fed with runoff from regionally extensive limestone bedrock in a summer-dry (Mediterranean) subhumid climate. Such factors may also have played a role in the formation of Carboniferous coal balls. They are found in paleoclimates intermediate between humid and semi-arid (Rowley *et al.* 1985). Coal balls also have been found in areas where there is earlier Carboniferous limestone and in coastal though not necessarily marine-influenced swamps (Scott & Rex 1985). Lycopods and horsetails of Carboniferous swamps may have been less acidifying than gymnosperms that subsequently invaded swamp habitats, and this may have restricted the formation of coal balls in rocks younger than Carboniferous. Comparable escalating effects of plant leachates on weathering through geological time also can be examined from the fossil record of Spodosols (Retallack 1986c) and of siderite nodules in paleosols (Retallack 1976).

Vertisols

Swelling clay soils are known to be at least as old as 2200 million years from an example near Waterval Onder, South Africa (see Fig. 9.9). Additional well preserved examples are Siluro-Devonian (Allen 1986a), Late Devonian (McPherson 1979), Jurassic (Goldbery 1982a, b), and Oligocene in age (McBride *et al.* 1968, Galloway 1978). The Siluro-Devonian examples show complex fracture patterns resembling lentil peds (Krishna & Perumal 1948) and the mukkara structure of modern Vertisols (Paton 1974). This complex structure contrasts with the simple blocky structure of the Waterval Onder clay. These Paleozoic Vertisols appear to have had a more unstable, smectitic clay than the illitic clays presumed to have dominated the Precambrian profile (Retallack 1986b). These differences in visible effects of physical properties of clay can be related to a number of factors such as the amount of rainfall, initial minerals available for weathering, and the greater demand for potash by vascular land plants compared to Precambrian microbial earths.

Aridisols

Alkaline, calcareous desert soils have a fairly continuous record from Precambrian time (ca. 1900 million years ago) to the present (Retallack 1986c). In Britain, fossil Aridisols are well known from Siluro-Devonian (Allen 1986a), Early Carboniferous (Wright 1982), Permo-Triassic (Steel 1974) and Late Jurassic rocks (Francis 1986). All of these paleosols contain burrows and root traces. The Late Jurassic paleosols include large stumps of trees (Fig. 1.2). The oldest large root traces in such paleosols now recorded are in the Early Carboniferous (Visean) Heatherslade Geosol near Cardiff, south Wales (Wright 1986b, 1987). If these are the most ancient woody root traces in Aridisols, then it would seem that trees colonized deserts in addition to other difficult sites, such as Spodosols and Histosols, later than better watered and nutrient-rich sites.

Some very ancient Aridisols are distinctive in that their carbonate is dolomitic and forms veins and pisolites (Campbell & Cecile 1981, Kalliokoski 1975) rather than nodular masses and layers of low-magnesian calcite most abundant in modern Aridisols. Some of this dolomite in paleosols is probably of diagenetic origin (Zenger *et al.* 1980). In other cases a diagenetic origin of dolomite is not evident (Retallack 1985). Dolomite is formed in few modern aridland soils, especially those of very high-base status (Dixon & Weed 1977). There may have been long-term changes in the mineralogy of soil carbonate in paleosols comparable to long-term changes in oceanic carbonates (Given & Wilkinson 1987).

Oxisols and duricrusts

These very strongly developed soils and indurated products of deep weathering extend well back into Precambrian time (Fig. 14.1), but may have been promoted by the stabilizing effect of land plants and of forests. Duricrusts occur below major geological unconformities of all ages, but the most impressive are associated with Devonian and younger unconformities. Phanerozoic lateritic crusts are commonly underlain by intensely leached masses of china clay (see Fig. 6.7). Such lateritic profiles are widespread on stable, low relief continents such as Africa and Australia (McFarlane 1976). The largest reserves of bauxites are along early Tertiary erosional landscapes in Guyana, Surinam, Guinea, Ivory Coast, Ghana, northern Queensland, and the Northern Territory. Some bauxites also are formed in closed depressions of karst topography (Bardossy 1982). Phanerozoic examples of this kind of bauxite are also much thicker and more deeply weathered than the few Precambrian examples known. Similarly, silcrete of the 1850-million-year-old basement beneath the Pitz Formation of the Canadian Northwest Territories (Ross & Chiarenzelli 1984) forms only thin veins, whereas thick ridges of silcrete are associated with latest Cretaceous and early Tertiary unconformities in central Australia (see Fig. 6.7). Deep continental weathering has become increasingly more profound since Precambrian time.

A near-modern world

Considering the three new orders of soils that appeared, together with notable changes in other kinds of soils and weathering products, the advent of forests can be seen as a time of accelerated diversification of soils on Earth. By Permian time vascular plants had established themselves in alpine to mangal, polar to tropical, and rainforest to desert environments. These late Paleozoic woody plants were mainly of kinds now extinct. Some features of Paleozoic soils, such as the calcareous nodules in Histosols, also are nearly extinct. Nevertheless, the world and its soils were becoming much more like those with which we are now familiar.

A finer web of life on land

The evolution of a terrestrial environment structured by trees profoundly affected numerous other creatures on land. The physical complexity of forests has been matched by increasingly complex interrelationships between forest organisms (Briand & Cohen 1987). Although forest ecosystems are complex, they have proven robust on geological time scales (Knoll 1986).

Forests were sources of evolutionary novelty, harboring the origin of many groups of plants and animals important to modern ecosystems such as gymnosperms, angiosperms, insects, amphibians, dinosaurs, mammals, and primates. Not all of these groups have left clear traces in forested paleosols, but their evolutionary history can be assessed against the paleoenvironmental information offered by paleosols.

Soil invertebrates

A general diversification of invertebrate trace fossils within paleosols through geological time is already apparent from the few available publications on this topic. A low-diversity assemblage of burrows and fecal pellets has been found in a thin paleosol (Darrenfelen Geosol) in the early Carboniferous (Visean) Cheltenham Limestone near Llanelly, south Wales, Britain (Wright 1983, 1987). Abundant burrows like those of cicadas (see Fig. 7.2) and earthworms (see Fig. 10.5) have been recognized in paleosols in the Early Triassic (late Scythian) Newport Formation near Sydney, Australia (Retallack 1976). Nine kinds of trace fossils, attributed to activity of crustaceans, earthworms, and insects, have been found in paleosols of the Early Eocene (Wasatchian) Willwood Formation of Wyoming, USA (Bown & Kraus 1983). Five kinds of trace fossils, including some attributed to dung beetles and sweat bees, were found in paleosols in the Oligocene Scenic Member of the Brule Formation in Badlands National Park, South Dakota, USA (Retallack 1983b, 1984b). Beautifully preserved nests of termites and 15 other kinds of animal burrows were found in paleosols of the Oligocene Jebel Qatrani Formation in the Fayum Depression of Egypt (Bown 1982). Whether this diversification of soil invertebrates of forest ecosystems has been steady or was achieved rapidly during late Paleozoic times is difficult to judge from present information.

Early land vertebrates

The oldest fossil amphibians are latest Devonian (Fammenian) and include *Ichthyostega* from East Greenland. Older Devonian (Frasnian) trackways have been found in eastern Victoria, Australia and Parana, Brazil (Bray 1985). A traditional view of these early amphibians, which emphasizes their similarities to lungfish, is that their limbs helped them overland to more permanent water during the dry seasons. A problem with this view is that modern lungfish burrow and aestivate rather than migrate during the dry season, as did ancient lungfish (Dubiel *et al.* 1987). The physiology and habitats of early ancestors of amphibians have promoted an alternative view that marine fish came out on land to avoid predation or to tap new food resources within the shelter of mangroves, like modern mudskippers (*Periopthalmus gracilis*, among others) of northeastern Australia (Nursall 1981). A seasonal or dry climate is not required by the second hypothesis but is essential to the first. Climatically sensitive features of paleosols may provide a test of these competing hypotheses. Paleosol features have been reported from each of these early amphibian-bearing sequences, and not all of them include calcareous nodules typical of seasonally dry climates.

Some of the earliest reptiles, such as the captorhinomorphs *Hylonomus lyelli* and *Archerpeton anthracos* and the pelycosaur *Protoclepsydrops haplous*, are found in paleosols in the Joggins Formation of Pennsylvanian age in the sea cliffs near Joggins, Nova Scotia (Carroll *et al.* 1972). They are preserved with remains of amphibians, millipedes, spiders, whip spiders, eurypterids, and land snails in sandstone casts of tree stumps. The hollows left by rotting of the stumps were traps for a variety of animals. The paleosols themselves have not been studied, but appear to have been weakly developed and waterlogged. The associated fossil plants are characteristic of Carboniferous swampy lowlands. Wet forests rich in insects and other invertebrates can be viewed as prerequisites for the conquest of land by carnivores such as the earliest amphibians and reptiles.

Dinosaurs

Because of their great size, dinosaurs should have had an effect on soil formation. It has been suggested, for example, that tracts of woodland may have been opened to herbaceous plants by migrating herds of dinosaurs, and that their piles of dung would have required an efficient fauna of decomposers (Bakker 1985). So far there has been no record of erosional scours, wooded grassland vegetation, or burrows of dung beetles in Mesozoic rocks, but this may only reflect a lack of attention to paleosols associated with dinosaurs. Paleosols associated with dinosaurs that I have seen include those of the Late Triassic (Carnian to Norian) Chinle Formation of Petrified Forest National Park, Arizona (Long & Padian 1986), the Late Jurassic to Early Cretaceous (mainly Tithonian) Morrison Formation in Dinosaur National Monument, Colorado (Dodson *et al.* 1980), the Early Cretaceous (Hauterivian to Barremian) Wealden Beds of the Isle of Wight, England (Alvin *et al.* 1981), the Early Cretaceous (Aptian to Albian) Otway Group of Victoria, Australia (Douglas & Williams 1982), the Late Cretaceous (Campanian) Judith River Formation in Dinosaur Provincial Park, Alberta (Dodson 1971), the Late Cretaceous (latest Campanian) Two Medicine Formation of the Landslide Butte area near Del Bonito, Montana (Horner 1984), and the Late Cretaceous (late Maastrichtian) Hell Creek Formation in Bug Creek near Fort Peck, Montana (Retallack *et al.* 1987).

Paleosols of the Wealden and Otway Groups are non-calcareous and probably formed under humid forests where bone is rare in paleosols but preserved in stream paleochannels. At the other extreme are paleosols of the Morrison and Two Medicine Formations with abundant caliche nodules formed in a subhumid, but not desertic, climate. Paleosols in both formations have abundant drab-haloed root traces and drab-surface horizons typical of paleosols formed under vegetation no sparser or smaller in stature than open woodland. Thus, dinosaurs appear to have lived in a variety of wooded ecosystems, as do southeast Asian elephants (*Elephas maximus* var. *sumatrensis*) and rhinoceroses (*Rhinoceros sondiacus* and *Didermoceras sumatrensis*) today. There is no evidence yet from paleosols that dinosaurs lived in desert or wooded grassland like some populations of African elephant (*Loxodonta africana*) and rhinoceros (*Ceratotherium simum*). This is not a preservational bias because desert and wooded grassland soils preserve bone well (Retallack 1984a). It may yet turn out, however, to be a study bias.

The rise to dominance and extinction of dinosaurs were both intriguingly abrupt events. Dinosaurs were a minor part of terrestrial faunas during Triassic time. At a level very close to the Triassic–Jurassic boundary most of the other large land animals such as rauisuchid thecodonts and procolophonid cotylosaurs disappeared. Dinosaurs diversified to take their place (Olsen *et al.* 1987). After flourishing as the dominant animals on land for 110 million years, dinosaurs themselves became extinct at the Cretaceous–Tertiary boundary (Russell 1979). Their role was taken by small mammals that hitherto had been only a minor part of terrestrial ecosystems. Both of these seemingly abrupt faunal overturns could profitably be re-examined with evidence from fossil soils. Some preliminary studies have already been made of the paleosols across the Cretaceous–Tertiary boundary (Retallack *et al.* 1987). This research is compatible with the idea that dinosaurs perished catastrophically during a major impact of a comet or asteroid with the Earth (Alvarez 1986).

Angiosperms

In contrast to dinosaurs, angiosperms rose gradually to dominance and are still prominent in most land vegetation. If angiosperms existed at all during mid-Triassic time, they were rare and geographically restricted plants. A number of possible early angiosperm fossils have been described over the years (Stewart 1983). Especially notable is a small plant of Late Triassic age (Carnian) which now (Cornet 1986) is known from superficially palm-like leaves (*Sanmiguelia lewisii*), ovuliferous (*Axelrodia burgeri*) and pollen organs (*Synangispadixis tidwellii*). This plant may be an ancestral gnetalean and evolutionarily close to early angiosperms. These well preserved fossils were found with a leaf litter of ferns rooted within a Fluvent in the swale of a meandering stream. Thus it was an early successional, streamside plant.

It is not until Early Cretaceous (Barremian) that fossil angiosperm flowers, fruits, pollen, and leaves are found in force and can be traced as a continuous fossil record up to the present. The earliest fossil remains in North America show a predilection for disturbed streamside and coastal environments (Retallack & Dilcher 1986). The earliest fossil angiosperms of the Early Cretaceous (Barremian) Potomac Group near Washington, DC, are associated with very weakly developed shaly and sandy Entisols of estuary margins, with pink clayey Inceptisols of levees and with Histosols of swamps. No fossil plants have been found in red, clayey Alfisols of well drained floodplains or in widespread, deeply weathered Ultisols or Oxisols developed on Paleozoic basement rocks throughout this region. Judging from the regional pollen rain into lake and river deposits which is dominated by conifer pollen, it is unlikely that angiosperms were present in these inland soils.

By mid-Cretaceous time (Cenomanian), angiosperm fossils were more common in many parts of the world, such as the Dakota Formation of central Kansas and Nebraska (Fig. 19.7). Beautifully preserved flowers, pollen, fruits, and leaves of angiosperms dominate leaf litters of moderately developed, drab, clayey, estuary margin Entisols and coastal swamp Histosols. Also dominated by fossil angiosperms are weakly developed, sandy paleosols (Psamments) of river levees. Curiously, fossil conifers have not been found within leaf litters of these fossil soils although their pollen, shoots, and cones are conspicuous in deposits of associated lakes and seas. Probably conifers dominated the vegetation of red, clayey Ultisols on better drained parts of the floodplain where fossil plants were not preserved (Retallack & Dilcher 1981a, b). By mid-Cretaceous time, then, angiosperms already dominated mangal, swamp, and levee vegetation, but had made limited gains in well drained forests.

This fossil record of soils associated with early angiosperms can be used to evaluate hypotheses concerning their evolutionary origins. A widely held view of the origin of angiosperms is that they were trees of dry tropical uplands with large insect-pollinated *Magnolia*-like flowers and fleshy, animal-dispersed fruits. Their rise to dominance according to this concept is explained by Cretaceous migrations of these plants into lowland sedimentary environments along with interdependent animal pollinators and dispersers. Evidence from paleosols and early angiosperm fossils does not support this view because the earliest angiosperms were lowland, weedy plants. Such plants today are small and produce numerous small (about 15–25 μm in diameter) pollen and (1 mm or less) seeds in order to be able to colonize widely scattered sites. Fossil leaves, fruits and pollen of the earliest Cretaceous angiosperms now known in various parts of the world are of this kind. They also are low in diversity and similar in regions as remote from one another as North America, Australia, and central Asiatic USSR. It is as if a small group of weedy plants were dispersing especially widely in coastal regions at this time (Truswell 1987).

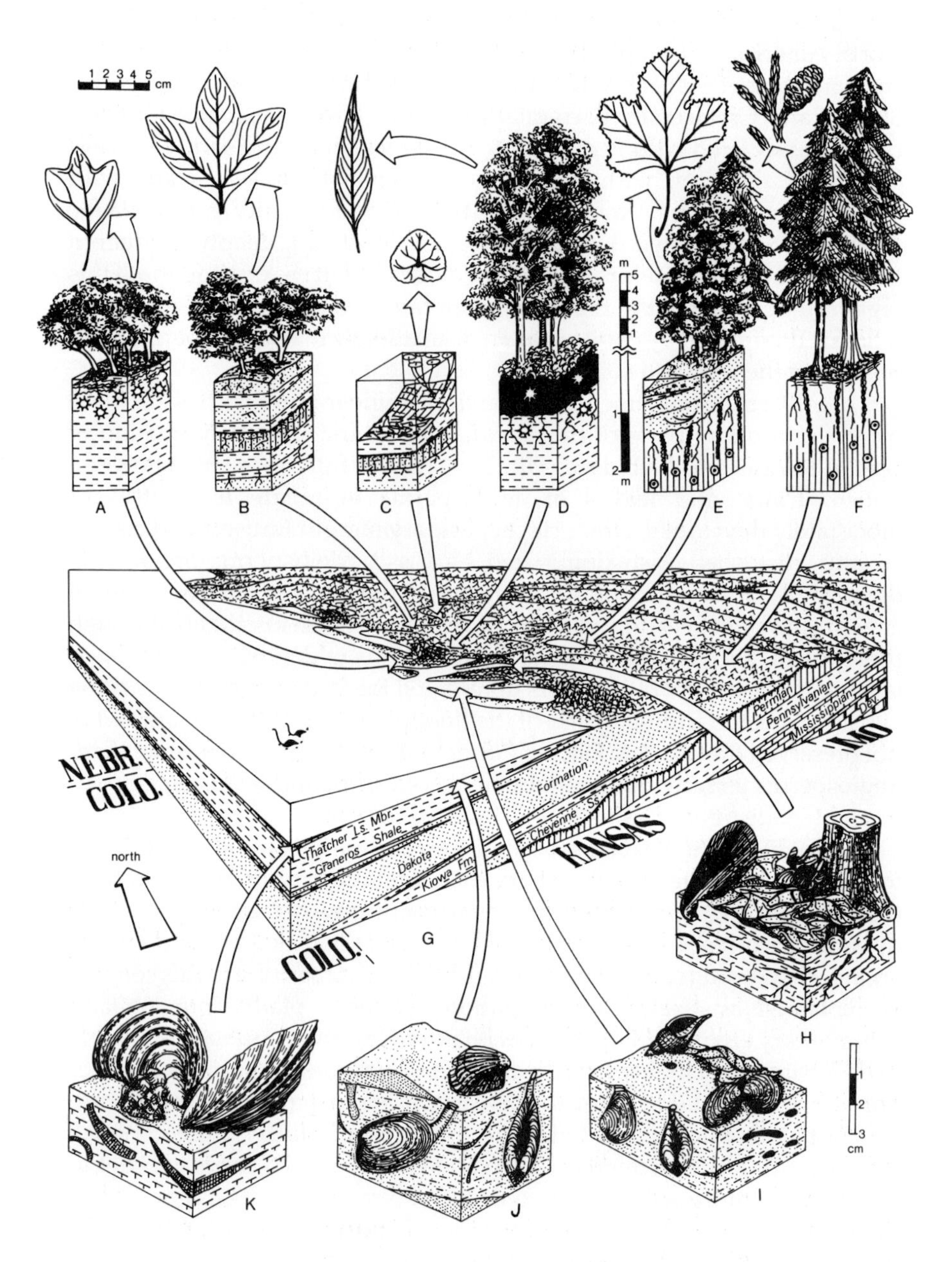

Figure 19.7 A reconstruction of soils and vegetation during mid-Cretaceous time in central Kansas, USA (data from Retallack & Dilcher 1981a, b).

Why did they migrate during Cretaceous time if allied Late Triassic plants were also streamside weeds? Wide dispersal and speciation of angiosperms may have been encouraged by fluctuations of the sea level in shallow continental seaways characteristic of mid-Cretaceous time. Such major paleogeographic changes would have caused widespread disturbance advantageous to pioneering plants. The slow steady gains of angiosperms and the equally gradual decline and extinction of pre-existing plants of disturbed habitats are indications that biological innovations also played a role. For weedy plants there would have been great advantage in an abbreviated life cycle in which pollination and fertilization followed one another in rapid succession as in modern angiosperms. In contrast, all modern gymnosperms and apparently also the most reproductively sophisticated of ancient seed ferns (Retallack & Dilcher 1988) were pollinated before the gametophyte was fully differentiated. Thus, fertilization is delayed. In pines, fertilization occurs many months after pollination. Near-synchronous pollination and fertilization followed by coordinated development of seeds and enclosing structures would have enabled weedy angiosperms to set seed faster than gymnospermous plants that were equally adapted in other ways to early successional habitats. Angiosperms remain the premier early successional plants on land, as can be seen from the rapid appearance of grasses and plantains in vacant lots.

Since mid-Cretaceous time angiosperms have diversified to include magnolias, orchids, and apples. In these plants, co-evolution with animal pollinators and dispersers has proceeded to extraordinary lengths. This subsequent evolution of angiosperms also can be addressed through the study of paleosols. The most important pollinators of angiosperms are bees, and most primitive groups of bees make nests in the ground. Fossil bees are rare in amber of Late Cretaceous age (Campanian) in New Jersey (Michener & Grimaldi, 1988). By Eocene time fossil bee nests, fossil bees in amber and lake deposits, and fossil flowers with the asymmetric shape characteristic of bee-pollinated blossoms are common (Crepet & Taylor 1985). Fossil nests in paleosols are a record of bees in well drained habitats where the bees themselves would not be preserved (Fig. 10.9). The evolutionary radiation of bees and their nests coincides with the dramatic evolutionary radiation of angiosperms during mid-Tertiary time (Retallack, 1984b).

The shape of evolution

The mid-Paleozoic advent of forests and the variety of soils that they formed was in some ways similar to major adaptive radiations of organisms, such as the diversification of vascular land plants. Soils, plant formations, and organisms all increased in diversity without extensive replacement of earlier kinds of soils or plants (Fig. 19.8). Ancient kinds of microbial ecosystems and ancient kinds of soils persisted in other parts of the landscape along with the new forested ecosystems and their soils. Forest ecosystems can be considered an adaptive breakthrough which presented new opportunities for both organisms and soil-forming processes. The shape of evolution of both soils and organisms in such times of innovation is one of diversification. Such spurts of innovation punctuate a longer term trend of diversification of life (Valentine 1986).

The increased diversity and biomass of forests more firmly established changes initiated under early land plants. Greater biomass and root penetration more effectively held the landscape against erosion. Afforestation of watersheds further slowed percolation of rain water into soils and its flow overland. Forests promoted meandering rather than braided streams and clayey rather than sandy soils (Schumm 1977). Not only was more oxygen photosynthetically produced by forests, but there was also

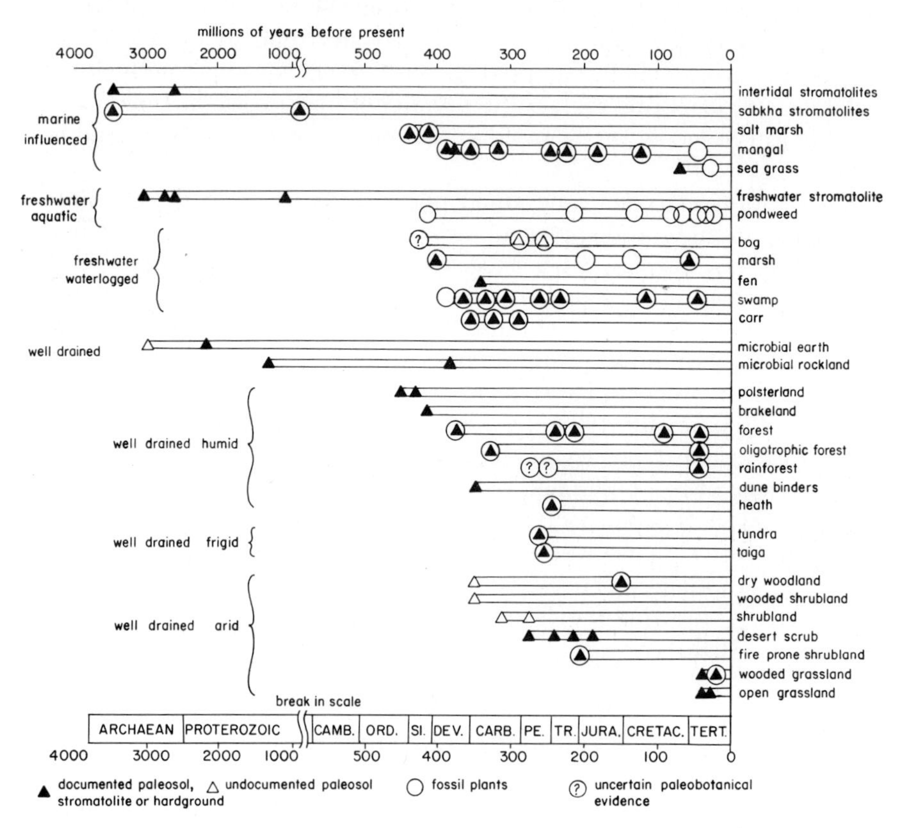

Figure 19.8 Geological ranges of plant formations, based on paleosols and associated fossils.

a greater potential for burying carbon in wooded swamps so that the atmosphere became more oxygenated as a result. The physical stability of Late Paleozoic woodland vegetation resulted in burial of carbon on an unprecedented scale in the form of coal seams that now fuel industrial economies.

With trees also appeared a new mechanism for burying carbon, i.e. the production of charcoal by wildfire. Fossil charcoal is known in the rock record at least as far back as Early Carboniferous (Tournaisian) in the Berea Sandstone near Amherst, Ohio (Cope & Chaloner 1985). Charcoal forms when the oxygen available is inadequate for complete burning, usually because it cannot diffuse into burning wood sufficiently fast. For wood to burn at all requires at least 13 vol. % of oxygen in the atmosphere (about 0.6 times the present level). If oxygen were present at such low levels, then wildfires would be expected to produce even more charcoal for a given volume of wood than they do at present levels of oxygenation. At higher than present levels of oxygenation, on the other hand, forests would burn more readily and completely, therefore diluting atmospheric oxygen with carbon dioxide. Charcoal is abundant in Late Paleozoic coals and red beds and would have been an additional way of oxygenating and regulating atmospheric composition by burying carbon in both well drained and swampy environments.

20 Grasses in dry continental interiors

Before the advent of grasses, dry continental interiors would have been vegetated by other kinds of plants. What these plants were like is difficult to say because well drained soils of desert regions preserve plant material only under exceptional circumstances (Retallack 1984a), such as the urine-impregnated middens of packrats (*Neotoma* spp.). There are few lakes and streams in deserts where plant fossils could be preserved and oases are surrounded by a lush growth of local plants completely different from those found in the desert beyond. Hence, the fossil record of plants in dry regions is both meager and biased. Some indications of ancient aridland ecosystems are provided by fossil root traces (Loope 1988), bones (Olson 1985), and burrows (Olson & Bolles 1975, Smith 1987) in calcareous red paleosols. Large plants and animals play a conspicuous role in desert ecosystems, but productivity is limited and there is much bare earth exposed. Aridisols forming in such environments do not appear greatly different from calcareous paleosols as old as 1900 million years (Campbell & Cecile 1981). For many millions of years after the mid-Paleozoic advent of trees, forests may have graded out through dry woodland into scrubby desert vegetation in dry continental interiors.

The mid-Tertiary advent of grasslands in subhumid to semi-arid regions signalled a new kind of ecosystem on Earth and new kinds of soils, Mollisols. Grasslands did not displace woodlands from humid regions or desert vegetation from very dry regions, but were interpolated between the two older kinds of plant formation. Grasslands are fast-cycling ecosystems of high productivity on short time scales. In woodland soils, nutrients are leached by rain water and bound up in standing wood, leaving little for large animals. In desert soils, nutrients are redistributed into duricrusts or remain unexploited in little weathered minerals because of inadequate soil water. Thus, they are denied to both plants and animals. Grasses, however, are fast-growing plants that are entirely edible both for large animals above ground and for soil invertebrates within. Grassland soils have dark, well structured surface horizons (mollic epipedons) that are rich in nutrients in a readily available form. As a result, they support a great diversity of mammals and other large vertebrates (Bell 1982). Rather than being communities intermediate between woodland and desert scrub in stature, density, and productivity, grasslands were a new and distinctive addition to the array of Earth's surface environments.

Grasses are unique among plants in the many adaptations that enable them to withstand grazing (Fig. 20.1). Some grasses grow most rapidly when grazed (McNaughton 1983), as I suspected as a teenager given the chore of mowing our suburban lawn. In addition to unfolding from a zone of dividing cells at the tip (apical meristem) of the stem, grass stems also elongate from zones of dividing cells above the nodes (intercalary meristems) of the plant. The uppermost parts can be eaten without totally

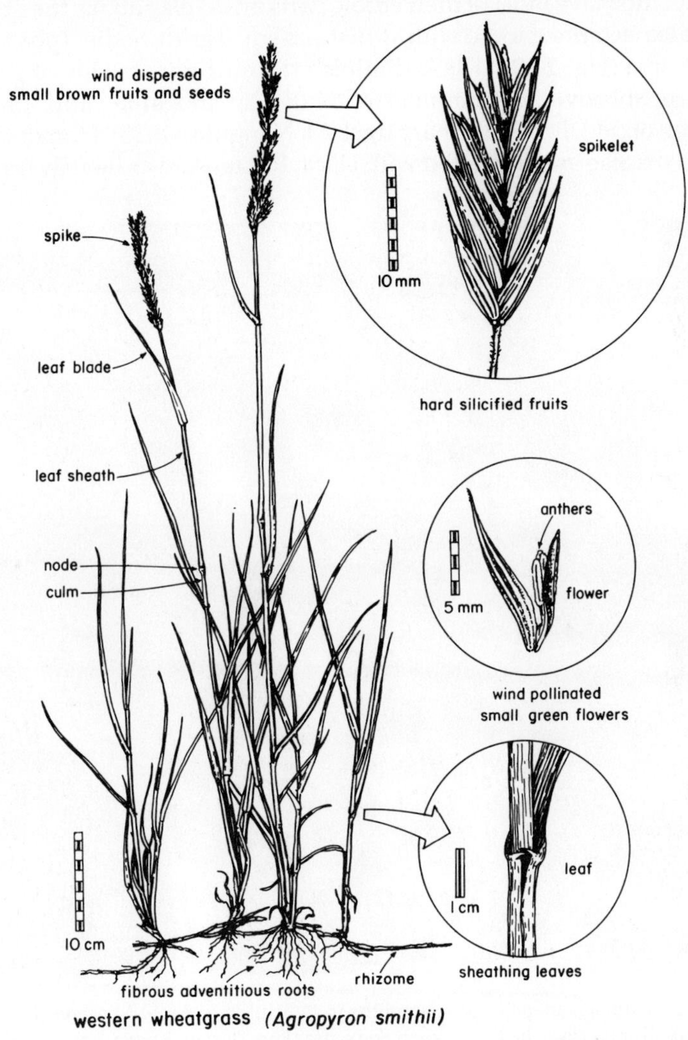

Figure 20.1 Adaptive features of grasses, as shown by western wheat grass *Agropyron smithii*.

interrupting growth. The apical meristem of many grass shoots is also protected from damage because it is hidden down within the sheathing bases of the leaves. In most plants, by contrast, the growing point is at the tip of the shoot and leaves arise and grow well behind it. By this arrangement the growth of the shoot is terminated if the tip is eaten.

Grasses also have a dense growth close to the ground. It is a modular kind of architecture spread by repetition of the same simple units of runners or rhizomes and their clumps of fibrous roots and linear leaves. Much of the plant's rhizomes and growing leaf bases are buried within the ground where they are protected from trampling and grazing. In contrast, woody plants have most of their edible parts on display above the ground.

Grass leaves are armored with a crust and small gritty bodies (phytoliths) of plant opal (Fig. 20.2). These diminish the nutritional value of grasses. They are abrasive to mammalian teeth, carcinogenic, and promote formation of calculi in the urinary tract (McNaughton 1985). Horsetails and some dicots also are encrusted with silica, but few are as heavily armored

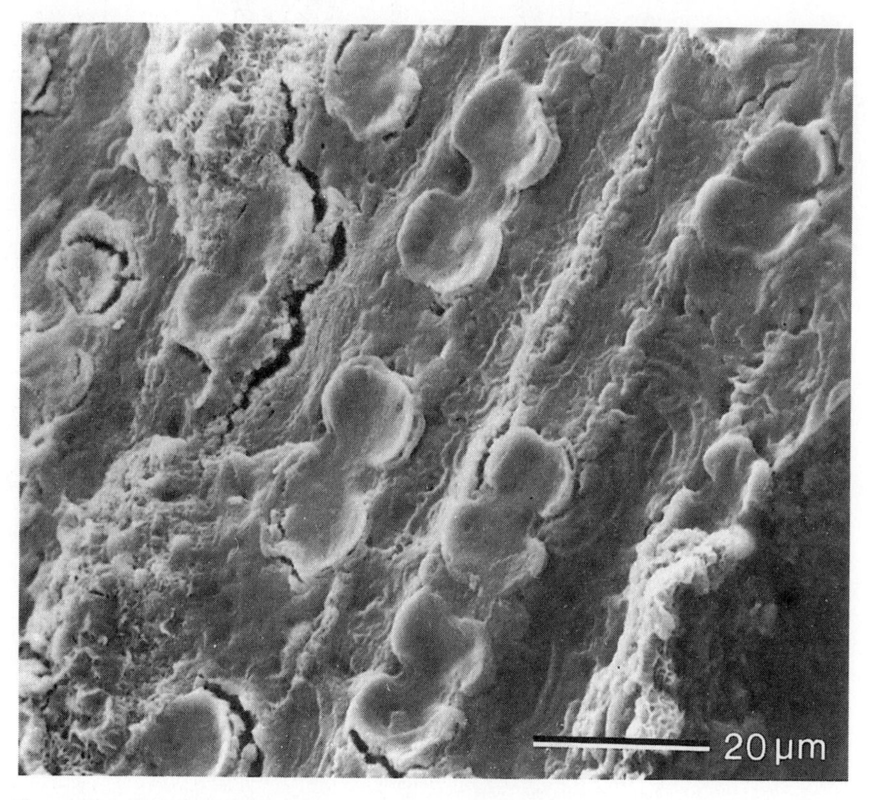

Figure 20.2 Dumbell-shaped opal phytoliths in the cuticle of a mid-Miocene (14-million-year-old) fossil grass from the Fort Ternan Beds, near Fort Ternan, Kenya. Scale bar is 20 μm (Kenyan National Museum specimen FT 13120).

as grasses. The siliceous skeletons strengthen bamboo and enable dead winter grass to remain standing long after associated herbaceous plants have withered and decayed.

If, despite all these defenses, grasses succumb to grazing and trampling, they have reproductive features that enable them to colonize scattered patches of bare ground. The plant body of grasses is small compared with other plants and their flowers are inconspicuous, small, and green or brown. They produce multitudes of small, smooth pollen grains and seeds that are dispersed by wind. This is very different from forest trees such as *Magnolia*, with their large, colorful flowers pollinated by beetles and their fleshy fruits dispersed by animals. It also is different from the showy bat-pollinated flowers of North American cacti. In contrast to these competitive and tolerant plants, grasses are weedy in their emphasis on reproduction independent of specific pollinating or dispersing animals (Grime 1979).

These features of grasses can also be found in their meager fossil record. The oldest fossils of grasses are simple monoporate pollen grains which can be securely traced as far back as Paleocene (Muller 1981). Less well documented are pollen grains in Cretaceous rocks. Parallel-veined fragments of leaves that could have belonged to sedges, rushes, or grasses have also been found as old as mid-Cretaceous (Cenomanian). It is not until Eocene time that there were surely grasses and sedges with rhizomes, sheathing leaf bases, basal node meristems, and the other distinctive features of modern grasses. Such fossils from the lignites of the Claiborne Formation of Tennessee, USA (Daghlian 1981) and from the upper Geiseltal Schichten southwest of Halle in East Germany (Krumbiegel *et al.* 1983) represent plants that lived as understory herbs in swamp woodland. The evolution of grasses in dry habitats is recorded by phytoliths in paleosols as old as Eocene in the Sarmiento Group of Argentina (Spalletti & Mazzoni 1978). Highly siliceous fossil grass leaves and fruits are found rarely in the Oligocene White River Group of the central Great Plains of North America (Galbreath 1974), but are much more diverse and abundant in Miocene and Pliocene rocks of this region (Thomasson 1979). The diversification of these fossils has been claimed to represent the spread of grassland vegetation. It could equally be interpreted as the acquisition of silica bodies as a defense against grazing (Thomasson 1985) within several evolutionary lineages of grasses that were already widespread.

The evolution of grasslands has also left an imprint in fossil and modern mammals (Bakker 1983). Adaptations to grassland among mammals are especially striking because they are similar in evolutionary lineages as independent as horses and antelopes. Grassland mammals have distinctive features of tooth and limb structure that set them apart from forest mammals and from their woodland ancestors in the fossil record. Their teeth have complexly infolded hard enamel and soft dentine that form low

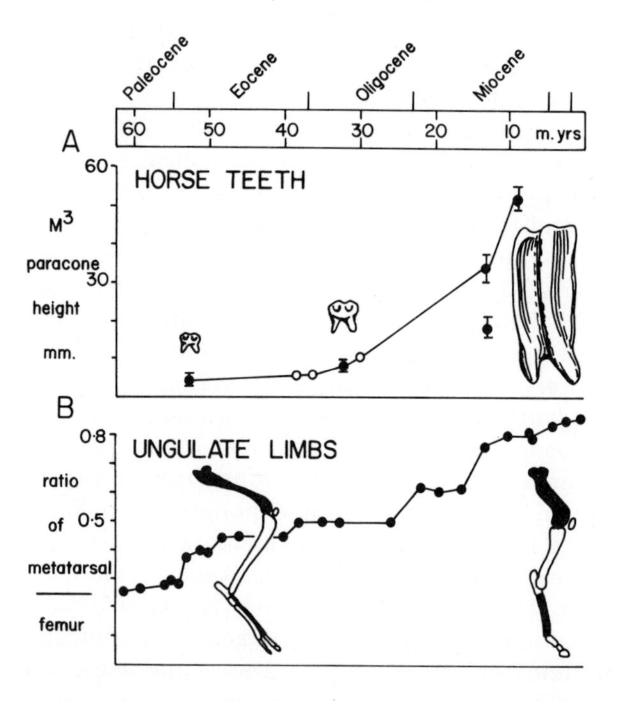

Figure 20.3 Indices of hypsodonty (height of M^3 paracone of horses) and cursoriality (metatarsal to femur ratios of ungulates) of Tertiary mammals of North America. (From Retallack 1988b, with permission of the Society of Economic Paleontologists and Mineralogists).

ridges on the flat, upper surface for grinding hard siliceous grasses. Their teeth are also high crowned (hypsodont) and in some cases continuously growing so that they continue to function despite wear. The magnificent grinding molars of modern grazing horses can be compared with the low-crowned molars of their ancestors thought to have browsed on leaves other than grasses (Fig. 20.3A). The ancestral tooth type for all mammals is unlike either of these, being jagged with sharp cusps. Mesozoic mammals had teeth of this kind, as do many modern insectivores.

The limbs of grassland herbivores are also distinctive. Their chief protection against predation is fleeing over open ground rather than climbing or hiding within woodlands. The elegant leggy appearance and increased running stride of antelopes and horses is very different from the generalized limb structure of woodland mammals. Their lower limbs (ulna and tibia) are elongated relative to the upper limbs (humerus and femur), which are heavily muscularized (Fig. 20.3B). The bones of the feet (metatarsals) and hands (metacarpals) are elongate so that they have become major limb bones. The bone (astragalus) connecting the hands and feet to their respective limb bones is ridged and pulley-like in order to restrict lateral motion. This contrasts with the smaller toe bones and rounded astragalus of raccoons and humans, designed for flexibility rather than speed. The hands and feet of grassland animals are held erect on their toes (unguligrade) rather than at an angle to the ground as in dogs (digitigrade) or flat to the ground as in bears and humans (plantigrade). The fingers and toes of grassland mammals are reduced in number compared with woodland creatures. Modern horses have only a single toe capped by a horny hoof. Three-toed horses and other forms relating them to the archaic mammalian design of five fingers and five toes are well represented in the fossil record (Simpson 1951). Grassland mammals appear totally redesigned compared with woodland prototypes.

In North America, these grassland adaptations are found in fossil mammals as old as Miocene (Fig. 20.3). Several lineages existed at that time with markedly hypsodont teeth and cursorial limb proportions. Grasslands may have existed earlier than this in North America or elsewhere as a selective pressure for the evolution of these animals. Hypsodonty and cursoriality of mammalian faunas in other parts of the world also provide only a minimum age for grasslands of Pliocene in Australia (Sanson 1982, Flannery 1982), Miocene in central Africa and Eurasia (Van Couvering 1980), and Eocene in Argentina (Webb 1978).

Neither the fossil record of grasses nor that of mammals is an entirely satisfactory guide to when and how grassland ecosystems evolved. The earliest known fossil grasses appear to have been minor parts of woodland and swamp vegetation and the earliest hypsodont teeth and cursorial limbs were adaptations to grasslands that existed earlier. These records do, however, constrain the origin of grasses and grasslands to some time during the Tertiary geological period and this can be assessed against independent evidence of paleosols.

Early grassland soils

The fossil record of soils is in general agreement with evidence from fossil mammals and grasses that grassland biomes emerged no earlier than Eocene. No definite grassland paleosols older than this have been reported. This is not to say that dry soils were not colonized by herbaceous vegetation until the advent of grasslands. Some of the red calcareous paleosols of Ordovician and Silurian age (see Figs. 18.4, 18.5, 18.8, 18.9) probably supported early land plants and include burrows filled with fecal pellets. These early herbaceous ecosystems of well drained soils lacked the intimate admixture of clay and organic matter, the granular structure and pervasive pellets and burrows of soils formed under grassland ecosystems. Considering evidence from the fossil record and also from modern soils, grasslands evolved as a unique kind of ecosystem during the Tertiary period.

North American Great Plains

A transition from forest to woodland, then to wooded grassland, and ultimately to open grassland has been documented from paleosols in volcaniclastic Late Eocene to Oligocene (Chadronian to Arikareean) alluvium of the White River and Arikaree Groups (see Fig. 6.2) in Badlands National Park, South Dakota (Retallack 1983a, b). This is an area of spectacular exposures (Fig. 2.6) about 100 km east of the Black Hills and near the geographic center of the Great Plains of North America.

Red, clayey, non-calcareous paleosols of Late Eocene age throughout much of this region (see Fig. 13.11) are evidence that it was heavily forested. In Badlands National Park, this time is represented by a thick, very strongly developed, non-calcareous, kaolinitic, clayey paleosol, the Yellow Mounds silty clay loam. This formed on Late Cretaceous (Maastrichtian) smectitic marine sediments (Pierre Shale and Fox Hills Formation) and a thin layer of Late Eocene alluvial sediments (Slim Buttes Formation). Climate during the Eocene–Oligocene transition (earliest Chadronian) may have been drier, perhaps subhumid. The Interior clay paleosol developed on the succeeding volcaniclastic alluvium (Chadron Formation) is a Paleudalf, also thick, clayey, and non-calcareous, but with more smectite than kaolinite.

Overlying these two basal, strongly developed paleosols is clayey volcaniclastic alluvium (Chadron Formation) with numerous early Oligocene (Chadronian) Paleustalf paleosols (Fig. 20.4A). Judging from their laterally continuous, drab, surface (A) horizons and abundant large, drab-haloed root traces, these Gleska Series paleosols are thought to have supported woodland. Only one other kind of paleosol of the Ohaka Series has been reported from this stratigraphic level. These are thin, gray Fluvents with relict bedding and fine root traces. They probably supported early successional herbaceous vegetation of disturbed streamsides. Apart from these ephemeral grassy areas, woodland appears to have been extensive in this region during Early Oligocene time.

During mid-Oligocene time (Orellan) deposition of alluvium (Scenic Member, Brule Formation) there was a greater variety of soils. Calcareous nodules and occasional pseudomorphs of evaporite minerals in the paleosols are evidence of a drier and perhaps more seasonal climate. The strongly developed pink and green Paleustalfs of the Gleska Series at this stratigraphic level have iron-stained calcareous stringers (petrocalcic horizons) and are closely associated with sandstone paleochannels. Red Fluvents of the Zisa Series, showing relict bedding and only small root traces of early successional vegetation, are interbedded with stream deposits. Away from paleochannels are Conata Series paleosols: Andic Ustochrepts with limited differentiation of a surface (A) and subsurface (Bt) horizon, abundant small root traces, and only scattered large, drab-

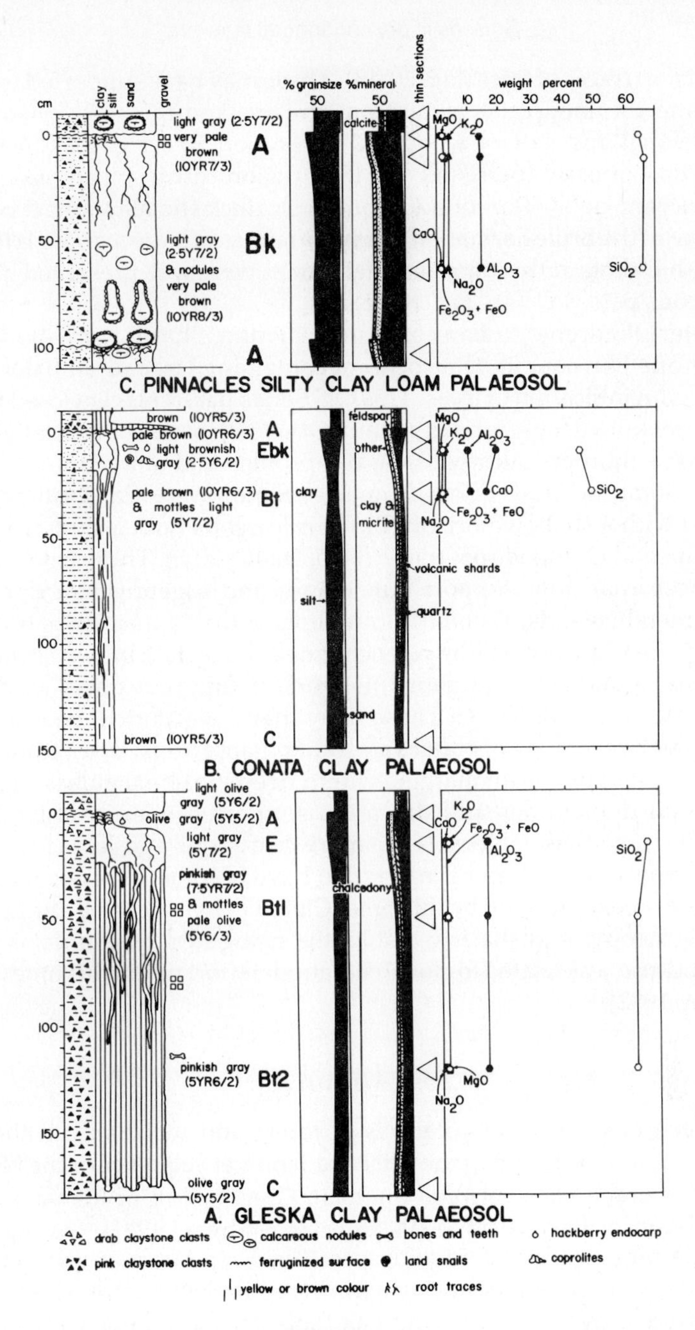

Figure 20.4 Columnar sections (measured in field) and mineralogical and chemical composition (by point counting and ICP chemical analysis) of (A) the type Gleska clay paleosol, a Udic Paleustalf in Early Oligocene Chadron Formation; B, Conata clay paleosol, an Andic Ustrochrept in the mid-Oligocene Scenic member of the Brule Formation; and C, the type Pinnacles silty clay loam paleosol, a Calciorthid in the Late Oligocene Sharps Formation, all in the Pinnacles area of Badlands National Park, South Dakota, USA (from Retallack 1986c, reprinted with permission of Blackwell Scientific Publications).

haloed root traces of trees (Fig. 20.4B). These may have supported wooded grassland on floodplains away from streams and their associated early successional and gallery woodland vegetation (see Fig. 6.3). A similar vegetation mosaic persisted in this region during Late Oligocene (Whitneyan) deposition of additional volcaniclastic alluvium (Poleslide Member of the Brule Formation). Diverse faunas of large mammals thrived in South Dakota at this time and their bones were well preserved in these calcareous paleosols (see Fig. 10.12).

In later Oligocene (Arikareean) ashy alluvium (Sharps Formation) there is only one Fluvaquentic Eutrochrept containing large, drab-haloed root traces as an indication of trees. This Ogi Series paleosol is enclosed within sandstones of a deeply incised stream. Other paleosols at this stratigraphic level have thinner, calcareous (A–Bk) profiles and abundant small root traces. Some of these of the Pinnacles Series were Calciorthids, light colored with a shallow horizon (Bk) of calcareous nodules, platy surface structure and clumped root traces (Figs. 20.4C, 20.5). These paleosols are like those that now support bunchgrass and sagebrush in dry intermontane rangelands. Ustollic Eutrandepts of the Samna Series higher in the sequence, away from the paleochannels, have dark brown surface (A) horizons, a laterally contiguous network of fine root traces, and thin, calcareous (Bk) horizons. These may have supported sod-forming, herbaceous vegetation of grasses and other prairie forbs. By this time, trees were confined to stream margins within deep erosional gullies, whereas floodplain depressions and dry areas supported open grassland (Fig. 20.6). Because these paleosols are more calcareous and less clayey than those below, climate can be inferred to have become increasingly dry. It may have been semi-arid by this time. Climatic drying can be attributed to the lengthening rain shadow cast by the rising Rocky Mountains to the west and also was related to global climatic deterioration at this time (Wolfe & Poore 1982).

East African Rift

Another transition from paleosols of forest and woodland to those of grasslands can be found in the Miocene Tinderet sequence of the Nyanza Rift. This graben lies between the main Gregory Rift of Kenya and the broad basin of Lake Victoria (Pickford & Andrews 1981). During early Miocene time, forests were widespread in East Africa as an equatorial extension east across the continent from the jungles of Zaire. Their deeply weathered, clayey, red lateritic paleosols can be mapped throughout Kenya and Uganda (McFarlane 1976). Early Miocene fossil localities of Songhor, Koru, and Rusinga Island in southwest Kenya are within sequences of ashes and lava flows from local carbonatite–nephelinite volcanoes. Paleosols in these sequences have preserved a variety of fossil

Figure 20.5 The type Pinnacles silty clay loam paleosol in Late Oligocene Sharps Formation of Badlands National Park, showing the irregular thickness of the finely structured surface (A) horizon. Hammer handle is 28 cm long.

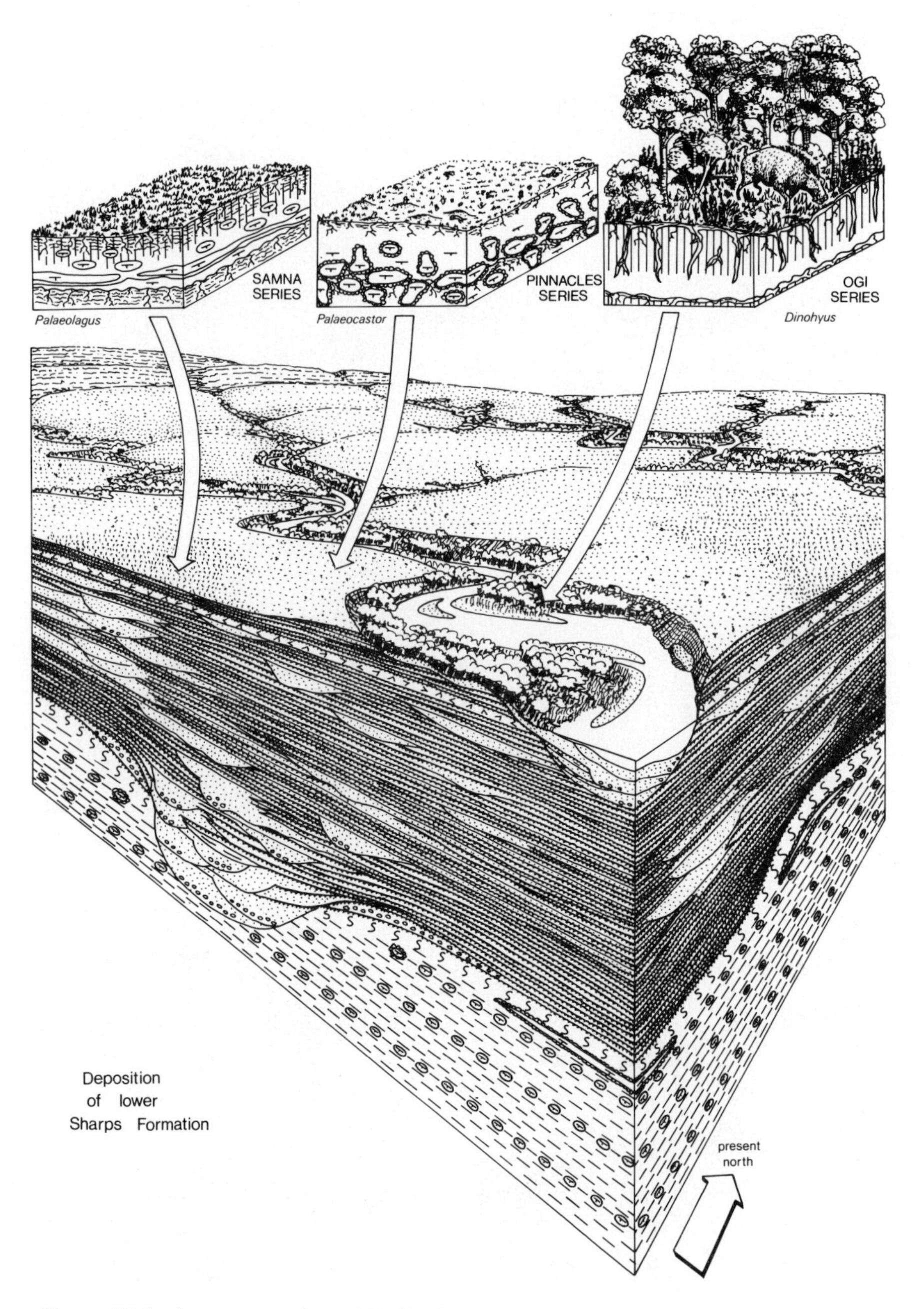

Figure 20.6 A reconstruction of Badlands National Park, South Dakota, during Late Oligocene (Arikareean) deposition of the lower Sharps Formation (from Retallack 1983b, with permission of the Geological Society of America).

apes, snails and fruits typical of tropical forest (Pickford 1986b). Normally these phosphatic and calcareous fossil hard parts would be destroyed in forest soils of humid climate (Retallack 1984a). The Miocene paleosols containing these fossils have remained calcareous despite their thick, clayey red profiles resembling tropical forest soils (see Fig. 21.4). Comparison of paleosols of different degree of development has shown that soil formation depleted carbonate content, but few became non-calcareous because the carbonatite parent material was very calcareous at the outset. Soils on comparably calcareous parent materials of shell and coral under wet forest in coastal East Africa are now quite decalcified (FAO 1971–81). Hence the Early Miocene red carbonatite paleosols probably supported dry tropical forest, similar to those now found around the periphery of the Guineo–Congolian basin (White 1983).

New kinds of soil formed in this region after the faulting of the western margin of the East African Rift and extrusion of plateau phonolites by Middle Miocene time (ca. 14 million years ago). These are well known in the sequence of paleosols in carbonatite–nephelinite tuffs sandwiched between thick phonolite flows near the village of Fort Ternan, south-western Kenya (Fig. 20.7). The top of the lower phonolite is red and weathered to clay, but is weakly calcareous and contains weakly expressed mukkara structure. This has been called the Lwanda Series paleosol, a Vertic Eutropept. The overlying brown claystone of the lower paleontological excavations here includes several weakly to moderately developed paleosols which also show evidence of cracking. These paleosols of the Rairo and Rabuor Series have been identified as Vertic Ustropepts and Vertic Argiustolls. Their mixture of small and large root traces is evidence of wooded grassland. Little bone was found in these paleosols.

The main paleontological excavation several meters higher in the sequence at Fort Ternan exposed two superimposed paleosols called the Chogo Series (Figs. 20.8, 20.9). These have a thin (10–12 cm), dark, surface (A) horizons and a shallow (18–26 cm) horizon (Bk) of calcareous nodules and stringers. These may be identified as Haplustolls. Their vegetation was predominantly grassy considering the abundance of fine root traces, fossil grass pollen and calcareous, grass-like stem casts. There also were numerous trees as indicated by stump casts up to 16 cm in diameter, by fossil fragments of thorn-bearing twigs and by pitted fruit stones. These were mainly found in the upper paleosol in the western part of the excavations where there may have been an ecotone into a patch of grassy woodland around a small ephemeral watercourse that was excavated there (Shipman 1981). The eastern part of the upper paleosol does not show such conspicuous evidence of trees and may have supported wooded grassland.

A thicker but otherwise similar paleosol of the Onuria Series, a

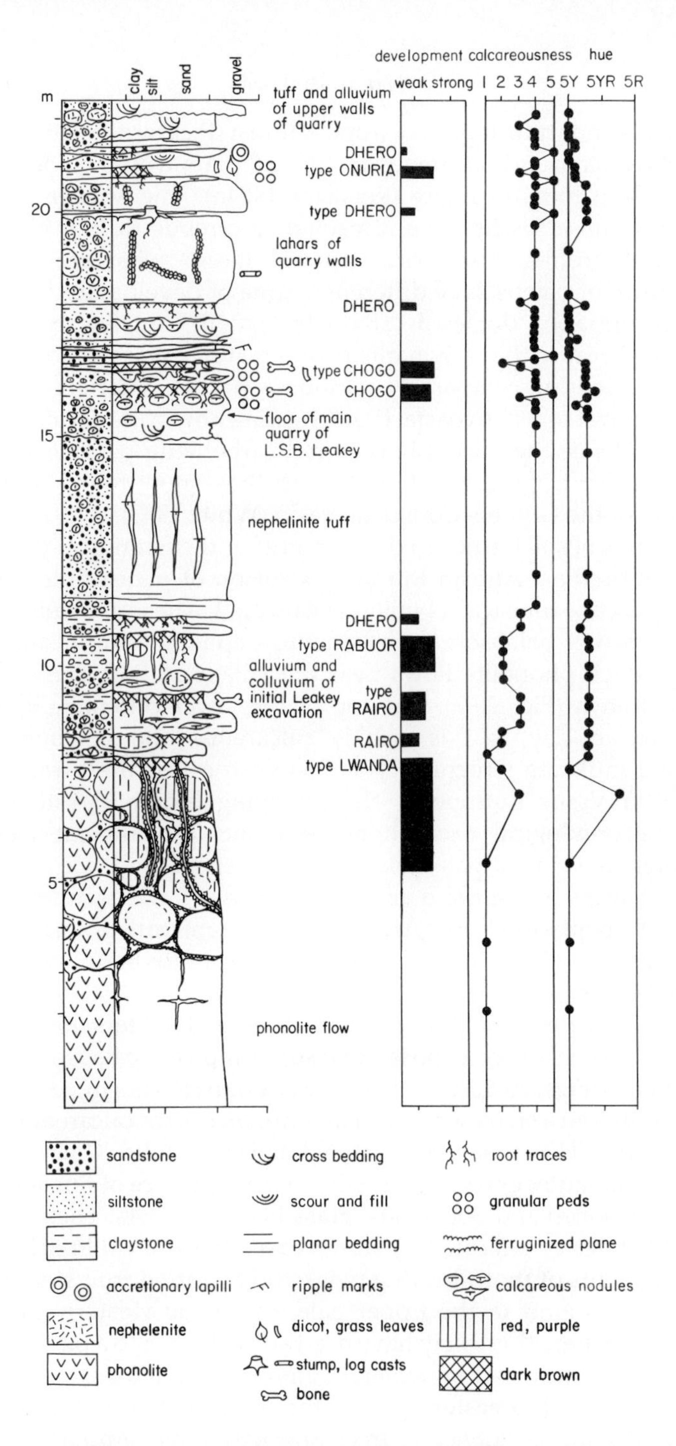

Figure 20.7 Columnar section (measured in field) of mid-Miocene (14-million-year-old) paleosols in the paleontological excavation at Fort Ternan National Monument, Kenya. Position of paleosols marked by boxes whose width corresponds to degree of development (Table 13.1). Calcareousness was estimated by reaction with dilute acid (Table 3.2) and hue with a Munsell color chart.

Figure 20.8 Middle Miocene (14-million-year-old) paleosols of the main fossiliferous layer in the excavation at Fort Ternan National Monument, Kenya. The top of the upper paleosol, the Chogo clay eroded phase paleosol, a Haplustoll, is at the hammer head and only the top of the Chogo clay ferruginized nodule variant paleosol is exposed at the base of the excavation. The hammer handle is 28 cm long.

Calciustoll, has been found capping a mudflow in the headwall of the large quarry at Fort Ternan National Monument. This has yielded rare dicotyledonous leaf fragments as well as abundant and exquisitely preserved fossil grasses (Fig. 20.2) that are close to their position of growth. Other thin paleosols with abundant relict bedding are the Dhero Series. These Orthents contain scattered small stumps and logs of early successional vegetation.

Climate during formation of these earliest known paleosols of wooded grassland in East Africa was drier than earlier in the Miocene. This can be related to the local building of volcanic cones and plateaus within the Nyanza Rift. General uplift of the western margin of the Gregory Rift cast a rain shadow that broke the equatorial band of tropical forest.

Other regions

Similar records of early grassland paleosols may be discerned on most continents. There is a superb record of early grasslands in Eocene and Oligocene volcaniclastic deposits of Argentina. These contain abundant

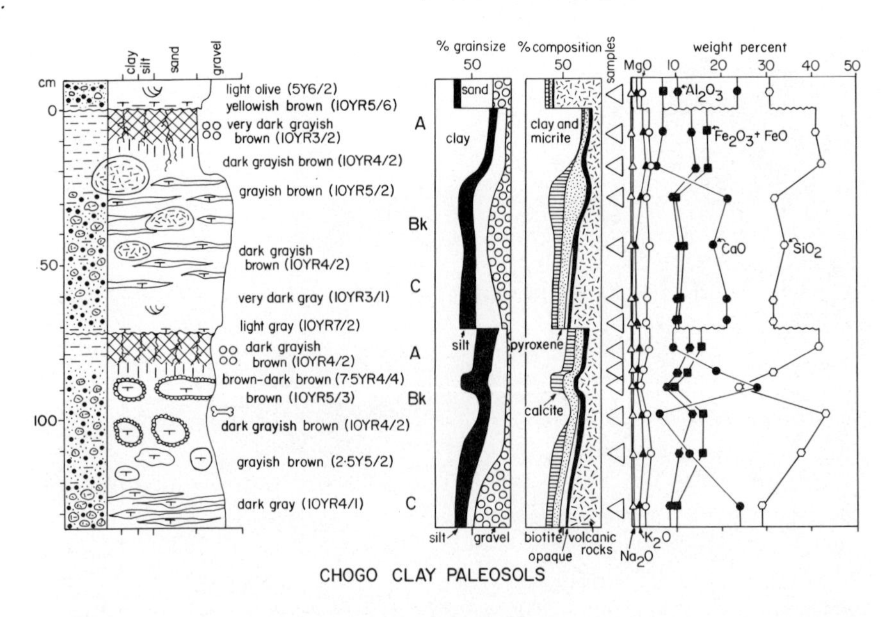

Figure 20.9 Columnar section (measured in field) and petrographic and chemical composition (estimated by point counting and atomic absorption spectrometric chemical analysis by C. McBirney) of the Chogo clay paleosols in mid-Miocene Fort Ternan Beds, Fort Ternan National Monument, Kenya. Lithological symbols as for Figure 20.7.

fossil mammals, phytoliths, and trace fossils of invertebrates, including dung beetles. Oligocene (Deseadean) paleosols near Paso Flores, Neuquen, have been briefly characterized (Frenguelli 1939) and appear to have been Inceptisols or Alfisols. Other paleosols extending back in age to Eocene (Mustersian) in the Sarmiento Group near Lago Colhue Huapi and Laguna del Mate in Chubut have been identified as Udolls and Aquolls (Spalletti & Mazzoni 1978). These are the oldest Mollisols yet recorded.

Paleosols similar to those formed under wooded grassland have been found in Late Miocene (8.3 million years old) alluvium of the Dhok Pathan Formation in northern Pakistan (see Figs. 11.8, 11.9). Probably they will be found in older deposits of the Asiatic interior because this was the likely source region of mammalian faunas adapted to grasslands that migrated into North America, Africa, and Europe during Miocene time.

Miocene (16 million years old) to modern vertebrate-bearing sequences in South Australia include thick duricrusts and numerous alluvial paleosols (Callen & Tedford 1976), but these have not yet been studied in detail. Grasslands appeared in Australia at least by Plio-Pleistocene, judging from the tooth morphology and limb proportions of fossil kangaroos (Sanson 1982, Flannery 1982).

Antarctica is one of the few continents unlikely to have a fossil record of grassland. Its Oligocene fossil floras were dominated by southern beech (*Nothofagus*) forests. By late Oligocene time glaciers already extended to sea level (Kemp 1978).

How did grasslands arise?

Grasslands have several advantages over other kinds of vegetation (Vogl 1974, Bell 1982) and these can be used to assess the paleopedological record of early grasslands.

Grasses flourish over trees in dry or otherwise unfavorable, highly seasonal and unpredictable environments. This includes the cool temperate, semi-arid Russian steppe, North American prairie, and Patagonian pampas and also seasonally dry subtropical and tropical wooded grasslands, salt marshes, coastal dunes, streamsides, and alpine meadows. Grasses survive in these difficult environments because of their small stature, unspecialized pollination and dispersal mechanisms, and protection of their tissues in rhizomes or other underground structures. Few plants are better suited than grasses to impersistent habitats such as seasonally snowy plains, floodprone streamsides, and human construction sites. There is, however, a limit to the ability of grasses to colonize such hostile habitats as barren rocks, salt pans and desert dunes. Nutrients and water are in too short supply for grasses in these habitats which tend to be vegetated by slow-growing, tolerant plants such as lichens and cacti.

Grasses recover quickly from fire because their fire-sensitive meristems and rhizomes may be underground and because they have little woody tissue that will burn long and hot. Grass leaves that have died back during winter or drought are especially flammable. Ground temperatures 3 cm below grass fires are only 3°C hotter than usual for as little as a few seconds, as fire rapidly consumes the available fuel and moves on (Gillon 1983). In contrast, grass fires are lethal for small shrubs and trees. Many modern grasslands are thought to be maintained by periodic fire (Johannessen *et al.* 1971).

Animal activity also may play a role in maintaining grasslands. The grazing and trampling activity of large herds of antelope and zebra keep East African game parks as well groomed as a city park (McNaughton 1984). Elephants knock down and tear apart trees in the dry season in order to consume their bark and leaves (Owen-Smith 1987). Plagues of grasshoppers and other insect pests may defoliate large areas, leaving them open for colonization by grasses.

A final factor encouraging the development of grasslands is competition among plants. Although trees may have an advantage in shading out and chemically poisoning lower growing plants, grasses can persist by shading out small seedlings of trees. Few kinds of plants create such a dense and continuous ground cover as grasses.

North American Great Plains

The Late Eocene to Oligocene sequence of paleosols in Badlands National Park, South Dakota, includes abundant evidence that the climate became drier with the appearance of grasslands (Retallack 1983a,b). Paleosols higher in the sequence are less clayey, more calcareous, and their calcic (Bk) horizons are higher within the profiles (see Fig. 9.5). Chalcedony pseudomorphs of evaporite minerals provide evidence of a dry season. Seasonality also is reflected in growth rings of rare fossil wood from here (Retallack 1986d).

Grasslands also appeared at a time when the landscape was becoming more disturbance prone. The degree of development of paleosols declines upwards in the sequence, with a lesser trend of increased development toward the top of each rock unit (see Fig. 13.14). Each unit represents an episode of erosional gully cutting followed by an episode of sediment accumulation approaching a dynamic equilibrium (see Figs 13.15, 13.16). Each of the four periods of downcutting ushered in more open vegetation and associated climatically sensitive features of the paleosols indicate drier climate as a cause of the erosional episodes. In only two of the erosional episodes can a large influx of volcanic ash or uplift of the source terrain be identified as a contributing factor. Thus, disturbances of vegetation evident from paleosols can be traced back to climatic effects.

Fire is a difficult factor to assess from grassland paleosols because phytoliths and ash left by grass fires may appear no different from organic and mineral matter produced by ordinary decay. The burning of trees, on the other hand, creates charcoal, which is a very persistent material in paleosols. I have seen charcoal in many paleosols, but none has yet been found in paleosols of Badlands National Park, South Dakota. This does not mean that fires never occurred there, but rather that the frequency of fires was so low that charcoal did not accumulate in soils. For this reason, it is unlikely that fire was important in the development of open grassland from wooded grassland in this region. It could, however, have played a role in the maintenance of open grassland once it had appeared.

Also of doubtful importance in the origin of open grassland in Badlands National Park was the role of mammals. Large titanotheres (*Menodus giganteus*) were potential "bulldozer herbivores" like modern elephants. However, they became extinct at the very time (near the Orellan–Chadronian boundary) when wooded grassland appeared for the first time. Another large mammal (*Metamynodon planifrons*) at this stratigraphic level and a little higher is found only in paleochannels and is thought to have lived more like a hippopotamus than an elephant (Prothero 1987). It could have locally demolished vegetation along streamsides, but paleosols are an indication that wooded grassland appeared first away from streams which were flanked by gallery woodlands. All the fossil mammalian evolutionary lineages became smaller in body mass in younger rocks of the Badlands. This and the increased abundance of burrowing mammals may have been a response to scarce resources in an increasingly dry climate. Despite these changes, however, most lineages of mammals persisted, including high browsers such as camels. Their teeth remained low-crowned and their limb proportions remained intermediate between those of archaic and of geologically younger mammals (see Fig. 20.3). Most of the mammals, such as oreodons (*Merycoidodon culbertsoni*), retained a semi-plantigrade stance that could be interpreted as evidence of soft paws rather than a hard narrow hoof. Although this fauna was living first in wooded grassland and then open grassland during Late Oligocene time, it remained better suited to woodland (Van Valkenburgh 1985, Prothero 1985). This mammalian fauna would have been gentler on grassland landscapes than modern grassland faunas and there is no evidence of adaptive changes in concert with the thinning of trees. This situation changed near the Oligocene–Miocene boundary which was a period of extreme desertification and ushered in an immigrant mammalian fauna much more suited to grasslands (Webb 1977).

A final factor also unlikely to have played a role in the emergence of grasslands in the Great Plains is competition among plants. The oldest grassland paleosols formed in a dry climate. This would have been marginal for grasses, as in the intermontane range lands of the North American west. The well structured surface horizon of these paleosols is variable in thickness (Fig. 20.5), perhaps because grasses and forbs of this early steppe were clumped. Plant competition was less important than being able to tolerate harsh conditions. Modern grasses compete not only for ground cover, but for reproductive success also by armoring their propagules with silica. In Late Miocene and Pliocene rocks there is evidence of such a competitive effort in the variety of silica-encrusted fossil fruits found (Thomasson 1979). Such fossils are extremely rare in Oligocene deposits (Galbreath 1974). Plant competition probably became important much later in the evolution of North American grasslands.

East African Rift

These same factors reconsidered for Miocene paleosols near Fort Ternan, Kenya, reveal different conditions for the advent of grasslands in East Africa. Climatic drying also is in evidence here from the shallower calcic (Bk) horizon of paleosols in the excavations at Fort Ternan compared with its depth in geologically older paleosols at Koru and Songhor in Kenya. All of these were volcaniclastic deposits close to carbonatite–nephelinite volcanoes. There is no indication that eruption frequency or style was different at Fort Ternan than in these other deposits, nor is there any charcoal as evidence for frequent disturbance by fire.

In contrast to the Badlands, both mammalian activity and plant competition could have been significant in promoting grassland at Fort Ternan. The large hoe-tusk elephants (*Prodeinotherium hobleyi*) found there may have demolished the small trunks (up to 16 cm in diameter) of trees both as "bulldozer" and "backhoe" herbivores. Other elephants (*Choerolophodon ngorora* and *Protanancus macinnesi*) and rhinoceroses (*Paradiceros mukirii*) could also have played a role in destroying trees. The mammalian fauna includes abundant hypsodont and hard-hooved antelopes (*Oioceros tanycerus, Kipsigicerus labidotus*) and giraffe ancestors (*Paleotragus primaevus*). Micro-wear on antelope teeth indicate that they were grazers and mixed grazer–browsers (Shipman 1986), but their limb and tooth proportions were most like those of living Indian woodland antelope (Gentry 1970). The density of root traces and the remains of fossil grasses indicate a luxuriant growth of grass. Fossil grass fragments have been found up to 24 cm long and 8 mm in diameter. They could easily have been 1 m or so tall. Such dense growth could have restricted seedling growth of trees. The grasses also were strongly silica encrusted (Fig. 20.2). Fossil twigs with narrow thorns are evidence that trees also expended resources in defence against herbivores.

Origin of grassland ecosystems

The earliest grasslands now recognized from paleosols in the Great Plains of North America and the western flanks of the Gregory Rift of East Africa were completely different from each other. In the Great Plains, Aridisols and Inceptisols supported an early grassland of indigenous flora and fauna little prepared for climatic drying that thinned out the trees. These earliest grasslands were like those of dry intermontane range lands in the western United States or the drier parts of Tsavo National Park in Kenya. Such early stages in the evolution of a major kind of ecosystem may have been less integrated than now, after millions of years of coevolution.

In East Africa, however, the earliest paleosols of grassland were Inceptisols and Mollisols associated with fossils of an immigrant mammalian fauna already partly adapted to open country. These grasslands were lush, like the tall grass prairie of the North American Great Plains or parts of Serengeti National Park, Tanzania. This ecosystem was already integrated to some extent and could be established under conditions of less severe climatic drying.

It seems unlikely that paleosols of Badlands National Park or Fort Ternan are entirely representative of early grassland evolution. However, what is known about them reveals a complex interplay of environmental and biotic factors. The biological innovations of Early Tertiary grasses enabled them

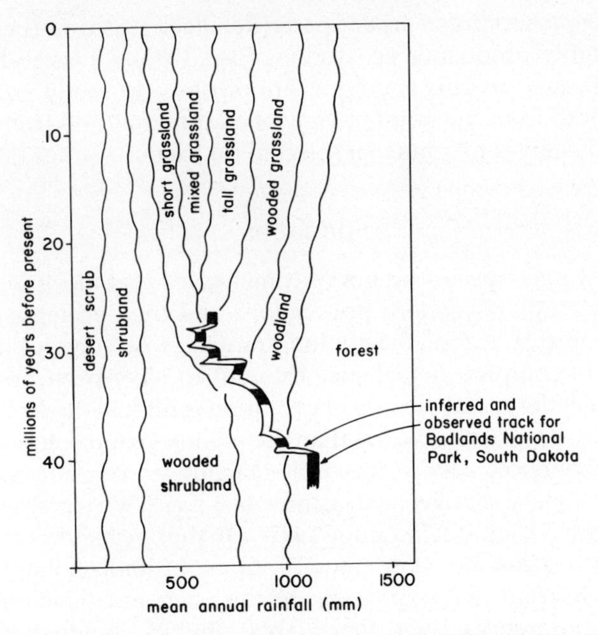

Figure 20.10 Hypothetical climatic range of vegetation formations in the Great Plains of North America during Tertiary time. This could be refined by additional tracks of climate and vegetation change like that shown for Badlands National Park.

to colonize disturbed habitats. At this time they were found as a minor part of vegetation of streamsides and swamps. The earliest steppes that formed at various times and places of drier climate during the early Tertiary may initially have been simply an opportunistic colonization of bare ground. Because grasslands appeared so much later than grasses, climatic drying is not in itself a sufficient explanation of grassland ecosystems. Grasslands as we know them today, with grazers maintaining disturbance and grasses sustaining grazers, took much longer to emerge. It may not have been so much a question of the evolution of the adaptations, which in itself may have been rapid, but rather the coming together of complementary parts of grassland ecosystems. Modern grassland ecosystems were then able to establish and maintain themselves within a widening window of climatic and edaphic conditions intermediate between those of woodlands and deserts (Fig. 20.10).

Evolutionary processes

Compared with other kinds of paleosols, those formed under grasslands are unusually fossiliferous with bones, snails, phytoliths, and stony fruits. This is because the calcareous composition of these soils is especially favorable for preserving such hard parts (Retallack 1984a) and because they are such highly productive ecosystems (Bell 1982). Grassland paleosols and their fossils present a unique opportunity to study evolutionary processes both from the point of view of adaptations and from the point of view of the environments that selected for them.

Speciation

Evolution of new species occurs on time scales that are too long to be revealed by study of modern populations and their ecology. The fossil record, in contrast, is replete with long-ranging fossil species that can be grouped into complex genealogies thought to have been produced by evolution. Yet the actual process of evolution is difficult to detect because it operates on time scales shorter than the million year resolution of many geological sequences. Hence, it is difficult to judge even simple questions about evolutionary change, such as how fast it is. Two extreme positions can be imagined (Gould & Eldredge 1977). On the one hand it could be that evolution proceeds by slow modifications of morphology that are imperceptible from one population to the next, but which amount to distinctive differences over time. This phyletic gradualism can be contrasted with punctuated equilibrium, by which evolution occurs in spurts of genetic change over a geologically short period of time that is nevertheless long by human standards. Species produced in this way then persist for millions of years virtually unchanged. Whether either of these views or some intermediate one represents the more general tempo of evolution can be studied using the fossil record, provided that there is some way of estimating the time elapsed between fossil samples. In non-marine vertebrate-bearing sequences paleosols may fill this role because the time scales of soil formation are intermediate between those of ecology and geology.

From this perspective, the Oligocene sequence of paleosols in Badlands National Park, South Dakota, is incomplete in a variety of ways (see Table 13.5). Not only are many of its paleosols well developed, but this aggrading sequence of paleosols is also punctuated by several episodes of erosion that may have lasted several million years (see Fig. 13.15). The vertebrate fossils of this region are not an adequate test of the evolutionary models of punctuated equilibrium or phyletic gradualism. Fossil mammals found there do seem to be fairly stable in shape and size within each depositional unit and this could be interpreted as evidence for evolutionary stasis following rapid evolutionary origin (Prothero 1985). However, it could also have been the result of a period of phyletic gradualism during the long periods of time unrecorded. Evidence from paleosols for the temporal resolution of this sequence compromises its relevance for fine scale evolutionary studies.

A more promising study area has been identified in the paleosol sequence of the Early Eocene (Wasatchian) Willwood Formation of Wyoming (Bown & Kraus 1981b). Each sample of vertebrates used here for evolutionary studies has been collected from as many as four paleosols (Gingerich 1980), which together may represent 50 000 years or so of soil formation. This resolution can be improved by sampling paleosol by paleosol (Winkler 1983). Existing data on mammalian evolution from this sequence are suggestive of both phyletic gradualism and punctuated equilibrium for different features and evolutionary lineages. Further studies of this kind are needed to determine which is the more usual evolutionary pattern and under what circumstances each occurs.

Natural selection

In studying mammalian evolution from the fossil record, it is natural to focus on what was selected: the fossil mammals. In such studies, it is common to assume that their various features such as high-crowned teeth and elongate limbs were optimally designed for a particular purpose. In this case, the purpose would be eating and avoiding being eaten in open grasslands. It is this assumption that enables reconstruction of former vegetation from presumed adaptive features of mammalian fossils. This assumption of optimal adaptation has been lampooned (by Gould & Lewontin 1979) as the Panglossian paradigm after the character in Voltaire's novel "Candide" who declaimed through a series of misfortunes that, "all is for the best in this best of all possible worlds." What is needed to avoid such false optimism is independent evidence of paleoenvironment against which the degree of adaptation or maladaptation of species or communities can be assessed. In some cases, paleosols can play that role.

A Panglossian view of Oligocene mammalian faunas of Badlands National Park, South Dakota, is that they lived in forest, because they resemble modern forest faunas in a variety of adaptive features (Van Valkenburgh 1985). From the new perspective offered by paleosols, Late Oligocene faunas there can now be seen as maladapted to grasslands compared with the Miocene faunas of Fort Ternan in Kenya or of Nebraska, USA, or to modern faunas of these same regions. Nevertheless, these Late Oligocene faunas persisted for millions of years despite their apparent inadequacy. Similarly, an antiquated technology such as hand copying of manuscripts can remain viable in the context of its own times. It is only with the hindsight of printing or photocopying that it seems archaic.

If vegetation and climate are not selecting for near constant adaptive grades in mammalian populations over geological time scales, then what is? One possibility is that evolution is constrained by difficulties in altering highly coordinated biological systems for reproduction, growth and development. However, this also seems unlikely considering the evidence for the occasional rapid evolution of individual species (Gingerich 1980) and faunal overturn of mammalian communities (Prothero 1985). Could it be, then, that evolutionary novelties are shouldered aside by the web of ecological interactions in natural mammalian ecosystems? One way of approaching this question is to compare geological stability with ecological integration of fossil mammalian assemblages as in the following section.

Coevolution

Assemblages of organisms that live together are often called communities and their various component organisms labeled as consumers, producers, or similar terms that reflect their role in maintaining the community. An extreme version of this marketplace view of ecology is that communities can be regarded as "superorganisms" in which particular species are as indispensable to the function of the community as organ systems of a single creature (Clements 1928). Extending this concept, extreme versions of the Gaia hypothesis view the Earth itself as a "superorganism." If communities really are so highly integrated, then presumably their evolution has more to do with biological forces such as competition (Van Valen 1973), independent of all but the most profound environmental perturbations (Olson 1985). A contrasting, equally extreme, and more Ereban view is that communities are only accidental associations of species whose abundance is determined by gradients in environmental factors, such as moisture, rather than by interaction with associated organisms (Whittaker 1978). In this view, associations of organisms are difficult to define in both space and time. Their evolution and migration is a direct reaction to environmental perturbations such as climatic change (Bernabo & Webb 1977).

Grassland ecosystems, as currently understood from the fossil record of mammals, grasses, and paleosols, appear to have evolved from something a little like an accidental association to something a little more like a superorganism. It is remarkable that similar kinds of mammals persisted in South Dakota throughout Oligocene time despite profound changes in vegetation from woodland to grassland. Adaptive features of both carnivores and their prey ungulates were stable over this interval (Fig. 20.3). In this case, mammalian assemblages were remarkably robust in the face of environmental change and were not especially dependent on vegetation. Greater interaction between mammals and vegetation is evident from the immigrant Miocene grassland ecosystem of East Africa, where this mammalian fauna also forms a recognizable dynasty (Pickford & Andrews 1981). Similarly, in North America a new Miocene grassland ecosystem established itself as a recognizable dynasty that was severely tested by later environmental deterioration (Webb 1977). Climatic and environmental change were extremely trying during Pleistocene Ice Ages with surprisingly little effect on mammalian community evolution. In the long term, mammalian communities appear to have become more robust in the face of external environmental change, owing to evolutionary improvements in the design of individual species and their web of interactions.

Whether or not these tentative conclusions withstand future scrutiny, the fossil record of mammals and of vegetation revealed by paleosols can be seen as a fertile testing ground for basic ideas about evolution.

21 Human impact on landscapes

Even if our bones and the debris of civilization soon become diagnostic fossils of one of the briefest biostratigraphic zones in geological history, our effect on land surfaces of the world already is conspicuous and irreversible. Modern cities, dams, parking lots, and highways are reshaping the landscape. The pace and scope of human activity are now altering elemental cycles that have sustained life on this planet for millions of years. The burning of oil and coal, for example, is reintroducing carbon into the atmosphere as carbon dioxide, diluting breathable oxygen in the atmosphere, and creating a global warming by the greenhouse effect.

Such activities have long been of concern to intellectuals. In his *Timaeus*, Plato wrote of the deforestation and desertification of Greece during classical times compared with the parkland enjoyed by his ancestors. He blamed these changes on the Iron Age civilization of his times and viewed the past as a Golden Age when humans lived in harmony with their environment. Similarly, the pollution of the Industrial Revolution of the 19th century spurred Romantic poets of England such as Wordsworth, Shelley, and Byron to find solace in the Lake district of northwestern England or the rustic landscapes of Greece and Italy of that time. An accurate reconstruction of environmental degradation attributable to human impact is needed in order to assess how damage can be minimized in the future. Historical records, artifacts, and bones all are useful for this task, but it is the landscape that suffers most. The record of human impact on paleosols promises to be a useful source of information on early human use and abuse of the landscape.

Fossil bones and stone tools provide a record of early human evolution that can guide the search for paleosols relevant to human origins and subsequent human dominion over the landscape. A variety of apelike fossils from Miocene rocks of Europe, Asia, and Africa have been regarded as possible human ancestors. Some of these called dryopithecines are generalized apes ancestral to gibbons, gorillas, and chimpanzees (Kelley 1986). One of the best known of these is *Proconsul nyanzae*, a medium-sized (10 kg), quadrupedal ape from the early Miocene (17 million years old) Hiwegi Formation on Rusinga Island (Fig. 21.1). Other Miocene apes,

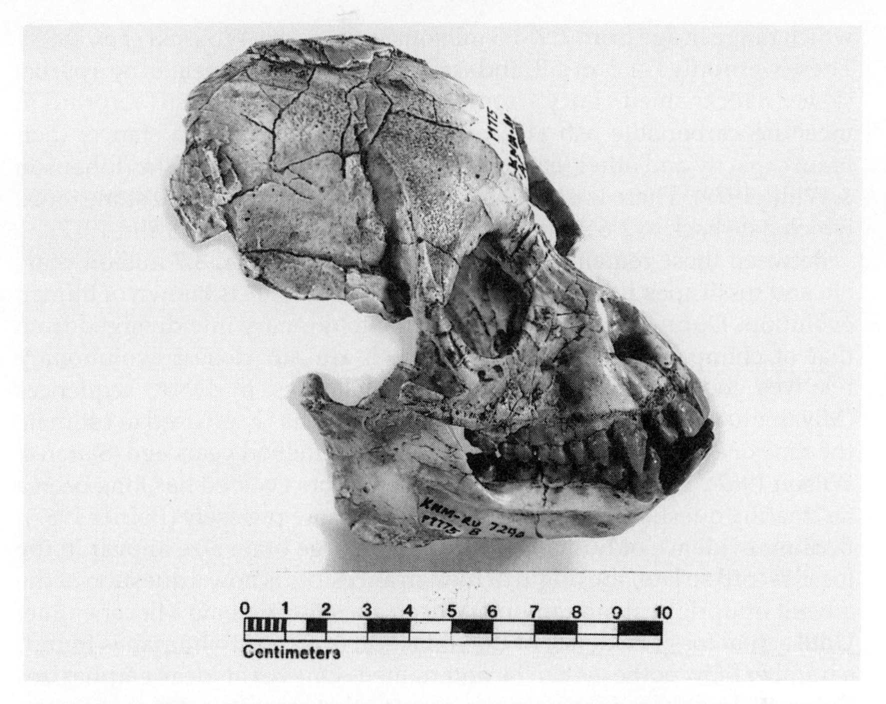

Figure 21.1 Skull of the dryopithecine primate *Proconsul nyanzae* from the Early Miocene (Kisingirian or 17-million-year-old) Hiwegi Formation, near Kiahera Hill, Rusinga Island (Kenyan National Museum specimen RU 7290; photograph by Sean Poston).

often called ramapithecines, had reduced canines and divergent rows of molars, unlike the prominent canines and parallel rows of molars in other apes. These features of ramapithecines are more like the arrangement in humans, and these fossils were once considered promising human ancestors. However, ramapithecine skull remains of *Sivapithecus indicus* from the Late Miocene (8.3 million years old) Dhok Pathan Formation of northern Pakistan are more like those of orangutans than of chimpanzees or humans (Pilbeam 1982). Other small-canined apes such as *Kenyapithecus wickeri* from the mid-Miocene (14 million years old) Fort Ternan Beds near Fort Ternan, Kenya, are known from more fragmentary remains (Pickford 1985, Senut 1988). Other kinds of African Miocene apes still are being found (Leakey & Leakey 1986a,b) and their interrelationships and diversity remain far from clear.

African rather than Asian apes remain the most promising human ancestors because the most geologically ancient fossils that are even remotely human are known from the East African Rift. Pliocene (3.5–3.7 million years old) remains from near Laetoli in Tanzania (Leakey & Harris 1987) have been referred to the same species (*Australopithecus afarensis*) as more abundant remains from Hadar, Ethiopia (Johanson *et al*. 1982), which range in age from 2.9–3.3 million years (Sarna-Wojcicki *et al.* 1985). These were only 1–1.5 m tall, and walked fully erect as revealed by a partial skeleton nicknamed "Lucy" from Hadar and by a set of fossil footprints in melilitite–carbonatite ash at Laetoli. Despite their human stance, their brain capacity and other features of the skull were very apelike (Johanson & White 1979). There is no evidence that these creatures used stone tools, which date back to 2.6 million years at Hadar (Roche & Tiercelin 1977).

Between these remains of early human ancestors ca. 3.7 million years old and fossil apes 10 million years old in Kenya, little is known of human evolution. During this time the human evolutionary line diverged from that of chimpanzees and gorillas, which are our closest evolutionary relatives judging from evidence of similarities in DNA sequences (Miyamoto *et al.* 1987). Similar genetic evidence has been used to estimate the time of evolutionary divergence at about 5 million years ago (Sarich & Wilson 1967). How the earliest human ancestors evolved has long been a fascinating question that now can be posed more precisely (Poirier 1987). Because evidence of human culture and of large brain size appear in the fossil record so late, the origin of human ancestors is now a question of the advent of upright stance among African apes during Late Miocene time. Until actual fossil evidence of the transition from ape to human is found, a number of hypotheses can be entertained. One set of ideas emphasizes the martial side of human nature in arguing that erect stance arose because of the need to have hands free for throwing projectiles at predators in increasingly open grassland habitats. A related idea is that erect stance enabled our ancestors to see predators over tall grass, an ability that was not needed in the complex three-dimensional habitats of forests. Another set of ideas emphasizes the opportunistic side of human nature: ready to exploit new niches in ways unusual for primates as a whole. Erect stance can be seen as a means of freeing hands to manipulate small seeds of grasslands as opposed to large and locally abundant fruits of tropical forests. Humans of early grasslands also can be imagined as scavengers like hyenas. Erect stance could have been a solution for the problem of searching large areas of open country for scattered carcasses in a way that was energetically economical. A final set of theories envisages humans as domestic, family-oriented creatures. By this view, hands would have been useful for carrying food back to a home base or for carrying infants while doing domestic chores.

Similar questions of lifestyle also have been posed for the various fossil australopithecine species. These consistently are found in pairs: a robust species and a gracile species (Fig. 21.2). The most ancient lightly built species (*Australopithecus africanus*) is found in cave deposits near Makapansgat and Sterkfontein in South Africa dated at 2.3–3 million years old. A contemporaneous thickset form (*Australopithecus boisei*) has been found recently near Lake Turkana, Kenya (Walker *et al.* 1986). During the

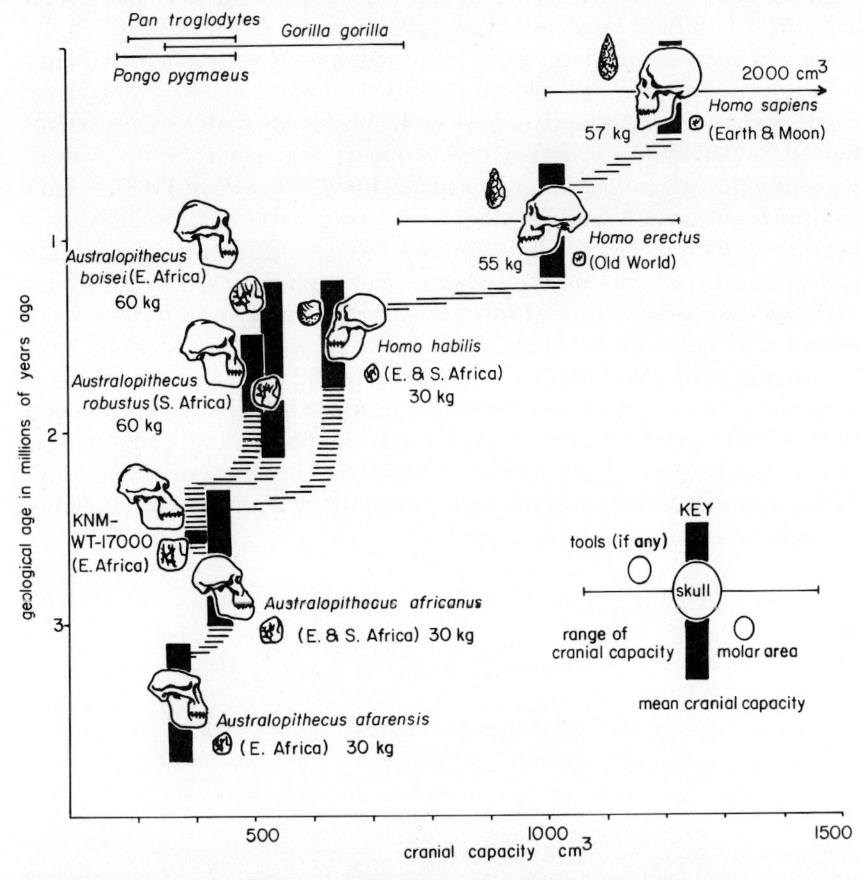

Figure 21.2 The stratigraphic range, cranial capacity, body weight, molar volume, and stone tools of early hominid species, showing a dramatic increase in tooth size in *Australopithecus* and in brain size in *Homo*, both compared with body size (data from Skelton *et al.* 1986, Poirier 1987).

interval 1.3–1.9 million years ago, robust and gracile forms shared a wide area of Africa. From Lake Turkana in northern Kenya to Olduvai Gorge in northern Tanzania, the robust *Australopithecus boisei* was sympatric with gracile *Homo habilis*. In the South African caves during this time interval the robust *Australopithecus robustus* was sympatric with *Homo habilis* (Vrba 1985). When two closely related species of animals live in the same general area, it is usually because they exploit the environment in slightly different ways. The robust species with their huge grinding teeth and heavily muscled skulls may have eaten large quantities of tough seeds and tubers of grasslands or low-grade forest browse. The gracile species may have been more omnivorous. They have been imagined as open country scavengers, as big game hunters, and as mixed hunters and gatherers.

Homo habilis is the earliest likely representative of our own genus and lived ca. 1.8 million years ago near Lake Turkana, Kenya (Fig. 21.3). Its brain was markedly enlarged (Fig. 21.2) compared with its small (about 1 m tall) and distinctly australopithecine body (Johanson *et al.* 1987). These creatures have been thought to be responsible for associated stone tools of the Oldowan tradition, but it now appears that some South African *Australopithecus robustus* could also use tools (Brain *et al.* 1988). A later cultural tradition of Acheulian stone tools dating back ca. 1.5 million years in Olduvai Gorge (Bed II; Hay 1976) can more confidently be linked to a later species of our genus, *Homo erectus*. Remains of these tall (1.6–1.8 m) and large-brained creatures (Brown *et al.* 1985a) and their stone tools have been found not only in Africa, but also in Eurasia from Spain to Java. Archeological excavations of butchery sites reveal that *Homo erectus* was in part a big game hunter. They may have used fire for cooking, as indicated by fossil charcoal and burnt earth as old as 1.4 million years at Chesowonja, Kenya (Gowlett *et al.* 1981). Fire can be used to move and corral game and is one way that humans could have substantially altered vegetation and landscape very early in their history.

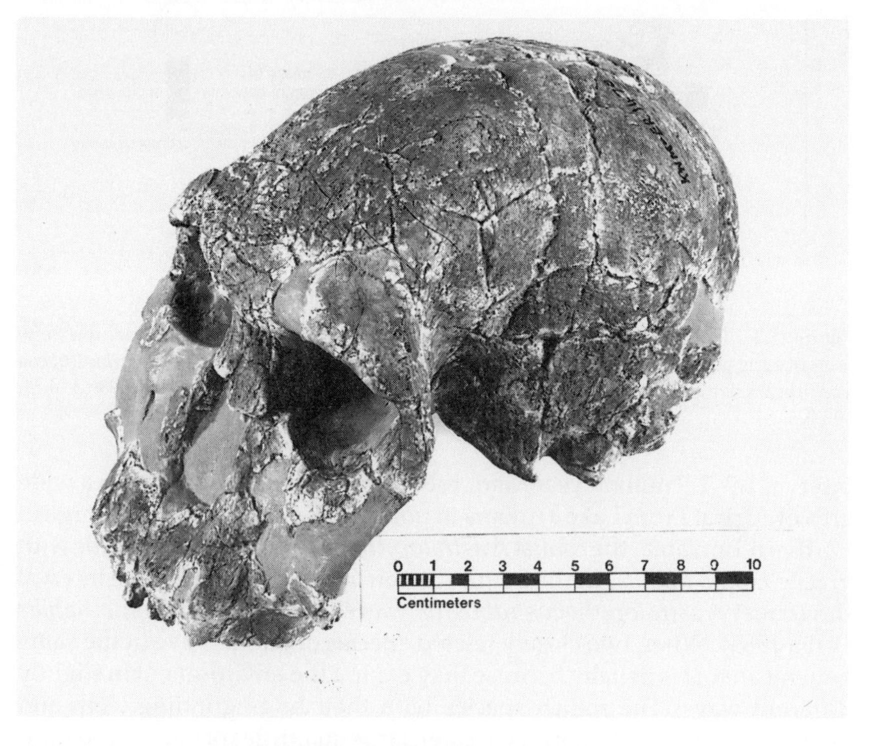

Figure 21.3 Restored skull of *Homo habilis* from the late Pliocene (1.8 million years ago) Koobi Fora Formation, 36 km northeast of Koobi Fora Spit, Kenya (Kenyan National Museum specimen ER 1470; photograph by Sean Poston).

The potential environmental impact of humanity became more and more pronounced with each succeeding culture. A more elaborate tool kit and indications of art and religion are found along with fossils of our own species, *Homo sapiens*. The Neanderthalers were a stocky race best known from the region of the European alps from about 300 000 to 35 000 years ago. Morphologically modern humans may have originated in Africa ca. 100 000 years ago (Stringer & Andrews 1988). Eurasia and Australia were colonized by modern humans ca. 50 000 years ago (White & O'Connell 1979). North and South America may have been invaded as long ago as 30 000 years via Siberia and Alaska when lowered sea level of the last glacial advance created the Bering land bridge. There is some doubt about such an early colonization, but few dispute the peopling of the New World 11 500 years ago by big game hunters with Clovis-style arrowheads. About this time many large animals such as mammoth, mastodon, and groundsloth became extinct. The extinction of large native marsupials such as *Diprotodon* in Australia and of giant flightless birds such as *Dinorthis* in New Zealand at the hands of hunters also would have had significant consequences for vegetation and landscapes (Owen-Smith 1987).

More obvious and far reaching effects of human cultures on the soil were associated with the domestication of animals and the advent of agriculture. Fossil horse incisors with the characteristic wear pattern produced by gnawing poles when tethered have been found in southern France as old as 14 000 years and perhaps as old as 30 000 years. Small carvings and cave paintings of what appear to be bridles on horses are known as old as 14 000 years (Bahn 1984). It is possible that these early Europeans managed herds of horse and other animals in a similar way to modern Lapps who follow reindeer. The rise of city states and their dependence on agriculture can be documented in archeological excavations as old as 10 000 years ago in the Middle East, 9000 years in southeast Asia, and 7500 years in Mesoamerica (Heiser 1981). These apparently independent transitions from hunter–gatherer nomadic societies to sedentary agricultural civilizations on several continents have sparked speculation on more general causes of agriculture (Cohen & Armelagos 1984). Taking a martial view, it could be argued that there was population growth in good years beyond the sustainable carrying capacity of the landscape for hunter–gatherer communities. Extended nomadic wanderings during droughts could have begun to produce intertribal conflicts. Within such a tense political environment, agriculture could have promoted the power and economic influence of particular tribes and villages. Opportunistic views on the origin of agricultural civilizations are based on the idea that human ingenuity exploited new resources in areas that were marginal for traditional kinds of hunting and gathering. In deserts, for example, some oasis date palms may have been an important source of food that later was cultivated. In woodlands also, game is scarce and difficult to hunt. Corn and other plants with large seeds that allow them to germinate in woodland could have assumed increased importance to humans in such regions. Alternatively, the advent of agriculture can be seen as a way of nurturing increased populations that was less arduous than nomadic wandering. Cultivation could have been a consequence of deliberate conservation of particular food plants at the expense of others that were uprooted. Planting also could have arisen from religious ceremonies associated with human burial or fertility rites. By these views agricultural use of plants arose out of cultural forces rather than because of changing or marginal environmental conditions.

This outline of critical events in human prehistory is based in large part on evidence of the traditional materials of physical anthropology and archeology: fossil bones, tools, and other debris of civilization. The questions themselves are too far reaching to be answered on these bases alone. Radiometric dating, interpretation of past sedimentary environments, and other kinds of scientific investigation have played an invaluable role in reconstructing early stages in human evolution. To this increasingly multidisciplinary approach can now be added the evidence of paleosols.

Human origins

What little is known of human evolution tends to reinforce the view that the earliest human ancestors were uncommon creatures living under environmental constraints similar to other animals and having no especially distinctive impact on their environment. If there was no direct impact of the earliest humans on soils, evidence of paleoenvironment from paleosols may prove useful in assessing general theories of human origins. Many ideas about the origin of upright stance involve specific kinds of past vegetation such as wooded grassland. Reconstructions of past vegetation from paleosols can be compared with the adaptations of mammalian faunas and putative human ancestors in order to assess selection pressures on their evolution. If the earliest fossil human ancestors with erect stance were found in woodland paleosols, rather than wooded grassland paleosols, then grassland-dependent martial and opportunistic hypotheses would be falsified in favor of environmentally independent nurturing hypotheses. For the most part, however, the value of such studies will be in providing additional information in a field where any kind of evidence is sparse and difficult to obtain.

Early Miocene forests of southwestern Kenya

The badlands of Koru, Songhor, and Rusinga Island in southwestern Kenya have long been known as a source of fossil dryopithecines, such as the quadrupeds *Proconsul africanus* and *P. nyanzae*, the large *P. major*,

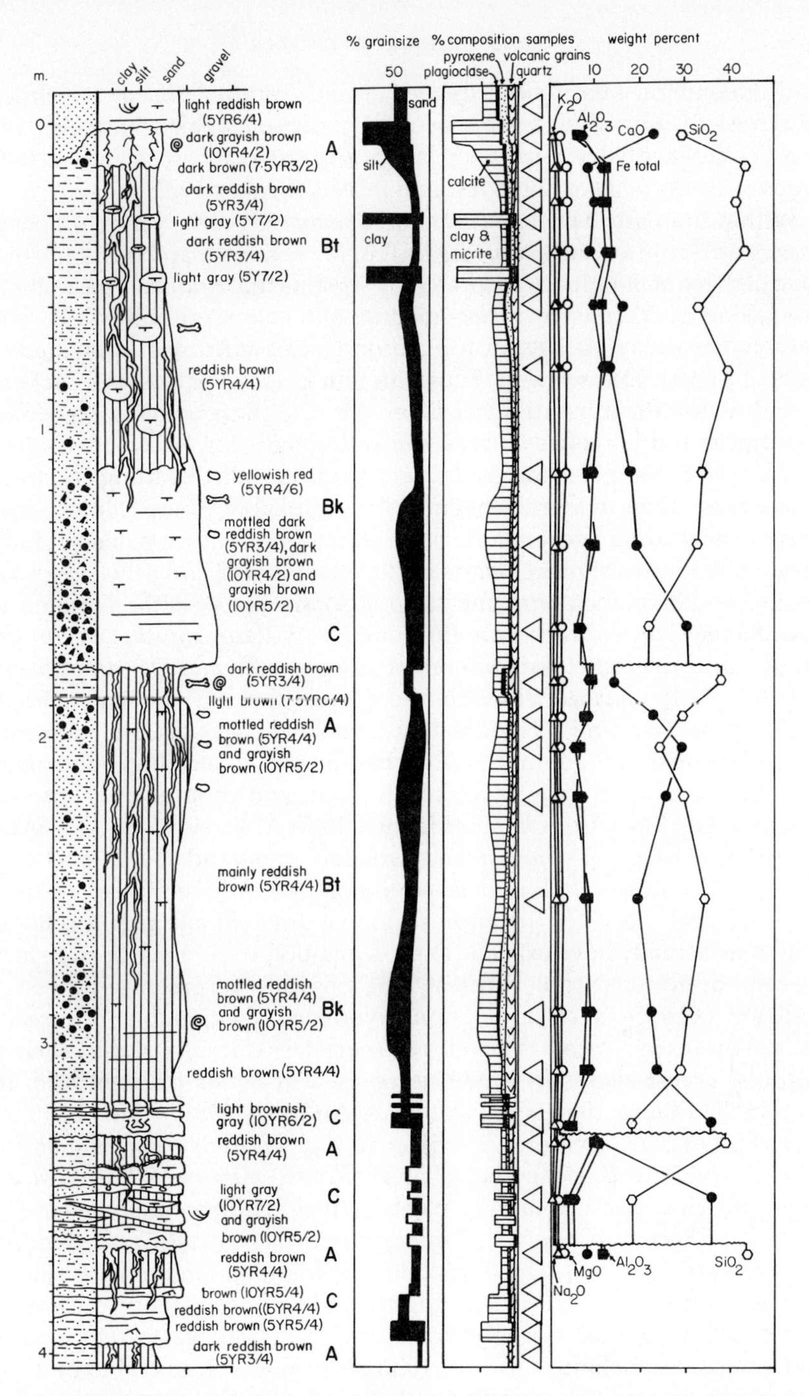

Figure 21.4 Columnar section (measured in field) and petrographic and chemical composition (by point counting by G. S. Smith, and atomic absorption spectrometric chemical analysis by C. McBirney) of the early Miocene (19 million years) Tut (above) and Choka clay paleosols (below), both Haplustalfs, in the Kapurtay Agglomerate, near Songhor, Kenya. Lithological symbols as for Figure 20.7.

and the aboreal *Limnopithecus legetet* and *Dendropithecus macinnesi* (Pickford 1986b). These sequences of paleosols, paleochannels, and carbonatite–nephelinite tuffs and lahars are remarkably fossiliferous with wood, leaves, snails, millipedes, and insects (Leakey 1952).

Within the Early Miocene (19 million years old) carbonatite tuffs near Koru and Songhor, two separate sedimentary facies can be recognized. One of these at Songhor was formed by streams that drained Precambrian crystalline rocks nearby, thus mixing non-calcareous alluvium with carbonatite and nephelinite tuff. A second facies at Koru is dominated by tuffs of a small volcanic vent. Four different kinds of paleosols have been found within the mixed alluvial facies. Most of these contain large fossil root traces and branches of trees, indicating that all were wooded. Buru and Kiewo Series paleosols contain conspicuous relict bedding and fossil root traces, as in modern Fluvents. They probably formed under early successional woody vegetation near watercourses. Kiewo paleosols have yielded 30 species of fossil mammals (listed for beds 5 and 6 by Pickford & Andrews 1981), including the dryopithecine apes (with numbers of specimens) *Proconsul africanus* (3), *P. major* (4), *Rangwapithecus gordoni* (4), *R. vancouveringi* (1), *Limnopithecus legetet* (8), and *Dendropithecus macinnesi* (1). Associated with these well drained lowland alluvial paleosols are two kinds of moderately developed paleosols. The Tut Series formed on mixed alluvium and ash. It has deeply penetrating drab-haloed root traces, a subsurface clayey (Bt) horizon and prominent calcareous nodules (Fig. 21.4). Comparable Haplustalfs in Africa now support small-leaved rain-green dry forest and wooded grassland (FAO 1971–81). These paleosols were calcareous because of their carbonatitic ashy parent material, and this gives the impression of a drier climate and vegetation than may actually have existed. Their vegetation could have been as lush as semi-deciduous tropical forest, but tropical evergreen rainforest is unlikely because such vegetation colonizing comparably calcareous coastal dunes in Africa has developed much less calcareous soils, such as Ultisols. Forest also is in evidence for the Tut Series paleosol from its mammalian fauna (bed 9 of Pickford & Andrews 1981) which includes forest-dependent creatures such as flying squirrels (*Paranomalurus bishopi* and *P. walkeri*) and bush babies (*Komba minor* and *K. robustus*). Dryopithecine apes from this paleosol include *Proconsul major* (2), *Rangwapithecus gordoni* (1), *Limnopithecus legetet* (4), and *Dendropithecus macinnesi* (1). The third kind of lowland moderately developed paleosol, the Choka Series, is distinguished by its consistently calcareous composition, abundant large burrows, and cocoon-like casts. It formed on carbonatite–nephelinite tuff little diluted by non-volcanic alluvium. Its vegetation was probably similar to that of the Tut Series and it contains similar snail, turtle, snake, bird, rodent, and ungulate remains (from bed 8 of Pickford & Andrews 1981).

Other kinds of paleosols have been found in an abandoned quarry 2 km north of Koru. These paleosols fill and cover paleokarst etched into the tops of carbonatite welded tuffs that presumably were on the flanks of a small volcanic vent. Weakly developed paleosols with large boulders of weathered carbonatite are called the Mobaw Series. They represent soils developed on colluvial fill of the paleokarst. Mobaw paleosols probably supported early successional woody vegetation. The fauna of the Mobaw Series is best known from a locality closer to Koru ("maize crib" or number 14 of Pickford 1986b). Included are millipedes, chameleons, varanid lizards, insectivores, chevrotains, and deer-like ruminants. The most fossiliferous level in the abandoned quarry to the north is another kind of paleosol, a brilliant red, uniform profile of the Kwar Series. These paleosols show the beginnings of a clayey (Bt) and calcareous (Bk) subsurface horizon that do not qualify as argillic or calcic. These Ustropepts probably supported dry forest similar to better developed profiles near Songhor. This also is evident from their fossil fauna (locality 10 of Pickford 1986b) of snails, lizards, snakes, birds, insectivores, gomphothere elephants, chalicotheres, chevrotains, deer-like ruminants, and the primates *Proconsul africanus, P. major, Limnopithecus legetet, Micropithecus clarki,* and *M. songhorensis*. A similar fauna has been found in larger collections from a similar paleosol 3 km north of Koru (locality 29 of Pickford & Andrews 1981).

Thus, apart from locally disturbed paleosols under early successional vegetation, these carbonatite–nephelinite volcanoes and nearby lowlands were extensively forested during early Miocene time. These forests were home to a variety of early apes. No evidence of wooded grassland has been discovered this old in Kenya.

Mid-Miocene grassland mosaic of southwestern Kenya

Fort Ternan National Monument, near the railway siding of Fort Ternan in southwestern Kenya, is now a large quarry excavated over many seasons by Louis S. B. Leakey and others in search of the rare fossil ape now known as *Kenyapithecus wickeri*. Compared with older Kenyan sites, few other ape species are found at Fort Ternan, and there is a very different assemblage of fossil antelopes. Paleosols provide clues to environmental changes behind this faunal overturn.

A mosaic of habitats is represented by paleosols in the upper levels of the excavation at Fort Ternan (see Fig. 20.7). The main fossiliferous layer is two superimposed brown paleosols (Chogo Series) with granular-structured, dark surface (A) horizons over calcareous subsurface (Bk) horizons (see Figs. 20.8, 20.9). These are Haplustolls and typical of grassland vegetation. The upper paleosol at the eastern end is the only confirmed source of ramapithecine apes in place. It contains small stump casts (7.5 and 16 cm in diameter and 2 m apart). Also found in this upper paleosol are large nodular masses from which articulated remains of genet-like mongooses (*Kanuites leakeyi*) were found (Savage & Long 1986), the only fully articulated skeletons in this excavation. A small gully-like feature also was unearthed near here (Shipman 1981). This part of the excavation may represent an ecotone into a small, wooded, ephemeral watercourse burrowed by dens of mongooses. This interpretation from paleosols of a mosaic of woodland and wooded grassland is compatible with evidence from fossil plants including fossil grasses, thorny twigs, and fruits of trees and from fossil antelopes that dominate the vertebrate assemblage (Shipman 1986).

Dhero Series paleosols in the excavation here are thin (only 12 cm) Orthents developed on lahars. These supported early successional wooded grassland, as indicated by scattered stump casts up to 31 cm in diameter. An Onuria Series profile has a thicker (19 cm), granular-structured, yellow surface (A) horizon over a thin calcareous (Bk) horizon like Calciustolls of the central Serengeti Plains of today. Remarkably, there are fossil grasses (see Fig. 20.2) arching out of this paleosol and into overlying alluvial grits. Rare finds of dicotyledonous leaf fragments in the surface of this paleosol are an indication that it may also have supported some trees.

Remains of *Kenyapithecus* also are known from other Miocene sites in western Kenya ranging in age from 17–10 million years (Pickford 1985). The paleosols yielding these fossils at Maboko (Andrews *et al.* 1981) appear to have supported seasonally waterlogged, woody vegetation. At Nachola these fossil apes have been found in brown paleosols with stump casts of *Acacia*, a setting suggestive of grassy woodland (Pickford *et al.* 1984). Wooded grassland existed in Africa at the time of *Kenyapithecus*, but these apes were fossilized more often in wooded parts of the vegetation mosaic.

Late Miocene woodlands of northern Pakistan

Throughout much of northern Pakistan and India an enormously thick sequence of alluvial outwash of the Siwalik Group from the rising Himalaya, Karakorum, and Hindu Kush Mountains has preserved a remarkably complete fossil record of Miocene to modern mammalian faunas. That part of the sequence (Dhok Pathan Formation) of late Miocene age (8.3 million years old) has yielded abundant remains of fossils dominated by three-toed horses (*Cormohipparion theobaldi*). Ape fossils are rare, but the recently described fossil face of *Sivapithecus indicus* (Pilbeam 1982) and some limb bones (Rose 1986) provide indications of arboreal habits and an evolutionary relationship to living orangutans. The paleosols in which they are found also provide clues to their ecology.

Near the small village of Kaulial in northern Pakistan, at least eight kinds of paleosols have been found in alluvial sediments about 8.3 million years old (see Figs. 11.7–9). These were deposited by streams draining Himalayan foothills and were fed by an ancestral Ganges River reaching deeply into the metamorphic core of the Himalaya. Ancestral Ganges swampy lowlands were forested, both in clayey swales (Khakistari Series) and on sandy levees (Pandu Series). A greater variety of paleosols were found in deposits of the former foothills drainage. These represent early successional stream-margin vegetation (Sarang and Kala Series), gallery forest (Bhura Series), floodplain forest (Lal and Sonita Series) and wooded grassland (Pila Series). Similar soils and vegetation now are found in the moist tropical monsoon forest belt of the Gangetic plain of Uttar Pradesh. This concept of late Miocene vegetation is well in accord with the vertebrate fauna known from these rocks which includes riverine creatures such as crocodiles, forest browsers such as chevrotains and pigs, and open woodland creatures such as antelope and three-toed horse (Badgley & Behrensmeyer 1980). Ramapithecine primates were rare elements of these diverse tropical faunas and have been found well preserved only in paleochannel deposits and paleosols of associated levees (Sarang Series). They may have lived also in the adjacent gallery forests, but they have not been found in vertebrate assemblages of the other paleoenvironments farther from streams.

Additional support for the idea that ramapithecine primates were riverside woodland or forest apes can be gained from studies of paleosols at higher stratigraphic levels of the Dhok Pathan Formation paleomagnetically dated at about 7.4 million years (see Fig. 18.10). At this level ramapithecines were locally close to extinction, and several immigrant mammalian species such as the antelope *Selenoportax lydekkeri* came to dominate less diverse faunas than existed before (Barry *et al.* 1982). These changes can be related to paleoenvironmental changes evident from paleosols. Only four kinds of paleosols were found, two representing open woodland and two streamside early successional vegetation. Similar soils and vegetation now are found in the dry tropical monsoon forest belt of Punjab.

It is likely that *Ramapithecus* and *Sivapithecus* were dependent on moist forested environments which retreated eastward across the Indian subcontinent as the Himalayan and other ranges cast a lengthening rain shadow. Chinese ramapithecines are found in deposits of swamp woodlands (Etler 1984). The living orangutan (*Pongo pygmaeus*) appears to be the last remnant of a once diverse group of Asian ramapithecine apes.

Initial steps to humanity

From the limited evidence now at hand, wooded grasslands were part of the East African landscape as long ago as 14 million years. This new kind of vegetation could have played a role in early human evolution, but whether it did or not will be better understood when the life habits and stance of fossil apes such as *Kenyapithecus* are better understood. If one takes the view that *Kenyapithecus* walked erect, then their presence near the earliest wooded grassland paleosols in Africa would be supportive of martial or opportunistic hypotheses for the origin of erect stance. On the other hand, *Kenyapithecus*, like *Proconsul* before, and like its probable Asiatic derivatives *Ramapithecus*, *Sivapithecus*, and *Pongo*, more likely remained a partly arboreal, quadrupedal, sexually dimorphic ape (Senut 1988). Indifference to wooded grassland habitats apparent from these preliminary paleosol studies, thus supports nurturing hypotheses for human erect stance. Although paleosols are abundant, widespread, and can be used to constrain hypotheses toward a unique answer, more bones that inform us of the lifestyle of creatures that made the transition from ape to human still are badly needed.

Early human ecology

Ideas concerning the ecology of early human fossils in Africa are anything but consistent. The oldest of these remains of Pliocene age (3–4 million years) are usually referred to a single species *Australopithecus afarensis*. There is evidence from associated fossil antelope and pigs (Johanson *et al.* 1982) and from fossil pollen (Bonnefille *et al.* 1987) that these early australopithecines lived in the Afar region of Ethiopia initially in montane forest, then successively in wooded grassland and woodland. The same species known farther south in Tanzania is thought to have lived in dry wooded grassland, to judge from evidence of fossil footprints, pollen, and vertebrates (Leakey & Harris 1987). Opinion is divided on the ecology of the various sympatric species of Plio-Pleistocene (1.3–1.9 million years) hominids of Africa. One view based on the occurrence of bones in different sedimentary facies near Lake Turkana in Kenya (Behrensmeyer 1982b) is that early humans (*Homo habilis*) lived in wooded grasslands and lake margins, but the robust australopithecines (*Australopithecus boisei*) were restricted to streamside gallery forests. Another view based on fossil antelope associated with early hominid remains in bone breccias of South African caves (Vrba 1985) is that early gracile australopithecines (*A. africanus*) lived in woodlands, the geologically younger robust australopithecines (*A. robustus*) in moist wooded grassland and early humans (*Homo erectus*) in dry wooded grassland. There are distinctive colors and textures of different parts of the South African cave deposits that may reflect changes in vegetation and climate (Butzer 1983). The study of paleosols is a comparable approach for reconstructing alluvial sedimentary environments of Kenya, Tanzania, and Ethiopia.

Pliocene grassland mosaic of the East African Rift

Spectacular evidence of the earliest known erect-walking, human ancestor fossils have been found at two distant locations in East Africa: Laetoli in Tanzania and Hadar in Ethiopia. At Laetoli, melilitite–carbonatite ashes of Pliocene age (3.5–3.7 million years old) have yielded footprints of erect walking early human ancestors and bones and teeth referred to *Australopithecus afarensis* (Leakey & Harris 1987). The footprint-bearing tuffs are a weakly developed Andept paleosol that supported early successional wooded grassland. This is indicated also by footprints of guinea fowl and giraffe. Moderately developed paleosols have also been found at Laetoli and these include extensive termitaries and subterranean nests of bees and wasps. Some of the large blocks of matrix containing these bee nests in the National Museum of Kenya have the structure and dark color of grassland soil surface horizons. These preliminary observations

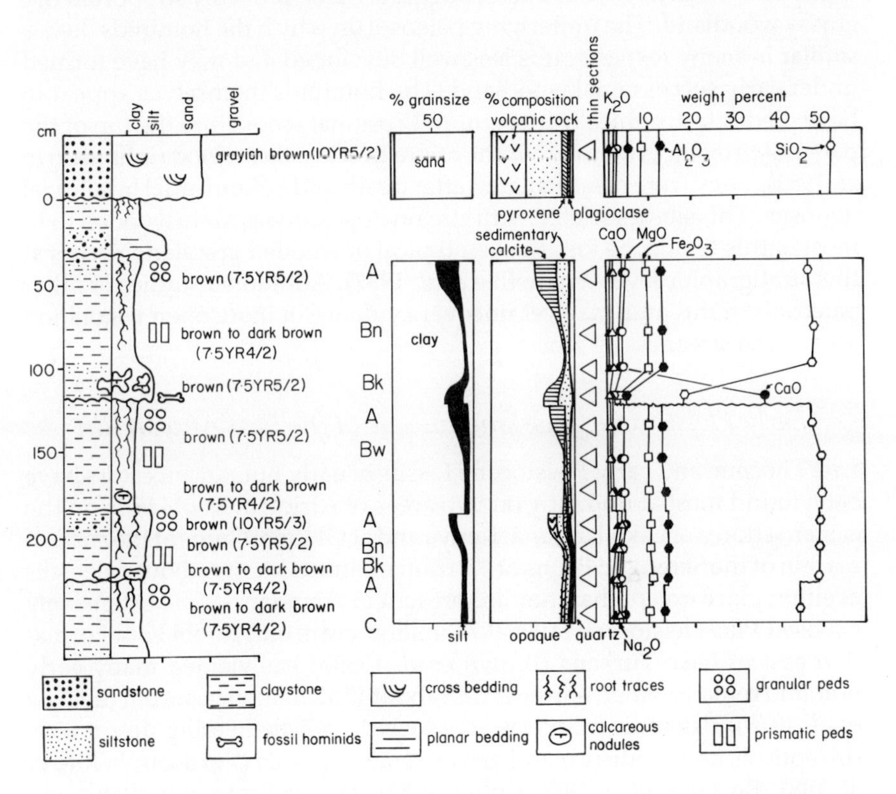

Figure 21.5 Columnar section (measured in field) and petrographic and chemical composition (by point counting by G. S. Smith and atomic absorption spectrometric chemical analysis by C. McBirney) of paleosols in the late Pliocene (3.0 million year old) Denen Dora member of the Hadar Formation at a locality (AL #333) 4 km northwest of the junction of Kada Hadar and the Awash River, Afar region, Ethiopia.

are in accord with evidence from vertebrates. Additional studies of paleosols could provide a clearer picture of non-catastrophic environments undisturbed by volcanism.

Near Hadar in the Afar Triangle of Ethiopia, in the Deren Dora Member of the Hadar Formation of Pliocene age (about 3.0 million years old), have been found a remarkable concentration of skeletal remains of *Australopithecus afarensis*. At one of the most fossiliferous localities (AL #333) there are remains of at least 13 individuals, including three juveniles, within the calcic (Bk) horizon of a paleosol (Johanson *et al*. 1982). The paleosol, developed on alluvial deposits entombing the fossils, has a granular surface (A) horizon and a clayey, prismatic structured subsurface (Bn) horizon (Fig. 21.5). The well formed calcic horizon (Fig. 3.16C) and slight enrichment of sodium in this profile are evidence that it was a Natrustoll (work in progress with S. Radosevich & M. Taieb). This, together with its large root traces, suggests that it probably supported dry grassy woodland. The underlying paleosol on which the hominids died is similar in many respects. It is less well developed and may have formed under early successional woodland. The hominids themselves appear to have died catastrophically in a group. Erosional scours into the top of the paleosol provide evidence that they perished in a flash flood (Johanson *et al*. 1982). They were disarticulated after death and before burial by alluvial deposits. This site associated with stream deposits was more wooded than most of this region, as fossil pollen typical of wooded grassland occurs at this stratigraphic level (Bonnefille *et al*. 1987). Additional studies of other paleosols in this area may yet uncover evidence of more open vegetation away from streams.

Early Pleistocene grassland mosaic of the East African Rift

Late Pliocene and early Pleistocene fossils of early human ancestors have been found most abundantly on two areas of Africa: at Koobi Fora on the eastern shores of Lake Turkana, Kenya and at Olduvai Gorge at the eastern margin of the Serengeti Plains of Tanzania. Little is known about paleosols at either place except that they are present in abundance. The extensively exposed Plio-Pleistocene (4.35–0.74 million years old) Koobi Fora Formation east of Lake Turkana (Brown *et al*. 1985b) has yielded many early hominid remains ranging in age from about 1 to 2 million years old (Leakey *et al*. 1978). Paleosols from this area include caliche-bearing desert soils (Inceptisols and Aridisols) and clayey, cracking soils (Vertisols; White *et al*. 1981, Burggraf *et al*. 1981, Cohen 1982). This indicates woodland and grassland habitats more lush than are there now, but the relative extent of habitats could be refined by detailed study.

A little more is known about paleosols at Olduvai Gorge. This spectacularly exposed sequence ranges in age from Late Pliocene to mid-

Pleistocene (1.9–0.4 million years old). Geological mapping of sedimentary facies (Hay 1976) has shown that the sequence accumulated in a large restricted lacustrine basin like modern Amboseli National Park of Kenya. From my own brief examination of Olduvai Gorge, the paleosols are very similar to modern soils of the Amboseli Basin (as described by Sombroek *et al.* 1982). Lake margin sequences include weakly developed zeolitic paleosols which were presumably alkaline soils (Entisols and Inceptisols) supporting scrubby, salt-tolerant vegetation. Also near lake margin sequences are paleosols with dark-brown, granular surface (A) horizons and calcareous nodular subsurface (Bk) horizons. These may have been grassland soils (Mollisols). A prominent red unit (Bed III) in the middle of the sequence contains numerous well drained, calcareous paleosols (perhaps Inceptisols) with large root traces. These probably were woodland soils normally confined to elevated regions around the lake. They occupied the entire lake basin during a short interval of unusually rainy and freely drained conditions. The waxing and waning importance of nearby montane forests compared with lowland grassland in this sequence is also apparent from studies of fossil pollen (Bonnefille 1984). The gray

Figure 21.6 The quarry face at the 1959 discovery site (FLK) of a robust australopithecine skull (*Australopithecus boisei*) in the lower part of the sequence (Bed I below darker Bed II exposed above) of central Olduvai Gorge, Tanzania. The lower bedded unit includes moderately developed caliche bearing paleosols.

lacustrine facies with brownish gray lowland paleosols returned again during deposition of the uppermost unit of the main sequence (Bed IV).

Within these varied environments, a robust australopithecine skull (*Australopithecus boisei*) was found by Mary Leakey in 1959 from weakly developed zeolitic lakeside paleosols (at site FLK; Fig. 21.6). A stone circle and tools of Oldowan culture (at site DK) were farther from the lake but in similar paleosols formed on sediment filling the bouldery top of a basalt flow. Such observations need to be extended to more sites and related to a more comprehensive accounting of paleosols here before a clear picture of early human ecology will emerge.

Homo erectus is found below the band of red paleosols in Olduvai Gorge at stratigraphic levels as old as 1.5 million years. *Homo erectus* is thought to have used fire in Kenya as early as 1.4 million years ago (Gowlett *et al.* 1981). Early human use of fire for hunting could have had profound effects on the environment. In some ways these effects mimic climatic drying. A more arid and less stable environment is indicated by caliche-bearing paleosols of the Masek, Ndutu and Naisiusiu Beds, ca. 0.5 million years old, that cap the Olduvai sequence (Hay & Reeder 1978). Perhaps this and other thick sequences of Pleistocene paleosols will some day yield evidence of the antiquity and extent of early human dominion over the landscape.

Early human land use

Studies of paleosols associated with early human fossils are still few and far between. So far, however, they have tended to support evidence from associated pollen and vertebrates for a variety of habitats for early humans. Early erect-walking forms such as *Australopithecus afarensis* ranged through woodland and wooded grassland and, to judge from palynological evidence, also through forest. If this pattern of habitat indifference can be substantiated by further studies of the kind outlined here, then hypotheses of human evolution based on family dynamics will look better than the various grassland-dependent martial and opportunistic hypotheses. Late Pliocene and early Pleistocene species of *Australopithecus* and *Homo* evidently lived within similar mosaic habitats represented by a variety of paleosols, but whether they were restricted to particular habitats within the mosaic and what these might have been remain unclear. There is evidence that some of them had stone tools. These have been preserved in a variety of habitats including paleosols of woodland and also of grassland and scrub. Hunting on the scale indicated by the abundance of these tools, and the use of fire by *Homo erectus*, could have had a marked impact on local ecosystems. Like so many other features of early human evolution, however, this also remains to be read from the copious fossil record of paleosols.

A tamed landscape

Some effects of human use of soils are obvious. Traces of walls and roads may remain for millenia. The towns and temples of antiquity have been well and carefully studied for some time. So too have cultivated fields and garbage dumps of ancient civilizations (Butzer 1982), which can reveal aspects of everyday life more readily than more spectacular archeological finds like the tomb of Tutankhamen.

Cultivation and waste disposal leave distinctive traces in soils. Plowing physically destroys natural soil structure creating a disturbed zone of massive structure (plaggen epipedon) down to a sharp base over undisturbed soil. Where the soil is well penetrated and overturned by plowing, the washing down of clay near the base of the plow layer produces a distinctive clayey horizon (agric horizon of Soil Survey Staff 1975). Plowing also can be seen from surface furrowing of the soil, which can be persistent. Waste disposal has the effect of adding not only artifacts to the soil, but also food scraps, bone, shells, and charcoal. Even when these specific waste products are not clearly preserved, surfaces associated with human settlement (anthropic epipedon) can be recognized by their compacted and footworn appearance and by their enrichment in phosphate and organic matter (Eidt 1985).

Such features in soils and paleosols can be used to document the transition from scattered farms to the tailored agricultural landscapes of England or Illinois, or to the urban jungles of Athens or New York. Such changes are also a matter for history. The study of paleosols is of special use as a record of earlier and more subtle phases in human exploitaon of the land. Have densely populated regions been tamed mainly by the Industrial Revolution? Or is this the culmination of millenia of human abuse? These kinds of questions have proven difficult and controversial, but paleopedological approaches can now be applied to them.

Desertification of Greece

It is a sobering experience to visit the sites of ancient Mediterranean civilizations. Where ancient history informs us of the great cedar forests of Phoenicia, there are barren hillsides terraced by farmers in an attempt to curb further erosion (Thirgood 1981). Such observations could be multiplied throughout the region and have fueled suspicions that the deserts of this region were largely the products of human abuse. On the other hand, it could be that the advance of rocky and sandy deserts around the Mediterranean Sea is due to dry climate induced by global changes in wind patterns and temperature. In this view, human settlement was incidental to or hastened desertification that already was in progress

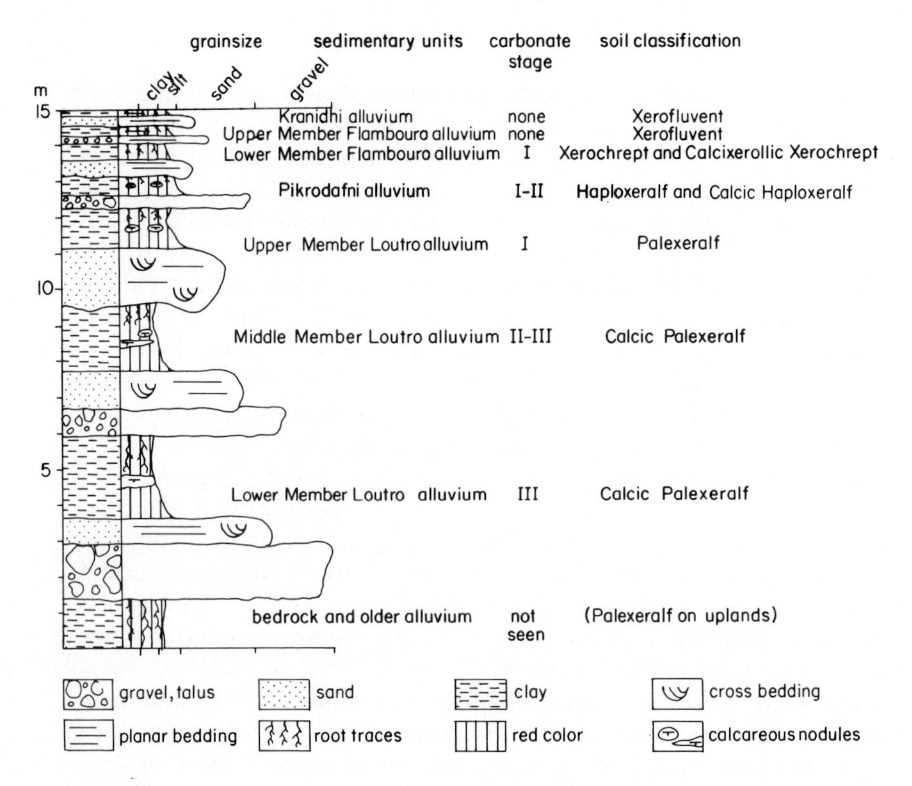

Figure 21.7 Composite diagram of alluvial and soil stratigraphy of the southern Argolid, Greece (from Pope & Van Andel 1984, reprinted with permission of Academic Press).

(Williams 1979). The effects of climate and humans in creating deserts are difficult to disentangle.

One place where human modification of soils has been documented is in the southern Argolid near Kranidhion, Greece (Pope & Van Andel 1984). This is a mountainous region with relief of up to 1000 m, enjoying a warm (mean July temperature 23°C and January 5°C), summer dry, subhumid (mean annual precipitation 932 mm) climate. By means of geomorphic mapping, with attention to associated artifacts and radiocarbon dating, it has been possible to subdivide alluvial valley fills here into several sedimentary units formed during brief periods of destabilization. Each unit is separated by paleosols of varying development (Fig. 21.7). The periods of landscape erosion do not seem to have been caused by climatic, sea-level, or tectonic changes. They each can be related to the waxing and waning of different human cultures in the area (Fig. 21.8).

During Neolithic times (about 4500 years ago) the land was stable (as indicated by a well developed paleosol) and populations were low (low density of archeological sites). From pollen records in coastal swamps and

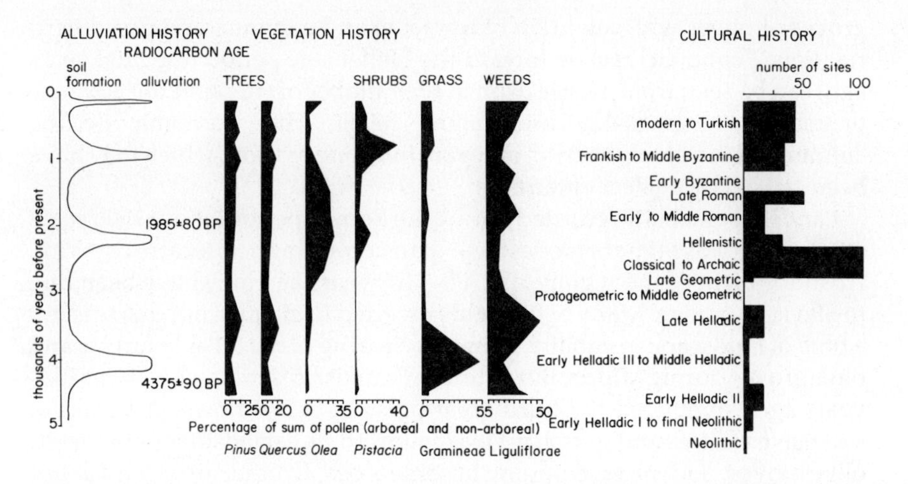

Figure 21.8 History of alluviation, vegetation and culture of the past 5000 years in southern Argolid, Greece. Periods of soil formation and alluviation are generalized from Figure 21.10. Pollen data are from a coastal bog (from Pope & Van Andel 1984, reprinted with permission of Academic Press).

the nature of these soils and paleosols, the vegetation of this time can be reconstructed as maquis (indicated by *Pistacia lentiscus* and *P. terebinthus*) on the hilltops, with oak (mainly *Quercus ilex*) parkland and oak–pine (*Pinus halepensis* and *Quercus ilex*) streamside gallery woodland at lower elevations. At this time the cultivation of wheat, barley, oats and lentils was limited to scattered, well watered lowland sites. Sheep and goats also were herded.

With the advent of bronze and pottery styles of the Early Helladic period (ca. 4000 years ago) there was considerable population growth, and cultivation and deforestation expanded into the hills. This early land abuse stimulated catastrophic slope erosion and a crisis for human populations, which consequently decreased. During the ensuing slow repopulation of the Middle and Late Helladic period, forests returned as grassland and maquis shrunk. Cultivation during this period may have been limited to terraces. Olive (*Olea europaea*) groves were extended. Multiple cropping of olives, vines, figs, cereals, and other crops, with pasturing of long-fallow fields of weeds (indicated by Liguliflorae pollen), reduced soil erosion. This was the time of Mycenae and the Homeric legends. The dark ages that followed the collapse of these late Helladic cultures (ca. 3200 years ago) may have been politically caused. They were not a response to environmental deterioration in this area and do not coincide with a period of alluviation.

During the ensuing repopulation of the Archaic to early Classical period (about 3200–2300 years ago), larger areas of woodland were converted to groves of olives with retention of terracing and continued crop and herd rotation. Economic decline toward the Hellenistic period (ca. 2200 years ago) can be seen from a large drop in the number of sites and the advance of maquis, presumably into fallow fields. This economic decline culminated in a second period of severe hillslope erosion, which may have been caused by neglect of terraces.

Landscape stability returned during the Roman period (about 1500 years ago), which was a time of modest re-expansion and agricultural prosperity. Abandonment of the region ca. 1300–1400 years ago may have been due to Slavic invasions. Many of the fields reverted to maquis, but this was not a time of landscape instability. Careless clearing of the hillside forests and olive groves during Middle Byzantine to Frankish resettlement (about 1000 years ago) stimulated the third great pulse of debris flows. A period of extensive maquis and grassland was followed by establishment of forest, olive groves, and mixed cropping in some areas. In other areas land abuse and alluviation have continued up to the present.

This reconstruction of human impact on a Mediterranean landscape over the past five millenia reveals that population pressures can have severe consequences for landscape stability. Forest clearance and neglect of soil conservation practices such as terracing are most damaging. Abandonment of the landscape for political reasons, as at the end of the Late Helladic and Late Roman periods, had little impact on landscape stability. There also were long periods of moderately intense land use and stability during Archaic to Classical and Roman times. Although the Argolid, like many other parts of the Mediterranean, is now more rocky and barren than it was in Neolithic times, its tamed landscape is more than ever dependent on human care for conserving whatever productivity remains.

Acidification of the British Lake District

The cool and humid highlands of the British Isles and Scandinavia are barren and wild landscapes that have inspired generations of poets and musicians with their rocky crags, sweeping slopes of purple-flowered heather (*Calluna vulgaris*), and spreading bogs of moss (*Sphagnum* spp.). These low-diversity plant assemblages have been regarded as the final phases of a series of vegetation changes following retreat of continental ice caps from these regions ca. 13 000 years ago (Table 21.1). The till and clay left by the glaciers was essentially unevenly ground rock, rich in weatherable minerals. With climatic improvement, these raw soils (Entisols) were colonized by grasses and other herbaceous plants (such as *Rumex acetosella*), forming vegetation like Arctic tundra. This was followed by Inceptisols supporting woodlands of juniper (*Juniperus communis*) and birch (*Betula nana*) like the taiga of northern Siberia. With

Table 21.1 Traditional view of north-west European pedogenic trends during post-glacial times, 13,000 years to present (from Macphail 1986)

Period	Climate	Pedogenic process	Archaeology
Sub-Atlantic	Cool, wet oceanic	Hydromorphism (peat formation)	Medieval Saxon Romano-British Iron Age
2,500			
Sub-Boreal	Warm, dry	Podzolization	Bronze Age (Copper Age) Neolithic
4,500			
Atlantic	Warm, wet "climatic optimum"	Lessivage	Mesolithic
7,500			
Boreal	Relatively warm, dry	Decalcification	
9,000			
Pre-Boreal	Sub-Arctic	Raw Soils	
10,000 (cool temperate Alleröd Interstadial)			Upper Palaeolithic
Late Glacial	Sub-Arctic		
13,000 yrs BP			

further climatic amelioration and weathering, brown, clayey soils (Alfisols) were formed under oak (*Quercus petraea*) forest. Decalcification in these soils under the cool, wet climatic regime proceeded until few weatherable minerals remained. Acidic podzolized soils (Spodosols) were then invaded by pines (*Pinus sylvestris*) and heather (*Calluna vulgaris*). In boggy depressions, moss (*Sphagnum*) formed peaty accumulations (Histosols). In this view, acidification of upland soils is part of a natural cycle of weathering and the highlands are true primeval wilderness.

There is, however, another view that podzolization and hydromorphism of upland regions are products of human forest clearance, and catastrophic loss of clay and nutrients, in soils that otherwise might have continued to support montane woodlands (Macphail 1986). In this view, the highlands are less wild than they may seem.

These issues have been addressed by studies of soil and vegetation change in the Langdale and Coniston Fells in the heart of the Lake District of northwestern England (Pennington 1965, Pennington *et al.* 1977, Hall & Folland 1970). This is a mountainous region with elevations of up to

963 m. Its present climate is cold (mean January temperature 4°C and July 15°C at Keswick) and rainy (889 mm at Penrith to 4699 mm annually in the central mountains).

Human settlement of coastal lowlands around the Lake District dates back ca. 7000 years. Neolithic settlements and agriculture were widespread in coastal lowlands by 5000 years ago (Walker, D. 1966). The mountains, however, remained forested with pine (*Pinus*), birch (*Betula*), oak (*Quercus*) and elm (*Ulmus glabra*) to elevations of 600 m, as revealed by fossil pollen and by old forest soils (Dystrochrepts and Hapludalfs) developed on glacial till and outwash (Fig. 21.9). During this time the hard volcanic ashes high in the hills of the Great Langdale, in the heart of the Lake District, were quarried for hand axes. Few complete hand axes have been found, perhaps because the quarries were used by summer visitors who collected only rough axe heads for final polishing on their return home to the lowlands (Plint 1962). A strong decline in elm (*Ulmus*) pollen is found in the highlands ca. 4200 years ago, as in many other settled parts of Neolithic Europe. The decline has been attributed to the cutting of elm as feed for stock, even in highland regions remote from settlements. The decline also could be due to a virulent pathogen similar to Dutch elm disease (Watts 1985) or to a change to cooler and wetter paleoclimate (Khotinskiy 1984). It is possible that all three agents were significant.

Forest clearance of the uplands in the Lake District began sporadically from about 5000 years ago as revealed by pollen of weeds associated with human settlement, such as plantain (*Plantago lanceolata*). The stone circles of the area, such as Castlerigg, are thought to date from the Beaker culture about 4000 years ago. Large areas had been cleared by the Bronze Age some 3000 years ago. Forest clearance can be detected in lowland bogs (Fig. 21.10) by a marked influx of illitic clay, rich in potassium, that diluted

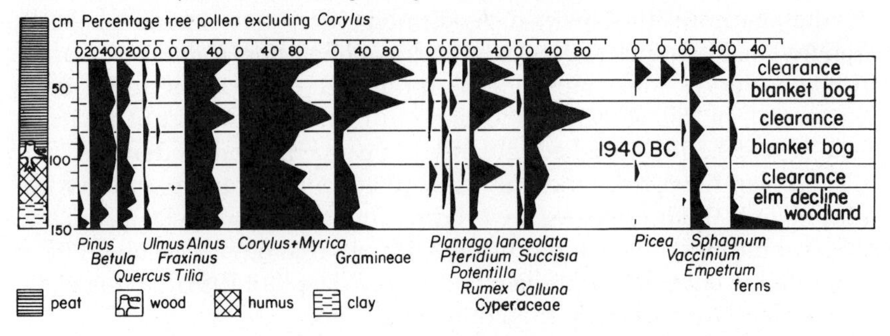

Figure 21.9 Pollen diagram from peat (Fibrist) overlying an old woodland soil (Dystrochrept) in an upland bog (Red Tarn Moss), high in the hills near Great Langdale, Lake District, England. Heather (*Calluna*) and moss (*Sphagnum*) expanded after each episode of forest clearance and invasion of weeds (*Plantago, Rumex, Potentilla* and Cyperaceae) (from Pennington 1965, with permission of the Royal Society of London).

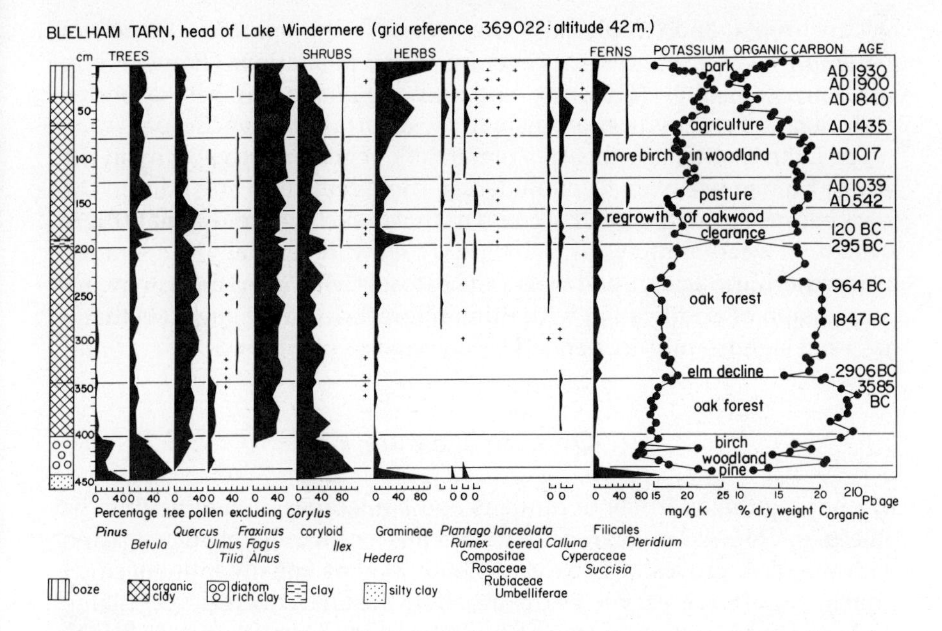

Figure 21.10 Pollen diagram of a lowland lake (Blelham Tarn) near Lake Windermere in the Lake District, England. Influx of clayey sediment (mainly illite, and so reflected in increased potassium in analysis) is correlated with the elm (*Ulmus*) decline and with subsequent episodes of forest clearance indicated by pollen of weeds (*Plantago*) and crops (cereals) (modified from Pennington 1965, Pennington *et al.* 1977).

the normal accumulation of peat. The cleared uplands mostly were invaded by heath (*Calluna*) on Aquods in fairly well drained sites or by moss (*Sphagnum*) on Fibrists in waterlogged depressions. A record of both kinds of vegetation is well preserved in moss peat that overlies the stumps of these former forests. These upland soils may already have been deeply weathered, but the loss of clays following clearance left them permanently acidic. Had forests been left untouched, recycling of nutrients through leaf fall may have enabled forests to persist for longer, as they appear to have done in other areas of the Lake District. The onset of blanket peat coincides in many bogs with pollen of weeds (*Plantago lanceolata*) that implicate forest clearance rather than climatic change as the cause.

Under Roman domination ca. 1700–1200 years ago, much of the easily cleared uplands were under cultivation for cereals. Lowland forests were not cleared, perhaps because they were too difficult to drain, plow, or clear. By the time of the Viking raids about 1000 years ago, Norse-speaking immigrants began to clear and cultivate the mountain valleys. Upland rangelands were turned over to sheep grazing. This pattern of agriculture persisted until about 200 years ago when cheaper imported cereals stimulated a shift entirely to grazing. Now uncultivated lowland fields on

Dystrochrepts and Hapludalfs are invaded by bracken (*Pteridium aquilinum*). Their progressive cover by timber plantations (*Pinus, Picea*, and *Larix*), heather (*Calluna*), and moss (*Sphagnum*) has extended podzolization and hydromorphism to most parts of the landscape.

The Lake District has shown a remarkable regenerative capacity in the face of human use over seven millenia. There remains a splendor in the grass because of high rainfall. It could be that in such a humid climate there is a natural weathering cycle in which raw soils will eventually be degraded by weathering to acidic Spodosols and Histosols. However, the patchwork progression of acidification with human forest clearance suggests that it has been significantly accelerated by human use of the land.

On human nature

In studying the interplay of humans and landscapes, it is difficult to be objective. The wild moorlands of Britain may seem untouched compared with the olive groves of Greece, yet both may be equally influenced by human impact. It is easy to be disturbed by the modern erosion of hiking trails and proliferation of tourist facilities in the Lake District or by the expansion of heavy industry and slum housing in Greece. But what of the generations of accumulated environmental degradation that preceded these modern excesses? The long-term behavior of landscapes and soils needs to be assessed in establishing the carrying capacity of landscapes for human societies.

Objectivity also has been difficult to attain in studies of human origins. During the war years of the early 1940s and the colonial era before that, martial hypotheses of human origins were popular. Technological innovations of the 1950s encouraged opportunistic theories of human origins. With the rise of the women's movement and the study of modern hunter–gatherer societies in the 1960s, nurturing hypotheses have become more popular. Political currents appear to have guided views on human origins and early ecology as much as the scientific evidence itself.

Perhaps we humans will always be too myopic or timid to accept a true view of our place in nature. There is promise, however, that the study of paleosols added to an ever expanding array of other approaches to assessing our prehistory will converge upon a consistent view of our past. Perhaps by that time our biological nature may be seen more as an archaic heritage than as an imperative that constrains our freedom of choice and survival.

References

Aandahl, A. R. 1982. *Soils of the Great Plains*. Lincoln: University of Nebraska Press.

Ahmad, N. & R. L. Jones 1969. Genesis, chemical properties and mineralogy of limestone-derived soils, Barbados, West Indies. *Tropical Agriculture* **46**, 1–15.

Al-Janabi, A. M. & J. H. Drew 1967. Characterization and genesis of a Sharpsburg–Wymore soil sequence in southwestern Nebraska. *Proceedings of the Soil Science Society of America* **31**, 238–44.

Allaart, J. H. 1976. The pre-3760 m.y. old supracrustal rocks of the Isua area, central West Greenland, and the occurrence of quartz-banded ironstones. In *The early history of the Earth*, B. F. Windley (ed.), 177–89. London: Wily-Interscience.

Allan, J. R. & R. K. Matthews 1982. Isotopic signatures associated with early meteoric diagenesis. *Sedimentology* **29**, 797–817.

Allen, J. R. L. 1965. Fining upwards cycles in alluvial successions. *Geological Journal* **4**, 229–46.

Allen, J. R. L. 1973. Compressional structures (patterned ground) in Devonian pedogenic limestones. *Nature* **243**, 84–6.

Allen, J. R. L. 1974. Geomorphology of Siluro-Devonian alluvial plains. *Nature* **249**, 644–5.

Allen, J. R. L. 1986a. Pedogenic calcretes in the Old Red Sandstone facies (Late Silurian–Early Carboniferous) of the Anglo–Welsh area, southern Britain. In *Paleosols: their recognition and interpretation*, V. P. Wright (ed.), 58–86. Oxford: Blackwell.

Allen, J. R. L. 1986b. Time scales of colour change in late Flandrian intertidal muddy sediments of the Severn Estuary. *Proceedings of the Geological Association* **97**, 23–8.

Allen, K. C. 1980. A review of *in situ* Late Silurian and Devonian spores. *Review of Palaeobotany and Palynology* **29**, 253–70.

Allen, P. 1947. Notes on Wealden fossil soil beds. *Proceedings of the Geologists' Association* **57**, 303–14.

Almond, J. E. 1985. The Silurian–Devonian fossil record of the Myriapoda. *Philosophical Transactions of the Royal Society of London* **B309**, 227–37.

Altschuler, Z. S., M. M. Schnepfe, C. C. Silber & F. O. Simon 1983. Sulfur diagenesis in Everglades peat and the origin of pyrite in coal. *Science* **221**, 221–7.

Alvarez, W. R. 1986. Toward a theory of impact crises. *Eos* **67**, 649–58.

Alvin, K. L. 1971. *Weichselia reticulata* (Stokes et Webb) Fontaine from the Wealden of Belgium. *Memoirs de l'Institut Royal des Sciences Naturelles de Belgique* No. 166.

Alvin, K. L., C. J. Fraser & R. A. Spicer 1981. Anatomy and palaeoecology of *Pseudofrenelopsis* and associated conifers in the English Wealden. *Palaeontology* **24**, 759–78.

American Commission of Stratigraphic Nomenclature 1961. Code of stratigraphic nomenclature. *Bulletin of the American Association of Petroleum Geologists* **45**, 645–65.

Anderson, D. M. & A. Banin 1974. Soil and water and its relationship to the origin of life. In *Cosmochemical evolution and the origins of life*, J. Oró, S. L. Miller, C. Ponnamperuma & R. S. Young (eds.), vol. 2, 23–36. Dordrecht: Reidel.

Anderson, J. M. & H. M. Anderson 1985. *Palaeoflora of southern Africa. Prodromus of South Africa megafloras: Devonian to Lower Cretaceous*. Rotterdam: Balkema.

Andreis, R. R. 1981. *Identificacion e importancia geologica de los paleosuelos*. Porto Alegre: Universitaria Federal do Rio Grande do Sul.

Andreis, R. R., R. Leguizamon & S. Archangelsky 1986. El paleovalle de Malanzan: nuevos criterios para la estratigraphia del Neopaleozoico do la Sierra de Los Llanos, La Rioja, Republica Argentina. *Boletin de la Academia Nacional de Ciencias Cordoba*, Argentina **57**, 1–119.

Andrews, P. W., G. E. Meyer, D. R. Pilbeam, J. A. Van Couvering & J. A. H. Van Couvering 1981. The Miocene fossil beds of Maboko Island, Kenya: geology, age, taphonomy and palaeontology. *Journal of Human Evolution* **10**, 35–48.

Arkley, R. J. 1963. Calculation of carbonate and water movement in soil from climatic data. *Soil Science* **96**, 239–48.

Arrhenius, S. A. 1909. *The life of the universe as conceived by man from the earliest ages to the present* (translated by H. Borns), 2 vols. London: Harper & Row.

Ascaso, C., J. Galvan & C. Ortega 1976. The pedogenic action of *Parmelia conspersa, Rhizocarpon geographicum* and *Umbilicaria pustulata. Lichenologist* **8**, 151–71.

Aubert, H. & M. Pinta 1977. *Trace elements in soils* (translated by L. Zuckerman and P. Segalen). Amsterdam: Elsevier.

Augustithis, S. S. & J. Otteman 1966. On diffusion rings and sphaeroidal weathering. *Chemical Geology* **1**, 201–9.

Awramik, S. M. 1971. Precambrian columnar stromatolite diversity: reflection of metazoan appearance. *Science* **174**, 825–6.

Baas-Becking, L. G. M., I. R. Kaplan & D. Moore 1960. Limits of the natural environment in terms of pH and oxidation–reduction potentials. *Journal of Geology* **68**, 243–84.

Badgley, C. 1986. Taphonomy of mammalian fossil remains from Siwalik rocks of Pakistan. *Paleobiology* **12**, 119–42.

Badgley, C. & A. K. Behrensmeyer 1980. Paleoecology of Middle Siwalik sediments and faunas, northern Pakistan. *Palaeogeography, Palaeoclimatology, Palaeoecology* **30**, 133–55.

Bahn, P. G. 1984. *Pyreneean prehistory*. Warminster: Aris & Phillips.

Baker, V. R. 1982. *The channels of Mars*. Austin: University of Texas Press.

Bakker, R. T. 1983. The deer flees, the wolf pursues: incongruencies in predator–prey coevolution. In *Coevolution*, D. J. Futuyma & M. Slatkin (eds.), 350–82. Sunderland: Sinauer.

Bakker, R. T. 1985. *The dinosaur heresies*. New York: William Morrow.

Banin, A. 1986. Clays on Mars. In *Clay minerals and the origin of life*, A. G. Cairns-Smith & H. Hartman (eds.), 106–15. Cambridge: Cambridge University Press.

Banks, H. P. 1969. Occurrence of *Cooksonia*, the oldest vascular land plant macrofossil, in the Upper Silurian of New York state. *Journal of the Indian Botanical Society, Golden Jubilee Volume* **50A**, 227–35.

Barbieri, M. 1985. *The semantic theory of evolution*. London: Harwood Scientific.

Barbour, E. H. 1897. Nature, structure and phylogeny of *Daemonelix. Bulletin of the Geological Society of America* **8**, 305–14.

Bardossy, G. 1982. *Karst bauxites*. Amsterdam: Elsevier.

Barghoorn, E. S. 1981. Aspects of Precambrian paleobiology. In *Paleobotany, paleoecology and evolution*, vol. 1, K. J. Niklas (ed.), 1–16. New York: Praeger.

Barghoorn, E. S. & S. A. Tyler 1965. Microorganisms from the Gunflint Chert. *Science* **147**, 563–77.

Barker, R. J., R. E. McDole & G. H. Logan 1983. *Idaho soils atlas*. Moscow: University of Idaho Press.

Barley, K. P. 1959. Earthworms and soil fertility. IV. The influence of earthworms on the physical properties of a red brown earth. *Australian Journal of Agricultural Research* **10**, 371–6.

Barrell, J. 1913. The Upper Devonian delta of the Appalachian Geosyncline. *American Journal of Science* **36**, 429–72.

Barrientos, X. & J. Selverstone 1987. Metamorphosed soils as stratigraphic indicators in deformed terranes: an example from the eastern Alps. *Geology* **15**, 841–4.

Barry, J. C., E. H. Lindsay & L. L. Jacobs 1982. A biostratigraphic zonation of the middle and upper Siwaliks of the Potwar Plateau of northern Pakistan. *Palaeogeography, Palaeoclimatology, Palaeoecology* **37**, 95–130.

Barshad, I. 1966. The effect of a variation in precipitation on the nature of clay mineral formation in soils from acid and basic igneous rocks. *Proceedings of International Clay Conference, Jerusalem* **1**, 167–73.

Barsukov, V. L., V. P. Volkhov & I. L. Khodakovsky 1982. The crust of Venus: theoretical models of chemical and mineral composition. *Journal of Geophysical Research* Suppl. 87A, 3–9.

Basan, P. 1979. Trace fossil nomenclature: the developing picture. *Palaeogeography, Palaeoclimatology, Paleoecology* **28**, 143–67.

Basilevsky, A. T., R. O. Kuzmin, O. V. Nikolaeva, A. A. Pronin, A. B. Ronca, V. S. Avdvesky, G. R. Uspensky, Z. P. Cheremukhina, V. V. Semenchenko & V. M. Ladygin 1985. The surface of Venus as revealed by the Venera landings: Part II. *Bulletin of the Geological Society of America* **96**, 137–44.

Baskin, Y. 1956. A study of authigenic feldspars. *Journal of Geology* **64**, 132–55.

Basu, A. 1981. Weathering before the advent of land plants: evidence from unaltered K-feldspars in Cambrian–Ordovician arenites. *Geology* **9**, 132–3.

Basu, A., D. S. McKay & T. Gerke 1987. Petrology and provenance of Apollo 15 drive tube 15007/8. In *Proceedings of the 18th Lunar and Planetary Science Conference*, G. Ryder (ed.), 283–98. Cambridge: Cambridge University Press.

Bathurst, R. G. C. 1975. *Carbonate sediments and their diagenesis*, 2nd edn. Amsterdam: Elsevier.

Batten, D. J. 1973. Palynology of early Cretaceous soil beds and associated strata. *Palaeontology* **16**, 399–424.

Beadle, N. C. W. 1981. *The vegetation of Australia*. Cambridge: Cambridge University Press.

Beck, C. B. 1960. The identity of *Archaeopteris* and *Callixylon*. *Britttonia* **12**, 351–68.

Beck, C. B. 1967. *Eddya sullivanensis* gen. et sp. nov., a plant of gymnospermous morphology from the Upper Devonian of New York. *Palaeontographica* **B121**, 1–22.

Beck, C. B. 1981. *Archaeopteris* and its role in vascular plant evolution. In *Paleobotany, paleoecology and evolution*, K. J. Niklas (ed.), 193–230. New York: Praeger.

Beerbower, R. 1985. Early development of continental ecosystems. In *Geologic factors and the evolution of plants*, B. H. Tiffney (ed.), 47–91. New Haven: Yale University Press.

Beeunas, M. A. & L. P. Knauth 1985. Preserved stable isotopic signature of subaerial diagenesis in the 1.2 b.y. Mescal Limestone, central Arizona: implications for the timing and development of a terrestrial plant cover. *Bulletin of the Geological Society of America* **96**, 737–45.

Behrensmeyer, A. K. 1982a. Time resolution in fluvial vertebrate assemblages. *Paleobiology* **8**, 211–27.

Behrensmeyer, A. K. 1982b. The geological context of human evolution. *Annual Reviews of Earth and Planetary Sciences* **10**, 39–60.

Behrensmeyer, A. K. & L. Tauxe 1982. Isochronous fluvial systems in Miocene deposits of northern Pakistan. *Sedimentology* **29**, 331–52.

Behrensmeyer, A. K., D. Western & D. E. Dechant-Boaz 1979. New perspectives in vertebrate paleoecology from a bone bed assemblage. *Paleobiology* **5**, 12–21.

Bell, R. H. V. 1982. The effect of soil nutrient availability on community structure in African ecosystems. In *Ecology of tropical savannas*, B. J. Huntley & B. J. Walker (eds.), 193–216. Berlin: Springer.

Bequaert, J. & F. M. Carpenter 1941. The antiquity of social insects. *Psyche* **48**, 50–5.

Bergstrom, J. 1979. Morphology of fossil arthropods as a guide to phylogenetic relationships. In *Arthropod phylogeny*, A. P. Gupta (ed.), 3–56. New York: Van Nostrand-Reinhold.

Bernabo, J. C. & T. Webb 1977. Changing patterns in the Holocene pollen record from northeastern North America: a mapped summary. *Quaternary Research* **8**, 64–96.

Berry, E. W. 1923. Pathological conditions among fossil plants. In *Paleopathology*, R. L. Moodie (ed.), 99–108. Chicago: University of Chicago Press.

Bertrand-Sarfati, J. & A. Moussine-Pouchkine 1983. Pedogenetic and diagenetic fabrics in upper Proterozoic Sarnyéré Formation (Gourma, Mali). *Precambrian Research* **20**, 225–42.

Besly, B. M. & P. Turner 1983. Origin of red beds in a moist tropical climate (Etruria Formation, Upper Carboniferous, U.K.). In *Residual deposits: surface related weathering processes and materials*, R. C. L. Wilson (ed.), 131–47. Oxford: Blackwell.

Bethke, C. M. & S. P. Altaner 1986. Layer-by-layer mechanism of smectite illitization and application of a new rate law. *Clays and Clay Minerals* **34**, 136–42.

Biju-Duval, B., M. Deynoux & P. Rognon 1981. Late Ordovician tillites of the central Sahara. In *Earth's pre-Pleistocene glacial record*, M. J. Hambrey & W. B. Harland (eds.), 99–107. Cambridge: Cambridge University Press.

Binder, A. B., R. E. Arvidson, E. A. Guiness, K. L. Jones, E. C. Morris, T. A. Mutch, D. C. Pieri & C. Sagan 1977. The geology of the Viking lander 1 site. *Journal of Geophysical Research* **82**, 4439–51.

Birkeland, P. W. 1984. *Soils and geomorphology*. New York: Oxford University Press.

Blatt, H. 1979. Diagenetic processes in sandstones. In *Aspects of diagenesis*, P. A. Scholle & P. R. Schluger (eds.), 141–57. *Special Publication of the Society of Economic Paleontologists and Mineralogists*, No. 26.

Blodgett, R. H. 1988. Calcareous paleosols in the Triassic Dolores Formation, southwestern Colorado. In *Paleosols and weathering through geologic time: principles and applications*, J. Reinhardt & W. R. Sigleo (eds.), 103–21. *Special Papers of the Geological Society of America* No. 216.

Bodman, G. B. & A. J. Mahmud 1932. The use of moisture-equivalents in the textural classification of soils. *Soil Science* **33**, 363–74.

Bohn, H., B. McNeal & G. O'Connor 1985. *Soil chemistry*, 2nd edn. New York: Wiley.

Bonnefille, R. 1984. Palynological research at Olduvai Gorge. *National Geographic Research Report* **17**, 227–43.

Bonnefille, R., A. Vincens & G. Buchet 1987. Palynology, stratigraphy and paleoenvironment of a Pliocene hominid site (2.9–3.3 m.yr) at Hadar, Ethiopia. *Palaeogeography, Palaeoclimatology, Palaeoecology* **60**, 249–81.

Booth, W. E. 1941. Algae as pioneers in plant succession and their importance in erosion control. *Ecology* **22**, 38–46.

Boucot, A. J., J. F. Dewey, D. L. Dineley, R. Fletcher, W. K. Fyson, J. G. Griffin, C. F. Hickox, W. S. McKerrow & A. M. Zeigler 1974. The geology of the Arisaig area, Antigonish County, Nova Scotia. *Special Papers of the Geological Society* No. 139.

Bouma, A. H. 1969. *Methods for the study of sedimentary structure*. New York: Wiley-Interscience.

Bowen, H. J. M. 1979. *Environmental chemistry of the elements*. London: Academic Press.

Bowles, C. G. & W. A. Braddock 1963. Solution breccies of the Minnelusa Formation in the Black Hills, South Dakota and Wyoming. *Professional Paper of the U.S. Geological Survey* **475C**, 91–5.

Bown, T. M. 1982. Ichnofossils and rhizoliths of the nearshore Jebel Qatrani Formation (Oligocene), Fayum Province, Egypt. *Palaeogeography, Palaeoclimatology, Palaeoecology* **40**, 255–309.

Bown, T. M. & M. J. Kraus 1981a. Lower Eocene alluvial paleosols (Willwood Formation, northwest Wyoming, U.S.A.) and their significance for paleoecology, paleoclimatology and basin analysis. *Palaeogeography, Palaeoclimatology, Palaeoecology* **34**, 1–30.

Bown, T. M. & M. J. Kraus 1981b. Vertebrate fossil-bearing paleosol units (Willwood Formation, northwest Wyoming, U.S.A.): implications for taphonomy, biostratigraphy and assemblage analysis. *Palaeogeography, Palaeoclimatology, Palaeoecology* **34**, 31–56.

Bown, T. M. & M. J. Kraus 1983. Ichnofossils of the alluvial Willwood Formation (Lower Eocene), Bighorn Basin, northwest Wyoming, U.S.A. *Palaeogeography, Palaeoclimatology Palaeoecology* **43**, 95–128.

Bown, T. M. & M. J. Kraus 1987. Integration of channel and floodplain suites in aggrading fluvial systems. I. Developmental sequence and lateral relations of lower Eocene alluvial paleosols, Willwood Formation, Bighorn Basin, Wyoming. *Journal of Sedimentary Petrology* **57**, 587–601.

Brady, L. F. 1947. Invertebrate tracks from the Coconino Sandstone of northern Arizona. *Journal of Paleontology* **21**, 466–72.

Brain, C. K. 1981. *The hunters or the hunted? An introduction to African cave taphonomy*. Chicago: University of Chicago Press.

Brain, C. K., C. S. Churcher, J. D. Clark, F. E. Grine, P. Shipman, R. L. Sosman, A. Turner & V. Watson 1988. New evidence of early hominids and their culture and environment from the Swartkrans Cave, South Africa. *South African Journal of Science* **84**, 828–36.

Braterman, P. S., A. G. Cairns-Smith & R. W. Sloper 1983. Photooxidation of hydrated Fe^{2+} enhanced by deprotonation: significance for banded iron formations. *Nature* **303**, 163–4.

Bray, A. A. 1985. The evolution of terrestrial vertebrates: environmental and physiological considerations. *Philosophical Transactions of the Royal Society of London* **B309**, 289–322.

Breithaupt, B. H. & D. Duvall 1986. The oldest record of serpent aggregation. *Lethaia* **19**, 181–5.

Brewer, R. 1976. *Fabric and mineral analysis of soils*, 2nd edn. New York: Krieger.

Brewer, R. & J. R. Sleeman 1969. The arrangement of constituents in Quaternary soils. *Soil Science* **107**, 435–41.

Brewer, R., K. A. W. Crook & J. G. Speight 1970. Proposal for soil stratigraphic units in the Australian stratigraphic code. *Journal of the Geological Society of Australia* **17**, 103–9.

Briand, F. & J. E. Cohen 1987. Environmental correlates of food chain length. *Science* **238**, 956–60.

Bridge, J. S. 1985. Paleochannel patterns inferred from alluvial deposits: a critical evaluation. *Journal of Sedimentary Petrology* **55**, 579–89.

Brook, G. A., M. E. Folkoff & E. O. Box 1983. A world model of soil carbon dioxide. *Earth Surface Processes Landforms* **8**, 79–88.

Brooks, H. K. 1955. Healed wounds and galls on fossil leaves from the Wilcox deposits (Eocene) of western Tennessee. *Psyche* **62**, 1–9.

Brooks, R. R. 1987. *Serpentine and its vegetation*. Portland: Dioscorides.

Brown, F. H., R. Harris, R. E. Leakey & A. C. Walker 1985a. Early *Homo erectus* skeleton from west Lake Turkana, Kenya. *Nature* **316**, 788–92.

Brown, F. H., I. McDougall, T. Davies & R. Maier 1985b. An integrated Plio-Pleistocene chronology for the Turkana Basin. In *Ancestors: the hard evidence*, E. Delson (ed.), 82–90. New York: Alan R. Liss.

Brown, R. W. 1934. *Celliforma spirifer*, the fossil larval chambers of mining bees. *Journal of the Washington Academy of Sciences* **24**, 532–9.

Brown, R. W. 1941a. The comb of a wasp nest from the Upper Cretaceous of Utah. *American Journal of Science* **239**, 54–6.

Brown, R. W. 1941b. Concerning the antiquity of social insects. *Psyche* **48**, 105–10.

Bryan, K. & C. C. Albritton 1943. Soil phenomena as evidence of climatic changes. *American Journal of Science* **241**, 467–90.

Buckland, W. 1837. *Geology and mineralogy considered with reference to natural theology*, 2 vols. London: W. Pickering.

Buechner, H. K. & H. C. Dawkins 1961. Vegetation changes induced by elephants and fire in Murchison Falls National Park, Uganda. *Ecology* **42**, 752–66.

Bullock, P., N. Fedoroff, A. Jongerius, T. Tursina & U. Babel 1985. *Handbook of soil thin section description*. Albrighton: Waine Research.

Bunch, T. E. 1975. Petrography and petrology of basaltic achondritic polymict breccias (howardites). *Geochimica et Cosmochimica Acta, Supplement* **6**, 469–92.

Bunch, T. E. & S. Chang 1980. Carbonaceous chondrites. II. Carbonaceous chondrite phyllosilicates and light element geochemistry as indicators of parent body processes and surface conditions. *Geochimica et Cosmochimica Acta* **44**, 1543–77.

Buol, S. W. & F. D. Hole 1961. Clay skin genesis in Wisconsin soils. *Proceedings of the Soil Science Society of America* **25**, 377–9.

Buol, S. W., F. D. Hole & R. D. McCracken 1980. *Soil genesis and classification*, 2nd edn. Ames: Iowa State University Press.

Burggraf, D. R., H. J. White, H. J. Frank & C. F. Vondra 1981. Hominid habitats in the Rift Valley, part 2. In *Hominid sites: their geologic settings*, G. Rapp & C. F. Vondra (eds.), 115–47. Boulder: Westview.

Burke, R. M. & P. W. Birkeland 1979. Reevaluation of multiparameter relative dating techniques and their application to the glacial sequence along the eastern escarpment of the Sierra Nevada, California. *Quaternary Research* **11**, 21–51.

Button, A. & N. Tyler 1981. The character and significance of Precambrian paleoweathering and erosion surfaces in southern Africa. In *Economic geology: 75th anniversary volume 1905–1980*, B. J. Skinner (ed.), 686–709. El Paso: Economic Geology.

Butzer, K. W. 1982. *Archaeology as human ecology: method and theory for a contextual approach*. Cambridge: Cambridge University Press.

Butzer, K. W. 1983. Sediment matrices of the South African australopithecines: a review. In *Hominid origins*, K. J. Reichs (ed.), 101–8. Washington: University Press of America.

Byrnes, J. G., T. D. Rice & D. Karaolis 1977. *Guilielmites* formed from phosphatized concretions in the Ashfield Shale of the Sydney Basin. *Records of the Geological Survey of New South Wales* **18**, 169–200.

Cady, J. G. & R. B. Daniels 1968. Genesis of some very old soils – the Paleudults. *Transactions of the 9th International Soil Science Congress, Adelaide* **4**, 103–12.

Cairns-Smith, A. G. 1971. *The life puzzle*. Toronto: University of Toronto Press.

Cairns-Smith, A. G. 1982. *Genetic takeover*. Cambridge: Cambridge University Press.

Cairns-Smith, A. G. & H. Hartman (eds.) 1986. *Clay minerals and the origin of life*. Cambridge: Cambridge University Press.

Calkin, P. E. & J. M. Ellis 1984. Development and application of a lichenometric dating curve, Brooks Range, Alaska. In *Quaternary dating methods*, W. C. Mahaney (ed.), 227–46. Oxford: Elsevier.

Callen, R. A. & R. H. Tedford 1976. New late Cenozoic rock units and depositional environments, Lake Frome area, South Australia. *Transactions of the Royal Society of South Australia* **100**, 125–67.

Campbell, F. H. A. & M. P. Cecile 1981. Evolution of the early Proterozoic Kilohigok Basin, Bathurst Inlet–Victoria Island, North West Territories. In *Proterozoic basins of Canada*, F. H. A. Campbell (ed.), 103–31. *Papers of the Geological Survey of Canada* No. 81–10.

Campbell, I. H. & G. G. C. Claridge 1987. *Antarctica: soils, weathering processes and environment*. Amsterdam: Elsevier.

Campbell, S. E. 1979. Soil stabilization by a prokaryotic desert crust: implications for Precambrian land biota. *Origins of Life* **9**, 335–48.

Carpenter, K. 1982. Baby dinosaurs from the Late Cretaceous Lance and Hell Creek Formation and a description of a new species of theropod. *Contributions to Geology of the University of Wyoming* **20**, 123–34.

Carr, M. H. 1981. *The surface of Mars*. New Haven: Yale University Press.

Carroll, R. L., E. S. Belt, D. L. Dineley, D. Baird & D. C. McGregor 1972. Vertebrate paleontology of eastern Canada. *Excursion Guidebook for the 24th International Geological Congress, Montreal* No. A59.

Carroll, R. L. 1988. *Vertebrate paleontology and evolution*. New York: W. H. Freeman.

Chaloner, W. G. & M. Muir 1968. Spores and floras. In *Coal and coal-bearing strata*, D. Murchison & T. S. Westoll (eds.), 127–46. New York: Elsevier.

Chaloner, W. G. & A. Sheerin 1979. Devonian macrofloras. In *The Devonian System*, M. R. House, C. T. Scrutton & M. G. Bassett (eds.), 145–61. *Special Papers in Palaeontology* No. 23.

Chaloner, W. G. & A. Sheerin 1981. The evolution of reproductive strategies in early land plants. In *Evolution today*, G. G. E. Scudder & J. Reveal (eds.), 93–100. Pittsburg: Hunt Institute for Botanical Documentation.

Chamberlain, T. C. 1895. The classification of American glacial deposits. *Journal of Geology* **3**, 270–7.

Chang, S. & T. E. Bunch 1986. Clays and organic matter in meteorites. In *Clays and the origin of life*, A. G. Cairns-Smith & H. Hartman (eds.), 116–29. Cambridge: Cambridge University Press.

Chapman, D. J. 1985. Geological factors and biochemical aspects of the origin of land plants. In *Geologic factors and the evolution of plants*, B. H. Tiffney (ed.), 23–45. New Haven: Yale University Press.

Chapman, V. J. 1977. *Wet coastal ecosystems. Ecosystems of the world* No. 1. Amsterdam: Elsevier.

Chen, J. H. & G. R. Tilton 1976. Isotopic lead investigations on the Allende carbonaceous chondrite. *Geochimica et Cosmochimica Acta* **40**, 635–43.

Chesworth, W. 1973. The parent rock effect in the genesis of soils. *Geoderma* **10**, 215–25.

Chiarenzelli, J. R., J. A. Donaldson & M. Best 1983. Sedimentology and stratigraphy of the Thelon Formation and the sub-Thelon regolith. *Papers of the Geological Survey of Canada* 83-1A, 443–5.

Chilingarian, G. V. & K. H. Wolf (eds.) 1976. *Compaction of coarse-grained sediments, II.* Amsterdam: Elsevier.

Chou, C.-L., W. V. Boynton, R. W. Bild, J. Kimberlin & J. T. Wasson 1976. Trace element evidence regarding a chondritic component in howardite meteorites. *Proceedings of the 7th Lunar Science Conference. Geochimica et Cosmochimica Acta, Suppl.* **7**, 3501–8.

Clark, G. R. & R. A. Lutz 1980. Pyritization in the shells of living bivalves. *Geology* **8**, 268–71.

Clarke, R. S., E. Jarosewich, B. Mason, J. Nelen, M. Gomez & J. R. Hyde 1970. The Allende, Mexico, meteorite shower. *Smithsonian Contributions to the Earth Sciences* **5**, 1–53.

Clements, F. E. 1928. *Plant succession and indicators*. New York: H. W. Wilson.

Clemmey, H. & N. Badham 1982. Oxygen in the Precambrian atmosphere: an evaluation of the geological evidence. *Geology* **10**, 141–6.

Cline, M. G. 1953. Major kinds of profiles and their relationships in New York. *Proceedings of the Soil Science Society of America* **17**, 123–7.

Cline, M. G. 1955. Soil survey of the territory of Hawaii. *U.S. Department of Agriculture Series* 1939, No. 25.

Cloud, P. E. 1976. The beginnings of biospheric evolution and their biochemical consequences. *Paleobiology* **2**, 351–87.

Cloud, P. & K. R. Lajoie 1980. Calcite-impregnated defluidization structures in littoral sands of Mono Lake, California. *Science* **210**, 1009–12.

Cohen, A. S. 1982. Paleoenvironments of root casts from the Koobi Fora Formation, Kenya. *Journal of Sedimentary Petrology* **52**, 401–14.

Cohen, A. S. 1985. The Okefenokee Swamp: a low sulfur end-member of a depositional model for coastal plain coals. In *Economic geology: coal, oil and gas*, A. T. Cross (ed.), 189–298. *Comptes Rendus Neuvième Congrès International de Stratigraphie et de Géologie du Carbonifère* 4.

Cohen, M. N. & G. J. Armelagos (eds.) 1984. *Paleopathology at the origins of agriculture*. Orlando: Academic Press.

Cole, A. C. 1968. Pogonomyrmex *harvester ants*. Knoxville: University of Tennessee Press.

Cole, M. M. 1986. *The savannas*. London: Academic Press.

Coleman, J. M. 1988. Dynamic changes and processes in the Mississippi River Delta. *Bulletin of the Geological Society of America* **100**, 999–1015.

Coleman, M. L. 1985. Geochemistry of diagenetic non-silicate minerals: kinetic considerations. *Philosophical Transactions of the Royal Society of London* **A315**, 39–56.

Collins, W. H. 1925. North shore of Lake Huron. *Memoirs of the Geological Survey of Canada* 143.

Colman, S. M. 1986. Levels of time information in weathering measurements, with examples from weathering rinds on volcanic clasts in the western United States. In *Rates of chemical weathering of rocks and minerals*, S. M. Colman & D. P. Dethier (eds.), 379–93. Orlando: Academic Press.

Compston, W. & R. T. Pidgeon 1986. Jack Hills, evidence of more very old detrital zircons in Western Australia. *Nature* **321**, 766–9.

Cooke, W. B. 1979. *The ecology of fungi*. Boca Raton: CRC Press.

Cooper, K. W. 1964. The first fossil tardigrade: *Beorn leggi* Cooper, from Cretaceous amber. *Psyche* **71**, 41–8.

Cope, M. J. & Chaloner, W. G. 1985. Wildfire: an interaction of biological and physical processes. In *Geologic factors and the evolution of plants*, B. H. Tiffney (ed.), 257–77. New Haven: Yale University Press.

Corliss, J. B., J. A. Baross & S. E. Hoffman 1981. An hypothesis concerning the relationship between submarine hot springs and the origin of life on Earth. *Oceanologica Acta* No. S-P, *Proceedings of the 26th International Geological Congress, Paris, 1980* 59–69.

Cornet, B. 1986. The leaf venation and reproductive structures of a Late Triassic angiosperm, *Sanmiguelia lewisii*. *Evolutionary Theory* **7**, 231–309.

Cotter, E. 1978. The evolution of fluvial style, with special reference to the central Appalachian Paleozoic. In *Fluvial sedimentology*, A. D. Miall (ed.), 361–83. *Memoir of the Canadian Society of Petroleum Geologists* 5.

Coyne, L., G. Pollack & R. Kloepping 1984. Room-temperature luminescence from kaolin induced by organic amines. *Clays and Clay Minerals* **32**, 58–66.

Cracraft, H. J. & N. Eldredge (eds.) 1979. *Phylogenetic analysis and paleontology*. New York: Columbia University Press.

Crane, P. R. & E. A. Jarzembowski 1980. Insect leaf mines from the Palaeocene of southern England. *Journal of Natural History* **14**, 629–36.

Creber, G. T. & W. G. Chaloner 1985. Tree growth in the Mesozoic and early Tertiary and the reconstruction of palaeoclimates. *Palaeogeography, Palaeoclimatology, Palaeoecology* **52**, 35–60.

Cremeens, D. L. & D. L. Mokma 1986. Argillic horizon expression and classification in soils of two Michigan hydrosequences. *Journal of the Soil Science Society of America* **50**, 1002–7.

Crepet, W. L. & D. W. Taylor 1985. The diversification of the Leguminosae: first fossil evidence of the Mimosoideae and Papilionoideae. *Science* **228**, 1087–9.

Crick, F. H. C. 1981. *Life itself: its origin and nature*. Touchstone: Simon & Schuster.

Crow, J. H. & Cooper, A. F. 1971. Crytobiosis. *Scientific American* 225 (12), 30–6.

Cruikshank, D. P. & R. H. Brown 1987. Organic matter on asteroid 130 Elektra. *Science* **238**, 183–4.

Crumpler, L. S., J. W. Head & D. B. Campbell 1986. Orogenic belts on Venus. *Geology* **14**, 1031–4.

Cummings, M. L. & J. V. Scrivner 1980. The saprolite at the Precambrian–Cambrian contact, Irvine Park, Chippewa Falls, Wisconsin. *Transactions of the Wisconsin Academy of Sciences, Arts and Letters* **68**, 22–9.

Curtis, C. D. 1985. Clay mineral precipitation and transportation during burial diagenesis. *Philosophical Transactions of the Royal Society of London* **A315**, 91–105.

Daghlian, C. P. 1981. A review of the fossil record of monocotyledons. *Botanical Review* **47**, 517–55.

Daily, B. &. M. R. Cooper 1976. Clastic wedges and patterned ground in the Late Ordovician–Early Silurian tillites of South Africa. *Sedimentology* **23**, 271–83.

Damuth, J. 1982. Analysis of the preservation of community structure in assemblages of fossil mammals. *Paleobiology* **8**, 434–46.

Dan, J. & D. H. Yaalon 1982. Automorphic soils in Israel. In *Aridic soils and geomorphic processes*, D. H. Yaalon (ed.), 103–15. *Catena, Supplement* No. 1.

Dash, B., K. N. Sahni & D. A. Bowes 1987. Geochemistry and original nature of Precambrian khondalites in the eastern Ghats, Orissa, India. *Transactions of the Royal Society of Edinburgh, Earth Sciences* **78**, 115–27.

Davis, W. M. 1899. The geographic cycle. *Geographical Journal* **14**, 481–504.

de Coninck, F., D. Righi, J. Maucorps & A. M. Robin 1974. Origin and micromorphological nomenclature of organic matter in sandy spodosols. In *Soil microscopy*, G. K. Rutherford (ed.), 263–80. Kingston: Limestone Press.

Decourten, F. L. & B. Bovee 1986. *Stigmaria* from the upper Carboniferous Rest Spring Shale and a possible Lepidodendrale rooting ground, Death Valley region, California. *Abstracts of the 82nd Annual Meeting of the Cordilleran Section, Geological Society of America* **18**, 100.

Deer, W. A., R. A. Howie & J. Zussman 1962–3. *Rock forming minerals*, 5 vols. New York: Wiley.

de Wit, H. A. 1978. *Soils and grassland types of the Serengeti Plain (Tanzania)*. Wageningen: Mededelingen Landbouwhogeschool.

de Wit, M. J., R. Hart, A. Martin & P. Abbott 1982. Archaean biogenic structures associated with mineralized hydrothermal vent system and regional metasomatism, with implication for greenstone belt studies. *Economic Geology* **77**, 1783–802.

di Castri, F., D. W. Goodall & R. L. Specht 1981. *Mediterranean shrublands. Ecosystems of the world*, No. 11. Amsterdam: Elsevier.

Dickinson, W. R. & C. A. Suczek 1979. Plate tectonics and sandstone composition. *Bulletin of the American Association of Petroleum Geologists* **63**, 2164–82.

Dickson, J. A. D. & M. L. Coleman 1980. Changes in carbon and oxygen isotopic composition during limestone diagenesis. *Sedimentology* **27**, 107–18.

Dimichele, W. A., T. L. Phillips & R. G. Olmstead 1987, Opportunistic evolution: abiotic environmental stress and the fossil record of plants. *Review of Palaeobotany and Palynology* **50**, 151–78.

Dixon, J. B. & S. B. Weed (eds.) 1977. *Minerals in soil environments*. Madison: Soil Science Society of America.

Dixon, J. C. & R. C. Young 1981. Character and origin of deep arenaceous weathering mantles on the Bega batholith, southwestern Australia. *Catena* **8**, 97–109.

Dodson, P. 1971. Sedimentology and taphonomy of the Oldman Formation (Campanian), Dinosaur Provincial Park, Alberta (Canada). *Palaeogeography, Palaeoclimatology, Palaeoecology* **10**, 21–74.

Dodson, P., A. K. Behrensmeyer, R. T. Bakker & J. McIntosh 1980. Paleoecology of the Jurassic Morrison Formation, western North America. *Paleobiology* **6**, 208–32.

Dokuchaev, V. V. 1883. Russian chernozem (Russkii chernozem). In *Collected writings (Sochineye)*, vol. 3 (translated by N. Kaner, 1967). Jerusalem: Israel Program for Scientific Translations.

Dorn, R. I. & T. M. Oberlander 1982. Rock varnish. *Progress in Physical Geography* **6**, 317–67.

Dott, R. H. 1974. Cambrian tropical storm waves in Wisconsin. *Geology* **2**, 243–6.

Douglas, J. G. & G. E. Williams 1982. Southern polar forests: the Early Cretaceous floras of Victoria and their paleoclimatic significance. *Palaeogeography, Palaeoclimatology, Palaeoecology* **39**, 171–85.

Droser, M. L. & D. J. Bottjer 1986. A semiquantitative field classification of ichnofabric. *Journal of Sedimentary Petrology* **56**, 558–9.

Dubiel, R. F., R. H. Blodgett & T. M. Bown 1987. Lungfish burrows in the Upper Triassic Chinle and Dolores Formations, Colorado Plateau. *Journal of Sedimentary Petrology* **57**, 512–21.

Duchač, K. C. & J. S. Hanor 1987. Origin and timing of metasomatic silicification of an Early Archaean komatiite sequence, Barberton Mountain Land, South Africa. *Precambrian Research* **37**, 125–46.

Dumanski, J. & R. J. St. Arnaud 1966. A micropedological study of eluvial soil horizons. *Canadian Journal of Soil Science* **46**, 287–92.

Dunham, R. J. 1969. Vadose pisolites in the Capitan reef (Permian), New Mexico and Texas. In *Depositional environments in carbonate rocks*, G. M. Friedman (ed.), 182–91. *Special Publication, Society of Economic Paleontologists and Mineralogists, Tulsa* No. 14.

Dunn, J. R. 1953. The origin of deposits of tufa in Mono Lake. *Journal of Sedimentary Petrology* **23**, 18–23.

Eagleman, J. C. 1980. *Meteorology*. New York: Van Nostrand.

Edwards, D. & D. S. Edwards 1986. A reconsideration of the Rhyniophytina Banks. In *Systematic and taxonomic approaches in palaeobotany*, R. A. Spicer & B. A. Thomas (eds.), 201–22. *Special Volume Systematics Association* No. 31. Oxford: Oxford University Press.

Edwards, P. & R. L. Folk 1979. Coprolites. In *The encyclopedia of paleontology*, R. W. Fairbridge & D. Jablonski (eds.) 224–5. Stroudsburg: Dowden, Hutchinson & Ross.

Edwards, R. & K. Atkinson 1986. *Ore deposit geology*. London: Chapman & Hall.

Eggert, D. A. 1961. The ontogeny of Carboniferous arborescent Lycopsida. *Palaeontographica* **B108**, 43–92.

Eggert, D. A. 1962. The ontogeny of Carboniferous Sphenopsida. *Paleontographica* **B110**, 99–127.

Eggleton, R. A. 1986. The relation between crystal structure and silicate weathering rates. In *Rates of chemical weathering of rocks and minerals*, S. M. Colman & D. P. Dethier (eds.), 21–40. New York: Academic Press.

Eglinton, G., J. R. Maxwell & C. T. Pillinger 1972. The carbon chemistry of the Moon. *Scientific American* **227** (4), 81–90.

Eidt, R. C. 1985. Theoretical and practical considerations in the analysis of anthrosols. In *Archaeological geology*, G. Rapp & J. A. Gifford (eds.), 155–90. New Haven: Yale University Press.

Eigen, M., W. Gardiner, P. Schuster & R. Winkler-Oswatitsch 1981. The origin of genetic information. *Scientific American* **244** (4), 88–115.

Elliott, R. E. 1985. Quantification of peat to coal compaction stages, based especially on phenomena in the East Pennine coalfield, England. *Proceedings of the Yorkshire Geological Society* **45**, 163–72.

Endrody-Younga, S. & S. B. Beck 1983. Onychophora from mesic grassland in South Africa (Onychophora: Peripatopsidae). *Annals of the Transvaal Museum* **33**, 347–52.

Engel, A. E., S. P. Itson, C. G. Engel, D. M. Stickney & E. J. Cray 1974. Crustal evolution and global tectonics: a petrogenic view. *Bulletin of the Geological Society of America* **85**, 843–58.

Erhart, H. 1965. Le temoinage paleoclimatique de quelques formations paleopediques dans leur rapport avec la sedimentologie *Geologische Rundschau* **54**, 15–23.

Essene, E. J. & D. C. Fisher 1986. Lightning strike fusion: extreme reduction and metal–silicate liquid immiscibility. *Science* **234**, 189–93.

Esteban, M. & C. F. Klappa 1983. Subaerial exposure environment. In *Carbonate depositional environments*, P. A. Scholle, D. G. Bebout & C. H. Moore (eds.), 1–54. Tulsa: American Asociation of Petroleum Geologists.

Etler, D. 1984. The fossil hominoids of Lufeng, Yunnan Province, The People's Republic of China: a series of translations. *Yearbook of Physical Anthropology* **27**, 1–56.

Ettensohn, F. R., G. R. Dever &. T. S. Grow 1988. A paleosol interpretation for profiles exhibiting subaerial exposure "crusts" from the Mississippian of the Appalachian Basin. In *Paleosols and weathering through geologic time: principles and applications*, J. Reinhardt & W. R. Sigleo (eds.), 35–48. *Special Paper of the Geological Society of America* No. 216.

Eugster, H. P. 1969. Inorganic bedded cherts from the Magadi area, Kenya. *Contributions to Mineralogy and Petrology*, 1–31.

Eugster, H. P. & L. A. Hardie 1975. Sedimentation in an ancient playa lake complex: the Wilkins Peak Member of the Green River Formation of Wyoming. *Bulletin of the Geological Society of America* **86**, 319–34.

Evans, G., J. W. Murray, H. E. J. Biggs, R. Bate & P. Bush 1973. The oceanography, ecology, sedimentology and geomorphology of parts of the Trucial Coast barrier island complex, Persian Gulf. In *The Persian Gulf*, B. H. Purser (ed.), 233–77. Berlin: Springer.

Evans, H. E. & M. J. W. Eberhard 1970. *The wasps*. Ann Arbor: University of Michigan Press.

Evans, J. G. 1972. *Land snails in archaeology*. New York: Seminar.

Evenari, M., I. Noy-Meir & D. W. Goodall (eds.) 1986. *Hot deserts and arid shrublands. Ecosystems of the world*, Nos. 12A & 12B. Amsterdam: Elsevier.

Falini, F. 1965. On the formation of coal deposits of lacustrine origin. *Bulletin of the Geological Society of America* **76**, 1317–46.

Fallou, F. A. 1862. *Pedologie: oder allgemeine und besondere Bodenkunde*. Dresden: G. Schonfeld.

Faniran, A. 1971. The parent material of Sydney laterites. *Journal of the Geological Society of Australia* **18**, 159–64.

FAO 1971–81. *Soil map of the world 1 : 5,000,000*, 10 vols. Paris: UNESCO.

Farrow, C. E. G. & D. J. Mossman 1988. Geology of Precambrian paleosols at the base of the Huronian Supergroup, Elliot Lake, Ontario, Canada. *Precambrian Research* **42**, 107–39.

Feakes, C. R. & G. J. Retallack 1988. Recognition and characterization of fossil soils developed on alluvium: a Late Ordovician example. In *Paleosols and weathering through geologic time: principles and applications*, J. Reinhardt and W. R. Sigleo (eds.), 35–48. *Special Paper, Geological Society of America* No. 216.

Feduccia. A. 1980. *The age of birds*. Cambridge: Harvard University Press.

Fehrenbacher, J. B., I. J. Jansen & K. R. Olson 1986. Loess thickness and its effect on soils in Illinois. *Bulletin of the Illinois Agricultural Experimental Station* No. 782.

Ferguson, D. K. 1985. The origin of leaf assemblages – new light on an old problem. *Review of Palaeobotany and Palynology* **46**, 117–88.

Fischer, W. A. 1978. The habitat of the early vertebrates: trace and body fossil evidence from the Harding Formation (Middle Ordovician), Colorado. *Mountain Geologist* **15**, 1–26.

Fisher, D. C. 1979. Evidence for subaerial activity of *Euproops danae* (Merostoma, Xiphosurida). In *Mazon Creek fossils*, M. H. Nitecki (ed.), 379–447. New York: Academic Press.

Fisher, G. C. & O.-L. Yam 1984. Iron mobilization by heathland plant extracts. *Geoderma* **32**, 339–45.

Fisher, R. V. & H.-U. Schminke 1984. *Pyroclastic rocks*. New York: Springer.

Fitzpatrick, E. A. 1980. *Soils*. London: Longman.

Fitzpatrick, E. A. 1984. *Micromorphology of soils*. New York: Chapman & Hall.

Flach, K. W., W. D. Nettleton, L. H. Gile & J. G. Cady 1969. Pedocementation: induration by silica, carbonates and sesquioxides in the Quaternary. *Soil Science* **107**, 442–53.

Flannery, T. 1982. Hindlimb structure and evolution in the kangaroos (Marsupialia: Macropodoidea). In *The fossil vertebrate record of Australasia*, P. V. Rich & E. M. Thompson (eds.), 507–23. Clayton: Monash University Press.

Fogel, R. & F. Hunt 1983. Contribution of mycorrhizae and soil fungi to nutrient cycling in a Douglas fir ecosystem. *Canadian Journal of Forest Research* **13**, 219–32.

Folk, R. L. 1965. Some aspects of recrystallization in ancient limestones. In *Dolomitization and limestone genesis*, L. C. Pray & R. C. Murray (eds.), 14–48. *Special Publication, Society of Economic Paleontologists and Mineralogists* 13.

Folk, R. L. & E. F. McBride 1976. The Caballos Novaculite revisited. Part 1. Origin of novaculite members. *Journal of Sedimentary Petrology* **46**, 659–69.

Folk, R. L., H. H. Roberts & C. H. Moore 1973. Black phytokarst from Hell, Cayman Islands, British West Indies. *Bulletin of the Geological Society of America* **84**, 2351–60.

Follmer, L. R. 1978. The Sangamon soil in its type area – a review. In *Quaternary soils*, W. C. Mahaney (ed.), 125–65. Norwich: Geoabstracts.

Follmer, L. R., E. D. McKay, J. A. Lineback & D. L. Gross 1979. Wisconsinan, Sangamonian and Illinoian stratigraphy in central Illinois. *Guidebook of the Illinois State Geological Survey* 13.

Foster, R. C. 1981. Polysaccharides in soil fabrics. *Science* **214**, 665–7.

Fox, S. W. and K. Dose 1972. *Molecular evolution and the origin of life*. San Francisco: Freeman.

Frakes, L. A. 1979. *Climates through geologic time*. Amsterdam: Elsevier.

Francis, J. E. 1986. The calcareous paleosols of the basal Purbeck Formation (Upper Jurassic), southern England. In *Paleosols: their recognition and interpretation*, V. P. Wright (ed.), 112–38. Oxford: Blackwell.

Frank, J. R., A. B. Carpenter & T. W. Oglesby 1982. Cathodoluminescence and composition of calcite cement in the Taum Sauk Limestone (Upper Cambrian), southeast Missouri. *Journal of Sedimentary Petrology* **52**, 631–8.

Franklin, J. F. & C. T. Dyrness 1973. Natural vegetation of Oregon and Washington. *General Technical Report of the U.S. Department of Agriculture Forest Service* No. PNW-8.

Frenguelli, J. 1939. Nidos fosiles de insectos en el Terciario del Neuquen y Rio Negro. *Notas del Museo de La Plata* 4 *Paleontologia* **18**, 379–402.

Frey, M. (ed.) 1987. *Low temperature metamorphism*. Glasgow: Blackie.

Frey, R. W., J. D. Howard & W. A. Pryor 1978. *Ophiomorphia*: its morphologic, taxonomic and environmental significance. *Palaeogeography, Palaeoclimatology, Palaeoecology* **23**, 199–299.

Frey, R. W., S. G. Pemberton & J. R. Fagerstrom 1984. Morphological, ethological and environmental significance of the ichnogenera *Scoyenia* and *Ancorichnus*. *Journal of Paleontology* **58**, 511–28.

Freytet, P. & J.-C. Plaziat 1982. *Continental carbonate sedimentation and pedogenesis – Late Cretaceous and Early Tertiary of sourthern France*. Stuttgart: E. Schwiezerbart'sche.

Friedmann, E. I. & R. Weed 1987. Microbial trace-fossil formation, biogenous and abiotic weathering in the Antarctic cold desert. *Science* **236**, 703–5.

Fritsch, A. 1899. *Fauna der Gaskohle und der Kalksteine der Permformation Böhmens. Vol. 4(1) Arthropoda (Hexapoda, Myriapoda), and vol. 4(2) Myriapoda pars II, Arachnida*. Prague: F. Rivnač.

Fuchs, L. H., E. Olsen & K. J. Jensen 1973. Mineralogy, mineral chemistry and composition of the Murchison (C2) meteorite. *Smithsonian Contributions to the Earth Sciences* No. 10.

Galbreath, E. C. 1974. Stipid grass "seeds" from the Oligocene and Miocene deposits of northeastern Colorado. *Transactions of the Illinois State Academy of Science* **67**, 366–8.

Gall, L. F. & B. H. Tiffney 1983. A fossil noctuid moth egg from the Late Cretaceous of eastern North America. *Science* **219**, 507–9.

Galloway, W. E. 1974. Deposition and diagenetic alteration of sandstone in northeast Pacific arc-related basins: implications for greywacke genesis. *Bulletin of the Geological Society of America* **85**, 379–90.

Galloway, W. E. 1978. Uranium mineralization in a coastal plain fluvial aquifer system: Catahoula Formation, Texas. *Economic Geology* **73**, 1655–76.

Gardner, L. R. 1980. Mobilization of Al and Ti during weathering – isovolumetric chemical evidence. *Chemical Geology* **30**, 151–65.

Gardner, T. W., E. G. Williams & P. W. Holbrook 1988. Pedogenesis of some Pennsylvanian underclays: ground-water, topographic and tectonic controls. In *Paleosols and weathering through geologic time: principles and applications*, J. Reinhardt & W. R. Sigleo (eds.), 81–101. *Special Paper, Geological Society of America* No. 216.

Garrels, R. M. & F. T. Mackenzie 1971. *Evolution of sedimentary rocks*. New York: W. W. Norton.

Gastaldo, R. A. 1986. Implications on the paleoecology of autochthonous Carboniferous lycopods in clastic sedimentary environments. *Palaeogeography, Palaeoclimatology, Palaeoecology* **53**, 191–222.

Gay, A. L. & D. E. Grandstaff 1980. Chemistry and mineralogy of Precambrian paleosols at Elliot Lake, Ontario, Canada. *Precambrian Research* **12**, 349–73.

Gensel, P. G. & H. N. Andrews 1984. *Devonian paleobotany*. New York: Praeger.

Gentry, A. W. 1970. The Bovidae (Mammalia) of the Fort Ternan fossil fauna. In *Fossil vertebrates of Africa*, L. S. B. Leakey & R. J. G. Savage (eds.), Vol. 2, 243–324. London: Academic Press.

Gerasimov, I. P. 1971. Nature and originality of paleosols. In *Paleopedology: origin, nature and dating of paleosols*, D. H. Yaalon (ed.), 15–27. Jerusalem: Israel University Press.

Gerhard, S. & W. Rietschel 1968. Ein Stuck Bernstein und seine Enschlusse. *Natur und Museum* **98**, 515–20.

Gerrard, J. (ed.) 1987. *Alluvial soils*. New York: Van Nostrand Reinhold.

Gibson, E. K., S. T. Wentworth & D. S. McKay 1983. Chemical weathering and diagenesis of a cold desert soil from Wright Valley, Antarctica: an analog of Martian weathering processes. *Proceedings 13th Lunar and Planetary Science Conference 2, Journal of Geophysical Research Supplement* 88A, 912–18.

Gile, L. H., F. F. Peterson & R. B. Grossman 1966. Morphological and genetic sequences of carbonate accumulation in desert soils. *Soil Science* **101**, 347–60.

Gile, L. H., J. W. Hawley & J. B. Grossman 1980. Soils and geomorphology in the Basin and Range area of southern New Mexico. Guidebook to the Desert Project. *Memoirs of the New Mexico Bureau of Mines and Mineral Resources* No. 39.

Gillespie, W. H., G. W. Rothwell & S. E. Scheckler 1981. The earliest seeds. *Nature* **293**, 462–4.

Gillon, D. 1983. The fire problem in tropical savannas. In *Tropical savannas, ecosystems of the world*, No. 13, F. Bourliere (ed.), 617–41. Amsterdam: Elsevier.

Gingerich, P. D. 1980. Evolutionary patterns in early Cenozoic mammals. *Annual Reviews of Earth and Planetary Sciences* **8**, 407–24.

Given, R. K. & B. H. Wilkinson 1987. Dolomite abundance and stratigraphic age: constraints on rates and mechanisms of Phanerozoic dolostone formation. *Journal of Sedimentary Petrology* **57**, 1068–78.

Glinka, K. D. 1927. *The great soil groups of the world* (translated by C. F. Marbut). Ann Arbor: Edwards Bros.

Glinka, K. D. 1931. *Treatise on soil science (Pochvovedeniye)* (translated by A. Gourevich, 1963). Jerusalem: Israel Program for Scientific Translations.

Glob, P. V. 1969. *The bog people: Iron Age man preserved* (translated by R. Bruce-Mitford). Ithaca: Cornell University Press.

Goldbery, R. 1982a. Paleosols of the Lower Jurassic Mishhor and Ardon Formations ("Laterite Derivative Facies"), Makhtesh Ramon, Israel. *Sedimentology* **29**, 669–90.

Goldbery, R. 1982b. Structural analysis of soil microrelief in paleosols of the Lower Jurassic "Laterite Derivative Facies" (Mishhor and Ardon Formations), Makhtesh Ramon, Israel. *Sedimentary Geology* **31**, 119–40.

Goldich, S. S. 1938. A study in rock weathering. *Journal of Geology* **46**, 17–58.

Goldring, R. & A. Seilacher 1971. Limulid undertracks and their sedimentological implications. *Neues Jahrbuch für Paläontologie Abhandlung* **137**, 422–42.

Golubič, S. & S. E. Campbell 1979. Analogous microbial forms in recent subaerial habitats and in Precambrian cherts, *Gloeothece caerulea* Geitler and *Eosynechococcus moorei* Hofmann. *Precambrian Research* **8**, 201–17.

Goodwin, A. M. 1981. Archaean plates and greenstone belts. In *Precambrian plate tectonics*, A. Kroner (ed.), 105–35. Amsterdam: Elsevier.

Gordon, C. C. & J. E. Buikstra 1981. Soil pH, bone preservation and sampling bias at mortuary sites. *American Antiquity* **46**, 566–71.

Gore, A. J. P. (ed.) 1983. *Mires – swamp, bog, fen and moor. Ecosystems of the World*, Nos. 4A and 4B. Amsterdam: Elsevier.

Gottsberger, G. 1978. Seed dispersal by fish in the inundated regions of Humaita, Amazonia. *Biotropica* **10**, 170–83.

Goudie, A. 1973. *Duricrust in tropical and subtropical landscapes.* Oxford: Clarendon Press.

Gould, S. J. & N. Eldredge 1977. Punctuated equilibria: the tempo and mode of evolution reconsidered. *Paleobiology* **3**, 115–51.

Gould, S. J. & R. C. Lewontin 1979. The spandrels of San Marco and the Panglossian paradigm: a critique of the adaptationist program. *Proceedings of the Royal Society of London* **205**, 581–98.

Gowlett, J., J. Harris, D. Walton & B. Wood 1981. Early archaeological sites, hominid remains and traces of fire from Chesowonja, Kenya. *Nature* **294**, 125–9.

Grabau, A. W. 1940. *The rhythm of the ages*. Peking: Henri Vetch.

Grande, L. 1980. Paleontology of the Green River Formation, with a review of the fish fauna. *Bulletin of the Geological Survey of Wyoming* No. 63.

Grandstaff, D. E., M. J. Edelman, R. W. Foster, E. Zbinden & M. M. Kimberley 1986. Chemistry and mineralogy of Precambrian paleosols at the base of the Dominion and Pongola Groups. *Precambrian Research* **32**, 97–131.

Gray, J. 1985a. The microfossil record of early land plants: advances in understanding of early terrestrialization. *Philosophical Transactions of the Royal Society of London* **B309**, 167–95.

Gray, J. 1985b. Early terrestrial ecosystems: the animal evidence. *Abstracts, Geological Society of America* **17**, 596.

Greeley, R. 1985. *Planetary landscapes*. London: George Allen & Unwin.

Griffin, J. J., H. Windom & E. D. Goldberg 1968. The distribution of clay minerals in the world ocean. *Deep-Sea Research* **15**, 433–59.

Griffiths, E. 1965. Microorganisms and soil structure. *Biological Review* **40**, 129–42.

Grim, R. E. 1968. *Clay mineralogy*, 2nd edn. New York: McGraw-Hill.

Grime, J. P. 1979. *Plant strategies and vegetation processes*. New York: Wiley.

Grotzinger, J. P. 1986. Cyclicity and paleoenvironmental dynamics, Rocknest Platform, northwest Canada. *Bulletin of the Geological Society of America* **97**, 1208–31.

Groves, D. I., J. S. R. Dunlop & R. Buick 1981. An early habitat of life. *Scientific American* **245**(4), 64–73.

Guthrie, R. L. & J. E. Witty 1982. New designations for soil horizons and the new Soil Survey Manual. *Journal of the Soil Science Society of America* **46**, 443–4.

Haasis, F. W. 1921. Relations between soil type and root form of western yellow pine seedlings. *Ecology* **2**, 291–303.

Habicht, J. K. A. 1979. Paleoclimate, paleomagnetism and continental drift. *Studies in Geology of the American Association of Petroleum Geologists* No. 9.

Haldane, J. B. S. 1929. The origin of life. *Rationalist Annual* **148**, 3–10.

Halffter, G. & W. D. Edmonds 1982. *The nesting behavior of dung beetles (Scarabaeinae)*. Mexico City: Instituto de Ecologia.

Halffter, G. & E. G. Matthews 1966. The natural history of dung beetles of the subfamily Scarabaeinae (Coleoptera, Scarabaeidae). *Folia Entomologica Mexicana* No. 12–14.

Hall, B. R. & C. J. Folland 1970. Soils of Lancashire. *Bulletin of the Soil Survey of Great Britain* No. 5.

Hall, R. D. & D. Michaud 1988. The use of hornblende etching, clast weathering and soils to date alpine glacial and periglacial deposits: a study from southwestern Montana. *Bulletin of the Geological Society of America* **100**, 458–67.

Hallam, A. 1981. *Facies interpretation and the stratigraphic record*. Oxford: Freeman.

Hallbauer, D. K., H. M. Jahns & H. A. Beltmann 1977. Morphological and anatomical observations on some Precambrian plants from the Witwatersrand, South Africa. *Geologische Rundschau* **66**, 477–91.

Hallberg, G. R., N. C. Wollenhaupt & G. A. Miller 1978. A century of soil formation on soil derived from loess in Iowa. *Journal of the Soil Science Society of America* **42**, 339–43.

Handlirsch, A. 1910. Fossile Wespennester. *Berichte Senckenberg Naturforschung Gesellschaft* **41**, 265–6.

Häntzschel, W. 1975. Trace fossils and problematica. In *Treatise on invertebrate paleontology. Part W. Miscellanea* Suppl. 1, 2nd edn, R. C. Moore and C. Teichert (eds.), W1–W269. Boulder & Lawrence: Geological Society of America and University of Kansas Press.

Hardcastle, J. 1889. Origin of loess deposit of the Timaru Plateau. *Transactions of the New Zealand Institute, Wellington* **22**, 406–16.

Harden, J. W. 1982a. A quantitative index of soil development from field descriptions: examples from a chronosequence in central California. *Geoderma* **28**, 1–28.

Harden, J. W. 1982b. A study of soil development using the geochronology of Merced River deposits, California. Unpub. PhD thesis, University of California, Berkeley.

Harris, T. M. 1957. A Liasso-Rhaetic flora in South Wales. *Proceedings of the Royal Society of London* **B147**, 289–308.

Harris, T. M. 1981. Burnt ferns in the English Wealden. *Proceedings of the Geologist's Association* **92**, 47–58.

Hartmann, W. K., R. G. Strom, S. J. Weidenschilling, K. R. Blasius, A. Woronow, M. R. Pence, R. A. F. Grieve, J. Diaz, C. R. Chapman, E. M. Shoemaker & K. L. Jones 1981. Chronology of planetary volcanism by comparative studies of planetary cratering. In *Basaltic volcanism on the terrestrial planets*, W. M. Kaula (ed.), 1048–127. New York: Pergamon Press.

Haude, R. 1970. Die Entstehung von Steinsalz-Pseudomorphosen. *Neues Jahrbuch für Geologie und Paläontologie Jahrgang* 1970, 1–10.

Hay, R. L. 1976. *Geology of Olduvai Gorge*. Berkeley: University of California Press.

Hay, R. L. & R. J. Reeder 1978. Calcretes of Olduvai Gorge and the Ndolanya Beds of northern Tanzania. *Sedimentology* **25**, 649–73.

Hays, J. D., J. Imbrie & N. J. Shackleton 1976. Variations in the Earth's orbit: pacemaker of the Ice Ages. *Science* **194**, 1121–32.

Heckel, P. H. 1986. Sea-level curve for Pennsylvanian eustatic marine transgressive–regressive depositional cycles along midcontinent outcrop belt, North American. *Geology* **14**, 330–4.

Hedberg, H. D. 1976. *International stratigraphic guide*. New York: Wiley.

Heiken, G., M. Duke, D. S. McKay, U. S. Clanton, R. Fryxell, J. S. Nagle, R. Scott & G. A. Sellers 1973. Preliminary stratigraphy of the Apollo 15 drill core. *Proceedings of the 4th Lunar Science Conference Geochimica et Cosmochimica Acta, Supplement* **4**(1), 191–213.

Heiken, G., R. V. Morris, D. S. McKay & R. M. Fruland 1976. Petrographic and ferromagnetic resonance studies of the Apollo 15 deep drill core. *Proceedings of the 7th Lunar Science Conference Geochemica et Cosmochimica Acta, Supplement* 7(1), 93–111.

Heiser, C. B. 1981. *Seed to civilization*. 2nd edn. San Francisco: Freeman.

Heizer, R. F. & L. K. Napton 1969. Biological and cultural evidence from prehistoric human coprolites. *Science* **165**, 563–8.

Herbert, C. 1972. Palaeodrainage patterns in the southern Sydney Basin. *Records of the Geological Survey of New South Wales* **14**, 5–18.

Hilgard, E. W. 1892. A report on the relations of soil to climate. *U.S. Department of Agriculture Weather Bulletin* **3**, 1–59.

Hillel, D. 1980. *Fundamentals of soil physics*. New York: Academic Press.

Ho, C. & J. M. Coleman 1969. Consolidation and cementation of recent sediments in the Atchafalaya Basin. *Bulletin of the Geological Society of America* **80**, 183–92.

Hoffman, R. L. 1969. Myriapoda, exclusive of Insecta. In *Treatise of invertebrate paleontology. Arthropoda. Part R4*, R. C. Moore (ed.), 573–620. Boulder and Lawrence: Geological Society of America and University of Kansas Press.

Hofmann, H. J. 1985. Precambrian carbonaceous megafossils. In *Paleoalgology*, D. F. Toomey & M. H. Nitecki (eds.), 20–33. Berlin: Springer.

Hofmann, H. J. & G. D. Jackson 1987. Proterozoic ministromatolites with radial-fibrous fabric. *Sedimentology* **34**, 963–71.

Holdridge, L. R. 1947. Determination of world plant formations from simple climatic data. *Science* **105**, 367–8.

Holdridge, L. R., W. C. Grenke, W. H. Hatheway, T. Liang & J. A. Tosi 1971. *Forest environments in tropical life zones*. Oxford: Pergamon Press.

Holland, H. D. 1984. *The chemical evolution of the atmosphere and oceans*. Princeton: Princeton University Press.

Holm, N. G. 1985. New evidence for a tubular structure of β-iron(III) oxide hydroxide – akaganeite. *Origins of Life* **15**, 131–9.

Holzhey, C. S., R. D. Yeck & W. D. Nettleton 1974. Microfabric of some argillic horizons in udic, xeric and torric soil environments of the United States. In *Soil microscopy*, G. K. Rutherford (ed.), 747–60. Kingston: Limestone.

Horner, J. R. 1982. Evidence of colonial nesting and "site fidelity" among ornithischian dinosaurs. *Nature* **297**, 675–6.

Horner, J. R. 1984. Three ecologically distinct vertebrate faunal communities from the Late Cretaceous Two Medicine Formation of Montana, with discussion of evolutionary pressures induced by interior seaway fluctuations. In *Northwest Montana and adjacent Canada*, J. D. McBane and P. R. Garrison (eds.), 299–303. *Guidebook and Proceedings of the Montana Geological Society Field Conference Symposium*.

Houghton, H. F. 1980. Refined techniques for staining plagioclase and alkali feldspars. *Journal of Sedimentary Petrology* **50**, 629–31.

Hower, J., E. V. Eslinger, M. E. Hower & E. A. Perry 1976. Mechanism of burial metamorphism of argillaceous sediment: 1, mineralogical and chemical evidence. *Bulletin of the Geological Society of America* **87**, 725–37.

Huddle, J. W. & S. H. Patterson 1961. Origin of Pennsylvanian underclays and related seat rocks. *Bulletin of the Geological Society of America* **72**, 1643–60.

Hunt, R. M., X. X. Xue & J. Kaufman 1983. Miocene burrows of extant bear dogs: indication of early denning behavior of large mammalian carnivores. *Science* **221**, 364–6.

Hutton, J. 1795. *Theory of the Earth, with proofs and illustrations*. Edinburgh: W. Creech (facsimile edn. 1959, Codicote: Wheldon & Wesley).

Isherwood, D. & A. Street 1976. Biotite-induced grusification of the Boulder Creek Granodiorite, Boulder County, Colorado. *Bulletin of the Geological Society of America* **87**, 366–70.

Jackson, M. L., S. A. Tyler, A. L. Willis, G. A. Bourbeau & R. P. Pennington 1948. Weathering sequence of clay size minerals in soils and sediments. I. Fundamental generalizations. *Journal of Physical and Colloid Chemistry* **52**, 1237–61.

Jackson, T. A. & W. D. Keller 1970. A comparative study of the role of lichens and "inorganic" processes in the chemical weathering of recent Hawaiian lava flows. *American Journal of Science* **269**, 446–66.

Jain, A. V. & M. E. Lipshutz 1973. Shock history of mesosiderites. *Nature* **242**, 26–8.

James, N. P. & P. W. Choquette (eds.) 1987. *Paleokarst*. New York: Springer.

Jeanson, C. 1967. Migrations chimiques dans un sol artificiel: étude micromorphologique. *Geoderma* **1**, 325–45.

Jell, P. A. 1982. An early Jurassic millipede from the Evergreen Formation in Queensland. *Alcheringa* **7**, 195–9.

Jenik, J. 1978. Roots and root systems in tropical trees: morphologic and ecological aspects. In *Tropical trees as living systems*, P. B. Tomlinson & M. H. Zimmerman (eds.), 323–49. Cambridge: Cambridge University Press.

Jennings, J. N. 1985. *Karst geomorphology*. Oxford: Blackwell.

Jenny, H. 1941. *Factors in soil formation*. New York: McGraw-Hill.

Joe, H., K. Kuma, W. Paplawsky, B. Rea & G. Arrhenius 1986. Abiotic photosynthesis from ferrous carbonate (siderite) and water. *Origins of Life* **16**, 369–70.

Johannessen, C. L., W. A. Davenport & S. McWilliams 1971. The vegetation of the Willamette Valley. *Annals of the Association of American Geographers* **61**, 286–302.

Johanson, D. C. & T. D. White, 1979. A systematic assessment of early African hominids. *Science* **203**, 321–30.

Johanson, D. C., M. Taieb and Y. Coppens 1982. Pliocene hominids from the Hadar Formation, Ethiopia (1973–1977): stratigraphic, chronologic and paleoenvironmental contexts, with notes on hominid morphology and systematics. *American Journal of Physical Anthropology* **57**, 373–402.

Johanson, D. C., F. T. Masao, G. G. Eck, T. D. White, R. C. Walter, W. Y. Kimbel, B. Asfaw, P. Manega, P. Ndessokia & G. Suwa 1987. New partial skeleton of *Homo habilis* from Olduvai Gorge, Tanzania. *Nature* **327**, 205–9.

Johnson, D. L. & D. Watson-Stegner 1987. Evolution model of pedogenesis. *Soil Science* **143**, 349–66.

Johnson, D. L., D. Watson-Stegner, D. N. Johnson & R. J. Schaetzl 1987. Proisotropic and proanisotropic processes of pedoturbation. *Soil Science* **143**, 278–91.

Johnson, K. G. 1972. Evidence for tidal origin of Late Devonian clastics in eastern New York State. *Proceedings of the 24th International Geological Congress, Montreal* **6**, 285–93.

Jones, D., M. J. Wilson & J. M. Tait 1980. Weathering of a basalt by *Pertusaria corallina*. *Lichenologist* **12**, 277–89.

Jones, M. G. K. 1981. The development and function of plant cells modified by endoparasitic nematodes. In *Plant parasitic nematodes*, B. M. Zuckerman & R. A. Rohde (eds.), vol. 3, 255–79. New York: Academic Press.

Jones, R. L. & H. C. Hanson 1985. *Mineral licks, geophagy and the biogeochemistry of North American ungulates*. Ames: Iowa State University Press.

Kabata-Pendias, A. & H. Pendias 1984. *Trace elements in soils and plants*. Boca Raton: CRC Press.

Kalliokoski, J. 1975. Chemistry and mineralogy of Precambrian paleosols in northern Michigan. *Bulletin of the Geological Society of America* **86**, 371–6.

Kalliokoski, J. 1986. Calcium carbonate cement (caliche) in Keweenawan sedimentary rocks (1.1 Ga), Upper Peninsula of Michigan. *Precambrian Research* **32**, 243–59.

Kalliokoski, J. & E. J. Welch 1985. Keweenawan-age caliche paleosol in the lower part of the Calument and Hecla Conglomerate, Centennial Mine, Calument, Michigan. *Bulletin of the Geological Society of America* **96**, 1188–93.

Kasting, J. F. 1987. Theoretical constraints on oxygen and carbon dioxide concentrations in the Precambrian atmosphere. *Precambrian Research* **34**, 205–29.

Keller, E. A. 1976. *Environmental geology*. Columbus: Merrill.

Keller, W. D. 1954. Bonding energies of some silicate minerals. *American Mineralogist* **39**, 783–93.

Kelley, J. 1986. Species recognition and sexual dimorphism in *Proconsul* and *Rangwapithecus*. *Journal of Human Evolution* **15**, 461–95.

Kemp, E. M. 1978. Tertiary climatic evolution and vegetation history in the southeast Indian Ocean region. *Palaeogeography, Palaeoclimatology, Palaeoecology* **24**, 169–208.

Kerridge, J. F., A. L. Mackay & W. V. Boynton 1979. Magnetite in CI carbonaceous meteorites: origin by aqueous activity on a planetismal surface. *Science* **205**, 395–7.

Khotinskiy, N. A. 1984. Holocene vegetation history. In *Late Quaternary environments of the Soviet Union*, A. A. Velichko, H. E. Wright & C. W. Barnosky (eds.), 179–200. Minneapolis: University of Minnesota Press.

Kidston, R. & W. H. Lang 1917. On Old Red Sandstone plants showing structure from the Rhynie chert bed, Aberdeenshire. Part I, *Rhynia gwynne-vaughani* Kidston & Lang. *Transactions of the Royal Society of Edinburgh* **51**, 761–84.

Kidston, R. & W. H. Lang 1921. On Old Red Sandstone plants showing structure from the Rhynie chert bed, Aberdeenshire. Part V. The Thallophyta occurring in the peat bed, the succession of plants through a vertical section of the bed and the conditions of accumulation and preservation of the deposit. *Transactions of the Royal Society of Edinburgh* **52**, 855–702.

Kimberley, M. M., D. E. Grandstaff & R. T. Tanaka 1984. Topographic control on Precambrian weathering in the Elliot Lake uranium district, Canada. *Journal of the Geological Society of London* **141**, 229–33.

Kimmins, J. P. 1987. *Forest Ecology*. New York: Macmillan.

Kindle, E. M. 1934. Concerning "lake balls," "*Cladophora* balls" and "coal balls." *American Midland Naturalist* **15**, 752–60.

Kitching, J. W. 1980. On some fossil arthropoda from the limeworks, Makapansgat, Potgietersrus. *Palaeontologia Africana* **23**, 63–8.

Kjellesvig-Waering, E. N. 1961. The Silurian Eurypterida of the Welsh borderland. *Journal of Paleontology* **35**, 789–835.

Kjellesvig-Waering, E. N. 1986. A restudy of the fossil Scorpionida of the world. *Palaeontographica Americana* No. 55.

Klappa, C. F. 1978. Biolithogenesis of *Microcodium*: elucidation. *Sedimentology* **25**, 489–522.

Klappa, C. F. 1979a. Calcified filaments in Quaternary calcretes: organo-mineral interactions in the subaerial vadose environment. *Journal of Sedimentary Petrology* **49**, 955–68.

Klappa, C. F. 1979b. Lichen stromatolites: criterion for subaerial exposure and a mechanism for the formation of laminar calcretes (caliche). *Journal of Sedimentary Petrology* **49**, 387–400.

Klute, A. (ed.) 1986. *Methods of soils analysis. Part 1. Physical and mineralogical methods*, 2nd edn. Madison: American Society of Agronomy.

Knoll, A. H. 1984. The Archean/Proterozoic transition: a sedimentary and paleobiological perspective. In *Patterns of change in Earth evolution*, H. D. Holland & A. F. Trendall (eds.), 221–42. Berlin: Springer.

Knoll, A. H. 1985a. Exceptional preservation of photosynthetic organisms in silicified carbonates and silicified peats. *Philosophical Transactions of the Royal Society of London* **B311**, 111–22.

Knoll, A. H. 1985b. Patterns of evolution in the Archaean and Proterozoic eons. *Paleobiology* **11**, 53–64.

Knoll, A. H. 1986. Patterns of change in plant communities through geological time. In *Community ecology*, J. Diamond and T. J. Case (eds.), 126–41. New York: Harper & Row.

Knoll, A. H., S. Golubič, J. Green & K. Swett 1986. Organically preserved microbial endoliths from the Late Proterozoic of East Greenland. *Nature* **321**, 856–7.

Knoll, M. A. & W. C. James 1987. Effect of the advent and diversification of vascular plants on mieral weathering through geologic time. *Geology* **15**, 1099–102.

Kornberg, A. 1980. *DNA replication*. San Francisco: Freeman.

Krassilov, V. A. 1981. *Orestovia* and the origin of vascular plants. *Lethaia* **14**, 235–50.

Kraus, M. J. 1987. Integration of channel and floodplain suites in aggrading fluvial systems. II. Lateral relations of lower Eocene alluvial paleosols, Willwood Formation, Bighorn Basin, Wyoming. *Journal of Sedimentary Petrology* **57**, 602–12.

Kraus, M. J. 1988. Nodular remains of early Tertiary forests, Bighorn Basin, Wyoming. *Journal of Sedimentary Petrology* **58**, 888–93.

Kraus, M. J. & L. T. Middleton 1987. Dissected paleotopography and base-level changes in a Triassic fluvial sequence. *Geology* **15**, 18–21.

Krishna, P. C. & S. Perumal 1948. Structure in black cotton soils of the Nizamsager Project area, Hyderabad state, India. *Soil Science* **66**, 29–38.

Krumbein, W. C. & R. M. Garrels 1952. Origin and classification of chemical sediments in terms of pH and oxidation-reduction potentials. *Journal of Geology* **60**, 1–33.

Krumbein, W. E. & C. Giele 1979. Calcification in a coccoid cyanobacterium associated with the formation of desert stromatolites. *Sedimentology* **26**, 593–604.

Krumbein, W. E. & K. Jens 1981. Biogenic rock varnishes of the Negev Desert (Israel): an ecological study of iron and manganese transformation by cyanobacteria and fungi. *Oecologia* **50**, 25–38.

Krumbiegel, G., L. Rüffle & H. Haubold 1983. *Das eozäne Geiseltal*. Lutherstadt: A. Ziemsen.

Kukla, G. J. 1977. Pleistocene land-sea correlations. 1. Europe. *Earth-Science Reviews* **13**, 307–74.

Kühne, W. G. & T. Schlüter 1985. A fair deal for the Devonian arthropod fauna of Rhynie. *Entomologia Generalis* **11**, 91–6.

Kühnelt, W. 1976. *Soil biology*, 2 edn. East Lansing: Michigan State University.

Kunze, G. W. & E. H. Templin 1956. Houston Black Clay, the type Grumosol. II. Mineralogical and chemical characterization. *Proceedings of the Soil Science Society of America* **20**, 91–6.

Labandiera, C. C., B. S. Beall & F. M. Hueber 1988. Early insect diversification: evidence from a Lower Devonian bristletail from Quebec. *Science* **242**, 913–6.

Lambrecht, K. 1933. *Handbuch der Paläornithologie*. Berlin: Gebruder Bornträger.

Leakey, L. S. B. 1952. Lower Miocene invertebrates from Kenya. *Nature* **169**, 624–5.

Leakey, M. D. & J. M. Harris (eds.) 1987. *Laetoli, a Pliocene site in northern Tanzania*. Oxford: Clarendon Press.

Leakey, R. E. & M. G. Leakey 1986a. A new Miocene hominoid from Kenya. *Nature* **324**, 143–6.

Leakey, R. E. & M. G. Leakey 1986b. A second new Miocene hominoid from Kenya. *Nature* **324**, 146–8.

Leakey, R. E., M. G. Leakey & A. K. Behrensmeyer 1978. The hominid catalogue. In *Koobi Fora Research Project. Vol. 1. The fossil hominids and an introduction to their context 1968–1974*, M. G. Leakey & R. E. Leakey (eds.), 82–182. Oxford: Clarendon Press.

Leary, R. L. 1981. Early Pennsylvanian geology and paleobotany of the Rock Island County, Illinois, area. Part 1. Geology. *Reports of Investigations of the Illinois State Museum* No. 37.

Leclerq, S. & H. P. Banks 1962. *Pseudosporochnus nodosus* sp. nov., a Middle Devonian plant with cladoxylalean affinities. *Palaeontographica* **B110**, 1–34.

Lee, K. E. 1985. *Earthworms*. Sydney: Academic Press.

Leo, R. F. & E. S. Barghoorn 1976. Silica in the biosphere. *Acta Cientifica Venezolana* **27**, 231–4.

Lepsch, I. F. & S. W. Buol 1974. Investigations on an Oxisol–Ultisol toposequence – S. Paulo state, Brazil. *Proceedings of the Soil Science Society of America* **38**, 491–6.

Lessig, H. D. 1961. The soils developed on Wisconsin and Illinoian-age glacial outwash along Little Beaver Creek and the adjoining upper Ohio Valley, Columbiana County, Ohio. *Ohio Journal of Science* **6**, 286–94.

Levine, E. L. & E. J. Ciolkosz 1983. Soil development in till of various ages in northeastern Pennsylvania. *Quaternary Research* **19**, 85–99.

Lewis, R. S. & E. Anders 1975. Condensation time of solar nebula from extinct ^{129}I in primitive meteorites. *Proceedings of the National Academy of Sciences of the U.S.A.* **72**, 268–73.

Lewis, S. E. 1972. Fossil caddis fly (Trichoptera) cases from the Ruby River Basin (Oligocene) of southwestern Montana. *Annals of the Entomological Society of America* **65**, 518–9.

Lieth, H. & R. H. Whittaker 1975. *Primary productivity of the biosphere*. New York: Springer.

Lochman-Balk, C. 1971. The Cambrian of the craton of the United States. In *Cambrian of the New World*, C. H. Holland (ed.), 79–167. New York: Wiley.

Lockley, M. 1986. The paleobiological and paleoenvironmental importance of dinosaur footprints. *Palaios* **1**, 37–48.

Long, R. A. & K. Padian 1986. Vertebrate biostratigraphy of the Late Triassic Chinle Formation, Petrified Forest National Park, Arizona: preliminary results. In *The beginning of the age of dinosaurs*, K. Padian (ed.), 161–9. Cambridge: Cambridge University Press.

Loope, D. B. 1988. Rhizoliths in ancient eolianites. *Sedimentary Geology* **56**, 301–14.

Lovelock, J. E. 1979. *Gaia: a new look at life on Earth*. Oxford: Oxford University Press.

Lowe, D. R. 1983. Restricted shallow-water sedimentation of Early Archaean stromatolitic and evaporitic strata of the Strelley Pool Chert, Pilbara Block, Western Australia. *Precambrian Research* **19**, 239–83.

Lowe, D. R., G. R. Byerly, B. L. Ransom & B. W. Nocita 1985. Stratigraphic and sedimentological evidence bearing on structural repetition in Early Archaean rocks of the Barberton Greenstone Belt, South Africa. *Precambrian Research* **27**, 167–86.

Lucas, C. 1976. Vestiges de paleosols dans le Permien des Pyrenées et le Trias inferieur des Pyrénees et de l'Aquitaine. *Comptes Rendus Hebdomadaires des Sciences de l'Académie des Sciences, Paris, Série D* **282**, 1419–22.

Luttrell, G. W., M. L. Hubert & V. M. Jussen 1986. Lexicon of new formal geologic names of the United States 1976–1980. *Bulletin of the U.S. Geological Survey* No. 1564.

Lytle, S. A. 1968. The morphological characteristics and relief relationships of representative soils in Louisiana. *Bulletin of the Louisiana Agricultural Experiment Station* No. 631.

MacDougall, J. D., G. W. Lugmair & J. F. Kerridge 1984. Early solar system aqueous activity: Sr isotope evidence from the Orgueil CI meteorite. *Nature* **307**, 249–51.

MacFayden, W. A. 1950. Sandy gypsum crystals from Berbera, British Somaliland. *Geological Magazine* **87**, 409–20.

Machette, M. N. 1985. Calcic soils of the southwestern United States. In *Soils and Quaternary geology of the southwestern United States*, D. L. Weide (ed.), 1–21. *Special Paper, Geological Society of America* No. 203.

Macphail, R. I. 1986. Paleosols in archaeology: their role in understanding Flandrian pedogenesis. In *Paleosols: their recognition and interpretation*, V. P. Wright (ed.), 263–90. Oxford: Blackwell.

Maglione, G. 1981. An example of recent continental evaporitic sedimentation: the Chadean Basin (Africa). In *Evaporite deposits*, G. Busson (ed.), 5–9. Houston: Gulf.

Maher, K. A. & D. J. Stevenson 1988. Impact frustration of the origin of life. *Nature* **331**, 612–4.

Main, B. Y. 1985. Further studies on the systematics of Ctenizid trapdoor spiders: a review of the Australian genera (Araneae : Mylagomorphae : Ctenizidae). *Australian Journal of Zoology, Supplement Series* No. 108.

Mann, A. W. & R. D. Horwitz 1979. Groundwater calcrete deposits in Australia: some observations from Western Australia. *Journal of the Geological Society of Australia* **26**, 293–303.

Marbut, C. F. 1935. *Atlas of American agriculture. Part III. Soils of the United States*. Washington: Government Printer.

Margulis, L. 1981. *Symbiosis and cell evolution*. San Francisco: Freeman.

Margulis, L., B. D. D. Grosovsky, J. F. Stolz, E. J. Gong-Collins, S. Lenk & A. Lopez-Cortes 1983. Distinctive microbial structures and pre-Phanerozoic fossil record. *Precambrian Research* **20**, 443–77.

Marion, G. M., W. H. Schlesinger & P. J. Fonteyn 1985. CALDEP: a regional model for soil formation in southwestern deserts. *Soil Science* **139**, 468–81.

Martin, L. D. & D. K. Bennett 1977. The burrows of the Miocene beaver *Palaeocastor*, western Nebraska, U.S.A. *Palaeogeography, Palaeoclimatology, Palaeoecology* **22**, 173–93.

Martyn, J. E. & G. I. Johnson 1986. Geological setting and origin of fuchsite-bearing rocks near Menzies, Western Australia. *Australian Journal of Earth Science* **33**, 1–18.

Mason, B. & E. J. Jarosewich 1973. The Barea, Dyarrl Island and Emery meteorites, and a review of the mesosiderites. *Mineralogical Magazine* **39**, 204–15.

Mason, B. & C. B. Moore 1982. *Principles of geochemistry*, 4th edn. New York: Wiley.

Matten, L. C. 1973. Preparation of pyritized plant petrifactions: "a plea for pyrite." *Review of Palaeobotany and Palynology* **16**, 165–73.

Maxwell, J. A. 1968. *Rock and mineral analysis*. New York: Interscience.

Maxwell, J. C. 1964. Influence of depth, temperature and geologic age on porosity of quartzose sandstone. *Bulletin of the American Association of Petroleum Geologists* **48**, 697–709.

McBride, E. F., W. L. Lindemann & P. S. Freeman 1968. Lithology and petrology of the Gueydan (Catahoula) Formation in south Texas. *Report of Investigations, Texas Bureau of Economic Geology* No. 63.

McCarthy, T. S., A. J. Erlank, J. P. Willis & L. H. Ahrens 1974. New chemical analyses of six achondrites and one chondrite. *Meteoritics* **9**, 215–22.

McCormack, D. E. & L. P. Wilding 1974. Proposed origin of lattisepic fabric. In *Soil microscopy*, G. K. Rutherford (ed.), 761–71. Kingston: Limestone.

McDonnell, K. L. 1974. Depositional environments of the Triassic Gosford Formation, Sydney Basin. *Journal of the Geological Society of Australia* **21**, 107–32.

McFadden, L. D. & J. C. Tinsley 1985. Rate and depth of pedogenic carbonate accumulation in soils: formulation and testing of a compartment model. In *Soils and Quaternary geology of the southwestern United States*, D. L. Weide (ed.), 23–41. *Special Paper, Geological Society of America* No. 203.

McFarlane, M. J. 1976. *Laterite and landscape*. New York: Academic Press.

McGee, W. J. 1878. On the relative positions of the forest bed and associated drift formations in northeastern Iowa. *American Journal of Science* **27**, 189–213.

McKay, D. S. & A. Basu 1983. The production curve for agglutinates in planetary regoliths. *Proceedings of the 14th Lunar and Planetary Science Conference Journal of Geophysic Research, Supplement* **88**(1), 193–9.

McKay, D. S., R. M. Fruland & G. M. Heiken 1974. Grain size and evolution of lunar soils. *Proceedings of the 5th Lunar Science Conference Geochimica et Cosmochimica Acta, Supplement* **5**(1), 887–906.

McLean, D. M. 1978. Land floras: the major late Phanerozoic atmospheric carbon dioxide/oxygen control. *Science* **200**, 1060–2.

McNaughton, S. J. 1983. Compensatory plant growth as a response to herbivory. *Oikos* **40**, 329–36.

NcNaughton, S. J. 1984. Grazing lawns: animals in herds, plant form and coevolution. *American Naturalist* **124**, 863–6.

McNaughton, S. J. 1985. Silica as a defense against herbivory. *Ecology* **66**, 528–35.

McPherson, J. G. 1979. Calcrete (caliche) palaeosols in fluvial redbeds of the Aztec Siltstone (Upper Devonian), southern Victoria Land, Antarctica. *Sedimentary Geology* **22**, 319–20.

Mellett, J. S. 1974. Scatological origin of microvertebrate fossil accumulations. *Science* **185**, 349–50.

Mermut, A. R., M. A. Arshad & R. J. St. Arnaud 1984. Micropedological study of termite mounds of *Macrotermes* in Kenya. *Journal of the Soil Science Society of America* **48**, 613–20.

Metting, B. 1987. Dynamics of wet and dry aggregate stability from three year microalgal soil conditioning experiment in the field. *Soil Science* **143**, 139–43.

Meyen, S. V. 1982. The Carboniferous and Permian flora of Angaraland (a synthesis). *Biological Monographs Lucknow* 7.

Meyer, C. 1985. Ore metals through geologic history. *Science* **227**, 1421–8.

Meyer, K. O. 1961. Lumbricidea-Bauten aus Pleistozänen Sanden. *Verhandlungen Vereins Naturwissenschaft Heimatforschung Hamburg* **35**, 10–15.

Michener, C. D. 1974. *The social behavior of bees*. Cambridge: Belknap Press of Harvard University.

Michener, C. D. & D. Grimaldi 1988. A *Trigona* from Late Cretaceous amber of New Jersey (Hymenoptera, Apidae, Meliponinae). *American Museum Novitates* 2917, 1–10.

Mikulic, D. G., D. E. G. Brigs & J. Kluessendorf 1985. A new exceptionally well preserved biota from the Lower Silurian of Wisconsin, U.S.A. *Philosophical Transactions of the Royal Society of London* **B311**, 75–85.

Miller, S. L. & J. L. Bada 1988. Submarine hot springs and the origin of life. *Nature* **334**, 609–11.

Miller, S. L. & L. E. Orgel 1974. *The origins of life on Earth*. Englewood Cliffs: Prentice-Hall.

Mishler, B. D. & S. P. Churchill 1985. Transition to a land flora: phylogenetic relationships of the green algae and bryophtyes. *Cladistics* **1**, 305–28.

Miyamoto, M. M., J. L. Sloghtom & M. Goodman 1987. Phylogenetic relations of humans and African apes from DNA sequences in the $\psi\eta$-globin region. *Science* **238**, 369–73.

Monod, J. 1971. *Chance and necessity* (translated by A. Wainhouse). New York: Knopf.

Mooney, H. A., P. M. Vitousek & P. A. Matson 1987. Exchange of materials between terrestrial ecosystems and the atmosphere. *Science* **238**, 926–32.

Morgan, J. 1959. The morphology and anatomy of American species of the genus *Psaronius. Illinois Biological Monographs* No. 27.

Morris, R. C. 1985. Genesis of iron ore in banded iron-formation by supergene and supergene-metamorphic processes. In *Handbook of strata-bound and stratiform ore deposits. Vol. 13. Regional studies and specific deposits*, K. H. Wolf (ed.), 73–235. Amsterdam: Elsevier.

Morris, S. C., R. K. Pickerill & T. L. Harland 1982. A possible annelid from the Trenton Limeston (Ordovician) of Quebec, with a review of fossil oligochaetes and other annulate worms. *Canadian Journal of Earth Sciences* **19**, 2150–7.

Morris, S. F. 1979. A new fossil terrestrial isopod with implications for East African Miocene landform. *Bulletin of the British Museum (Natural History), Geology* **32**, 71–5.

Morrison, R. B. 1967. Principles of Quaternary stratigraphy. In *Quaternary soils*, R. B. Morrison & H. E. Wright (eds.), 1–69. *Proceedings of the 7th Congress of the International Association of Quaternary Research* Vol. 9.

Morrison, R. B. 1978. Quaternary soil stratigraphy – concepts, methods and problems. In *Quaternary soils*, W. C. Mahaney (ed.), 77–108. Norwich: Geoabstracts.

Morton, J. P. 1985. Rb–Sr evidence for punctuated illite/smectite diagenesis in the Oligocene Frio Formation, Texas Gulf Coast. *Bulletin of the Geological Society of America* **96**, 114–22.

Moussa, M. T. 1970. Nematode fossil trails from the Green River Formation (Eocene) of the Unita Basin, Utah. *Journal of Paleontology* **44**, 304–7.

Mueller, G. 1968. Genetic histories of nitrate deposits from Antarctica and Chile. *Nature* **219**, 1131–4.

Muhs, D. R. 1983. Airborne dust fall on the California Channel Islands, U.S.A. *Journal of Arid Environments* **6**, 223–38.

Muhs, D. R. 1984. Intrinsic thresholds in soil systems. *Physical Geography* **5**, 99–110.

Muir, J. W. & J. Logan 1982. Eluvial/illuvial coefficients of major elements and the corresponding losses and gains in three soil profiles. *Journal of Soil Science* **33**, 295–308.

Muller, J. 1981. Fossil pollen records of extant angiosperms. *Botanical Review* **47**, 1–142.

Mumpton, F. A. & W. C. Ormsby 1976. Morphology of zeolites in sedimentary rocks by scanning electron microscopy. *Clays and Clay Minerals* **24**, 1–23.

Munsell Color 1975. *Munsell color charts*. Baltimore: Munsell Color.

Murphy, C. P. 1983. Point-counting pores and illuvial clay in thin sections. *Geoderma* **31**, 133–50.

Murphy, C. P. 1986. *Thin section preparation of soils and sediments*. Berkhamstead: A.B. Academic.

Murray, R. C. 1964. Origin and diagenesis of gypsum and anhydrite. *Journal of Sedimentary Petrology* **34**, 512–23.

Mutch, T. A., R. E. Arvidson, A. B. Binder, E. A. Guiness & E. C. Morris 1977. The geology of the Viking Lander 2 site. *Journal of Geophysical Research* **82**, 4452–67.

Nagy, B. 1975. *Carbonaceous meteorites*. Amsterdam: Elsevier.

Nanninga, N. (ed.) 1985. *Molecular cytology of* Escherichia coli. London: Academic Press.

Neall, V. E. 1977. Genesis and weathering of Andosols in Taranaki, New Zealand. *Soil Science* **123**, 400–8.
Nesbitt, H. W. & G. M. Young 1989. Formation and diagenesis of weathering profiles. *Journal of Geology* **97**, 129–47.
Nettleton, W. D. & B. R. Brasher 1983. Correlation of clay minerals and properties of soils in the Western United States. *Soil Science Society of America Journal* **67**, 1032–6.
Neumann-Mahlkau, P. 1976. Recent sand volcanoes in the sand of a dike under construction. *Sedimentology* **23**, 421–5.
Neville, A. C. 1975. *Biology of the arthropod cuticle*. New York: Springer.
Nikiforoff, C. C. 1943. Introduction of paleopedology. *American Journal of Science* **241**, 194–200.
Nikiforoff, C. C. 1959. Reappraisal of the soil. *Science* **129**, 186–96.
Niklas, K. J. 1982. Chemical diversification and evolution of plants as inferred from paleobiochemical studies. In *Biochemical aspects of evolutionary biology*, M. H. Nitecki (ed.), 29–91. Chicago: University of Chicago Press.
Nilsson, T. 1964. Standardpollendiagramme und C^{14}-Datierungen aus dem Ageröds Mosse in mittlere Schonen. *Lund Universität Årsskrifter Neue Folge 2*, No. 59.
Nilsson, T. 1983. *The Pleistocene: geology and life in the Quaternary Ice Age*. Hingham: Reidel.
Nisbet, E. G. 1987. *The young Earth*. Boston: George Allen & Unwin.
X North American Commission on Stratigraphic Nomenclature 1982. North American stratigraphic code. *Bulletin of the American Association of Petroleum Geologists* **67**, 841–75.
Northcote, K. H. 1974. *A factual key for the recognition of Australian soils*. Adelaide: Rellim Technical.
Northcote, K. H. & J. K. M. Skene 1972. Australian soils with saline and sodic properties. *Soil Publications of the Commonwealth Scientific and Industrial Research Organization, Australia* No. 27.
Nursall, J. R. 1981. Behavior and habitat affecting the distribution of five species of sympatric mudskippers in Queensland. *Bulletin of Marine Science* **31**, 730–5.

O'Connor, F. B. 1967. The Enchytraeidae. In *Soil biology*, A. Burges & F. Raw (eds.), 213–57. London: Academic Press.
Ohmoto, H. & R. P. Felder 1987. Bacterial activity in the warmer sulfate-bearing Archaean oceans. *Nature* **328**, 244–6.
Ollier, C. 1969. *Weathering*. New York: Elsevier.
Olsen, P. E., N. H. Shubin & M. H. Anders 1987. New Early Jurassic tetrapod assemblages constrain Triassic–Jurassic tetrapod extinction event. *Science* **237**, 1025–9.
Olsen, R. A., R. B. Clark & J. H. Bennett, 1981. The enhancement of soil fertility by plant roots. *American Scientist* **69**, 378–84.
Olson, E. C. 1985. Nonmarine vertebrates and Late Paleozoic climates. In *Paleontology, paleoecology, paleogeography*, J. T. Dutro & H. W. Pfefferkorn (eds.), 403–14. *Comptes Rendus Neuvième Congrès International de Stratigraphie et de Géologie du Carbonifère* 5. Carbondale: Southern Illinois University Press.
Olson, E. C. & K. Bolles 1975. Permo-Carboniferous freshwater burrows. *Fieldiana, Geology* **33**, 271–90.
Oparin, A. I. 1924. *Proiskhozhdenie zhizni* (The origin of life). Moscow: Rabochii (translated in an appendix of J. D. Bernal, 1967, *The origin of life*. New York: World).
Ortlam, D. 1971. Paleosols and their significance in stratigraphy and applied geology in the Permian and Triassic of southern Germany. In *Paleopedology: origin, nature and dating of paleosols*, D. H. Yaalon (ed.), 321–7. Jerusalem: Israel University Press.
Ortoleva, P., G. Auchmuty, J. Chadam, J. Heftmer, E. Merino, C. H. Moore & E. Ripley 1986. Redox front propagation and banding modalities. *Physica* **19D**, 334–54.
Ostrom, J. H. 1974. *Archaeopteryx* and the origin of flight. *Quarterly Review of Biology* **49**, 27–47.

Owen-Smith, N. 1987. Pleistocene extinctions: the pivotal role of megaherbivores. *Paleobiology* **13**, 351–62.

Palmer, A. R. 1983. The decade of North American geology 1983 Geologic Time Scale. *Geology* **11**, 503–4.

Page, A. L. (ed.) 1982. *Methods of soil analysis. Part 2, Chemical and microbiological properties*, 2nd edn. Madison: American Society of Agronomy.

Patching, W. R. 1987. *Soil survey of Lane County area, Oregon*. Washington: U.S. Government.

Paton, T. R. 1974. Origin and terminology for gilgai in Australia. *Geoderma* **11**, 221–42.

Paulusse, J. H. M. & C. Y. Jeanson 1977. Structuration du sol par les diplopodes: étude experimentale et microscopique. In *Soil organisms as components of ecosystems*, U. Lohm & T. Persson (eds.), 484–8. *Ecological Bulletin* No. 25.

Pavich, M. J. & S. F. Obermeier 1985. Saprolite formation beneath coastal plain sediments near Washington D.C. *Bulletin of the Geological Society of America* **96**, 886–900.

Pavich, M. J., L. Brown, J. Harden, J. Klein & R. Middleton 1986. ^{10}Be distribution in soils from Merced River terraces, California. *Geochimica et Cosmochimica Acta* **50**, 1727–35.

Pelczar, M. J., E. C. S. Chan, N. R. Krieg & M. F. Pelczar 1986. *Microbiology*. New York: McGraw-Hill.

Pennington, W. 1965. The interpretation of some post-Glacial vegetation at different Lake District sites. *Proceedings of the Royal Society of London* **B161**, 310–23.

Pennington, W., P. A. Cranwell, E. Y. Haworth, A. P. Bonny & J. P. Lishman 1977. Interpreting the environmental record in the sediments of Blelham Tarn. *Annual Report of the Freshwater Biological Association* **45**, 37–47.

Percival, C. J. 1986. Paleosols containing an albic horizon: examples from the Upper Carboniferous of northern England. In *Paleosols: their recognition and interpretation*, P. V. Wright (ed.), 87–111. Oxford: Blackwell.

Perry, R. S. & J. B. Adams 1978. Desert varnish: evidence of cyclic deposition of manganese. *Nature* **276**, 488–91.

Pickford, M. 1985. A new look at *Kenyapithecus* based on recent discoveries in western Kenya. *Journal of Human Evolution* **14**, 113–44.

Pickford, M. 1986a. Sedimentation and fossil preservation in the Nyanza Rift System, Kenya. In *Sedimentation in the African Rifts*, L. E. Frostick, R. W. Renault, I. Reid & J. J. Tiercelin (eds.), 345–62. *Special Publication, Geological Society, London* No. 25.

Pickford, M. 1986b. Cainozoic paleontological sites of western Kenya. *Münchner Geowissenshaften Abhandlungen* A8.

Pickford, M. & P. Andrews 1981. The Tinderet Miocene sequence in Kenya. *Journal of Human Evolution* **10**, 11–33.

Pickford, M., H. Ishida, Y. Nakano & H. Nakaya 1984. Fossiliferous localities of the Nachola-Samburu Hills area, northern Kenya. *Supplementary Issue African Study Monographs Kyoto Univ.* **2**, 45–56.

Pilbeam, D. 1982. New hominoid skull material from the Miocene of Pakistan. *Nature* **295**, 232–4.

Pinto, J. & H. D. Holland 1988. Paleosols and the evolution of the atmosphere, Part II. In *Paleosols and weathering through geologic time: principles and applications*, J. Reinhardt & W. R. Sigleo (eds.), 21–34. *Special Paper, Geological Society of America* No. 216.

Piperno, D. 1987. *Phytolith analysis*. San Diego: Academic Press.

Pitt, M. D., H. M. Madeley & J. R. Robertson 1961. Concretionary stream bar deposits. *Oklahoma Geology Notes* **21**, 301–6.

Playfair, J. 1802. *Illustrations of the Huttonian theory of the Earth*. London and Edinburgh: Cadell & Davies and William Creech (facsimile edn., University of Illinois Press, Urbana, 1956).

Plaziat, J. C. 1970. Huitres de mangrove et peuplements littoraux de l'Eocene inferieur des Corbières. *Geobios* **3**, 7–27.

Plint, R. G. 1962. Stone axe factory sites in the Cumbrian fells. *Transactions of the Cumberland & Westmoreland Antiquarian and Archaeological Society, New Series* **62**, 1–26.

Poinar, G. O. 1983. *The natural history of nematodes*. Englewood Cliffs: Prentice-Hall.

Poirier, F. E. 1987. *Understanding human evolution*. Englewood Cliffs: Prentice-Hall.

Polley, H. W. & S. L. Collins 1984. Relationships of vegetation and environment in buffalo wallows. *American Midland Naturalist* **112**, 178–86.

Polynov, B. B. 1927. *Contributions of Russian scientists to paleopedology*. Leningrad: USSR Academy of Sciences.

Pomerol, C. 1964. Decouverte de paléosols de type podzol au sommet de l'Auversien (Bartonien inférieur) de Moisselles (Sein-et-Oise). *Comptes Rendus Hebdomadaires des Séances de l'Academie des Sciences, Série D* **258**, 974–6.

Pope, K. O. & T. H. Van Andel 1984. Late Quaternary alluviation and soil formation in southern Argolid: its history, causes and archaeological implications. *Journal of Archaeological Science* **11**, 281–306.

Powers, L. W. & D. E. Bliss 1983. Terrestrial adaptations. In *Biology of Crustacea*, Vol. 8, F. J. Vernberg and Q. B. Vernberg (eds.), 271–333. New York: Academic Press.

Pratt, L. M., T. L. Phillips & J. M. Dennison 1978. Evidence of non-vascular land plants from the Early Silurian (Llandoverian) of Virginia, U.S.A. *Review of Palaeobotany and Palynology* **25**, 121–49.

Price, L. W. 1981. *Mountains and man*. Berkeley: University of California Press.

Prothero, D. R. 1985. North American mammalian diversity and Eocene–Oligocene extinctions. *Paleobiology* **11**, 389–405.

Prothero, D. R. 1987. The rise and fall of the American rhino. *Natural History* **96**(8), 26–33.

Radwanski, A. 1977. Present-day types of traces in the Neogene sequence: their problems of nomenclature and preservation. In *Trace Fossils 2*, J. P. Crimes and J. C. Harper (eds.), 227–64. Liverpool: Seal House.

Rahn, P. 1971. The weathering of tombstones and its relationship to the topography of New England. *Journal of Geological Education* **19**, 112–18.

Rao, C. P. 1985. Origin of coal balls in the Illinois Basin. In *Economic geology: coal, oil and gas*, A, T. Cross (ed.), 393–406. *Comptes Rendus Neuvième Congrès International de Stratigraphie et de Géologie du Carbonifère* No. 4. Carbondale: Southern Illinois University Press.

Rao, M. D., D. G. Odom & J. Oró 1980. Clays in prebiological chemistry. *Journal of Molecular Evolution* **15**, 317–31.

Ratcliffe, B. C. & J. A. Fagerstrom 1980. Invertebrate lebensspuren of Holocene floodplains: their morphology, origin and paleoecological significance. *Journal of Paleontology* **54**, 614–30.

Raven, P. H., R. F. Evert & H. Curtis 1981. *Biology of plants*. New York: Worth.

Reams, M. W. 1989. Stromatolitic humid climate carbonates: a variety of calcrete? In *International working meeting on soil micromorphology*, L. A. Douglas (ed.), 381–6. Amsterdam: Elsevier.

Reimer, T. O. 1986. Alumina-rich rocks from the Early Precambrian of the Kaapvaal Craton as indicators of paleosols and as products of other decompositional reactions. *Precambrian Research* **32**, 155–79.

Renton, J. J., M. T. Heald & C. B. Cecil 1969. Experimental investigation of pressure solution of quartz. *Journal of Sedimentary Petrology* **39**, 1107–17.

Retallack, G. J. 1975. The life and times of a Triassic lycopod. *Alcheringa* **1**, 3–29.

Retallack, G. J. 1976. Triassic paleosols in the upper Narrabeen Group of New South Wales. Part I. Features of the paleosols. *Journal of the Geological Society of Australia* **23**, 383–99.

Retallack, G. J. 1977. Triassic paleosols in the upper Narrabeen Group of New South Wales. Part II. Classification and reconstruction. *Journal of the Geological Society of Australia* **24**, 19–35.

Retallack, G. J. 1979. Middle Triassic coastal outwash plain deposits in Tank Gully, Canterbury, New Zealand. *Journal of the Royal Society of New Zealand* **9**, 397–414.

Retallack, G. J. 1980. Late Carboniferous to Middle Triassic megafossil floras from Sydney Basin. In *A guide to the Sydney Basin*, C. Herbert & R. J. Helby (eds.), 384–430. *Bulletin of the Geological Survey of New South Wales* No. 26.

Retallack, G. J. 1981a. Preliminary observations on fossil soils in the Clarno Formation (Eocene to early Oligocene), near Clarno, Oregon. *Oregon Geology* **43**, 147–50.

Retallack, G. J. 1981b. Comment on "Reinterpretation of the depositional environment of the Yellowstone fossil forests." *Geology* **9**, 52–3.

Retallack, G. J. 1981c. Two new approaches for reconstructing fossil vegetation: with examples from the Triassic of eastern Australia. In *Communities of the past*, J. Gray, A. J. Boucot & W. B. N. Berry (eds.), 271–95. Stroudsburg: Hutchinson-Ross.

Retallack, G. J. 1982. Paleopedological perspectives on the development of grasslands during the Tertiary. In *Proceedings of the 3rd North American Paleontological Convention*, vol. 2, B. Mamet and M. J. Copland (eds.), 417–21. Toronto: Business & Economic Service.

Retallack, G. J. 1983a. A paleopedological approach to the interpretation of terrestrial sedimentary rocks: the mid-Tertiary fossil soils of Badlands National Park, South Dakota. *Bulletin of the Geological Society of America* **94**, 823–40.

Retallack, G. J. 1983b. Late Eocene and Oligocene paleosols from Badlands National Park, South Dakota. *Special Paper, Geological Society of America* No. 193.

Retallack, G. J. 1983c. Middle Triassic estuarine deposits near Benmore Dam, southern Canterbury and northern Otago, New Zealand. *Journal of the Royal Society of New Zealand* **13**, 107–27.

Retallack, G. J. 1984a. Completeness of the rock and fossil record: estimates using fossil soils. *Paleobiology* **10**, 59–78.

Retallack, G. J. 1984b. Trace fossils of burrowing beetles and bees in an Oligocene paleosol, Badlands National Park, South Dakota. *Journal of Paleontology* **58**, 571–92.

Retallack, G. J. 1985. Fossil soils as grounds for interpreting the advent of large plants and animals on land. *Philosophical Transactions of the Royal Society of London* **B309**, 105–42.

Retallack, G. J. (ed.) 1986a. Precambrian paleopedology. *Precambrian Research* **32**, 93–259.

Retallack, G. J. 1986b. Reappraisal of a 2200-Ma-old paleosol from near Waterval Onder, South Africa. *Precambrian Research* **32**, 195–252.

Retallack, G. J. 1986c. The fossil record of soils. In *Paleosols: their recognition and interpretation*, P. V. Wright (ed.), 1–57. Oxford: Blackwell.

Retallack, G. J. 1986d. Fossil soils as grounds for interpreting long term controls on ancient rivers. *Journal of Sedimentary Petrology* **56**, 1–18.

Retallack, G. J. 1988a. Field recognition of paleosols. In *Paleosols and weathering through geologic time: principles and applications*, J. Reinhardt & W. R. Sigleo (eds.), 1–20. *Special Paper, Geological Society of America* No. 216.

Retallack, 1988b. Down to earth approaches to vertebrate paleontology. *Palaios* **3**, 335–44.

Retallack, G. J. & D. L. Dilcher 1981a. A coastal hypothesis for the dispersal and rise to dominance of flowering plants. In *Paleobotany, paleoecology and evolution*, K. J. Niklas (ed.), vol. 2, 27–77. New York: Praeger.

Retallack, G. J. & D. L. Dilcher 1981b. Early angiosperm reproduction: *Prisca reynoldsii* gen. et sp. nov. from mid-Cretaceous coastal deposits in Kansas, U.S.A. *Palaeontographica* **B179**, 103–37.

Retallack, G. J. & D. L. Dilcher 1986. Cretaceous angiosperm invasion of North America. *Cretaceous Research* **7**, 227–52.

Retallack, G. J. & D. L. Dilcher 1988. Reconstruction of selected seed ferns. *Annals of the Missouri Botanical Garden* **75**, 1010–57.

Retallack, G. J. & C. R. Feakes 1987. Trace fossil evidence for Late Ordovician animals on land. *Science* **235**, 61–3.

Retallack, G. J., G. D. Leahy & M. D. Spoon 1987. Evidence from paleosols for ecosystem changes across the Cretaceous/Tertiary boundary in eastern Montana. *Geology* **15**, 1090–3.

Rex, G. M. & A. C. Scott 1987. The sedimentology, paleoecology and preservation of Lower Carboniferous plant deposits at Pettycur, Fife, Scotland. *Geological Magazine* **124**, 43–66.

Rhoads, D. C. & J. W. Morse 1970. Evolutionary and ecologic significance of oxygen deficient marine basins. *Lethaia* **4**, 413–28.

Richards, B. N. 1987. *The microbiology of terrestrial ecosystems*. Harlow: Longman Scientific and Technical.

Richards, P. W. 1952. *The tropical forest*. Cambridge: Cambridge University Press.

Richardson, D. H. S. 1981. *The biology of mosses*. New York: Wiley.

Richardson, S. M. 1978. Vein formation in the C1 carboniferous chondrites. *Meteoritics* **13**, 141–59.

Richmond, G. M. & D. S. Fullerton 1987. Summation of Quaternary glaciations in the United States of America. In *Quaternary glaciations in the northern hemisphere*, V. Sirrava, D. Q. Bowen & G. M. Richmond (eds.), 183–96. Oxford: Pergamon Press.

Ride, W. D. L., C. W. Sabrosky, G. Bernardi & R. V. Melville 1985. *International code of zoological nomenclature*, 3rd edn. Berkeley: University of California Press.

Riding, R. & L. Voronova 1985. Morphological groups and series in Cambrian calcareous algae. In *Paleoalgology*, D. F. Toomey & M. H. Nitecki (eds.), 56–78. Berlin: Springer.

Rieke, H. H. & G. V. Chilingarian 1974. *Compaction of argillaceous sediments*. Amsterdam: Elsevier.

Rinehart, J. S. 1980. *Geysers and geothermal energy*. New York: Springer.

Rinne, R. J. K. & P. Barclay-Estrup 1980. Heavy metals in a feather moss, *Pleurozium schreberi*, and in soils in NW Ontario, Canada. *Oikos* **34**, 59–67.

Robertson, D. S., J. E. Tilsley & G. M. Hogg 1978. The time-bound character of uranium deposits. *Economic Geology* **73**, 1409–19.

Robison, R. A. 1986. A marine myriapodlike fossil from the Middle Cambrian of Utah. *Abstracts, Geological Society of America* **19**, 823.

Roche, H. & J.-J. Tiercelin 1977. Découverte d'une industrie lithique ancienne *in situ* dans la formation d'Hadar, Afar central, Ethiopie. *Comptes Rendus Hebdomadaires des Séances de l'Academie des Sciences, Paris, Série D* **284**, 1871–4.

Rockwell, T. K., D. L. Johnson, E. A. Keller & G. R. Dembroff 1985a. A late Pleistocene–Holocene soil chronosequence in the Ventura Basin, southern California, U.S.A. In *Geomorphology and soils*, K. S. Richards, R. R. Arnett & S. Ellis (eds.), 309–27. London: George Allen & Unwin.

Rockwell, T. K., E. A. Keller & D. L. Johnson 1985b. Tectonic geomorphology of alluvial fans and mountain fronts near Ventura, California. In *Tectonic geomorphology*, M. Morisawa & J. T. Hack (eds.), 183–207. Boston: George Allen & Unwin.

Rohr, D. M., A. J. Boucot, J. Miller & M. Abbot 1986. Oldest termite nest from the Upper Cretaceous of Texas. *Geology* **14**, 87–8.

Romell, L. G. 1935. An example of myriapods as mull formers. *Ecology* **16**, 67–71.

Ronov, A. B. 1964. Common tendencies in the evolution of the Earth's crust, ocean and atmosphere. *Geochemistry* **8**, 715–43.

Ronov, A. B. & A. A. Migdisov 1971. Geochemical history of the crystalline basement and sedimentary cover of the Russian and North American platforms. *Sedimentology* **16**, 137–85.

Roscoe, S. M. 1968. Huronian rocks and uraniferous conglomerates in the Canadian Shield. *Geological Survey of Canada Paper* No. 68–40.

Rose, M. D. 1986. Further hominoid postcranial specimens from the Late Miocene Nagri Formation of Pakistan, *Journal of Human Evolution* **15**, 333–67.

Ross, G. M. & J. P. Chiarenzelli 1984. Paleoclimatic significance of widespread Proterozoic silcretes in the Bear and Churchill provinces of the northwestern Canadian Shield. *Journal of Sedimentary Petrology* **55**, 196–204.

Ross, K. A. & R. V. Fisher 1986. Biogenic grooving in glass shards. *Geology* **14**, 571–3.

Rothwell, G. W. 1975. The Callistophytaceae (Pteridospermopsida): I, vegetative structures. *Palaeontographica* **B151**, 171–96.

Rowley, D. B., A. Raymond, J. T. Parrish, A. L. Lottes, C. R. Scotese & A. M. Zeigler 1985. Carboniferous paleogeographic, phytogeographic and paleoclimatic reconstructions. *International Journal of Coal Geology* **5**, 7–42.

Rozen, O. M. 1967. Metamorphosed bauxite pebbles in conglomerate among Precambrian schists of the Kokchetov Massif. *Doklady Akademia Nauk USSR Earth Sciences Section* (American Geological Institute translation) **174**, 66–8.

Rucklidge, J. C. 1984. Radioisotope detection and dating with particle accelerators. In *Quaternary dating methods*, W. C. Mahaney (ed.), 17–32. Amsterdam: Elsevier.

Ruhe, R. V. 1959. Stone lines in soils. *Soil Science* **87**, 223–31.

Ruhe, R. V. 1969. *Quaternary landscapes in Iowa*. Ames: Iowa State University Press.

Ruhe, R. V. 1984. Soil–climate system across the prairies in the midwestern U.S.A. *Geoderma* **34**, 201–19.

Ruhling, A. & G. Tyler 1970. Sorption and retention of heavy metals in the woodland moss *Hylocomium splendens* (Hedev) Br. & Sh. *Oikos* **21**, 92–7.

Ruibal, R., L. Tevis & V. Roig 1969. The terrestrial ecology of the spadefoot toad, *Scaphiopus hammondi*. *Copeia* 571–84.

Runge, E. C. A. 1973. Soil development and energy models. *Soil Science* **115**, 183–93.

Runham, N. W. & P. T. Hunter 1970. *Terrestrial slugs*. London: Hutchinson.

Runnegar, B. 1982. Oxygen requirements, biology and phylogenetic significance of the late Precambrian worm *Dickinsonia*, and the evolution of the burrowing habit. *Alcheringa* **6**, 228–39.

Rushforth, S. R. 1971. A flora from the Dakota Sandstone Formation (Cenomanian), near Westwater, Grand County, Utah. *Science Bulletin, Brigham Young University, Biology Series* No. 14(3).

Russell, D. A. 1979. The enigma of the extinction of the dinosaurs. *Annual Review of Earth and Planetary Sciences* **7**, 163–82.

Russell, R. S. 1977. *Plant root systems: their function and interaction with the soil*. London: McGraw-Hill.

Ryer, T. A. & A. W. Langer 1980. Thickness change involved in the peat-to-coal transformation for a bituminous coal of Cretaceous age in central Utah. *Journal of Sedimentary Petrology* **50**, 987–92.

Sadler, P. M. 1981. Sediment accumulation rates and the completeness of stratigraphic sections. *Journal of Geology* **89**, 569–84.

Sagan, C. & J. B. Pollack 1974. Differential transmission of sunlight on Mars: biological implications. *Icarus* **21**, 490–5.

Sahni, B. 1932. On the genera *Clepsydropsis* and *Cladoxylon* of Unger and on a new genus *Austroclepsis*. *New Phytologist* **31**, 270–8.

Sakagami, S. F. & C. D. Michener 1962. *The nest architecture of the sweat bees (Halictinae). A comparative study of behavior*. Lawrence: University of Kansas Press.

Sanchez, P. A. & S. W. Buol 1974. Properties of some soils of the upper Amazon Basin of Peru. *Proceedings of the Soil Science Society of America* **38**, 117–21.

Sanford, R. L. 1987. Apogeotropic roots in an Amazon rain forest. *Science* **235**, 1062–4.

Sanson, G. D. 1982. Evolution of feeding adaptations in fossil and recent macropodids. In *The fossil vertebrate record of Australasia*, P. V. Rich & E. M. Thompson (eds.), 490–506. Clayton: Monash University Press.

Sarich, V. M. & A. C. Wilson 1967. Immunological time scale for human evolution. *Science* **158**, 1200–3.

Sarjeant, W. A. S. (ed.) 1983. *Terrestrial trace fossils*. Stroudsburg: Dowden, Hutchinson & Ross.

Sarna-Wojcicki, A. M., C. E. Meyer, P. H. Roth & F. T. Brown 1985. Ages of tuff beds at East African early hominid sites in the Gulf of Aden. *Nature* **313**, 306–8.

Savage, R. J. G. & M. R. Long 1986. *Mammal evolution*. New York: Facts on File.

Schau, M. K. & J. B. Henderson 1983. Archaean weathering at three localities on the Canadian Shield. *Precambrian Research* **20**, 189–202.

Schaetzl, R. J. & C. J. Sorenson 1987. The concept of "buried" versus "isolated" paleosols; examples from northeastern Kansas. *Soil Science* **143**, 426–35.

Scheckler, S. E. 1986. Floras of the Devonian–Mississippian transition. In *Land plants – notes for a short course*, R. A. Gastaldo (ed.), 81–96. *Science Studies, Department of Geology, University of Tennessee* No. 15.

Schenk, E. 1937. Insektenfrassgange oder Bohrlocher von Pholadiden in Ligniten aus dem Braunkohlenflotz bei Köln. *Neues Jahrbuch für Mineralogie, Geologie und Paläontologie* **77**, 392–401.

Schlee, D. 1980. *Bernstein-Raritaten*. Stuttgart: Staatliche Museum für Naturkunde.

Schlüter, T. 1984. Kretazische Lebensspuren von solitaren Hymnopteren. *Aufschluss Heidelberg* **35**, 423–30.

Schmidt, V. & D. A. McDonald 1979. The role of secondary porosity in the course of sandstone diagenesis. In *Aspects of diagenesis*, P. A. Scholle & P. R. Schluger (eds.), 175–207. *Special Publication, Society of Economic Paleontologists and Mineralogists* No. 26.

Scholle, P. A. 1978. A color illustrated guide to carbonate rock constituents, textures, cements and porosities. *Memoir of the American Association of Petroleum Geologists* No. 27.

Scholle, P. A. 1979. A color illustrated guide to constituents, textures, cements and porosities of sandstones and associated rocks. *Memoir of the American Association of Petroleum Geologists* No. 28.

Schopf, J. M. 1982. Forms and facies of *Vertebraria* in relation to Gondwana coal. In *Geology of the central Transantarctic Mountains*, M. D. Turner & J. E. Splettstoesser (eds.), 37–62. *Antarctic Research Series* No. 36. Washington: American Geophysical Union.

Schopf, J. M., E. Mencher, A. J. Boucot & H. N. Andrews 1966. Erect plants in the early Silurian of Maine. *Professional Paper, U.S. Geological Survey* No. 550D, 69–75.

Schopf, J. W. (ed.) 1983. *Earth's earliest biosphere*. Princeton: Princeton University Press.

Schopf, J. W. & B. M. Packer 1987. Early Archaean (3.3-billion-to-3.5-billion-year-old) microfossils from Warrawoona Group, Australia. *Science* **237**, 70–3.

Schreyer, W., G. Werding & K. Abraham 1981. Corundum–fuchsite rocks in greenstone belts of southern Africa: petrology, geochemistry, and possible origin. *Journal of Petrology* **22**, 191–231.

Schumm, S. A. 1956. The role of creep and rainwash on the retreat of badland slopes. *American Journal of Science* **254**, 693–706.

Schumm, S. A. 1977. *The fluvial system*. New York: Wiley.

Schumm, S. A. 1981. Evolution and response of the fluvial system: sedimentologic implications. In *Recent and ancient non-marine depositional environments: models for exploration*, F. C. Ethridge & R. M. Flores (eds.), 19–29. *Special Publication, Society of Economic Paleontologists and Mineralogists* No. 31.

Scott, A. C. & G. Rex 1985. The formation and significance of Carboniferous coal balls. *Philosophical Transactions of the Royal Society of London* **B311**, 123–37.

Sehgal, J. L. & G. Stoops 1972. Pedogenic calcic accumulation in arid and semi-arid regions of the Indo-Gangetic alluvial plain of the erstwhile Punjab (India). Their morphology and origin. *Geoderma* **8**, 59–72.

Sehgal, J. L., C. Sys & D. R. Bhumbla 1968. A climatic soil sequence from the Thar desert to the Himalayan Mountains in Punjab (India). *Pedologie* **18**, 351–73.

Seilacher, A. 1984. Late Precambrian and Early Cambrian metazoa: preservational or real extinctions? In *Patterns of change in Earth evolution*, H. D. Holland & A. F. Trendall (eds.), 159–68. Berlin: Springer.

Selden, P. A. 1985. Eurypterid respiration. *Philosophical Transactions of the Royal Society of London* **B309**, 219–26.

Senior, B. R. & J. A. Mabbutt 1979. A proposed method of defining deeply weathered rock units based on regional geological mapping in southwest Queensland. *Journal of the Geological Society of Australia* **26**, 237–54.

Senut, B. 1988. Taxonomie et fonction chez les Hominoidea miocenes Africains: example de l'articulation du crude. *Annales de Paleontologie* **74**, 128–54.

Sepkoski, J. J. & A. H. Knoll 1983. Precambrian–Cambrian boundary: the spike is driven and the monolith crumbles. *Paleobiology* **9**, 199–206.

Serdyuchenko, D. P. 1968. Metamorphosed weathering crusts of the Precambrian: their metallogenic and petrographic fabric. In *Precambrian geology*, B. Hejtman (ed.), 37–42. *Proceedings of the 13th International Geological Congress* No. 4. Prague: Academia.

Seward, A. C. 1898. *Fossil plants*, vol. 1. Cambridge: Cambridge University Press (facsimile edn., New York: Hafner, 1963).

Shachak, M., C. G. Jones & Y. Granot 1987. Herbivory in rocks and the weathering of a desert. *Science* **236**, 1098–9.

Shacklette, H. T. 1965. Element content of bryophytes. *Bulletin of the U.S. Geological Survey* No. 1198G.

Sharp, R. P. 1940. The ep-Archaean and ep-Algonkian erosion surfaces, Grand Canyon, Arizona. *Bulletin of the Geological Society of America* **51**, 1235–69.

Sharp, R. P. & M. C. Malin 1984. Surface geology from Viking landers on Mars: a second look. *Bulletin of the Geological Society of America* **95**, 1398–412.

Shear, W. A., P. M. Bonamo, J. D. Grierson, W. D. I. Rolfe, E. L. Smith & R. A. Norton 1984. Early land animals in North America: evidence from Devonian age arthropods from Gilboa, New York. *Science* **224**, 492–4.

Sheihing, M. H. & H. W. Pfefferkorn 1984. The taphonomy of land plants in the Orinoco Delta: a model for the incorporation of plant parts in clastic sediments of Late Carboniferous age of Euramerica. *Review of Palaeobotany and Palynology* **41**, 205–40.

Sherman, G. D. 1952. The genesis and morphology of the alumina-rich laterite clays. In *Problems of clay and laterite genesis*, A. F. Frederickson (ed.), 154–61. New York: American Institute of Mining and Metallurgical Engineering.

Sherwood-Pike, M. A. & J. Gray 1985. Silurian fungal remains: probable records of Ascomycetes. *Lethaia* **18**, 1–20.

Shipman, P. 1981. *Life history of a fossil*. Cambridge: Harvard University Press.

Shipman, P. 1986, Paleoecology of Fort Ternan reconsidered. *Journal of Human Evolution* **15**, 193–204.

Shlemon, R. J. 1985. Application of soil-stratigraphic techniques to engineering geology. *Bulletin of the Association of Engineering Geologists* **22**, 129–42.

Sidorenko, A. V. 1963. Problemy osadochnoi geologii dokembria (Problems in Precambrian sedimentary geology). *Sovetskaya Geologia (Soviet Geology)* (4), 3–23.

Siegel, B. Z. 1977. *Kakabekia*, a review of its physiological and environmental features and their relation to its possible ancient affinities. In *Chemical evolution of the early Precambrian*, C. Ponnamperuma (ed.), 143–54. New York: Academic Press.

Simkin, T. & R. S. Fiske 1983. *Krakatau 1883*. Washington: Smithsonian Institution Press.

Simonson, R. W. 1941. Studies of buried soils formed from till in Iowa. *Proceedings of the Soil Science Society of America* **6**, 373–81.

Simonson, R. W. 1978. A multiple-process model of soil genesis. In *Quaternary soils*, W. C. Mahaney (ed.), 1–25. Norwich: Geoabstracts.

Simpson, G. G. 1951. *Horses*. New York: Oxford University Press.

Singer, A. & E. Galan (eds.) 1984. *Palygorskite–sepiolite: occurrence, genesis, uses*. Amsterdam: Elsevier.

Skelton, R. R., M. M. McHenry & G. M. Drawhorn 1986. Phylogenetic analysis of early hominids. *Current Anthropology* **27**, 21–43.

Smart, P. & N. K. Tovey 1981. *Electron microscopy of soils and sediments: examples*. Oxford: Clarendon Press.

Smeck, N. E. 1973. Phosphorus: an indicator of pedogenetic weathering processes. *Soil Science* **115**, 199–206.

Smiley, C. J. & W. C. Rember 1981. Paleoecology of the Miocene Clarkia Lake (northern Idaho) and its environs. In *Communities of the past*, J. Gray, A. J. Boucot & W. B. N. Berry (eds.), 551–90. Stroudsburg: Dowden, Hutchinson & Ross.

Smith, R. M. H. 1987. Helical burrow casts of therapsid origin from the Beaufort Group (Permian) of South Africa. *Palaeogeography, Palaeoclimatology, Palaeoecology* **60**, 155–69.

Smoot, E. L. & T. N. Taylor 1986. Structurally preserved fossil plants from Antarctica. II. A Permian moss from the Transantarctic Mountains. *American Journal of Botany* **73**, 1683–91.

Soil Survey Staff 1951. Soil Survey Manual. *Handbook, U.S. Department of Agriculture* No. 18.

Soil Survey Staff 1962. *Supplement to U.S.D.A. Handbook 18, Soil Survey Manual* (replacing p. 173–88). Washington: Government Printer.

Soil Survey Staff 1975. Soil taxonomy. *Handbook, U.S. Department of Agriculture* No. 436.

Sokal, R. R. & P. H. Sneath 1963. *Principles of numerical taxonomy*. San Francisco: Freeman.

Solem, A. & E. L. Yochelson 1979. North American Paleozoic land snails, with a summary of other Paleozoic nonmarine snails. *Professional Paper, U.S. Geological Survey* No. 1072.

Sombroek, W. G., H. M. H. Braun & B. J. A. van der Pouw 1982. *Exploratory soil map and agroclimatic zone map of Kenya, 1980: scale 1 : 100,000*. Nairobi: Kenya Soil Survey.

Sorem, R. K. & R. H. Fewkes 1979. *Manganese nodules*. New York: I.F.I.-Plenum.

Southgate, P. N. 1986. Cambrian phoscrete profiles, coated grains and microbial processes in phosphogenesis: Georgina Basin, Australia. *Journal of Sedimentary Petrology* **56**, 429–41.

Spalletti, L. A. & M. M. Mazzoni 1978. Sedimentologia de Grupo Sarmiento en el perfil ubicado al sudeste del Lago Colhue Huapi, Provincia de Chubut. *Obras del Centenario del Museo de La Plata* **4**, 261–83.

Specht, R. L. (ed.) 1979. *Heathlands and related shrublands. Ecosystems of the world*, vols. 9A and 9B. Amsterdam: Elsevier.

Spicer, R. A. 1981. The sorting and deposition of allochthonous plant material in a modern environment at Silwood Lake, Silwood Park, Berkshire, England. *Professional Paper, U.S. Geological Survey* No. 1143.

Spiess, F. N., K. C. MacDonald, T. Atwater, R. Ballard, A. Carranza, D. Cordoba, C. Cox, V. M. Diaz Garcia, J. Francheteau, J. Guerrero, R. Hawkins, R. Haymon, R. Hessler, T. Juteau, M. Kastner, R. Larson, B. Luydendyk, J. D. MacDougall, S. Miller, W. Normark, J. Orcutt & C. Rangin 1980. East Pacific Rise: hot springs and geophysical experiments. *Science* **207**, 1421–33.

Stace, H. C. T., G. D. Hubble, R. Brewer, K. H. Northcote, J. R. Sleeman, M. J. Mulcahy & E. G. Hallsworth 1968. *A handbook of Australian soils*, Adelaide: Rellim Technical.

Stach, E., M.-Th. Mackowsky, M. Teichmüller & R. Teichmüller 1975. *Stach's textbook of coal petrology* (translated by D. G. Murchison, G. H. Taylor & F. Zierkie). Berlin: Borntraeger.

Staley, J. T., F. Palmer & J. B. Adams 1982. Micocolonial fungi: common inhabitants on desert rocks? *Science* **215**, 1093–5.

Stanley, K. O. & G. Faure 1979. Isotopic composition and sources of strontium in sandstone cements: the High Plains sequence of Wyoming and Nebraska. *Journal of Sedimentary Petrology* **49**, 45–54.

Stanworth, C. W. & J. P. N. Badham 1984. Lower Proterozoic red beds, evaporites and secondary sedimentary uranium deposits form East Arm, Great Slave Lake, Canada. *Journal of the Geological Society, London* **141**, 235–42.

Staples, L. W. 1965. Origin and history of the thunder egg. *The Ore Bin* **27**, 195–204.

Stauffer, P. 1979. A fossilized honey bee comb from late Cenozoic cave deposits at Batu Caves, Malay Peninsula. *Journal of Paleontology* **53**, 1416–21.

Stebbins, G. L. & G. J. C. Hill 1980. Did multicellular plants invade the land? *American Naturalist* **115**, 342–53.

Steel, R. J. 1974. Cornstone (fossil caliche) – its origin, stratigraphic and sedimentological importance in the New Red Sandstone, Scotland. *Journal of Geology* **82**, 351–69.

Steitz, T. A., P. H. Ohlendorf, D. B. McKay, W. F. Anderson & B. F. Matthews 1982. Structural similarity in the DNA-binding domains of catabolite gene activator and cro repressor proteins. *Proceedings of the National Academy of Sciences of the U.S.A.* **79**, 3097–100.

Stephens, C. G. 1943. The pedology of a South Australian fen. *Transactions of the Royal Society of South Australia* **67**, 191–9.

Stevenson, F. J. 1969. Pedohumus: accumulation and diagenesis during the Quaternary. *Soil Science* **107**, 470–9.

Stevenson, F. J. 1986. *Cycles of soil: carbon, nitrogen, phosphorus, sulfur, micronutrients.* New York: Wiley.

Stewart, A. J., D. H. Blake & C. D. Ollier 1986. Cambrian river terraces and ridgetops in central Australia: oldest persisting landforms. *Science* **233**, 758–61.

Stewart, W. N. 1983. *Paleobotany and the evolution of plants.* Cambridge: Cambridge University Press.

Stolper, E. 1977. Experimental petrology of eucrite meteorites. *Geochimica et Cosmochimica Acta* **41**, 587–611.

Stolper, E. & H. Y. McSween 1979. Petrology and origin of the shergottite meteorites. *Geochimica et Cosmochimica Acta* **43**, 1475–98.

Stoops, G. 1983. Micromorphology of oxic horizons. In *Soil micromorphology*, vol. 1, P. Bullock and C. P. Murphy (eds.), 419–40. Berkhamsted: A.B. Academic.

Størmer, L. 1977. Arthropod invasion of land during Late Silurian and Devonian. *Science* **197**, 1362–4.

Stratigraphic Nomenclature Committee 1973. Australian code of stratigraphic nomenclature. *Journal of the Geological Society of Australia* **20**, 105–12.

Straus, A. 1977. Gallen, Minen und andere Frasspuren in Pliozän von Willershausen am Harz. *Verhandlungen Vereins Botanische Province Brandenburg* **113**, 43–80.

Stringer, C. B. & P. Andrews 1988. Genetic and fossil evidence for the origin of modern humans. *Science* **239**, 1263–8.

Strother, P. K. & C. Lenk 1983. *Eohostimella* is not a plant. *American Journal of Botany* **70**, 80.

Strother, P. K. & A. Traverse 1979. Plant microfossils from Llandoverian and Wenlockian rocks of Pennsylvania. *Palynology* **3**, 1–21.

Stubblefield, S. P. & T. N. Taylor 1988. Recent advances in palaeomycology. *New Phytologist* **108**, 3–25.

Sudo, T., S. Shimoda, H. Yotsumota & S. Aita 1981. *Electron micrographs of clay minerals.* Amsterdam: Elsevier.

Sukacheva, I. D. 1980. Evolutsiya stroitelnogo povedeniya lichinok ruchinikov Trichoptera (Evolution of building behavior in the larvae of Trichoptera). *Zhurnal Obshchei Biologii*, 457–68.

Summerfield, M. A. 1983. Petrography and diagenesis of silcrete from the Kalahari Basin and Cape coastal zone, southern Africa. *Journal of Sedimentary Petrology* **53**, 895–909.

Suppe, J. 1985. *Principles of structural geology.* Englewood Cliffs: Prentice-Hall.

Suttner, L. J. & P. K. Dutta 1986. Alluvial sandstone composition and paleoclimate. I. Framework mineralogy. *Journal of Sedimentary Petrology* **56**, 329–45.

Syers, J. K. & I. K. Iskander 1973. Pedogenic significance of lichens. In *The lichens*, V. Ahmadjian & M. E. Hale (eds.), 225–48. New York: Academic Press.

Tan, J. H. (ed.) 1984. *Andosols.* New York: Van Nostrand Reinhold.

Tauxe, L. & C. Badgley 1984. Transition stratigraphy and the problem of remanence lock-in-time in the Siwalik red beds. *Geophysical Research Letters* **11**, 611–13.

Taylor, J. M. 1950. Pore-space reduction in sandstone. *Bulletin of the American Association of Petroleum Geologists* **34**, 701–6.

Taylor, S. R. 1982. *Planetary science: a lunar perspective*. Houston: Lunar and Planetary Institute.

Taylor, T. N. 1988. The origin of land plants: some answers, more questions. *Taxon* **37**, 805–33.

Thaer, A. D. 1857. *The principles of practical agriculture* (translated by W. Shaw & C. W. Johnson). New York: C. M. Saxton.

Thirgood, G. W. 1981. *Man and the Mediterranean forest*. London: Academic Press.

Thomasson, J. R. 1979. Late Cenozoic grasses and other angiosperms from Kansas, Nebraska and Colorado: biostratigraphy and relationships to living taxa. *Bulletin of the Geological Survey of Kansas* No. 218.

Thomasson, J. R. 1985. Miocene fossil grasses: possible adaptation in reproductive bracts (lemma and palea). *Annals of the Missouri Botanical Garden* **72**, 843–51.

Thompson, I. & D. S. Jones 1980. A possible onychophoran from the Middle Pennsylvanian Mazon Creek Beds of northern Illinois. *Journal of Paleontology* **54**, 588–96.

Thompson, J. B. 1972. Oxides and sulfides in regional metamorphism of pelitic schists. In *Geochemistry*, J. E. Gill (ed.), 27–35. *Proceedings of the 24th International Geological Congress, Montreal, Sect.* 10. Gardenvale: Harpells.

Thorp, J. 1949. Effects of certain animals that live in soils. *Scientific Monthly* **68**, 180–91.

Thorp, J. & E. C. Reed 1949. Is there laterite in rocks of the Dakota Group? *Science* **109**, 69.

Tite, M. S. 1972. *Methods of physical examination in archaeology*. London: Seminar.

Tomeoka, K. & P. R. Busek 1988. Matrix mineralogy of the Orgueil CI carbonaceous chondrite. *Geochimica et Cosmochimica Acta* **52**, 1622–40.

Toulmin, P., A. K. Baird, B. C. Clark, K. Keil, H. J. Rose, R. P. Christian, P. H. Evans & W. C. Kelliher 1977. Geochemical and mineralogical interpretation of the Viking inorganic chemical results. *Journal of Geophysical Research* **82**, 4625–34.

Traverse, A. 1988 *Paleopalynology*. London: Unwin Hyman.

Trewartha, G. T. 1982. *Earth's problem climates*. Madison: University of Wisconsin Press.

Truswell, E. M. 1987. The initial radiation and rise to dominance of angiosperms. In *Rates of evolution*, K. S. W. Campbell & M. F. Day (eds.), 101–28. London: Allen & Unwin.

Ugolini, F. C. 1986. Processes and rates of weathering in cold and polar desert environments. In *Rates of chemical weathering of rocks and minerals*, S. M. Colman & D. P. Dethier (eds.), 193–235. New York: Academic Press.

Vahrenkamp, V. C. & V. Rossinsky 1987. Preserved isotopic signature of subaerial diagenesis in the Mescal Limestone, central Arizona: discussion and reply. *Bulletin of the Geological Society of America* **99**, 595–7.

Valentine, J. W. (ed.) 1986. *Phanerozoic diversity patterns*. Princeton: Princeton University Press.

Valentine, K. W. G. & J. B. Dalrymple 1976. Quaternary buried paleosols: a critical review. *Quaternary Research* **6**, 209–22.

Van Couvering, J. A. H. 1980. Community evolution in East Africa during the Late Cenozoic. In *Fossils in the making*. A. K. Behrensmeyer and A. P. Hill (eds.), 272–98. Chicago: University of Chicago Press.

Van Donselaar-ten Bokkel Huinink, W. A. E. 1966. *Structure, root systems and periodicity of savanna plants and vegetation in northern Surinam*. Amsterdam: North-Holland.

Van Valen, L. 1973. A new evolutionary law. *Evolutionary Theory* **1**, 1–30.

Van Valkenburgh, B. 1985. Locomotor diversity within past and present guilds of large predatory mammals. *Paleobiology* **11**, 406–28.

Vassoevich, N. B. 1960. Opit postroeniya tipovoi krivoi gravitashionnogo uplotneniya glinnistikh osadkov (Experiment in constructing typical curve of gravitational compaction of clayey sediments). *Novosti Neftyanoi Tekhniki, Geologiya* (4), 11–15.

Veizer, J. & W. Compston 1976. $^{87}Sr/^{86}Sr$ in Precambrian carbonates as an index of crustal evolution. *Geochimica et Cosmochimica Acta* **40**, 905–14.

Veizer, J., W. Compston, J. Hoefs & H. Neilson 1982. Mantle buffering of the early oceans. *Naturwissenschaften* **69**, 173–80.

Vesey-Fitzgerald, D. F. 1973. Animal impact on vegetation and plant succession in Lake Manyara National Park, Tanzania. *Oikos* **24**, 314–25.

Vickery, A. M. & H. J. Melosh 1987. The large crater origin of SNC meterorites. *Science* **237**, 738–43.

Vinogradov, A. P. 1959. *The geochemistry of rare and dispersed chemical elements in soils*. New York: Consultants Bureau.

Vogl, R. J. 1974. Effects of fire on grasslands. In *Fire and ecosystems*, T. T. Kozlowski and C. E. Ahlgren (eds.), 139–94. New York: Academic Press.

Vokes, H. E., P. D. Snavely & D. A. Myers 1951. Geology of the southern and southwestern border areas of the Willamette Valley, Oregon. *U.S. Geological Survey Oil & Gas Investigation Map* No. OM110.

Von Damm, K. L., J. M. Edmond, B. Grant, C. I. Measures, B. Walden & P. F. Weiss 1985. Chemistry of submarine hydrothermal solutions at 21°N, East Pacific Rise. *Geochimica et Cosmochimica Acta* **49**, 2197–220.

Von Lengerken, H. 1954. *Die Brutfursorge- und Brutpflegeinstinkte der Kafer*. Leipzig: Geest & Portig.

Voorhies, M. R. 1975. Vertebrate burrows. In *The study of trace fossils*, R. W. Frey (ed.), 269–94. New York: Springer.

Voss, E. G., H. M. Burdet, W. G. Chaloner, P. Hiepko, J. McNeill, R. D. Meikle, D. H. Nicolson, R. C. Rollins, P. C. Silva & W. Greuter 1983. *International code of botanical nomenclature*. Hague: Junk.

Vrba, E. S. 1985. Ecological and adaptive changes associated with early hominid evolution. In *Ancestors: the hard evidence*, E. Delson (ed.), 63–71. New York: Alan R. Liss.

Walker, A. C., R. E. Leakey, J. M. Harris & F. H. Brown 1986. 2.5-Myr *Australopithecus boisei* from west of Lake Turkana, Kenya. *Nature* **322**, 517–22.

Walker, D. 1966. The Late Quaternary history of the Cumberland lowland. *Philosophical Transactions of the Royal Society of London* **B251**, 1–210.

Walker, P. H. 1966. Postglacial environments in relation to landscape and soils on the Cary Drift, Iowa. *Research Bulletin, Iowa State University* **549**, 838–75.

Walker, P. H. & B. E. Butler 1983. Fluvial processes. In *Soils: an Australian viewpoint*, Division of Soils CSIRO (ed.), 83–90. London: Academic Press.

Walker, T. R. 1967. Formation of red beds in modern and ancient deserts. *Bulletin of the Geological Society of America* **78**, 353–68.

Walton, J. 1936. On the factors which influence the external form of fossil plants; with descriptions of some species of the Paleozoic equisetalean genus *Annularia* Sternberg. *Philosophical Transactions of the Royal Society of London* **B226**, 219–37.

Warburg, M. R. 1968. Behavioral adaptations of terrestrial isopods. *American Zoologist* **8**, 545–59.

Warner, J. L. 1983. Sedimentary processes and crustal cycling on Venus. *Proceedings of the 13th Lunar Planetary Science Conference Journal of Geophysical Research Supplement* **88**(2), 495–500.

Washburn, A. L. 1980. *Geocryology*. New York: Wiley.

Wasson, J. T. 1984. *Meteorites: their record of early solar system history*. San Francisco: Freeman.

Watts, N. L. 1976. Paleopedogenic palygorskite from the basal Permo-Triassic of northwest Scotland. *American Mineralogists* **61**, 299–302.

Watts, W. A. 1985. Quaternary vegetation cycles. In *The Quaternary history of Ireland*, K. J. Edwards & W. P. Warren (eds.), 155–85. London: Academic Press.

Watson, J. P. 1967. A termite mound in an Iron Age burial ground in Rhodesia. *Journal of Ecology* **55**, 663–9.
Weaver, C. E. 1967. Potassium, illite and the ocean. *Geochimica et Cosmochimica Acta* **31**, 2181–96.
Weaver, J. E. 1919. The ecological relations of roots. *Carnegie Institution of Washington Publication* No. 286.
Weaver, J. E. 1920. Root development in the grassland formation. *Carnegie Institution of Washington Publication* No. 292.
Webb, S. D. 1972. Locomotor evolution in camels. *Forma et Functio* **5**, 99–112.
Webb, S. D. 1977. A history of savanna vertebrates in the New World. Part I. North America. *Annual Reviews of Ecology Systematics* **8**, 355–80.
Webb, S. D. 1978. A history of savanna vertebrates in the New World. Part II. South America and the Great Interchange. *Annual Reviews of Ecology and Systematics* **9**, 393–426.
Webster, T. 1826. Observations on the Purbeck and Portland Beds. *Transactions of the Geological Society, London* **2**, 37–44.
Wedepohl, K. H. 1969–78. *Handbook of geochemistry*, 2 vols. Berlin: Springer.
Weiss, A. 1981. Replication and evolution in inorganic systems. *Angewandte Chemie, International Edition in English* **20**, 850–60.
Weller, J. M. 1959. Compaction of sediments. *Bulletin of the American Association of Petroleum Geologists* **43**, 273–310.
West, I. M. 1975. Evaporites and associated sediments of the basal Purbeck Formation (Upper Jurassic) of Dorset. *Proceedings of the Geologist's Association* **86**, 205–25.
White, F. 1983. *The vegetation of Africa*. Paris: UNESCO.
White, H. J., D. R. Burggraf, R. B. Bainbridge & C. F. Vondra 1981. Hominid habitats in the Rift Valley. Part I. In *Hominid sites: their geologic settings*, G. Rapp & C. F. Vondra (eds.), 57–113. Boulder: Westview.
White, P. & J. O'Connell 1979. Australian prehistory: new aspects of antiquity. *Science* **203**, 21–8.
Whitney, M. 1909. Soils of the United States. *Bulletin of the Bureau of Soils, U.S. Department of Agriculture* No. 55.
Whittaker, R. H. 1978. Approaches to classifying vegetation. In *Classification of plant communities*, R. H. Whittaker (ed.), 1–31. The Hague: Junk.
Whittington, H. B. 1985. *The Burgess Shale*. New Haven: Yale University Press.
Whybrow, P. J. & H. H. McClure 1981. Fossil mangrove roots and paleoenvironments of the Miocene of the eastern Arabian peninsula. *Palaeogeography, Palaeoclimatology, Palaeoecology* **32**, 213–35.
Wieder, M. & D. H. Yaalon 1982. Micromorphological fabrics and developmental stages of carbonate nodular forms related to soil characteristics. *Geoderma* **28**, 203–20.
Willard, B. 1938. Evidence of Silurian land plants in Pennsylvania. *Proceedings of the Pennsylvania Academy of Science* **12**, 121–4.
Williams, E. H. 1899. *Manual of lithology*, 2nd edn. New York: Wiley.
Williams, G. E. 1968. Torridonian weathering and its bearing on Torridonian palaeoclimate and source. *Scottish Journal of Geology* **4**, 164–84.
Williams, G. E. 1969. Characteristics and origin of a Precambrian pediment. *Journal of Geology* **77**, 183–207.
Williams, G. E. 1986. Precambrian permafrost horizons as indicators of palaeoclimate. *Precambrian Research* **32**, 233–42.
Williams, M. A. J. 1979. Droughts and long-term climatic change: recent French research in arid north Africa. *Geographical Bulletin* **11**, 82–96.
Wilson, E. O. 1971. *The insect societies*. Cambridge: Harvard.
Wilson, E. O. & R. W. Taylor 1964. A fossil ant colony: new evidence of social antiquity. *Psyche* **71**, 93–103.
Wing, S. L. 1984. Relation of paleovegetation to geometry and cyclicity of some fluvial carbonaceous deposits. *Journal of Sedimentary Petrology* **54**, 52–66.

Winkler, D. A. 1983. Paleoecology of an Early Eocene mammalian fauna from paleosols in the Clarks Fork Basin, northwestern Wyoming. *Palaeogeography, Palaeoclimatology, Palaeoecology* **43**, 261–98.

Winkler, H. G. F. 1976. *Petrogenesis of metamorphic rocks*, 3rd edn. New York: Springer.

Woese, C. R. 1980. An alternative to the Oparin view of the primeval sequence. In *The origins of life and evolution*, H.-O. Halvorson and K. E. Van Holde (eds.), 65–76. New York: Alan R. Liss.

Wolfe, J. A. & R. Z. Poore 1982. Tertiary marine and non-marine climatic trends. In *Climate in Earth history*, Geophysics Study Committee (ed.), 154–8. Washington: National Academy of Sciences.

Wood, C. A. & L. D. Ashwal 1981. SNC meteorites: igneous rocks from Mars? *Proceedings of the 12th Lunar and Planetary Conference Geochimica et Cosmochimica Acta* **12**(2), 1359–75.

Woodburne, M. O. (ed.) 1987. *Cenozoic mammals of North America*. Berkeley: University of California Press.

Wright, V. P. 1981. The recognition and interpretation of paleokarsts: two examples from the Lower Carboniferous of South Wales. *Journal of Sedimentary Petrology* **52**, 83–94.

Wright, V. P. 1982. Calcrete paleosols from the Lower Carboniferous Llanelly Formation, South Wales. *Sedimentary Geology* **33**, 1–33.

Wright, V. P. 1983. A rendzina from the Lower Carboniferous of South Wales. *Sedimentology* **30**, 159–79.

Wright, V. P. 1985. The precursor environment for vascular plant colonization. *Philosophical Transactions of the Royal Society of London* **B309**, 143–5.

Wright, V. P. (ed.) 1986a. *Paleosols: their recognition and interpretation*. Oxford: Blackwell.

Wright, V. P. 1986b. The role of fungal biomineralization in the formation of early Carboniferous soil fabrics. *Sedimentology* **33**, 831–8.

Wright, V. P. 1987. The ecology of two early Carboniferous paleosols. In *European Dinantian environments*, J. Miller, A. E. Adams & V. P. Wright (eds.), 345–58. *Geological Journal Special Issue*, No. 12.

Wright V. P. & R. C. L. Wilson 1987. A Terra Rossa-like paleosol complex from the Upper Jurassic of Portugal. *Sedimentology* **34**, 259–73.

Yaalon, D. H. (ed.) 1971. *Paleopedology: origin, nature and dating of paleosols*. Jerusalem: Israel University Press.

Yaalon, D. H. 1975. Conceptual models in pedogenesis. Can soil-forming functions be solved? *Geoderma* **14**, 189–205.

Yanovsky, E., E. K. Nelson & R. M. Kingsbury 1932. Berries rich in calcium. *Science* **75**, 565–6.

Young, A. 1972. *Slopes*. Edinburgh: Oliver & Boyd.

Young, G. C. 1982. Devonian sharks from southeastern Australia and Antarctica. *Palaeontology* **25**, 817–43.

Young, G. M. & D. G. F. Long 1976. Ice wedge casts from the Huronian, Ramsay Lake Formation (2300 m.y. old), near Espanola, northern Canada. *Palaeogeography, Palaeoclimatology, Palaeoecology* **19**, 191–200.

Zaug, A. J. & T. R. Cech 1986. The intervening RNA sequence of *Tetrahymena* is an enzyme. *Science* **231**, 470–6.

Zenger, D. H., J. B. Dunham & R. L. Ethington 1980. Concepts and models of dolomitization. *Special Publication, Society of Economic Paleontologists and Mineralogists* No. 28.

Zeuner, F. 1938. Die Insektenfauna des Mainzer Hydrobienkalks. *Paläontologische Zeitschrift* **20**, 104–59.

Ziegler, A. M., C. R. Scotese, W. S. McKerrow, M. E. Johnson & R. K. Bambach 1979. Paleozoic paleogeography. *Annual Review of Earth and Planetary Sciences* **7**, 473–502.

Zinke, P. J. 1962. The pattern of individual forest trees on soil properties. *Ecology* **43**, 130–3.

Index